Engineering Mathematics

Her-Terng Yau

Yung-Dann Yu

Chao-Kuang Chen

Engineering Mathematics

Author／Her-Terng Yau, Yung-Dann Yu, Chao-Kuang Chen

Publisher／Ben-yuan Chen

Editor／Hao-Wei Huang

Cover Designer／Qiao-yun Dai

Imprint／CHUAN HWA BOOK CO., LTD.

Postal Savings Account／0100836-1

Book Identification Number／06526

First Edition, First Printing／2024

ISBN／978-626-328-912-3

ISBN／978-626-328-907-9(PDF)

Website／www.chwa.com.tw

CHUAN HWA Online Bookstore Open Tech／www.opentech.com.tw

If you have any questions about this book, please feel free to email us for guidance.

book@chwa.com.tw

Taipei Head Office (Northern District Branch)
Address: No. 21, Zhongyi Rd., Tucheng Dist., New Taipei City 23671, Taiwan (R.O.C.)
Phone: (02) 2262-5666
Fax: (02) 6637-3695、6637-3696

Central District Branch
Address: No. 26, Shuyi 1st Ln., South Dist., Taichung City 40256, Taiwan (R.O.C.)
Phone: (04) 2261-8485
Fax: (04) 3601-8600

Southern District Branch
Address: No. 12, Ying'an St., Sanmin Dist., Kaohsiung City 80769, Taiwan (R.O.C.)
Phone: (07) 381-1377
Fax: (07) 862-5562

PREFACE

"Engineering Mathematics" is a subject that is essential in engineering and electrical and computer science departments of universities worldwide. It serves as a foundational discipline for mathematical analysis shared across relevant programs. Consequently, it forms a fundamental skillset for all students in engineering and electrical and computer science disciplines, and is considered essential knowledge for becoming an engineer. However, due to its extensive content, spanning from fundamental engineering problem modeling, simplification, analysis, to solving, the textbook covers multiple significant mathematical topics. As a result, it is a challenging and complex subject to learn, encompassing various domains. This complexity sometimes leads students majoring in engineering, electrical, and computer science to feel fear, reluctant, or even abandonment towards the subject, which is indeed regrettable. Nevertheless, mastering the subject of "Engineering Mathematics" provides insight into the mysteries and principles of engineering and related fields, and serves as a foundational stepping stone for those who plan to pursue further studies in master's or doctoral programs and engage in thesis research. **With the support of numerous Chinese readers, this book is launched in its English edition. The key feature of this English edition is improved readability through enhanced text cohesion. Additionally, more engineering and electrical and computer science-related physical concepts have been introduced, making theoretical concepts easier to comprehend and integrate.**

Textbook content

This textbook is rich in content and is organized into five major sections, based on the foundation of mathematics used in engineering and electrical and computer science fields. The sections are "Ordinary Differential Equations," "Linear Algebra," "Vector Function Analysis," "Fourier Analysis and Partial Differential Equations," and "Complex Analysis." The material is suitable for a one-year course of engineering mathematics for undergraduate students in four-year programs, totaling six credits over two semesters. The key points of each section are briefly introduced as follows.

First part :
Ordinary Differential Equations (Chapter1~4)

In almost all universities, the content of the first semester of Engineering Mathematics in engineering and electrical and computer science departments is primarily focused on "Ordinary Differential Equations." This content typically covers four main themes: "First-Order Ordinary Differential Equations," "Higher-Order Linear Ordinary Differential Equations," "Laplace Transforms," and "Power Series Solutions of Ordinary Differential Equations."

Second part :
Linear Algebra (Chapter5~7)

This section of the textbook covers three main topics: "Vector Operations and Vector Spaces," "Matrix Analysis," and "Systems of Linear Differential Equations."

Third part :
Vector Function Analysis (Chapter8)

This section covers vector differentiation, the Del operator, line integrals, surface integrals, and the three major integral theorems (Green's theorem, Gauss's divergence theorem, and Stokes' theorem). These concepts and calculations are extensively applied in electromagnetics (electrical and information engineering field) and fluid mechanics (engineering field), making it a highly crucial unit.

Fourth part :
Fourier Analysis and Partial Differential Equations (Chapter9~10)

Chapter nine of this book introduces orthogonal function sets and Fourier analysis. This unit is used in the field of engineering to solve the partial differential equations in chapter ten. In addition to this, the field of electrical and information engineering extensively applies it to signal analysis and processing.

Fifth part :
Complex Analysis (Chapter11)

This chapter covers topics including complex arithmetic, complex variable functions and differentiation, complex variable function integration, Taylor and Laurent series, residue theorem, and definite integration of real variable functions. This chapter is applicable in the fields of engineering and electrical and information engineering for courses such as general physics, thermodynamics, fluid mechanics, automatic control, circuit theory, signals and systems, and electrical, mechanical, and control engineering.

Features of this book

This book is an important work written by the author, and its characteristics can be summarized as follows:

1. The content is detailed and presented in an easily understandable manner, making it engaging for beginners in engineering mathematics.

2. The text is well-organized with clear arrangement and highlighted reminders, making it easy for engineering mathematics beginners to grasp at a glance

3. Abundant and appropriately leveled conceptual examples help engineering mathematics beginners build confidence.

4. The problems-solving process is comprehensive and reviews calculus techniques, enabling engineering mathematics beginners to approach problems with ease.

5. The exercises are rich and cover various question types, allowing engineering mathematics beginners to familiarize themselves with concepts and formulas.

6. Relevant knowledge from engineering and electrical and information fields is introduced appropriately and explained thoroughly, aiding engineering mathematics beginners in applying concepts to other specialized courses.

Instructions

Video modeling: Guidance on modeling problems in engineering, physics, or electrical engineering, coupled with audio-visual explanations, provides a clearer understanding of the concepts.

https://www.youtube.com/htyauiem

1-4-2   Reduction to First-Order Linear ODEs

Some ODEs may resemble linear ODEs but have additional functions mixed in the terms such as $P(x)$, $Q(x)$, or the derivative term y'. In such cases, we can use the technique of variable transformation to rearrange and consolidate these "extras" in order to transform the original equation into a linear ODE, making it easier to solve. Let's illustrate this with the logistic growth model. Consider a population size $y(x)$ of a certain species, where x represents time. The population has a maximum size A, and the rate of change of the population size is proportional to the product of the current population size and the remaining growth space available. In other words, the function $y(x)$ describing the population size must satisfy the following equation: $\dfrac{dy}{dx} = ky(A-y)$

Table organization: Formulating important formulas in a tabulated format allows for quick reference and application.

3-4-2 The Basic Formula and Theorem for Laplace Inverse Transform

Since the Laplace inverse transform is the inverse operation of the Laplace transform, by referring to the results of the Laplace transform, we can naturally determine the original function from which it was transformed. This is what the Laplace inverse transform requires. Therefore, referring to the formulas in the beginning sections of this chapter, we can construct Table 1. We will use this as a basis to calculate the Laplace inverse transform of various types of functions. For the Laplace inverse transform of less common functions, you can refer to Appendix II. Additionally, after understanding the content of Chapter 11 on complex functions, readers can try to directly compute the inverse transform using the definition, skipping this part.

Table 1

function	Laplace transform	Laplace inverse transform
$u(t)$	$\mathcal{L}\{u(t)\} = \dfrac{1}{s}$	$\mathcal{L}^{-1}\{\dfrac{1}{s}\} = u(t)$ or 1
e^{at}	$\mathcal{L}\{e^{at}\} = \dfrac{1}{s-a}$	$\mathcal{L}^{-1}\{\dfrac{1}{s-a}\} = e^{at}$
$\sin wt$	$\mathcal{L}\{\sin wt\} = \dfrac{w}{s^2+w^2}$	$\mathcal{L}^{-1}\{\dfrac{1}{s^2+w^2}\} = \dfrac{1}{w}\sin wt$
$\cos wt$	$\mathcal{L}\{\cos wt\} = \dfrac{s}{s^2+w^2}$	$\mathcal{L}^{-1}\{\dfrac{s}{s^2+w^2}\} = \cos wt$
$\sinh wt$	$\mathcal{L}\{\sinh wt\} = \dfrac{w}{s^2-w^2}$	$\mathcal{L}^{-1}\{\dfrac{1}{s^2-w^2}\} = \dfrac{1}{w}\sinh wt$
$\cosh wt$	$\mathcal{L}\{\cosh wt\} = \dfrac{s}{s^2-w^2}$	$\mathcal{L}^{-1}\{\dfrac{s}{s^2-w^2}\} = \cosh wt$
t^n	$\mathcal{L}\{t^n\} = \dfrac{n!}{s^{n+1}}$	$\mathcal{L}^{-1}\{\dfrac{1}{s^{n+1}}\} = \dfrac{t^n}{n!}$; n is positive integer

Varying Levels of Difficulty: Exercises are categorized into foundational and advanced levels, progressively enhancing proficiency from basic to more complex concepts.

2-3 Exercises

Basic questions

1. Find the general solution of the following homogeneous ODE.
   (1) $y'' + y' - 2y = 0$.
   (2) $y' + 6y' + 9y = 0$.
   (3) $y'' - 36y = 0$.
   (4) $y'' - y' - 6y = 0$.
   (5) $y'' - 3y' + 2y = 0$.
   (6) $y'' + 8y' + 16y = 0$.

Solve the initial value problems for the following ODEs in questions 2 to 4.

2. $y'' + 3y' + 2y = 0$, $y(0) = 1$, $y'(0) = 0$.

Advanced questions

1. Find the general solution of the following homogeneous ODE
   (1) $y'' - y' + 10y = 0$.
   (2) $4y'' + y' = 0$.
   (3) $y'' - 10y' + 25y = 0$.
   (4) $y'' + 9y = 0$.
   (5) $y'' - 4y' + 5y = 0$.
   (6) $12y'' - 5y' - 2y = 0$.
   (7) $2y'' + 2y' + y = 0$.
   (8) $y''' + 3y'' + y' + 3y = 0$.
   (9) $\dfrac{d^4 y}{dx^4} - y = 0$

Video Instruction: Each chapter begins with QR codes linking to video tutorials by the authors of classic examples, transforming the textbook into a mobile classroom.

First-Order Ordinary Differential Equations

Epilogue

The book appropriately incorporates examples to establish and clarify key concepts in engineering mathematics. It also offers a comprehensive collection of practice exercises for teachers to assign as homework assignments and includes detailed solutions for all exercises in the instructor's manual for reference. **Additionally, the author has uploaded Chinese video content from classroom lectures on YouTube (https://www.youtube.com/c/htyauiem, or refer to the QR code on the back cover of the book), readers are welcome to subscribe to this channel and access the audio-visual content that complements the book according to their needs.**

Finally, this book will be dedicated to my advisor, Professor Chao-Kuang Chen. I would like to express my gratitude for his guidance in completing this engineering mathematics textbook. I dedicate this book to my advisor and extend my warmest wishes for his 90th birthday.

Her-Terng Yau

CONTENTS

Chapter 9   Orthogonal Functions and Fourier Analysis

Chapter 10   Partial Differential Equation

Chapter 11   Complex Analysis

Appendix

Index

1

First-Order Ordinary Differential Equations

Huygens
(1629~1695, The
Netherlands)

In order to accurately describe celestial motion and other physical phenomena, both differential equations and calculus emerged around the late 17th century. Ordinary differential equations (ODEs) are generally believed to have been proposed by Huygens (1629-1695) in 1693. By the mid-18th century, with the efforts of many mathematicians, ODEs became an independent discipline and found widespread applications in various fields such as mathematics and engineering today.

Learning Objectives

1-1
Introduction to Differential Equations

1-1-1 Understanding the origin of differential equations through applied problems

1-1-2 Understanding basic terms of differential equations

1-1-3 Capable of distinguishing between general solution, particular solution, and singular solution

1-1-4 Capable of distinguishing between initial value and boundary value problems

1-2
Separable First-Order ODEs

1-2-1 Mastering the method of separation of variables

1-2-2 Master the technique of using a change of variables to transform into a separable form

1-3
Exact ODEs and Integration Factor

1-3-1 Learn to determine and find particular solutions for exact ODEs

1-3-2 Learn to derive and use integrating factors

1-4
Linear ODEs

1-4-1 Master the techniques for solving first-order linear ODEs

1-4-2 Learn to solve various first-order ODEs that can be linearized

1-4-3 Master the methods for solving Bernoulli and Riccati equations

1-5
Solving First-Order ODEs with the Grouping Method

1-5-1 Learn to solve using common total differential formulas

1-5-2 Understand the framework for solving first-order ODEs

1-6
Application of First-Order ODEs

1-6-1 ODE engineering, physical modeling, and solutions

ExampleVideo

In engineering problems, utilizing mathematical modeling to describe a physical phenomenon is an extremely useful research method, and the most common approach is through differential equations. Differential equations can be broadly classified into two categories: ordinary differential equations (ODEs) and partial differential equations (PDEs), depending on whether the function to be solved is a single variable or multiple variables.

This chapter begins by introducing the overall concept of differential equations, followed by discussions on the representation and geometric interpretation of their solutions. The main focus of this chapter is to explore the methods for solving various types of first-order ordinary differential equations.

1-1   Introduction to Differential Equations

1-1-1   Modeling Physical Problems and the Production of Differential Equations

In general, mathematical modeling of physical problems often involves using derivatives and partial derivatives of functions, along with relevant physical properties and theorems, to establish the mathematical equations for the model. For example, in physics textbooks, we learn about projectile motion, and in terms of geometry, we can use differentiation to model the original curve based on the tangent of the curve. We can also use Newton's second law of motion to model a spring system, as described in detail below.

1. Modeling Projectile Motion

If an object is thrown upward from someone's hand, and its upward height is denoted as $y(t)$, which is a function of time, then according to kinematics, we can know the acceleration $a(t) = y''(t) = -g$. Additionally, if the initial height of the object above the ground when it is in the hand is y_0, and the initial upward velocity is v_0, neglecting air resistance, this physical problem can be modeled as a differential equation:

$$\begin{cases} y'' + g = 0 \\ y(0) = y_0 \text{ , } y'(0) = v_0 \end{cases}$$

As the figure 1-1 shows.

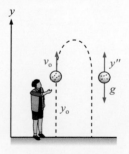

Figure 1-1

2. Modeling the Slope of a Tangent Line

We can utilize calculus if we want to model a curve in the xy-plane that passes through the point (1, 2) and has a slope of $\dfrac{y}{x}$. In calculus, the first derivative of a function represents the slope of the function's curve. Therefore, according to the given conditions the slope of the tangent line is $\dfrac{dy}{dx} = y' = \dfrac{y}{x}$, where y is a function of x. Additionally, since the curve passes through the point (1, 2), we can model it as follows:

$$\begin{cases} \dfrac{dy}{dx} = y' = \dfrac{y}{x} & \cdots\text{differential equation} \\ y(1) = 2 & \cdots\text{limitation factor} \end{cases}$$

Figure 1-2(a) represents a schematic diagram of the solutions to this equation, where the blue lines represent all the tangent lines that satisfy $y' = \dfrac{y}{x}$. The black slanted line represents the curve that passes through (1, 2). The detailed methods for solving this equation will be discussed in subsequent chapters.

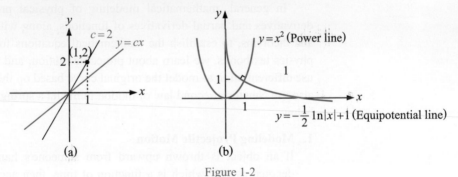

(a)　　　　　(b)

Figure 1-2

3. Modeling Orthogonal Curve Families

In electromagnetics, we know that electric field lines and equipotential lines exhibit orthogonal curves, If we want to model a curve in the xy-plane that passes through the point (1, 1) and is orthogonal to the electric field line $y = x^2$, we can utilize the fact that the slope of the tangent line to $y = x^2$ that is $y' = 2x$. We also know that the product of the slopes of two perpendicular lines is -1. Therefore, according to the given conditions, the product of the slopes of these two curves is -1. As a result, the slope of the orthogonal curve to $y = x^2$ is $\dfrac{dy}{dx} = y' = \dfrac{-1}{2x}$. Taking into account the constraints, we can model it as follows:

$$\begin{cases} \dfrac{dy}{dx} = -\dfrac{1}{2x} & \cdots\text{differential equation} \\ y(1) = 1 & \cdots\text{constraints} \end{cases}$$

Figure1-2(b) represents a schematic diagram of these two orthogonal curves.

4. Modeling of Mechanical Systems

Figure 1-3 represents a spring system without considering friction forces. In this system, an external force $f(t)$ acts on an object with mass m, and the spring has an elastic coefficient of k. The object is initially positioned at the origin and has an initial velocity of V_0. How can this spring system be modeled? Let $y(t)$ represent the

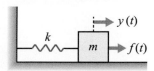

Figure 1-3   Mass-Spring System

displacement of this spring system. According to the given conditions, the total sum of forces ΣF is equal to $f(t) - k \cdot y$. Since the object's acceleration is $\dfrac{d^2 y}{dt^2} = y''$,

based on Newton's second law of motion, $\Sigma F = ma$, we obtain the dynamic equation for this spring system as follows: $f(t) - k \cdot y = my''$ that is $my'' + k \cdot y = f(t)$, with the initial conditions $y(0) = 0$; $y'(0) = V_0$, we can model it as follows:

$$\begin{cases} my'' + k \cdot y = f(t) & \cdots \text{differential equation} \\ y(0) = 0;\ y'(0) = V_0 & \cdots \text{constraints} \end{cases}$$

1-1-2   Definition and Classification of Differential Equations

In the previous discussion, we introduced the use of differential equations to model physical systems, where the unknown functions were single-variable functions. However, in many physical systems, there are also unknown functions that depend on multiple variables. For example, the temperature distribution T in a room may vary with different positions and over time. In this case, the physical quantity of temperature is a multivariable function T (position, time). Therefore, in the following, we will provide a detailed definition and classification of differential equations.

▶ Definition 1-1-1

Differential Equation, DE

Any equation that describes the relationship between an unknown function, its derivatives, and the independent variable is called a differential equation.   ◀

In the modeling of mechanical systems discussed earlier, for example, the unknown function of displacement $y(t)$ is described by the equation $my'' + ky = f(t)$ which represents the relationship between the unknown function $y(t)$, its derivative $y''(t)$ and the independent variable. This is known as a differential equation.

1. Classification

General differential equations can be classified into two types based on the number of independent variables in the unknown function, as listed in Table 1

Table 1

Equation	Unknown function	Symbol	Example
Ordinary Differential Equation, ODE	Single variable	$y = y(t)$	$my'' + cy' + ky = f(t)$
Partial Differential Equation, PDE	Multi-variable	$u = u(x, t)$	$\dfrac{\partial^2 u}{\partial x^2} = c^2 \dfrac{\partial^2 u}{\partial t^2}$

2. Basic Noun

When fully describing a differential equation, there are several common terms, including:

(1) **Order:** The highest order of the derivative in a differential equation is referred to as the order of the differential equation.

(2) **Degree:** In a differential equation where the orders are all integers (non-negative integers), the power of the highest order of derivative is the degree of the differential equation.

(3) **Linear:** A differential equation is considered linear when the unknown function and its derivatives satisfy the following properties

① The highest power of the unknown function and its derivatives is 1.

② There are no product terms where the unknown function and its derivatives are multiplied together.

③ The equation does not contain nonlinear functions such as trigonometric functions, exponential functions, etc.

Based on the given information, we can determine that in $\dfrac{dy}{dx} = y' = 2x$, the unknown function $y(x)$ is a single-variable function, its highest derivative is first order, and the highest derivative is first degree. There are no product terms or non-linear functions present. Therefore, it can be classified as a first-order linear ordinary differential equation.

Next are some examples of differential equations and their related information, as shown in Table 2.

Table 2

Equations	ODE or PDE	Order	Degree	Linear or Nonlinear
$(y''')^2 + (y')^3 + y' = t$	ODE	three	two	Nonlinear
$y'' + y' + 4y = \cos x$	ODE	two	one	Linear
$y'' + 3y' + 4y = \cos y$	ODE	two	one	Nonlinear
$xy' + (y'')^2 + xy^2$	ODE	two	two	Nonlinear
$\dfrac{\partial u}{\partial x} + \dfrac{\partial u}{\partial y} + 8x^5 + \sin y = u^2$	PDE	one	one	Nonlinear

Equations	ODE or PDE	Order	Degree	Linear or Nonlinear
$\dfrac{\partial^2 u}{\partial x^2} + \dfrac{\partial^2 u}{\partial y^2} = 0$	PDE	two	one	Linear
$\dfrac{dy}{dx} = \sqrt{1+y}$ [1]	ODE	one	two	Nonlinear

1-1-3 Solution of Ordinary Differential Equations

▶ **Definition 1-1-2**

Solution of Ordinary Differential Equations

A function y that satisfies the ordinary differential equation (ODE)
$f(x, y, y', y'', \cdots, y^{(n)}) = 0$ is called a solution of the ODE.   ◀

For example, $y_1 = \cos x$ satisfy $y_1'' + y_1 = -\cos x + \cos x = 0$, so y_1 is the solution of $y'' + y = 0$, and similarly, it can be proven $y_2 = \sin x$, $y_3(x) = 3\cos x + \sin x$ are also solutions of the equation $y'' + y = 0$.

If we further investigate, we will find that actually all expressions that can be written in the form $y_4(x) = c_1 \cos x + c_2 \sin x$ are solutions of the equation $y'' + y = 0$. Here, c_1 and c_2 are arbitrary independent constants. Moreover, this ODE is a second-order ODE, and two arbitrary independent constants in the solution $y_4(x) = c_1 \cos x + c_2 \sin x$ are precisely the two constants. As for the solutions $y_1 = \cos x$, $y_2 = \sin x$, and $y_3(x) = 3\cos x + \sin x$, they can be obtained by appropriately assigning values to c_1 and c_2. Based on this concept, we will classify and introduce the solutions of ODEs in the following section.

Here, we will categorize the solutions based on the concepts mentioned above. Let's consider an ODE represented by $f(x, y, y, \cdots, y^{(n)}) = 0$. The solutions and their corresponding geometric interpretations can be broadly classified into three categories, as described below.

---

[1] This formula needs to be transformed into $(\dfrac{dy}{dx})^2 = 1 + y$ and then determined.

1. **The Classification of Solutions**

    (1) **General Solution**

    The number of arbitrary independent constants in the solutions of an ODE is equal to the order of the ODE.

    (2) **Particular Solution**

    A solution that cannot be expressed in the general solution but still satisfies the ODE is called particular solutions.

    (3) **Singular Solution**

    The solutions of the ODE that cannot be expressed in the form of the general solution, but they are still solutions of ODE. When a singular solution occurs, the uniqueness of the solution to the initial value problem (see subsection 1-1-4) of the ODE will fail.

    So in $y'' + y = 0$, solution $y_4(x) = c_1 \cos x + c_2 \sin x$ is general solution, then $y_1(x) = \cos x$, $y_2(x) = \sin x$ and $y_3(x) = 3\cos x + \sin x$ are all particular solutions, As shown in Figure 1-4, these solutions have the following geometric interpretations:

2. **The Geometric Interpretations of Solutions**

    (1) **General Solution**

    A family of curves in the x-y plane.

    (2) **Particular Solution**

    A specific curve within the family of curves represented by the general solution.

    (3) **Singular Solution**

    The envelope or common tangent line of the general solution.[2]

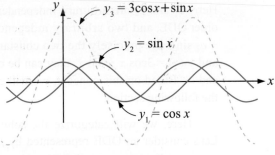

Figure 1-4

In a general ODE, it is not easy to identify singular solutions. By examining the following examples, we can better understand the characteristics of singular solutions and gain a clearer understanding of the geometric interpretations of general solutions and particular solutions.

---

[2] In general, the possibility of singular solutions arises more often in high degree nonlinear ODEs.

Example 1

Please confirm whether $y = x^2$ and $y = cx - \dfrac{1}{4}c^2$ (where c is an arbitrary constant) can satisfy the same differential equation $\dfrac{1}{4}(y')^2 - xy' + y = 0$. Plot both solutions on the xy-plane and discuss their relationship.

Solution

(1) Substituting directly into the original differential equation for verification, it can be confirmed that $y = x^2$ and $y = cx - \dfrac{1}{4}c^2$ are the solutions of ODE.

(2) Originally differential equation is first-order ODE, and in $y = cx - \dfrac{1}{4}c^2$ only has an arbitrary constant c, so $y = cx - \dfrac{1}{4}c^2$ is general solution. And $y = x^2$ cannot be solved by general solution, so is singular soliution, as well as the singular solution is the envelope of the general solution, as shown in the diagram on the below.

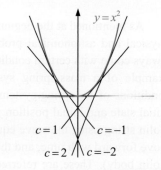

$$y = x^2$$

$$c = 1 \quad c = -1$$
$$c = 2 \quad c = -2$$

Q.E.D.

Example 2

Find the differential equation with $y(x) = ce^{\frac{x^2}{2}}$ as the general solution, where c is an arbitrary constant.

Solution   Let's differentiate it first and see what equation $y(x)$ satisfies:

$y'(x) = x \cdot ce^{\frac{x^2}{2}} = x \cdot y$, so y satisfies $y' - xy = 0$, now, y only has an arbitrary constant, Hence, from the classification of solutions, we know that the order of the differential equation with this as the general solution is 1. Therefore, the original equation is $y' - xy = 0$.

Q.E.D.

Example 3

Find the differential equation with $y(x) = A\cos x + B\sin x$ as the general solution, where A, B are arbitrary constants.

Solution   Similarly to the previous question, let's differentiate y and see what relationship it satisfies.

$$y(x)' = -A \cdot \sin x + B \cdot \cos x$$

$$y(x)'' = -A \cdot \cos x - B \cdot \sin x = -y(x)$$

So y satisfies $y'' + y = 0$. The question requires y to be the general solution, and since y contains only two arbitrary constants, it implies that the original equation should be a second-order ordinary differential equation (ODE). Therefore, the original equation is given by $y'' + y = 0$.

Q.E.D.

1-1-4   Problem Classification of Differential Equations

As mentioned at the beginning of the chapter, differential equations originated from physical and astronomical problems. Therefore, during the process of modeling, they always come with certain conditions of the system, such as initial conditions. Taking the example of a mass-spring system with the equation $my'' + ky = f(t)$, the initial conditions of the spring system ($t = 0$) can be specified as velocity $y'(0) = 0$ for static initial state and initial position $y(0) = 0$ at the origin. When modeling the vibration of a violin string using the wave equation, it is assumed that the two endpoints of the string move forward with time, and their displacements are 0 (because the string is fixed to the violin body). These are referred to as boundary conditions, meaning conditions at the boundaries. Based on this, differential equations can be classified into the following two types:

1. **Initial Value Problem, IVP**
    In a differential equation, when values are assigned for the same independent variable (conditions), we refer to it as an initial value problem (IVP),
    for instance: the mass-spring system mentioned above
    $my'' + ky = f(t)$, $y(0) = 0$, $y'(0) = 0$.

2. **Boundary Value Problem, BVP**
    In a differential equation, when values are assigned for two or more different independent variables (conditions), we refer to it as a boundary value problem (BVP),
    for instance: the given string vibration equation $y'' + y = 0$, $y(0) = 0$, $y(l) = 0$, where l represents the length of the string.

In the initial value problem of a differential equation, if the given initial conditions are sufficient, then the solution to that differential equation exists and is unique. Taking the example of the projectile motion equation $y'' + g = 0$ in Figure 1-1, when we know the initial height y_0, the initial upward velocity v_0, and the acceleration is $-g$, according to Newtonian kinematics, the complete trajectory $y(t)$ of the object can be determined, i.e., $y'' + g = 0$, $y(0) = y_0$, $y'(0) = v_0$. The solution to this ordinary differential equation (ODE) exists and is unique. However, for boundary value problems of differential equations, if a solution exists, it may not be unique. The relevant concepts will be introduced in Section 9-1 of this book.

Example 4

Decide whether the following ODE is an initial value problem or a boundary value problem.

$(1)\ y''(x) + y(x) = 0,\ \begin{cases} y(0) = 1 \\ y'(0) = -1 \end{cases}$
$\qquad (2)\ y''(x) + y(x) = 0,\ \begin{cases} y(0) = 1 \\ y(\dfrac{\pi}{2}) = 1 \end{cases}$

Solution

(1) This problem involves assigning a value to y for the same independent variable $x = 0$, so it is an initial value problem (IVP)

(2) This problem involves assigning values to y for two different independent variables, $x = 0$ and $x = \dfrac{\pi}{2}$, so it is a boundary value problem (BVP).   **Q.E.D.**

In fact, the boundary and initial conditions tell us which particular solution within the general solution of the ODE is applicable. In the case of Example 4, for part (1), Example 3 already tells us that $y(x) = A\cos x + B\sin x$ is the general solution to the equation. By using the initial condition $y(0) = 1$, we can solve for $A = 1$; and by using the condition $y'(0) = -1$, we can solve for $B = -1$. Therefore, the particular solution to this differential equation is $y(x) = \cos x - \sin x$. Similarly, by utilizing the boundary conditions, you can obtain the particular solution for part (2) as $y = \cos x + \sin x$.

1-1 Exercises

Basic questions

1. Determine the order, degree, linearity, and whether it is an ODE or a PDE for the following differential equations.
   (1) $y'' + y' + 4y = \cos x$.
   (2) $y' + 2y^4 + 3\sin x = x^5$.

2. Consider the ODE $y'' + 2y' + 3y = \sin x$, answer the following two questions:
   (1) The order of this ODE.
       (A)one (B)two (C)three (D)four.
   (2) The degree of this ODE.
       (A)one (B)two (C)three (D)four.

3. Which of the following is a linear ODE?
   (A) $yy' + 1 = \sin x$ (B) $y'' + \sin(y) = 0$
   (C) $y''' + 3y' = \cos x$ (D) $y' + e^y = x$.

4. Determine the ODE with $y = ce^x$ as its general solution.

5. Please describe the mathematical model of the curve that passes through the point (2, 3) and has a slope of x.

6. If an object is held in mid-air at an initial height of 1 meter and thrown upwards, the height of the object above the ground as a function of time is denoted by $y(t)$. Assuming there is no consideration for friction, describe the mathematical model for this scenario, given that the initial upward velocity is 5 meters per second.

Advanced questions

1. Determine the order, degree, linearity, and whether it is an ODE or PDE for the following differential equations.
   (1) $(1 + x^3)(4dy + 5dx) = 10xydx$.
   (2) $\left[1 + (y')^2\right]^{1/2} = 5y''$.
   (3) $x^2(y'')^2 + x(y')^3 + y = x^5$.
   (4) $d(yx) = y^2 dx$.
   (5) $\dfrac{\partial u}{\partial x} + \dfrac{\partial u}{\partial y} + 8x^5 + \sin y = u^2$.

(6) $\dfrac{\partial^2 u}{\partial x^2} \cdot \dfrac{\partial u}{\partial y} = \sin x$.

2. Please establish the mathematical model for the curve that passes through (1, 1) and has a tangent slope of $-\dfrac{x}{y}$.

3. Establish the mathematical model for the orthogonal curve of the equation $y - cx^2 = 0$.

4. Establish the mathematical model for the orthogonal curve of the equation $y = \dfrac{cx}{x+1}$.

5. Establish the mathematical model for the orthogonal curve of the equation $x^2 + y^2 = c^2$.

6. Using the concept of Newton's law of cooling, where the rate of temperature change of an object is proportional to the difference between the object's temperature and the ambient temperature, establish the mathematical model for the temperature of a cake at time t, given that the cake was just taken out of the oven at a temperature of 300°F while the room temperature was 70°F.

7. Consider a closed RL circuit with $R = 10\Omega$, $L = 2H$, and an electromotive force $E = 100V$. At $t = 0$, the current is 0, assume $I(t)$ represent the current magnitude at time t. Please establish the mathematical model for this circuit.

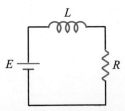

8. If the decay of ^{14}C is proportional to its current mass, there is a fossil bone with a ^{14}C content $M(t)$, and the half-life of ^{14}C is 5600 years. Please establish its mathematical model.

9. There was initially 60 liters of pure water in a tank. Starting at time $t = 0$, saltwater (with concentration $0.5\,\text{kg}/\ell$) is continuously injected $2\,\ell/\text{min}$ into the tank. It is assumed that the injected saltwater immediately mixes completely with the water in the tank, and the mixed saltwater simultaneously flows out of the tank at a rate of $3\,\ell/\text{min}$ liters per minute. As a result, after 60 minutes, all the water in the tank has completely drained. Please establish a mathematical model for the variation of salt content in the tank over time.

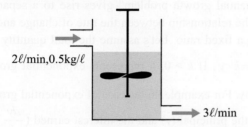

2ℓ/min,0.5kg/ℓ

3ℓ/min

10. Considering the mass, spring, and damping system as shown in the diagram below, assuming the initial displacement and velocity of the spring are both 0, please establish its mathematical model.

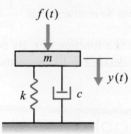

$f(t)$

m

$y(t)$

k c

11. There is a closed single-loop RLC circuit, where $R = 50\Omega$, $L = 30\text{H}$, $C = 0.025\text{F}$ and the electromotive force $E(t) = 200 \cdot \sin(4t)\text{V}$. Let $q(t)$ represent the charge across the capacitor, and at $t = 0$, the charge is 0 and the current is 0, Please establish the mathematical model for this circuit.

12. Try to find the ordinary differential equation to the following general solutions.
   (1) $y = c_1 e^{3x} + c_2 e^{-3x}$.
   (2) $y = c_1 \cos 2x + c_2 \sin 2x$.
   (3) $y = c_1 e^{5x} + c_2 x e^{5x}$.
   (4) $y = e^{-2x}(c_1 \cos x + c_2 \sin x)$.
   (5) $y = c_1 x + c_2 x^2$.
   (6) $y = c_1 x + c_2 x \cdot \ln x$.

1-2 Separable First-Order ODEs

We begin with the most fundamental method for solving first-order ODEs. In 1691, Leibniz discovered that by using the fundamental theorem of calculus, we can easily solve certain specific forms of first-order ODEs, which are now referred to as "separable variable type." However, there is a slightly weaker form known as "separable variable type with manipulation," which requires some additional steps to be performed in order to rearrange it into the "separable variable type."

In physical systems, the "exponential growth problem" gives rise to a separable variable type ODE, which describes the relationship between the rate of change and the total quantity of a species, exhibiting a fixed ratio. Let's assume the total quantity of a species is denoted by $y(t)$ then $\dfrac{dy}{dt} = k \cdot y$. If $k > 0$, it represents exponential growth, while $k < 0$ indicates exponential decay. For example, in the case of exponential growth, such as depositing money in a bank, the principal (y) and the interest earned ($\dfrac{dy}{dt}$) are related, where $k > 0$ represents the interest rate. Conversely, in the case of exponential decay, such as the luminosity of glass $y(t)$ and the absorbed luminosity $\dfrac{dy}{dt}$, $k < 0$ represents the absorption rate. Then $\dfrac{dy}{dt} = ky$ can be rearranged as $\dfrac{1}{y} dy = kdt$, where the left side is a function of y, and the right side is a function of t. By doing so, we can integrate the equation to find a solution. The detailed solution method is explained below.

1-2-1 Separation of Variables

If a first-order ODE has the form of $\dfrac{dy}{dx} = f_1(x) f_2(y)$, then with a little rearrangement, we can obtain:

$$\int F_2(y)dy = \int f_1(x)dx + c \;^3$$

---

[3] When performing indefinite integration on both sides of an equation, each side typically introduces an arbitrary constant of integration. However, when rearranging the equation and moving terms around, these constants on both sides can be combined into a single constant. Therefore, it is sufficient to add the constant of integration on one side only.

Among $F_2(y) = \dfrac{1}{f_2(y)}$, where c is a specific constant. In this case, we only need to apply the fundamental theorem of calculus and integrate both sides of the equation to obtain the general solution[4]. If we also have boundary or initial conditions, we can find the particular solution.

If $y = r$ is the root of $f_2(y) = 0$, in this case, directly using separation of variables to solve is not possible because $F_2(y)$ at $y = r$ diverges and does not exist. However, at this point, $y = r$ would be also a solution of the original ODE, for example, $\dfrac{dy}{dx} = y^2 - 1$, here, $y = \pm 1$ is also a solution of this ODE. Some scholars call this solution an additional solution or a constant solution. However, this type of solution generally has no special significance in physical systems, so this book focuses on solving general solutions and does not specifically discuss this type of solution.

Returning to the equation $\dfrac{dy}{dx} = ky$, then $\displaystyle\int \dfrac{1}{y}\,dy = \int k\,dx$. By applying the aforementioned concept, we obtain $\ln |y| = kx + c_1$, $y = ce^{kx}$ ($c = e^{c_1}$), If the initial quantity of the species is given by $y(0) = y_0$, then $c = y_0$, and the solution becomes $y = y_0 \cdot e^{kx}$. This is illustrated in Figure 1-5.

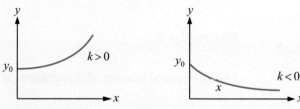

Figure 1-5

Here are a few examples to familiarize everyone with solving first-order ODEs using the method of separation of variables.

Example 1

Find the general solution of the equation $y' = x^2 + 2x + 3$.

Solution

$$\frac{dy}{dx} = x^2 + 2x + 3 \Rightarrow dy = (x^2 + 2x + 3)\,dx$$

$$\Rightarrow \int dy = \int (x^2 + 2x + 3)\,dx \Rightarrow y = \frac{1}{3}x^3 + x^2 + 3x + c,$$

the constant "c" is a constant that is related to the boundary or initial conditions.　Q.E.D.

---

[4] If ODE is $\dfrac{dy}{dx} = f'(x)$, then its general solution is $y = \displaystyle\int f(x)\,dx + c$.

Example 2

$y' = y^2\, e^x,\ y(0) = 1$.

Solution   After rearranging the original equation, we can obtain $\int y^{-2}\,dy = \int e^x\,dx$.

By applying the fundamental theorem of calculus to both sides of the equation, then

$$y = \frac{1}{c - e^x},$$

where c is a constant related to the boundary or initial conditions. With the initial condition $y(0) = 1$ then $1 = \dfrac{1}{c-1} \Rightarrow c = 2$, so the solution is

$$y(x) = \frac{1}{2 - e^x}$$

Note: In this case, the solution $y = \dfrac{1}{2 - e^x}$, that is $y = f(x)$, is called an explicit solution.
If the solution is written as $y(2 - e^x) - 1 = 0$, that is $F(x, y) = 0$, it is called an implicit solution.   Q.E.D.

Example 3

Find the general solution of the equation $y' = -\dfrac{2xy}{1+x^2}$.

Solution   Similarly, by rearranging the original equation slightly, we can obtain

$$\int \frac{1}{y}\,dy = \int \frac{-2x}{1+x^2}\,dx.$$

Then the fundamental theorem of calculus tells us that

$$\ln|y| + \ln|1 + x^2| = c.$$

Now, due to the logarithm property of "adding becomes multiplying,"

the equation can be simplified to $\ln|y(1 + x^2)| = c$, that is

$$y = \frac{k}{1+x^2}$$

where k is a constant related to the boundary or initial conditions.   Q.E.D.

Example 4

$e^y \sin x dx + 3dy = 0$.

Solution   After rearranging the original equation and integrating both sides, we obtain:

$$\int \sin x dx + \int 3e^{-y} dy = -c,$$

that is $-\cos x - 3e^{-y} = -c$, the genetal solution is

$$\cos x + 3e^{-y} = c.$$

Q.E.D.

The first-order ODE in Example 4 can be rearranged into the following general form:

$$M_1(x)M_2(y)dx = N_1(x)N_2(y)dy.$$

By dividing both sides of $M_1(y)N_1(x)$ and integrating, we obtain

$$\int \frac{M_1(x)}{N_1(x)} dx = \int \frac{N_2(y)}{M_2(y)} dy + c,$$

where c is a constant related to the boundary or initial conditions.

Example 5

$(y^2 + 1)dx = y \sec^2 x dy$.

Solution   In the given form, let's substitute $M_1(x) = 1$, $M_2(y) = y^2 + 1$, $N_1(x) = \sec^2 x$, $N_2(y) = y$, we can obtain

$$\int \frac{M_1(x)}{N_1(x)} dx = \int \frac{N_2(y)}{M_2(y)} dy + c$$

$$\Rightarrow \int \cos^2 x dx = \int \frac{y}{y^2 + 1} dy + c$$

$$\Rightarrow \int \frac{1 + \cos 2x}{2} dx = \frac{1}{2} \ln|y^2 + 1| + c$$

$$\Rightarrow \frac{1}{2}x + \frac{1}{4}\sin 2x = \frac{1}{2} \ln|y^2 + 1| + c.$$

Q.E.D.

1-2-2   Reduction to Separation of Variables

Some first-order ODEs cannot be directly solved using the method of separating variables. However, with suitable variable transformations, they can be converted into directly separable ODEs. Let's discuss "homogeneous" ODEs and "exact" ODEs as examples.

1. Homogeneous ODE

▶ **Definition 1-2-1**

First-Order Homogeneous ODEs

If for any pair of (x, y) in its domain and a constant λ,

$f(x, y)$ satisfies $f(\lambda x, \lambda y) = \lambda^k f(x, y)$, then $f(x, y)$ is called a homogeneous function of degree k. If $k = 0$, $f(x, y)$ is a homogeneous function of degree 0, then the ordinary differential equation (ODE) of the form $y' = f(x, y)$ is called a first-order homogeneous ODE.   ◀

For instance:

(1) $f(x, y) = x^2 + xy$ satisfies $f(\lambda x, \lambda y) = \lambda^2 (x^2 + xy)$, therefore, it is a second-order homogeneous function.

(2) $f(x, y) = \dfrac{x - y}{x + y}$ satisfies $f(\lambda x, \lambda y) = \lambda^0 \dfrac{x - y}{x + y}$ which is a zeroth-order homogeneous function.

For a general form of a differential equation $M(x, y)dx + N(x, y)dy = 0$, if $M(x, y)$ and $N(x, y)$ are both homogeneous functions of degree k, then $M(x, y)dx + N(x, y)dy = 0$

can be transformed into $y' = -\dfrac{M(x, y)}{N(x, y)} = \dfrac{x^k M_1(\dfrac{y}{x})}{x^k N_1(\dfrac{y}{x})} = f(x, y) = f(\dfrac{y}{x})$, which is a

first-order homogeneous ODE. Here, $f(x, y) = f(\dfrac{y}{x})$ represents a zeroth-order

homogeneous function. We will now solve this type of equation through variable transformations.

Assume $v = \dfrac{y}{x}$, we know that $dy = vdx + xdv$, so

$$y' = \frac{dy}{dx} = \frac{vdx + xdv}{dx} = v + x \cdot \frac{dv}{dx} = f(v).$$

Slightly arranged, we obtain

$$\int \frac{dv}{f(v) - v} = \int \frac{dx}{x} + c \cdots ①$$

(If we assume that $v = \dfrac{x}{y}$, then through a similar calculation, we obtain

$\int \dfrac{f(v)}{1 - vf(v)} dv = \int \dfrac{dy}{y} + c$). At this point, we can substitute the function f based on

the original boundary conditions into the equation to obtain the general solution. This will be illustrated in the following examples, Example 6 and Example 7.

Example 6

$$y' = \frac{y}{x+y}.$$

Solution   For any real number r, function $f(x,y) = \dfrac{y}{x+y}$ satisfies

$$f(rx,ry) = \frac{ry}{rx+ry} = \frac{y}{x+y} = f(x,y).$$

Therefore, f is a zero-degree homogeneous function. By performing a variable transformation,

$v = \dfrac{y}{x}$, we can obtain

$$f(v) = \frac{v}{1+v},$$

that is

$$\frac{1}{f(v)-v} = -\frac{1}{v^2} - \frac{1}{v}.$$

Substituting into equation ①, we obtain the general solution:

$$\ln|y| - \frac{x}{y} = c.$$

Example 7

$$y^2\, dx = (x^2 + xy)dy = 0.$$

Solution   After rearranging the original equation, we have:

$$\frac{dy}{dx} = \frac{-y^2}{x^2 + xy}.$$

As shown in Example 6, we can verify that function $f(x,y) = \dfrac{-y^2}{x^2+xy} = \dfrac{-(\frac{y}{x})^2}{1+\frac{y}{x}}$ is a

homogeneous function. Therefore, by performing the variable transformation $v = \dfrac{y}{x}$,

we obtain $f(v) = \dfrac{-v^2}{1+v}$. Substituting this into the formula ①, we get the general

solution

$$\frac{1}{2}\ln\left|\frac{2y}{x}+1\right| - \ln|y| = c.$$

2. Function-Type ODEs

If ODE has the form of $y' = \dfrac{dy}{dx} = f(ax + by + c)$, it is called Function-type ODE.

By applying the variable transformation $t = ax + by + c$ and differentiating both sides, we obtain $dt = adx + bdy$, that is

$$\frac{dt - adx}{b} = dy .$$

By substituting the original equation and integrating, we obtain the following result.

$$\int \frac{1}{a + bf(t)}\, dt = \int dx + c \cdots ②$$

At this point, substituting the boundary condition f and evaluating the indefinite integral will give us the general solution. Let's look at the following example.

Example 8

$\dfrac{dy}{dx} = (x + y + 1)^2 .$

Solution   By observing the original equation, we can make $x + y + 1 = t$, then $f(t) = t^2$, $a = b = c = 1$, substitute into ② obtain

$$\int \frac{1}{1 + t^2}\, dt = \int dx + c ,$$

after integrating, we obtain $\tan^{-1}(x + y + 1) = \tan^{-1} t = x + c$ [5], that is general solution

$$x + y + 1 = \tan(x + c).$$

Q.E.D.

Example 9

$y' = y - x - 1 + \dfrac{1}{x - y + 2} .$

---

[5] $\dfrac{d}{dt}\tan^{-1} t = \dfrac{1}{1 + t^2}$.

Solution   The right side of the equation in the original expression is equal to

$$-(x-y+2)+1+\frac{1}{x-y+2},$$

therefore, we can assume $f(t)=-t+1+\frac{1}{t}$, $a=1$, $b=-1$, $c=2$ in the formula ②, we

can obtain

$$\int\frac{t}{t^2-1}dt=\int dx.$$

After integrating, we obtain $\ln|t^2-1|=2x+k$, that is general solution

$$\ln|(x-y+2)^2-1|=2x+k.$$

Q.E.D.

1-2 Exercises

Basic questions

Solve the following ODE.

1. $y' = x + 1$.

2. $y' = \sin x$.

3. $y' = e^x$.

4. $dx + e^{3y}dy = 0$.

5. $e^y y' = x + x^3$.

6. $y' = x^2 + 1$, $y(0) = 1$.

7. $y' = \cos x$, $y(\frac{\pi}{2}) = 2$.

8. $y' = e^y$, $y(1) = 0$.

9. $y' = -4xy^2$, $y(0) = 1$.

10. $(\csc x)dy + e^{-y}dx = 0$, $y(0) = 0$.

Advanced questions

Solve the following first-order ODE.

1. $\dfrac{dy}{dx} = \sin 5x$.

2. $\dfrac{dy}{dx} = (x+1)^2$.

3. $y' + y^2 = xy^2$.

4. $y^3(2x^2 - 3x - 1)dx + 3y^2dy = 0$.

5. $2xy' - y^2 + 2y + 8 = 0$.

6. $(1 + x^2)(1 + y^2)dx - xydy = 0$.

7. $4yy' = e^{x-y^2}$, $y(1) = 2$.

8. $(3xy + y^2) + (x^2 + xy)y' = 0$.

9. $y' - (\dfrac{y}{x})^2 + 2(\dfrac{y}{x}) = 0$.

10. $2xyy' - y^2 + x^2 = 0$.

11. $y' = (x + y - 2)^2$.

12. $y' = \dfrac{x - y}{2x - 2y + 1}$.

13. $\dfrac{dp}{dt} = p - p^2$.

14. $\dfrac{dy}{dx} = (\dfrac{2y+3}{4x+5})^2$.

15. $(2y^2 - 6xy)dx + (3xy - 4x^2)dy = 0$.

16. $\dfrac{dy}{dx} + xy^3 \sec(\dfrac{1}{y^2}) = 0$.

17. $y' = \dfrac{y - x}{y + x}$.

$\quad$ (Hint : $\displaystyle\int \dfrac{1}{1+x^2}dx = \tan^{-1}(x) + c$)

18. $xy' = y^2 - y$.

19. $(1 + x^2)(1 + y^2)dx - xydy = 0$.

20. $\csc y dx + \sec^2 x dy = 0$.

Solve the initial value problem for the following first-order ODE.

21. $\dfrac{dy}{dx} = xye^{-x^2}$, $y(4) = 1$.

22. $x^2 \dfrac{dy}{dx} = y - xy$, $y(-1) = -1$.

23. $\dfrac{dy}{dx} = \dfrac{y^2 - 1}{x^2 - 1}$, $y(2) = 2$.

24. $\dfrac{dy}{dx} + 2y = 1$, $y(0) = \dfrac{5}{2}$.

1-3 Exact ODEs and Integration Factor

There are two common types of forms in which it is relatively easy to find the general solution for a first-order ODE. One is the separation of variables method discussed in the previous section, and the other is the exact ODE, which will be introduced in this section. We will first define and discuss the solution method for exact ODEs. For non-exact ODEs, we will explore how to find an integrating factor to transform them into exact form before solving them.

1-3-1 Exact ODEs

▶ **Definition 1-3-1**

Exact Ordinary Differential Equations (Exact ODEs)

For the differential equation $M(x, y)dx + N(x, y)dy = 0$, if there exists a smooth function [6]

$$\phi : R \times R \rightarrow R$$

such that $d\phi = M(x, y)dx + N(x, y)dy$, then the ODE is said to be exact. ◀

If we don't have any tools available, it can be impractical to guess whether an ODE is exact by trying to find the function ϕ. However, the following theorem provides a convenient condition for quickly determining if an ODE is exact. When it is confirmed that the original ODE is exact, it also provides a way to find the function ϕ.

▶ **Theorem 1-3-1**

Exact Conditions

$M(x, y)dx + N(x, y)dy = 0$ is exact if and only if
$$\frac{\partial M}{\partial y} = \frac{\partial N}{\partial x}.$$

Proof (selected readings)
【**Sufficient Conditions**】
If an ODE is exact, then according to the definition, we have

$$d\phi = \frac{\partial \phi}{\partial x}dx + \frac{\partial \phi}{\partial y}dy = M(x, y)dx + N(x, y)dy.$$

---

[6] Smooth function: A function for which all its derivatives of any order are continuous.

Therefore we obtain $\dfrac{\partial \phi}{\partial x} = M(x,y)$, $\dfrac{\partial \phi}{\partial y} = N(x,y)$.

According to the assumption, ϕ is a smooth function, which means its second derivatives exist and are continuous. And the order of partial differentiation does not affect the value of the derivative, so

$$\frac{\partial}{\partial x} N(x,y) = \frac{\partial^2 \phi}{\partial x \partial y} = \frac{\partial^2 \phi}{\partial y \partial x} = \frac{\partial}{\partial y} M(x,y).$$

【Necessary Conditions】

In $M(x,y)dx + N(x,y)dy = 0$, we assume that $\dfrac{\partial M}{\partial y} = \dfrac{\partial N}{\partial x}$. According to the conclusion we originally wanted, let's assume

$$\phi(x,y) = \int M(x,y)dx + \int N(x,y)dy,$$

then knowing by the hypothesis $\dfrac{\partial \phi}{\partial x} = M$, $\dfrac{\partial \phi}{\partial y} = N$,

$$\therefore \exists \phi : R \times R \to R \text{ causes } d\phi = \frac{\partial \phi}{\partial x} dx + \frac{\partial \phi}{\partial y} dy = 0. \qquad \blacktriangleleft$$

Therefore, when a first-order ODE is exact, there exists a smooth function $\phi(x, y)$ causes $d\phi = 0$, which means $\phi(x, y) = c$ is a general solution to the original ODE. Here, $\dfrac{\partial M}{\partial y} = \dfrac{\partial N}{\partial x}$ is called **the discriminant**. In fact, in the discussion of "necessary conditions," it has been shown how to find the function ϕ, then to find $\phi(x, y)$: by integrating M with respect to x and N with respect to y separately, and then adding the results, we can find $\varphi(x, y)$. However, in practice, it is common to combine terms to simplify the expression. Let's look at a few examples to illustrate this process.

Example 1

$(y^2 + x)dx + (2xy + 5)dy = 0$, $y(0) = 1$.

Solution　We take $M = y^2 + x$, $N = 2xy + 5$, $\dfrac{\partial M}{\partial y} = 2y = \dfrac{\partial N}{\partial x}$. Therefore, the original ODE is exact.
Because

$$\begin{cases} M = y^2 + x \\ N = 2xy + 5 \end{cases} \xrightarrow{\ \text{partial derivative}\ } \begin{cases} \phi(x,y) = xy^2 + \dfrac{1}{2}x^2 + g(y) \\ \phi(x,y) = xy^2 + 5y + f(x) \end{cases},$$

by comparing the two equations, we can take $f(x) = \frac{1}{2}x^2$, $g(y) = 5y$,

the general solution is

$$\phi(x, y) = xy^2 + \frac{1}{2}x^2 + 5y = c,$$

substituting the initial condition $y(0) = 1$, obtain $c = 5$, so the particular solution is

$xy^2 + \frac{1}{2}x^2 + 5y = 5$.   Q.E.D.

Example 2

$(2y + e^y + 6x^2)y' + 4 + 12xy = 0$.

Solution   We take $M = 4 + 12xy$, $N = 2y + e^y + 6x^2$, then $\dfrac{\partial M}{\partial y} = 12x = \dfrac{\partial N}{\partial x}$, so

ordinary ODE is exact.
Because

$$\begin{cases} M = 4 + 12xy \\ N = 2y + e^y + 6x^2 \end{cases} \xrightarrow{\text{partial derivative}} \begin{cases} \phi(x, y) = 4x + 6x^2y + g(y) \\ \phi(x, y) = y^2 + e^y + 6x^2y + f(x) \end{cases}.$$

We compare two equations and we take $g(y) = y^2 + e^y$, $f(x) = 4x$, the general solution is

$$\phi(x, y) = 6x^2y + 4x + y^2 + e^y + c.$$ Q.E.D.

1-3-2   Integration Factor

When a first-order ODE is not exact, we can multiply it by a suitable function to transform it into an exact form. Specifically, we assume the existence of a function I then using the discriminant formula $\dfrac{\partial(IM)}{\partial y} = \dfrac{\partial(IN)}{\partial x}$ to solve I. Then, by multiplying the original ODE by I, we obtain an exact ODE. The following steps provide a detailed explanation:

▶ Definition 1-3-2

Integration Factor

In the equation $M(x, y)dx + N(x, y)dy = 0$, if there exists a function $I(x, y)$ causes $IM(x, y)dx + IN(x, y)dy = 0$ be an exact differential equation, then this fuction is called an integrating factor.   ◀

1. **Calculation of Integration Factor**

As mentioned in the introduction, we assume that multiplying the equation by the integration factor I results in an exact equation. Then the discriminant tells us that $\dfrac{\partial IM}{\partial y} = \dfrac{\partial IN}{\partial x}$, and by applying the multiplication rule of differentials, we obtain

equation $N\dfrac{\partial I}{\partial x} - M\dfrac{\partial I}{\partial y} = I(\dfrac{\partial M}{\partial y} - \dfrac{\partial N}{\partial x})$ (referred to as equation ①), in other

words, if we can solve this partial differential equation, we can obtain the desired integration factor. However, the problem is that this partial differential equation may not always have a solution. Therefore, we need to impose some restrictions on I to simplify this PDE. For example, let's assume that $I(x, y)$ is a single-variable function:

(1) Assuming $I(x, y) = I(x)$

Then formula ① can simplified into $N\dfrac{dI}{dx} = I(\dfrac{\partial M}{\partial y} - \dfrac{\partial N}{\partial x})$, after

rearrangement and integrating, we obtain

$\ln|I| = \displaystyle\int \dfrac{dI}{I} = \int \dfrac{\dfrac{\partial M}{\partial y} - \dfrac{\partial N}{\partial x}}{N} dx$, if $\dfrac{\dfrac{\partial M}{\partial y} - \dfrac{\partial N}{\partial x}}{N} = f(x)$, then integration factor

$I = e^{\int f(x)dx}$.

(2) Assuming $I(x, y) = I(y)$

Then formula ① simplified into $-M\dfrac{dI}{dy} = I(\dfrac{\partial M}{\partial y} - \dfrac{\partial N}{\partial x})$, after rearrangement

and integrating, we obtain

$\ln|I| = \displaystyle\int \dfrac{dI}{I} = \int \dfrac{\dfrac{\partial M}{\partial y} - \dfrac{\partial N}{\partial x}}{-M} dy$, if $\dfrac{\dfrac{\partial M}{\partial y} - \dfrac{\partial N}{\partial x}}{-M} = f(y)$, then integration factor

$I = e^{\int f(y)dy}$.

Indeed, in addition to the two types of integrating factors mentioned above, there are two more types of integrating factors that can be solved using similar methods. The process of solving for these two types of integrating factors is omitted here. We can summarize the commonly used four integrating factors as shown in Table 3:[7] [8]

Table 3

$Mdx + Ndy = 0$ conditions	$\dfrac{\dfrac{\partial M}{\partial y} - \dfrac{\partial N}{\partial x}}{N} = f(x)$	$\dfrac{\dfrac{\partial M}{\partial y} - \dfrac{\partial N}{\partial x}}{-M} = f(y)$	$\dfrac{\dfrac{\partial M}{\partial y} - \dfrac{\partial N}{\partial x}}{N - M} = f(x+y)$	$\dfrac{\dfrac{\partial M}{\partial y} - \dfrac{\partial N}{\partial x}}{yN - xM} = f(xy)$
Integration factor	$e^{\int f(x)dx}$	$e^{\int f(y)dy}$	$e^{\int f(x+y)d(x+y)}$	$e^{\int f(xy)d(xy)}$

---

[7] The integrating factor given by equation ① represents the general solution, and if it exists, there are infinitely many integrating factors.

[8] Since differentiation is a linear operation, any constant multiple of the integrating factor I is still an integrating factor.

Example 3

$(2x + y^2)dx + xy\,dy = 0$.

Solution   Assume $M = 2x + y^2$, $N = xy$, then $\dfrac{\partial M}{\partial y} = 2y \neq \dfrac{\partial N}{\partial x} = y$. Therefore, the original ODE is not exact, and we need to find an integrating factor. Let's assume the integrating factor $I(x, y) = I(x)$ in this case, then

$$\frac{\dfrac{\partial M}{\partial y} - \dfrac{\partial N}{\partial x}}{N} = \frac{2y - y}{xy} = \frac{1}{x}.$$

Therefore through the integration factor of Table3 $I = e^{\int \frac{1}{x}dx} = e^{\ln|x|} = x$, the original ODE both multiplied x, then we obtain

$$(2x^2 + xy^2)dx + x^2\,x\,dy = 0 \text{ (Exact ODE)}.$$

Now, as demonstrated in Example 1. We obtain the general solution:

$$\phi(x, y) = \frac{1}{2}x^2y^2 + \frac{2}{3}x^3 = c.$$

Q.E.D.

Example 4

$y\,dx + (2x + 5 + \sin y)dy = 0$.

Solution   Assume $M = y$, $N = 2x + 5 + \sin y$, then $\dfrac{\partial M}{\partial y} = 1 \neq \dfrac{\partial N}{\partial x} = 2$. Therefore, the original ODE is not exact, and we need to find an integrating factor. Let's assume the integrating factor $I(x, y) = I(x)$ in this case, then

$$\frac{\dfrac{\partial M}{\partial y} - \dfrac{\partial N}{\partial x}}{-M} = \frac{1 - 2}{-y} = \frac{1}{y},$$

we obtain

$$I = e^{\int \frac{1}{y}dy} = e^{\ln|y|} = y.$$

After multiplying the original ODE by the integrating factor $I = y$,

$y^2 dx + (2yx + 5y + y \sin y)dy = 0$ is exact ODE,

using the method for solving exact ODEs, the general solution is

$$\phi(x, y) = xy^2 + \frac{5}{2}y^2 - y\cos y + \sin y = c \quad^9.$$

Q.E.D.

---

[9] In the process of solving Example 4, we need to perform integration by parts in the final step $\int y \sin y \, dy$.

Here, let's recall the formula for integration by parts:

$$\int f'(x)g(x)dx = f(x)g(x) - \int f(x)g'(x)dx ,$$

alternatively, we can use a shortcut method as shown in Figure 1-6.

We can obtain $\int y \sin y \, dy = -y \cos y + \sin y + k$.

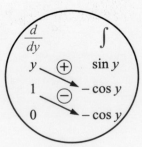

Figure 1-6

1-3   Exercises

Basic questions

1. Which one of the following mathematical expressions is an exact differential equation?

(1) $2xydx + (x^2 - 1)dy$.

(2) $(3x^2 + y^2 + 1)dx + (x^3 + 2xy^2 - 1)dy$.

(3) $(2 + x^2 y)dy + xy^2 dx$.

Find first-order ODE of following questions 2-8

2. $(5x + 4y)dx + (4x - 8y^3)dy = 0$.

3. $(2xy^2 - 3)dx + (2x^2 y + 4)dy = 0$.

4. $(2x^2 + 3x)dx + 2xydy = 0$.

5. $(x^2 - 2xy)dx + (\sin y - x^2)dy = 0$.

6. $(2 + x^2 y)y' + xy^2 = 0$, $y(1) = 2$.

7. $(3x + 2y)dx + xdy = 0$.

8. $ydx + (2x + 4)dy = 0$.

Advanced questions

1. Which one of the following mathematical expressions is an exact differential equation?

(1) $(\ln x + y)dx + (\ln x + x)dy$.

(2) $(\cos x \sin x - xy^2)dx + y(1 - x^2)dy$.

(3) $(\tan x - y)dx - (\sec x + y^2)dy$.

(4) $(1 + \ln x + \dfrac{y}{x})dx - (1 - \ln x + y)dy$.

(5) $(\tan x - \sin x \sin y)dx + \cos x \cos ydy$.

2. $[f(x) + g(y)]dx + [h(x) + p(y)]dy = 0$. In which condition, the ODE is exact?

3. Under what conditions, the equation $f(x)g(y)dx + h(x, y)dy = 0$ is exact?

Find first-order ODE of following questions 4-12.

4. $(\sin y - y \sin x)dx + (\cos x + x \cos y - y)dy = 0$.

5. $(x + y)^2 dx + (2xy + x^2 - 1)dy = 0$.

6. $(e^x + y)dx + (2 + x + ye^y)dy = 0$, $y(0) = 1$.

7. $xdx + (x^2 y + 4y)dy = 0$, $y(4) = 0$.

8. $xy^4 + e^x + 2x^2 y^3 y' = 0$.

9. $3y^4 - 1 + 12xy^3 \dfrac{dy}{dx} = 0$, $y(1) = 2$.

10. $(x^2 y^3 - \dfrac{1}{1 + 9x^2})dx + x^3 y^2 dy = 0$.

11. $2y^2 + ye^{xy} + (4xy + xe^{xy} + 2y)y' = 0$.

12. $(\cos x \cdot \sin x - xy^2)dx + y(1 - x^2)dy = 0$.

13. Find the value of k for which the equation
$\dfrac{dx}{dy} = \dfrac{1 + y^2 + kx^2 y}{1 - 2xy^2 - x^3}$ is exact, please solve it.

14. For the first-order differential equation $y - xy' = 0$,

(1) Prove that the above equation is not exact.

(2) Find a integrating factor $\mu(x)$.

(3) Find a integrating factor $v(y)$.

15. Determining whether
$(4xy + 2x^2 y) + (2x^2 + 3y^2)y' = 0$ is exact?

16. Find solution α causes ODE exact, and solve ODE
$2xy^3 - 3y - (3x + \alpha x^2 y^2 - 2\alpha y)y' = 0$.

Find first-order ODE of following questions 17-22

17. $(y^4 + 2y)dx + (xy^3 - 4x + 2y^4)dy = 0$.

18. $(3x - 2y)y' = 3y$.

19. $x^2 + y^2 + x + xyy' = 0$.

20. $1 + (3x - e^{-2y})y' = 0$.

21. $(y - x + 1)dx - (y - x + 5)dy = 0$, and $y(0) = 4$.

22. $(3y^2 + x + 1)dx + 2y(x + 1)dy = 0$, and $y(0) = 1$.

1-4 Linear ODEs

If an ODE has the form of $y' + P(x)y = Q(x)$, we call it a first-order linear ODE. This type of ODE commonly appears in various practical modeling scenarios, such as circuits, object motion, chemical dilution, radioactive substances, and thermodynamic systems, among others. It can be broadly categorized into two types: 1. $Q = 0$, referred to as homogeneous, 2. $Q \neq 0$, referred to as nonhomogeneous. The solution of these types of ODEs follows a similar approach to that of non-exact ODEs, where we transform them into finding integrating factors.

Let's use a simple example of a drug reaction to illustrate how to construct a first-order linear ODE. Suppose a drug, when consumed by a human body, decomposes at a rate of kx. Let $y(x)$ represent the concentration of the drug in the body at time x, and let the absorption rate by the human body be represented by $\dfrac{1}{x} y(x)$, the concentration of the drug in the body is $y'(x) = kx - \dfrac{1}{x} y(x)$, that is $y'(x) + \dfrac{1}{x} y(x) = kx$, now $P(x) = \dfrac{1}{x}$, $Q(x) = kx$, this ODE is a first-order linear ODE. Next, let's discuss how to solve it.

1-4-1 First Order Linear ODEs

▶ **Theorem 1-4-1**

Finding the General Solution for a Nonhomogeneous ODE

The general solution of $y' + P(x)y = Q(x)$ is $y = \dfrac{1}{I(x)} \int I(x)Q(x)dx + \dfrac{c}{I(x)}$, which $I(x)$ is integration factor.

proof

After rearrangement of the original equation, we obtain $[P(x)y - Q(x)]dx + dy = 0$, assuming that $M(x, y) = P(x)y - Q(x)$, $N(x, y) = 1$, finding that $\dfrac{\partial M}{\partial y} = P(x) \neq \dfrac{\partial N}{\partial x} = 0$,

so ordinary ODE is not exact, in order to find the general solution, we adopt the approach of using an integrating factor, where we let $I(x, y) = I(x)$, we can obtain

$$\frac{\dfrac{\partial M}{\partial y} - \dfrac{\partial N}{\partial x}}{N} = \frac{P(x) - 0}{1} = P(x).$$

Therefore, referring to Table 3 in Sections 1-2, we obtain the integrating factor as follows:

$$I = \exp\left[\int P(x)dx\right],$$

then we obtain

$$Iy' + IP(x)y = IQ(x) \quad \text{(exact ODE)} \cdots \text{①}$$

According to the product rule and chain rule of differentiation, the left-hand side of equation ① is equal to $(e^{\int P(x)dx}y)'$, so the formula ① will become $(Iy)' = IQ(x)$. Hence, according to the Fundamental Theorem of Calculus, integrating both sides of the equation, we obtain

$$Iy = \int IQ(x)dx + c.$$

The general solution is

$$y = \frac{1}{I(x)}\int IQ(x)dx + \frac{c}{I(x)}. \qquad \blacktriangleleft$$

Note that in the above general solution, $\dfrac{c}{I(x)}$ satisfies $y' + P(x)y = 0$, in other words, $\dfrac{c}{I(x)}$ is a homogeneous solution of the original ODE, $\dfrac{1}{I(x)}\int IQ(x)dx$ satisfies original ODE, so it is a particular solution. In other words, this general solution provides both the formula for the homogeneous solution and the particular solution simultaneously.

Example 1

If in the previously mentioned drug reaction model, $k = 3$, and the concentration at time $x = 1$ is 2, find the drug concentration function.

Solution　This ODE is $y' + \dfrac{1}{x}y = 3x$, $y(1) = 2$. In the original equation, let

$P(x) = \dfrac{1}{x}$, $Q(x) = 3x$. Calculating the integrating factor

$I = \exp\left(\int \dfrac{1}{x}dx\right) = \exp(\ln|x|) = x$, we obtain the general solution

$y = \dfrac{1}{x}\int x \times 3x\, dx + \dfrac{c}{x} = x^2 + \dfrac{c}{x}$. Substituting the condition $y(1) = 2$, we get $c = 1$.

Therefore, the solution is $y = x^2 + \dfrac{1}{x}$. 　　　Q.E.D.

We can follow a step-by-step procedure to solve this type of ODE as follows:

Step 1 Check $P(x)$, $Q(x)$.

Step 2 Calculate $I = e^{\int P(x)dx}$.

Step 3 Substitute $y = \dfrac{1}{I(x)}\int IQ(x)dx + \dfrac{c}{I(x)}$.

Step 4 Substitute the condition to find particular solution.

Example 2

$y' + y\tan x = \sin 2x$, $y(0) = 1$.

Solution

Step 1 In original equation, we assume $P(x) = \tan x$, $Q(x) = \sin 2x$.

Step 2 Calculate integration factor $I = \exp(\int \tan x dx) = \exp(-\ln|\cos x|) = \dfrac{1}{\cos x}$.

Step 3 We obtain the general solution $y = -2\cos^2 x + c\cos x$.

Step 4 Then substitute the condition $y(0) = 1$ finding that $c = 3$, so the solution is $y = -2\cos^2 x + 3\cos x$.   Q.E.D.

If we interchange the roles of x and y in the given differential equation, we can still apply the same steps 1 to 4 mentioned earlier.

Example 3

$dx + (3x - e^{2y})dy = 0$.

Solution The original ODE can be transformed into $\dfrac{dx}{dy} + 3x = e^{2y}$.

Step 1 $P(y) = 3$, $Q(y) = e^{2y}$.

Step 2 $I = e^{\int P(y)dy} = e^{\int 3dy} = e^{3y}$.

Step 3 $x = \dfrac{1}{e^{3y}}\int e^{3y}e^{2y}dy + \dfrac{c}{e^{3y}} = \dfrac{1}{5}e^{2y} + ce^{-3y}$.   Q.E.D.

1-4-2 Reduction to First-Order Linear ODEs

Some ODEs may resemble linear ODEs but have additional functions mixed in the terms such as $P(x)$, $Q(x)$, or the derivative term y'. In such cases, we can use the technique of variable transformation to rearrange and consolidate these "extras" in order to transform the original equation into a linear ODE, making it easier to solve. Let's illustrate this with the logistic growth model. Consider a population size $y(x)$ of a certain species, where x represents time. The population has a maximum size A, and the rate of change of the population size is proportional to the product of the current population size and the remaining growth space available. In other words, the function $y(x)$ describing the population size must satisfy the following equation:

$$\frac{dy}{dx} = ky(A - y)$$

$0 < y < A$. Where k represents the proportionality constant, namely $y' - kAy = -ky^2$, this ODE is similar to a first-order linear ODE, with the only difference being the additional y term. In this section, we will illustrate the solution techniques using examples of the "Bernoulli Equation, " "Function Differential Equation, " and "Riccati Equation."

1. **Bernoulli's Equation**

   Bernoulli, in 1695, mathematically modeled and analyzed the one-dimensional motion with the presence of drag force, providing a solution for the corresponding ODE. In modern times, we collectively refer to a first-order ODE with the following form as the Bernoulli equation:

   $$y' + P(x)y = Q(x)y^n,$$

   where n is any real number. For $n = 0, 1$, it is the linear case and has been discussed in subsection 1-4-1. Therefore we only consider $n \neq 0, 1$ here. Now, since the "junk" terms are purely composed of y, we can divide both sides of the equation by y^n. In doing so, the original equation becomes:

   $$y^{-n}y' + P(x)y^{1-n} = Q(x).$$

   Interestingly, the chain rule tells us that y^{1-n}, which its differential fuction is $(1 - n)y^{-n}y'$. Therefore, let's assume substitute $u - y^{1-n}$ then the original equation becomes:

   $$u' + (1 - n)P(x)u = (1 - n)Q(x).$$

   This is a first-order linear ODE, and the method used to solve it was invented by the calculus master Gottfried Wilhelm Leibniz (1646-1716). We can proceed to solve it as we have learned for linear ODEs. Let's see an example below:

Example 4

In the logistic growth model, if $k = 1$, $A = 1$, and $y(0) = 2$, solve the equation for $y(x)$.

Solution   The original ODE is $y' - y = -y^2$, it can know $P(x) = -1$, $Q(x) = -1$, $n = 2$. The original equation can be written as: $u' + u = 1$.

From the solution method for linear ODEs (integration factor $I = e^{\int 1 dx} = e^x$), we can obtain the general solution $u = y^{-1} = 1 + ce^{-x}$, so $y = \dfrac{1}{1 + ce^{-x}}$, then from $y(0) = 2$, we

obtain $c = -\dfrac{1}{2}$. The general solution is

$$y = \frac{1}{1 - \dfrac{1}{2}e^{-x}}.$$ Q.E.D.

Example 5

$3xy' + y + x^2y^4 = 0$.

Solution   By dividing both sides of the equation by $3x$ and comparing this with the general form of a linear ODE, we can assume $P(x) = \dfrac{1}{3x}$, $Q(x) = \dfrac{-x}{3}$, $n = 4$, then the original equation becomes

$$u' - \frac{1}{x}u = x.$$

From the solution method for linear ODEs (integration factor $I = e^{\int -\frac{1}{x} dx} = \dfrac{1}{x}$),

the general solution is

$$\frac{1}{xy^3} = x + c.$$ Q.E.D.

2. Function Differential Equation

Consider a first-order ODE of the form:

$$(\frac{dv}{dy})y' + P(x)v(y) = Q(x) \quad \cdots ①$$

Observing the two terms on the left side of the equation, the chain rule tells us that the derivative of $v(y)$ (with respect to x) is equal to $(\dfrac{dv}{dy})y'$.

This suggests that we choose $v(y)$ as a new variable. In fact: if assuming $u = v(y)$, then the original equation becomes

$$u' + P(x)u = Q(x).$$

We have returned to the standard form of a linear ODE.

Example 6

$(2\cosh y + 3x)dx + (x \sinh y)dy = 0$.

Solution   Let's first rearrange the original equation to the form of ①, then divide both sides by xdx and simplify. We obtain

$$\sinh(y)y' + \frac{2}{x}\cosh y = -3 .$$

Therefore, we can assume $u = \cosh y$ in formula ①, then the original equation becomes $u' + \frac{2}{x}u = -3$. From the solution method for linear ODEs (integration factor

$I = e^{\int \frac{2}{x}dx} = x^2$), the general solution is

$$x^2 \cosh(y) = -x^3 + c.$$

$\boxed{\text{Q.E.D.}}$

1-4-3 Riccati Equation (Selecting Reading)

We refer to equations of the following form as Riccati equations:

$$y' = P(x)y^2 + Q(x)y + R(x).$$

Mathematicians Riccati (1676-1754, Italy) and Bernoulli have conducted in-depth studies on this equation. Before attempting to solve it, let's make some observations. The most troublesome term in the original equation is $P(x)y^2$. Without it, we would be back to a linear case. The question is, how can we eliminate $P(x)y^2$?

Here, we borrow an interesting idea. Let's recall the simplest ODE mentioned at the beginning of this chapter: $y' = c$. Its solution represents a straight line with a slope of c. When we shift a line with a slope of c, the resulting line still has a slope of c. Therefore, all solutions can be obtained by shifting a certain line. We refer to this phenomenon as the translation invariance of solutions. Now, does the Riccati equation possess translation invariance?

Assume y_p is a particular solution. Let's perform a translation by $y_p + c$ and substitute it into the original equation. We obtain:

$$2P(x)y_p(x) + cP(x) + Q(x) = 0 \cdots ②$$

Surprisingly, the term $P(x)y^2$ is eliminated. Therefore, to eliminate $P(x)y^2$ we just need to perform a translation on the particular solution. There are two choices:

(1) Assume $y = y_p + \frac{1}{u}$, which leads back to a linear ODE.

(2) Assume $y = y_p + u$, which results in the Bernoulli equation
    Let's consider the following examples to illustrate these cases.

Example 7

$$y' = \frac{1}{x^2} y^2 - \frac{1}{x} y + 1 \,; y\,(1) = 3.$$

Solution We assume $c = 0$ in the formula ②, and assume the particular solution has the form of kx, substituting to the original equation then we can obtain $k^2 - 2k + 1 = 0$, the solution is $k = 1$. Therefore the particular solution is $y_p(x) = x$. Next, adopting the first choice, let's set $y = x + \frac{1}{u}$ to transform the original equation into a first-order linear

ODE: $u' + \frac{1}{x} u = \frac{-1}{x^2}$. The integrating factor can be obtained through Table 3 in

Sections 1-2 $I = e^{\int \frac{1}{x} dx} = x$, we obtain the general solution

$$y = x + \frac{x}{c - \ln|x|}.$$

Substituting the given condition, we obtain $c = \frac{1}{2}$, so the solution is

$$y = x + \frac{x}{\frac{1}{2} - \ln|x|}.$$

Q.E.D.

1-4 Exercises

Basic questions

Solve the following first-order ODE.

1. $\dfrac{dy}{dx} + y = e^{3x}$.

2. $y' - 2xy = x$, $y(0) = 1$.

3. $y' + 3x^2 y = x^2$.

4. $y' - xy = x^3$, $y(0) = 0$.

5. $y' - y = e^{2x}$.

6. $y' + 3y = 3$, $y(\dfrac{1}{3}) = 2$.

7. $y' + xy = xy^{-1}$.

8. $y' - \dfrac{1}{x}y = -xy^2$.

9. $y\dfrac{dx}{dy} - x = 2y^2$, $y(1) = 5$.

10. $y' + \dfrac{1}{x}y = 2$.

Advanced questions

Solve the following first-order ODE.

1. $x\dfrac{dy}{dx} - y = x^2 \sin x$.

2. $(x+1)\dfrac{dy}{dx} + y = \ln x$, $y(1) = 10$.

3. $y' + (\tan x)y = \cos^2 x$, $y(0) = 1$.

4. $y' + y\tan x = \sin x$, $y(0) = 1$.

5. $y' + y\tan x = \sec x$.

6. $y'\cos y + 2x\sin y = 2x$.

7. $y' + y(\dfrac{2}{t} - \dfrac{t}{4}) = \dfrac{1}{t}$.

8. $y' = \dfrac{1}{6e^y - 2x}$.

9. $e^y y' - e^y = x - 1$.

10. $xy' + (1+x)y = e^x$.

11. $x\dfrac{dy}{dx} + 2y = 3$.

12. $\dfrac{dr}{d\theta} + r\sec\theta = \cos\theta$.

13. $x\dfrac{dy}{dx} + 4y = x^3 - x$.

14. $\cos x\dfrac{dy}{dx} + (\sin x)y = 1$.

15. $dy + 2xydx = xe^{-x^2} y^3 dx$.

16. $y' + \dfrac{1}{x}y = 3x^2 y^3$.

17. $y^2 dx + (x^3 y + xy)dy = 0$.

18. $y' - \dfrac{2}{x}y = \dfrac{-1}{x}y^2$.

1-5 Solving First-Order ODEs with the Grouping Method

1-5-1 Total Differential Formula

In the previous sections, we discussed the process of transforming equations into linear ODEs through variable transformations and other methods, followed by solving them using integrating factors. In this section, we will discuss more complex nonlinear first-order ODEs, and the method of solving them is quite different: the total differential formula. Mathematically, the total differential is an operator that has the following form:

$$d(\phi) = \frac{\partial(\phi)}{\partial x} dx + \frac{\partial(\phi)}{\partial y} dy .$$

You can substitute any differentiable function (of two variables) into the operator ϕ to obtain a total differential formula (of course, you can extend the formula to the case of multiple variables as well). Table 4 summarizes some commonly used total differential formulas. Readers are encouraged to substitute functions into the definition of total differential and verify them using the rules of differentiation, chain rule, and other differential formulas.

Table 4

Number	Function	Total derivative	Function characteristics
①	$x \pm y$	$dx \pm dy$	Polynomial
②	$x^2 \pm y^2$	$xdx \pm ydy$	
③	xy	$ydx + xdy$	
④	$x^m y^n$	$x^{m-1} y^{n-1}[mydx + nxdy]^{10}$	
⑤-1	$\dfrac{x}{y}$	$\dfrac{ydx - xdy}{y^2}$	Total differential is related to $ydx - xdy$
⑤-2	$\dfrac{y}{x}$	$-\dfrac{ydx - xdy}{x^2}$	
⑤-3	$\ln(\dfrac{x}{y})$	$\dfrac{ydx - xdy}{xy}$	
⑤-4	$\tan^{-1}(\dfrac{x}{y})$	$\dfrac{ydx - xdy}{x^2 + y^2}$	

---

[10] $mydx + nxdy = \dfrac{d(x^m y^n)}{x^{m-1} y^{n-1}}$

In the section on exact ODEs, we used an integrating factor to transform the original equation into a total differential form in order to find the general solution. However, by directly using the total differential formula, even many non-linear ODEs can be solved directly. Take $(3y^2 + 3y)dx + (4xy + 3x)dy = 0$ for example, we combine polynomials with the same total degree to obtain:

$$y(3ydx + 4xdy) + 3(ydx + xdy) = 0.$$

Referring to the formulas in Table 4, this expression can be transformed into

$$\frac{d(x^3y^4)}{x^2y^2} + 3d(xy) = 0.$$

Therefore, after finding a common denominator and integrating both sides of the equation, the general solution is given by $x^3y^4 + x^3y^3 = c$. In fact, x^2y^2 is the integrating factor for the original equation. Indeed, if we can transform the equation into a total differential form using the full differential formula, then we no longer need to go through the process of linearization and searching for integrating factors. The full differential formula can directly help us find the integrating factor for the equation.

Example 1

$$\frac{dy}{dx} = \frac{x-y}{x+y}.$$

Solution   By rearranging the original equation as $(xdy + ydx) - xdx + ydy = 0$, Using the formula ③ from Table 4, we know $d(xy) = ydx + xdy$, so the original equation is equal to $d(xy) - xdx + ydy = 0$. Then integrating both sides of the equation

$$\int d(xy) - \int xdx + \int ydy = \int 0,$$

the general solution is

$$xy - \frac{1}{2}x^2 + \frac{1}{2}y^2 = c.$$

Example 2

$$\cos ydx - 2(x-y)\sin ydy - \cos ydy = 0.$$

Solution   By rearranging the original equation as $\cos y(dx - dy) - 2(x-y)\sin ydy = 0$. Through formula ①, we know $d(x - y) = dx - dy$, so the original equation is equal to $\cos yd(x - y) - 2(x-y)\sin ydy = 0$. Then integrating both sides of the equation

$$\int \frac{1}{(x-y)}d(x-y) = \int 2\frac{\sin y}{\cos y}dy,$$

the general solution is

$$(x-y)\cos^2 y = c.$$ Q.E.D.

Example 3

$$y' = \frac{2 + y\cos(xy)}{-x\cos(xy)} .$$

Solution   By rearranging the original equation as $\cos(xy)[xdy + ydx] + 2dx = 0$. Through formula ③, we know $d(xy) = ydx + xdy$, so the original equation is equal to $\cos(xy)d(xy) + 2dx = 0$. Then integrating both sides of the equation

$$\int \cos(xy)d(xy) + \int 2dx = \int 0 ,$$

the general solution is

$$\sin(xy) + 2x = c. \qquad \text{Q.E.D.}$$

Example 4

Solve $y' + \frac{1}{x}y = 3x^2y^3$.

Solution   By multiplying both sides of the equation by xdx, we can rearrange the original equation as $xdy + ydx = 3x^3y^3dx$. Using the formula ③ from Table 4, we know $d(xy) = 3(x^3y^3)dx$.
Then integrating both sides of the equation

$$\int \frac{1}{x^3y^3} \cdot d(xy) = \int 3dx ,$$

the general solution is

$$-\frac{1}{2}\frac{1}{(xy)^2} = 3x + c . \qquad \text{Q.E.D.}$$

Example 5

Solve $y^2 + (x^2 - xy)y' = 0$.

Solution   By rearranging the original equation as $y(ydx - xdy) + x^2dy = 0$, dividing both sides of the equation by yx^2 then we obtain $\frac{ydx - xdy}{x^2} + \frac{1}{y}dy = 0$.

Now through formula ⑤-2 from Table 4, we know $d(\frac{y}{x}) = -\frac{ydx - xdy}{x^2}$.

Therefore, we obtain $-d(\frac{y}{x}) + \frac{1}{y}dy = 0$. Then integrating both sides of the equation

$$-\int d(\frac{y}{x}) + \int \frac{1}{y}dy = \int 0 ,$$

the general solution is

$$-\frac{y}{x} + \ln|y| = c . \qquad \text{Q.E.D.}$$

Example 6

$(6x^2 - 3xy)\dfrac{dy}{dx} + 9xy - 2y^2 = 0$.

Solution By rearranging the original equation as

$$3x(3ydx + 2xdy) - y(2ydx + 3xdy) = 0,$$

now formula ④ from Table 4 tells us $\dfrac{d(x^m y^n)}{x^{m-1}y^{n-1}} = (my)dx + (nx)dy$. In this case,

we let $m = 3$, $n = 2$ (and $m = 2$, $n = 3$), then formula ④ can be rearranged as

$3\dfrac{d(x^3 y^2)}{xy} - \dfrac{d(x^2 y^3)}{xy} = 0$. Multiplying both sides of the equation by (xy) and

integrating both sides, the general solution is

$$3x^3 y^2 - x^2 y^3 = c.$$

Q.E.D.

1-5-2 Summary of First-Order ODE Solution Methods

We have learned various solution methods for first-order ODEs. In summary, the approach is to first determine if the equation is linear. If it is not linear, we then examine if it can be transformed into a linear form (Sections 1-1 to 1-4). If finding the integrating factor is challenging, we can even directly use the method of exact differentials (Section 1-5), by passing the linearization process. We have summarized the interpretation of these solution methods in Figure 1-7 to assist in quickly finding the appropriate method.

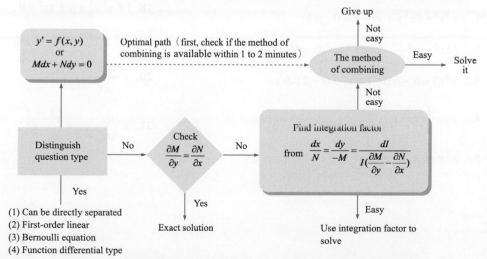

Figure 1-7 Flowchart for solving first-order ordinary differential equation

1-5 Exercises

Basic questions

Try using the grouping method to solve the following ODE.

1. $xdy + ydx = 0$.

2. $xdx + ydy + (x^2 + y^2)dy = 0$.

3. $xdy + ydx - xydy = 0$.

4. $4ydx + 3xdy = 0$.

5. $xy' = x + y$, $y(1) = 1$.

6. $y' + \dfrac{1}{x}y = xy^2$.

7. $xy' + 2y = e^{x^2}$.

8. $xy' + y + 4 = 0$.

9. $(y + x^2)dx - xdy = 0$.

10. $y'[2 - x\cos(xy)] - y\cos(xy) = 0$.

Advanced questions

Try using the grouping method to solve the following ODE.

1. $x^3y' + 3x^2y + x^2 - 1 = 0$, $y(1) = 1$.

2. $(2x^3 - y^3 - 3x)dx + 3xy^2dy = 0$.

3. $y^2(3ydx - 6xdy) = x(ydx + 2xdy)$.

4. $y' - \dfrac{1}{x}y = x^2 + 2$.

5. $x^3y' = x^2y - y^3$.

6. $xy' = \dfrac{2y^2}{x} + y$.

7. $3x^2y' - y^2 - 3xy = 0$.

8. $(x^2 + 3y^2)dx - 2xydy = 0$.

9. $ye^{xy}\dfrac{dx}{dy} + xe^{xy} = 12y^2$, $y(0) = -1$.

10. $(3x^2 - y^2)y' - 2xy = 0$.

11. $y = (y^4 + 3x)y'$.

12. $(2y^2 - 6xy)dx + (3xy - 4x^2)dy = 0$.

13. $y' - 2xy = x^2 + y^2$.

14. $xy' - y = \dfrac{y}{\ln y - \ln x}$.

15. $y' + \dfrac{y}{x} = (\ln x) \cdot y^2$, $y(1) = 1$.

16. $y' - \dfrac{2}{x}y = 4x$, $y(1) = 2$.

17. $x^2y' - xy = y^3$.

18. $1 + x^2y^2 + y + xy' = 0$.

19. $(x^2 + y^2 + x)dx + xydy = 0$.

20. $y' - \dfrac{y}{x} = x^3y^2$.

21. $x \cdot \dfrac{dy}{dx} - y = \dfrac{x^3}{y} \cdot e^{\frac{y}{x}}$.

1-6　Application of First-Order ODEs

ODEs have their origins in real-life problems, and in this section, we will showcase several modeling examples from practical life or physical systems to illustrate the origin of ODEs. By solving the corresponding ODEs, we can then explain the underlying physics, engineering, or geometric phenomena that led to their formulation.

1. Orthogonal Trajectories

Consider the function $F(x, y) = c$, where its graph represents a family of curves in the plane. As c varies, different curves are generated, as shown in Figure 1-8. We want to know how to find its orthogonal curve family?

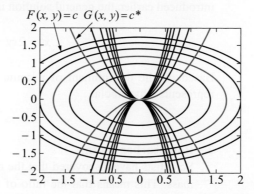

$F(x, y) = c$　$G(x, y) = c*$

Figure 1-8　Orthogonal Curve Family Illustration

First, we know that the product of the slopes of two perpendicular lines is -1. Therefore, if we can determine the slope of the tangent line to F, we can use it to formulate the equation for the orthogonal curves and solve them. By differentiating both sides of $F(x, y) = c$ with respect to x, we obtain $\dfrac{\partial F}{\partial x} + \dfrac{\partial F}{\partial y}\dfrac{dy}{dx} = 0$, thus, the

slope of the tangent line to $F(x, y) = c$ is $\dfrac{dy}{dx} = -\dfrac{\dfrac{\partial F}{\partial x}}{\dfrac{\partial F}{\partial y}}$. Therefore, the slope of the

orthogonal curves is

$$\frac{dy}{dx} = +\frac{F_y}{F_x} .$$

To solve this ODE, we can integrate it and obtain the equation for the orthogonal curves. Common engineering problems involving orthogonal curves include the relationship between electric field lines and equipotential lines in electromagnetics, as mentioned in Section 1-1. Here is an example:

Example 1

Find the orthogonal curve family of $x^2 + 2xy - y^2 + 4x - 4y = c$, and verify that the orthogonal curves are hyperbolas.[11]

Solution Assuming $F(x, y) = x^2 + 2xy - y^2 + 4x - 4y$, the slope of the orthogonal curves is

$$\frac{dy}{dx} = \frac{F_y}{F_x} = \frac{2x-2y-4}{2x+2y+4}.$$

We obtain $(2x + 2y + 4)dy = (2x - 2y - 4)dx$ after rearrangement. Using the methods introduced earlier, the general solution is

$$x^2 - 2xy - y^2 - 4x - 4y = k.$$

Based on the discriminant of a conic section, $(-2)^2 - 4 \cdot 1(-1) = 8 > 0$, the curve represents a hyperbola.

Q.E.D.

2. **Radiocarbon Dating**

Scientists have discovered that the rate of decay of ^{14}C is directly proportional to its mass. This means that the ratio of the mass of ^{14}C, $M(t)$, to its rate of change is a constant. This can be expressed using an ordinary differential equation (ODE) as:

$$\frac{dM}{dt} = -\alpha M ,$$

where α is a proportionality constant. For example, this can be used to determine the age of dinosaurs.

Example 2

In the Jurassic period, ^{14}C was used to estimate the age of dinosaurs. If it is known that the ^{14}C remaining of a dinosaur has only $\frac{1}{2000}$ of its initial amount, what is the approximate age of the dinosaur? (The half-life of ^{14}C is 5000 years.)

---

[11] In conic section $Ax^2 + Bxy + Cy^2 + Dx + Ey + F = 0$, we can determine the figure according to the value of $\Delta = B^2 - 4AC$:

(1) $\Delta > 0$ hyperbola (2) $\Delta = 0$ parabola (3) $\Delta < 0$ oval

Solution   Let $M(t)$ be the mass of ^{14}C, then $\dfrac{dM}{dt} = -\alpha M$, using the solution obtained

from the previous first-order ODE method, we obtain $M(t) = ke^{-\alpha t}$. Setting the initial mass as $M(0) = M_0$ and substituting the initial condition $M(0) = M_0$, it tell us $k = M_0$, that is $M(t) = M_0 e^{-\alpha t}$. Given that the half-life of ^{14}C is 5000 years, we have:

$M(5000) = \dfrac{1}{2} M_0$, that is $\dfrac{1}{2} M_0 = M_0 e^{-5000\alpha}$. We obtain

$$\alpha = \frac{\ln 2}{5000}$$

and the formula for $M(t)$ is $M(t) = M_0 \cdot \exp[-\dfrac{\ln 2}{5000} t]$.

From the given information, we know $\dfrac{1}{2000} M_0 = M_0 \cdot \exp[-\dfrac{\ln 2}{5000} t]$. The solution is the dinosaur lived approximately 54829 years ago.   Q.E.D.

3. Mixing Problem

As shown in Figure 1-9, if there is both inflow and outflow of a solution in a uniformly mixed container, the rate of change of solute concentration can be expressed as the inflow rate minus the outflow rate. Now, we will try to model this phenomenon. Let's assume that the container initially contains T_0 liters of saltwater with a concentration of $C_0 (kg/\ell)$. At a rate of V_i liters per minute, saltwater with a concentration of $C_i (kg/\ell)$ is injected into the container, and after sufficient mixing, it is withdrawn from the container at a rate of V_0 liters per minute. Assuming that the remaining salt content in the container is $x(t)$, we know that concentration is the ratio of salt content to volume. Therefore, we can write the equation as follows:

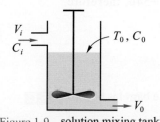

Figure 1-9   solution mixing tank

$$\frac{dx}{dt} = \text{Input rate} - \text{Output rate} = V_i C_i - \frac{V_o x}{(T_0 + (V_i - V_o)t)}$$

and $x(0) = T_0 \cdot C_0$, we can find $x(t)$ by sovling the ODE above.

Example 3

In the mixing problem of two different concentration saltwater, the total amount of salt in the mixer can be described by a first-order ordinary differential equation (ODE). Suppose there is a mixing container containing 300 gal of saltwater, and another saltwater with a flow rate of 3 gal per minute is injected into the mixing container. After sufficient mixing, the mixture is drained from the container at the same rate.

(1) If the concentration of the solution flowing into the mixing container is 2 lb/gal, please establish a model for the salt content in the mixing container at any time.

(2) Assuming initially there are 60 lb of salt dissolved in the initial 300 gal of the mixing container, how much salt will be present in the mixing container after a long period of time?

Solution

(1) Assuming when time is t, the salt content in the container is $x(t)$, so

$$\frac{dx}{dt} = 2 \times 3 - \frac{x(t)}{300} \times 3 ,$$

that is

$$\frac{dx}{dt} + \frac{1}{100} x(t) = 6 .$$

(2) The relationship between the salt content $x(t)$ and time can be obtained by solving the above equation. By using the solution method for linear ODEs (integration factor $I(t) = e^{\frac{t}{100}}$), the general solution is $x(t) = 600 + ce^{-t/100}$. From the initial condition $x(0) = 60 = 600 + c$ we find that $c = -540$, therefore $x(t) = 600 - 540e^{-t/100}$ and

$$\lim_{t \to \infty} x(t) = 600 .$$

Therefore, when a long time has passed, the salt content in the barrel is 600 lb. Q.E.D.

4. **Law of Cooling**
    To understand the effect of fluid temperature on the temperature of an object it contacts, we can use Newton's law of cooling, discovered by Newton himself. The rate of change of an object's temperature is proportional to the difference between the instantaneous temperature of the object and the temperature of the fluid it is in contact with. If we express this phenomenon in mathematical symbols, let T be the temperature of the object at time t, and let T_1 be the temperature of the object at $t = 0$, and T_0 be the temperature of the fluid (typically air). Therefore, according to Newton's law of cooling, the relationship between the temperature of the object and time can be expressed as:

$$\frac{dT}{dt} = k(T - T_0) ,$$

where k is a proportionality constant and $T(0) = T_1$.

Example 4

A cake was just taken out of the oven with an initial temperature of 300°F. After three minutes, the temperature drops to 200°F, and the room temperature is 70°F. How long (starting from the time it was taken out of the oven) does it take for the cake's temperature to reach the closest value to 70.5°F?

Solution   Let T represent the temperature of the cake at time t. According to Newton's law of cooling, we know $\dfrac{dT}{dt} = k(T-70)$ (k is a constant). Separating variables, we obtain $\displaystyle\int \dfrac{dT}{T-70} = \int kdt$.

The general solution is $T(t) = 70 + ce^{kt}$. Now through the known conditions:

① when $t = 0$, $T(0) = 300$, we obtain $c = 230$, that is $T(t) = 70 + 230e^{kt}$.

② when $t = 3$, $T(3) = 200$, we obtain $e^k = (\dfrac{130}{230})^{\frac{1}{3}}$.

Therefore, the relationship between temperature and time is:

$T(t) = 70 + 230(\dfrac{130}{230})^{\frac{1}{3}t}$. Assuming $70.5 = 70 + 230(\dfrac{130}{230})^{\frac{t}{3}}$, the solution is $t \approx 32.238$ minutes. At this time, the temperature of the cake is closest to 70.5°F.   Q.E.D.

5. **First Order Circuit**

First, let's introduce the voltage drop units for some common circuit components:

(1) **Resistor**

$ER = RI$ (R is resistor, the unit is ohms; I is electric current, the unit is ampere).

(2) **Inductance**

$E_L = L\dfrac{dI}{dt}$ (L is inductance, the unit is henrys).

(3) **Capacitance**

$E_C = \dfrac{1}{C}Q = \dfrac{1}{C}\displaystyle\int_0^t I(\tau)d\tau$,

where Q is electric charge (the unit is coulombs); C is capacitance (the unit is farads).

6. **The Nature of the Circuit**

According to Kirchhoff's voltage law, we know that the sum of voltage drops around a closed loop is equal to zero. Therefore, we can derive the following relationship:

(1) In an RL circuit (as shown in Figure 1-10), the relationship between current and time satisfies the following equation:

$$L\frac{dI}{dt} + RI = E(t).$$

(2) In an RC circuit (as shown in Figure 1-11), the relationship between current and capacitance satisfies the following equation:

$$RI + \frac{1}{C}\int_0^t I(\tau)d\tau = E(t) \quad \text{or} \quad R\frac{dI}{dt} + \frac{1}{C}I = \frac{dE}{dt}.$$

In the above three equations, $E(t)$ represents the electromotive force.

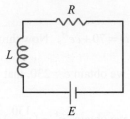

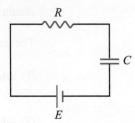

Figure 1-10 RL circuit   Figure 1-11 RC circuit

Example 5

In a closed RL circuit with $R = 10\Omega$, $L = 2H$, and a power supply of 100V, if the initial current is 0, please calculate the current at $t = 0.2$ seconds?

Solution According to Kirchhoff's voltage law, we know $2\frac{dI}{dt} + 10I = 100$ and $I(0) = 0$. The above equation is first-order linear ODE, where its integration factor is e^{5t}, so $e^{5t}\frac{dI}{dt} + 5e^{5t}I = 50e^{5t}$ is exact, that is $e^{5t}I = \int 50e^{5t}dt + k$. Through $I(0) = 0$ we know $k = -10$, so $I(t) = 10 - 10e^{-5t}$. Therefore, the current at $t = 0.2$ seconds can be obtained as $I(0.2) = 10 - 10e^{-1}$. Q.E.D.

1-6   Exercises

Basic questions

1. Given that the slope of a curve is $x + 1$ and it passes through the point (2, 5), find the equation of the curve.

2. Given a curve family represented by the equation $x^2 + y^2 = c$, find its orthogonal curve family.

3. Assume $F(x, y, a) = 0$, $G(x, y, b) = 0$ be two sets of mutually orthogonal curves.
   If $F(x, y, a) = x^2 - y^2 - a^2$, what is $G(x, y, b) = ?$

4. A water tank contains 200 liters of a saltwater solution with a concentration of 1.0 gram/liter. Clear water is added to the tank at a rate of 2 liters per minute, while the tank is simultaneously drained at the same rate. Assuming the solution is uniformly mixed during the addition and drainage processes, how long does it take for the concentration of salt in the solution to become 1% of its original concentration?

5. Physics states that electric field lines and equipotential lines are orthogonal to each other. In an experiment, two point charges of equal magnitude are located at $(-1, 0)$ and $(1, 0)$ respectively. It has been confirmed that the equation for the electric field lines established by these two charges in their vicinity is
   $x^2 + (y - c)^2 = 1 + c^2$ (where c is a constant).
   Find the equation for the equipotential lines in this case.

Advanced questions

1. Find the orthogonal family of curves to the curve family $x^2 + y^2 = cx$, where c is a constant.

2. Find the orthogonal family of curves to the curve family $x^2 + cy^2 = 4$, where c is a constant.

3. Find the orthogonal family of curves to the curve family $F(x, y, k) = 2x^2 - 3y = k$, where k is a constant.

4. Consider a single loop closed RC circuit with the following parameters: $R = 10\Omega$, $C = 10^{-3}F$ and a power supply of 100 V. If the initial current is 5 and the charge on the capacitor is $q(t)$, and if
   $i(t) = \dfrac{dq(t)}{dt}$, find the current in the circuit at any given time.

5. Experimental evidence shows that the rate of change of an object's temperature is directly proportional to the difference between its temperature and the ambient temperature (ignoring temperature gradients in the object). Now, consider a sample initially at 1000°C placed directly in a 25°C atmospheric cooling environment. After one hour, the temperature of the sample is measured to be 80°C.
   (1) Based on the given conditions, please define the variables and assign their physical meanings, along with their units.
   (2) Based on the given conditions, please write the differential equation describing the cooling behavior of the sample and solve it.
   (3) Try to find out from the solution function that the sample is cooled to 25°C (calculated at 25.01°C) required time.

6. The age of a fossil bone is determined to be how long? if its ^{14}C content is one-thousandth of the original content. (Given that the half-life of ^{14}C is 5600 years)

7. A mixing container with a capacity of 400 gallons is initially filled halfway and contains 50 grams of salt dissolved in it. If another solution of 2 grams per gallon is injected into the mixing container at a rate of 10 gallons per minute, and after thorough mixing, the solution is drained at a rate of 4 gallons per minute. What is the amount of salt in the mixing container when the solution overflows?

2

High-Order Linear Ordinary Differential Equation

Euler
(1707~1783,
Switzerland)

Leonhard Euler was one of the most outstanding mathematicians of the 18th century. He made significant contributions not only in the field of curved geometry but also in differential equations, mechanics, number theory, algebra, elementary functions, and complex variable functions. Euler was the first mathematician to systematically study higher-order differential equations and introduced the concept of the characteristic equation, which fundamentally provided a general form for solutions of linear differential equations.

Learning Objectives

2-1 Basic Theories	2-1-1	Learn to use differential operators	
	2-1-2	Learn to use Wronskian determinants	
	2-1-3	Master the techniques for solving regular ODE	

2-2 Solving Higher-Order ODE with the Reduction of Order Method	No section	Master the reduction of order method

2-3 Homogeneous Solutions of Higher-Order ODEs	2-3-1	Learn to use the characteristic equation to find homogeneous solutions

2-4 Finding Particular Solution Using the Method of Undetermined Coefficients	2-4-1	Understand the assumptions and principles of the method of undetermined coefficients
	2-4-2	Learn to apply the method of undetermined coefficients through examples

2-5 Finding Particular Solution Using the Method of Variation of Parameters	No section	Master the parametric variation method for finding particular solutions

2-6 Finding Particular Solution Using the Method of Inverse Differential Operators	2-6-1	Understanding the inverse operator
	2-6-2	Mastering the formulas and applications of commonly used inverse operators

2-7 Equidimensional Linear ODEs	2-7-1	Learn to solve Euler-Cauchy and other higher-order ODEs
	2-7-2	Learn to solve Legendre and other higher-order ODEs

2-8 The Applications of Higher-Order ODEs in Engineering	2-8-1	Master the modeling and solution of 'mass-damper-spring' systems
	2-8-2	Master the modeling and solution of single (or multiple) loop electrical circuit systems
	2-8-3	Master the modeling and solution of feedback-controlled solution mixing systems

ExampleVideo

In the first chapter of this book, we have already discussed the solution methods for first-order ordinary differential equations (ODEs) and their applications in engineering. However, in physical systems such as mass-spring-damper systems, RLC circuits, feedback-controlled mixing of solutions, and beam deformations, higher-order differential equations are needed to describe them. Therefore, this chapter will introduce the solution methods for higher-order linear ODEs and utilize them to solve some common engineering problems. Solving higher-order linear ODEs requires more theoretical background. **If the reader's goal is to focus on solving these equations, they can skip Sections 2-1 and 2-2 and directly proceed to Section 2-3 to learn how to solve higher-order linear ODEs.**

2-1 Basic Theories

Higher-order ODEs encompass both higher-order linear and higher-order nonlinear equations. However, most higher-order nonlinear problems require the use of numerical methods and simulation software such as MATLAB or Mathematica for their solution. Moreover, the dynamic behavior and phenomena associated with these equations can be quite complex, including the possibility of chaotic behavior. It is a highly intricate field of science.

Therefore, in this chapter, we will focus only on the study of idealized physical systems commonly encountered in engineering, for which the equations can be mostly simplified into higher-order linear ODEs. We will first investigate the form of solutions for higher-order linear ODEs and explore how to utilize linearly independent solutions to form the solution space. We will also examine the methods for solving both homogeneous and non-homogeneous linear ODEs. The ODEs with the following form will be referred to as higher-order linear ODEs:

$$a_n(x)y^{(n)} + a_{n-1}(x)y^{(n-1)} + \cdots + a_1(x)y' + a_0(x)y = R(x).$$

The solution of higher-order ODEs no longer relies on various calculus techniques, such as converting them into exact differentials. Therefore, we need to resort to a more abstract approach to interpret higher-order ODEs using "operators."

In the past, when studying calculus, we learned that differentiation follows the distributive law with respect to addition, subtraction, multiplication, and division. In other words, for any constant c, we have the relationship:

$$\frac{d}{dx}[f(x) \pm cg(x)] = \frac{d}{dx}f(x) \pm c\frac{d}{dx}g(x).$$

If we isolate a point and consider it, $\dfrac{d}{dx}$ it resembles an 'operation' on the function space. For convenience, let's define $D = \dfrac{d}{dx}$, we refer to this concept as an 'operator.'

Moreover, if we look at the first-order ODE from Chapter 1, $\dfrac{d}{dx}y + P(x)y = Q(x)$ can also be seen as $\dfrac{d}{dx} + P(x) = D + P(x)$ and interpret it as acting on y to yield $Q(x)$.

Next, we will further abstract and apply this concept to the solution of higher-order ODEs.

2-1-1 Differential Operator

▶ **Definition 2-1-1**

n-th Order Linear Differential Operator

We refer to the following sequence of expressions that involve both differentiation $\dfrac{d^i}{dx^i}$ and multiplication by functions $a_i(x)$ as an *n*-th order linear differential operator

$$a_n(x)\frac{d^n}{dx^n} + a_{n-1}(x)\frac{d^{n-1}}{dx^{n-1}} + \cdots + a_1(x)\frac{d}{dx} + a_0(x).$$ ◀

We use the symbol $L(D)$ to represent the aforementioned operator, where $D = \dfrac{d}{dx}$,

$D^2 = \dfrac{d^2}{dx^2}$, $\cdots$, $D^n = \dfrac{d^n}{dx^n}$. You can rewrite any linear ODE from Chapter 1 into the form of the example given in the introduction. In fact, higher-order linear ODEs can be directly represented as $L(D)y = R(x)$. Here, it is important to distinguish:

1. **R(x) = 0**
   The original equation $L(D)y = 0$ is called Homogeneous.

2. **R(x) ≠ 0**
   The original equation $L(D)y = R(x)$ is called Non-Homogeneous.

This is related to the subsequent solving process, where it is important for the readers to establish the concept that homogeneity is the foundation for solving non-homogeneous problems.

Example 1

Please rewrite the following ODE in the form of a differential operator $L(D)y = R(x)$, where $D = \dfrac{d}{dx}$.

(1) $y'' - 3y' + 2y = 0$.

(2) $5y''' + 2y'' - 3y' + 2y = 5\sin x$.

(3) $x^2 y'' - 3xy' + 2y = 0$.

(4) $(x+1)^2 y'' - 3(x+1)y' + 2y = 5(x+1)$.

Solution

(1) $y'' - 3y' + 2y = (D^2 - 3D + 2)y$; Second-order linear homogeneous constant coefficient ODE.

(2) $5y''' + 2y'' - 3y' + 2y = (5D^3 + 2D^2 - 3D + 2)y = 5\sin x$; Third-order linear non-homogeneous constant coefficient ODE.

(3) $x^2 y'' - 3xy' + 2y = (x^2 D^2 - 3xD + 2)y = 0$; Second-order linear homogeneous variable coefficient ODE.

(4) $[(x+1)^2 D^2 - 3(x+1)D + 2]y = 5(x+1)$; Second-order non-homogeneous linear variable coefficient ODE. $\boxed{\text{Q.E.D.}}$

2-1-2 Linear Independence and Linear Dependence of Functions

Recalling the Cartesian coordinate system learned in high school, on $\mathbf{R}^3$, a vector on any coordinate axis cannot be replaced by vectors on other axes. For example, the vector $i = (0, 0, 1)$ on the x-axis cannot be replaced by the vector $j = (0, 1, 0)$ on the y-axis. This inability to be replaced is what we refer to as "linear independence" (conversely, "linear dependence" is the term used when vectors can be replaced). In fact, the same concept applies to functions as well. Before discussing ODE solutions, let's talk about the independence and dependence of functions, which will help us understand the solution space of homogeneous ODEs. In this chapter, we will initially define linear dependence or independence based on functions, and delve deeper into the abstract definition in the chapter on linear algebra.

▶ **Definition 2-1-2**

Linear Independence and Linear Dependence of Functions

Assume $S = \{u_1, u_2, \cdots, u_n\}$ be a set of functions defined on the interval I.

(1) If a function u_k in S can be expressed as a linear combination of the other $(n-1)$ functions, then we say that $S = \{u_1, u_2, \cdots, u_n\}$ is linearly dependent over the interval I.

(2) If any function u_k in S cannot be expressed as a linear combination of the other $(n-1)$ functions, then we say that $S = \{u_1, u_2, \cdots, u_n\}$ is linearly independent over the interval I. ◀

The above definition is equivalent to the following operational definition using the symbol of linear combination. In fact, this is the form of definition that we often use.

▶ **Definition 2-1-3**

Linear Independence and Linear Dependence of Functions

Assume $S = \{u_1, u_2, \cdots, u_n\}$ be a set of functions defined on the interval I. If the linear combination $c_1 u_1 + c_2 u_2 + \cdots + c_n u_n$ (called formula ①) is zero function, that is $c_1 u_1 + c_2 u_2 + \cdots + c_n u_n = 0$, we have only $c_1 = c_2 = c_3 = \cdots = c_n = 0$, then we say that S is linearly independent on the interval I; Conversely, if there exists a $c_i \neq 0$ that causes formula ① become zero function, then we say that S is linear dependent on the interval I. ◀

Example 2

Please determine whether the following set of functions is linearly independent or linearly dependent.

(1) $S = \{x, x^2\}, x \in [0, \infty)$.

(2) $S = \{1, x, 2x, + 1\}, x \in [0, \infty)$.

Solution

(1) Let $u_1(x) = x$, $u_2(x) = x^2$, $x \in [0, \infty)$, if $c_1 x + c_2 x^2 = 0$ (zero function),
By substituting $x = 1$, $x = 2$, and solving the resulting system of equations, we get $c_1 = c_2 = 0$
$\therefore \{x, x^2\}$ at $[0, \infty)$ is linearly independent.

(2) Let $u_1(x) = 1$, $u_2(x) = x$, $u_3(x) = 2x + 1$, $x \in [0, \infty)$,
if $c_1 \cdot 1 + c_2 x + c_3(2x + 1) = 0 \Rightarrow (c_1 + c_3) + (c_2 + 2c_3)x = 0$ (zero function).
We know that the only linear function that is identically zero is $f(x) = 0$,
so $c_1 = -c_3$, $c_2 = -2c_3$, we obtain c_1, c_2, c_3 not all zeros,
we take $c_1 = -c_3 = -1$, $c_2 = -2$, $c_3 = 1$,
$\therefore \{1, x, 2x + 1\}$ at $[0, \infty)$ is linearly dependent.

From the above example, we can observe that in the set $S = \{x, x^2\}$, the function x cannot be expressed as a linear combination of x^2, and similarly, x^2 cannot be expressed as a linear combination of x. On the other hand, in the set $S = \{1, x, 2x + 1\}$, we can see that the function $2x + 1$ can be expressed as a linear combination of 1 and x. This validates the definition of linear independence and linear dependence

However, not all sets of functions are as easily determined for their dependence or independence as in the above example. In more complex cases, it is not practical to use the original definition. In such cases, we can adopt a matrix perspective. First, we differentiate $n - 1$ times to the linear combination of functions $c_1u_1 + c_2u_2 + \cdots + c_nu_n = 0$ at a specific point $x = a$. This leads to the following system of equations (referred to as system ②) :

$$\begin{bmatrix} u_1(a) & u_2(a) & \cdots & u_n(a) \\ u_1'(a) & u_2'(a) & \cdots & u_n'(a) \\ \vdots & & & \\ u_1^{(n-1)}(a) & u_2^{(n-1)}(a) & \cdots & u_n^{(n-1)}(a) \end{bmatrix} \begin{bmatrix} c_1 \\ c_2 \\ \vdots \\ c_n \end{bmatrix} = \begin{bmatrix} 0 \\ 0 \\ \vdots \\ 0 \end{bmatrix} \cdots ②$$

In the upcoming chapters on linear algebra, we will see that in a homogeneous system of linear equations $AX = 0$, where $X = \begin{bmatrix} c_1 \\ c_2 \\ \vdots \\ c_n \end{bmatrix}$, if the determinant of the coefficient matrix A is non-zero, then $X = 0$. This tells us that the set of functions $\{u_1, u_2, \cdots, u_n\}$ is linear independent. Conversely, if the set of functions $\{u_1, u_2, \cdots, u_n\}$ is linear dependent, then the determinant of the coefficient matrix of system ② is zero. Therefore, we obtain the following important tool called the Wronskian determinant. Using the Wronskian determinant, we can determine whether a set of functions is linearly dependent or independent, which is much more convenient than using the original definition.

▶ **Definition 2-1-4**

Wronskian Determinant

Let $S = \{u_1, u_2, \cdots, u_n\}$ be a set of functions defined on $[a, b]$. If all the functions are at least $n - 1$ times differentiable, we define the determinant

$$\begin{vmatrix} u_1 & u_2 & \cdots & u_n \\ u_1' & u_2' & \cdots & u_n' \\ \vdots & & & \\ u_1^{(n-1)} & u_2^{(n-1)} & \cdots & u_n^{(n-1)} \end{vmatrix}$$

as the Wronskian determinant of the function set S, denoted as $W(u_1, u_2, \cdots, u_n)$. ◀

Note that in the above definition 2-1-4, $W(u_1, u_2, \cdots, u_n)$ is a function, therefore

$$W(u_1, u_2, \cdots, u_n)(x) = \det \begin{bmatrix} u_1(x) & u_2(x) & \cdots & u_n(x) \\ u_1'(x) & u_2'(x) & \cdots & u_n'(v) \\ \vdots & \vdots & \ddots & \vdots \\ u_1^{(n-1)}(x) & u_2^{(n-1)}(x) & \cdots & u_n^{(n-1)}(x) \end{bmatrix}$$

$$= \begin{vmatrix} u_1(x) & \cdots & u_n(x) \\ u_1'(x) & \cdots & u_n'(x) \\ \vdots & \ddots & \vdots \\ u_1^{(n-1)}(x) & \cdots & u_n^{(n-1)}(x) \end{vmatrix}$$

▶ **Theorem 2-1-1**

The Relationship Between the Wronskian Determinant and the

Linear Independence or Dependence

Continuation of symbols in Definition 2-1-4

① If exists $x \in [a, b]$ causes $W(u_1, u_2, \cdots, u_n)(x) \neq 0$, then $S = \{u_1, u_2, \cdots, u_n\}$ is linear independent.

② If $S = \{u_1, u_2, \cdots, u_n\}$ on interval $[a, b]$ is linear dependent, then $W(u_1, u_2, \cdots, u_n) = 0$.

◀

We can check the independence of the function set $S = \{1, x, x^2\}$ for $0 \leq x < \infty$.

Through $W(1, x, x^2) = \begin{vmatrix} 1 & x & x^2 \\ 0 & 1 & 2x \\ 0 & 0 & 2 \end{vmatrix} = 2 \neq 0$, we know S is linear independence

combination by definition 2-1-1.

Example 3

Find the Wronskian determinant of $y_1 = e^{-x} \cos wx$ and $y_2 = e^{-x} \sin wx$ then discuss whether y_1 and y_2 is linear independent?

Solution   Here, assuming $w \neq 0$, then for all $x \in \mathbb{R}$, we have

$$W(y_1, y_2) = \begin{vmatrix} y_1 & y_2 \\ y_1' & y_2' \end{vmatrix}$$

$$= \begin{vmatrix} e^{-x} \cos wx & e^{-x} \sin wx \\ (-e^{-x} \cos wx - we^{-x} \sin wx) & (-e^{-x} \sin wx + we^{-x} \cos wx) \end{vmatrix}$$

$$= w \cdot e^{-2x} \neq 0,$$

we know $\{y_1, y_2\}$ is linear independent by definition 2-1-1 ①.

Example 4

Are the following functions linearly dependent or linearly independent within the given interval? Please explain the reason.

(1) e^{2x}, e^{-2x} ($-\infty < x < \infty$).

(2) $x + 1, x - 1$ ($0 < x < 1$).

(3) $\ln x, \ln x^2$ ($x > 0$).

Solution

(1) For all $x \in \mathbb{R}$, $W(e^{2x}, e^{-2x}) = \begin{vmatrix} e^{2x} & e^{-2x} \\ 2e^{2x} & -2e^{-2x} \end{vmatrix} = -4 \neq 0$,

by definition 2-1-1, we know $\{e^{2x}, e^{-2x}\}$ is linear independent.

(2) Same reason, because $W(x+1, x-1) = \begin{vmatrix} x+1 & x-1 \\ 1 & 1 \end{vmatrix} = 2 \neq 0$,

we know $\{x + 1, x - 1\}$ is linear independent by definition 2-1-1.

(3) Because $\ln x^2 = 2\ln x$, we know $\ln x, \ln x^2$ at $x > 0$ is linear dependent.   **Q.E.D.**

2-1-3   The Solution of Normal ODEs

The solutions of an nth-order linear ODE are closely related to its coefficient functions $a_i(x)$, $i = 0, 1, 2, \cdots, n$. In some cases, there may be special situations where the ODE has no solution or the solution is not unique. However, these cases are not commonly encountered in practical engineering. Here, we will start with the simpler condition where the solutions of the ODE exist and are unique.

▶ **Definition 2-1-5**

Normal ODE

If in nth-order linear ODE $a_n(x)y^{(n)} + a_{n-1}(x)y^{(n-1)} + \cdots + a_1(x)y' + a_0(x)y = R(x)$, variable coefficient $a_n(x)$, $a_{n-1}(x)$, $\cdots$, $a_1(x)$, $a_0(x)$, and $R(x)$ are continuous functions on I, then the ODE is said to be normal.   ◀

For instance, $y'' + 2y' + y = \sin x$, $-\infty < x < \infty$ is normal ODE. And at $y'' + \dfrac{2}{x}y' + y = \sin x$, $-\infty < x < \infty$ is not normal (because at the origin $x = 0$, $\dfrac{2}{x}$ is a discontinuity).

Example 5

Normal ODE $y'' + xy = 0$, the two solutions can be written in series form as follows:

$$y_1 = 1 - \frac{1}{6}x^3 + \frac{1}{180}x^6 - \frac{1}{12960}x^9 + \cdots,$$

$$y_2 = x - \frac{1}{12}x^4 + \frac{1}{504}x^7 - \frac{1}{45360}x^{10} + \cdots,$$

please determine if this ODE is a normal ODE and verify that these two solutions are linear independent.

Solution In this question $a_2(x) = 1$, $a_1(x) = 0$, $a_0(x) = x$ at interval $-\infty < x < \infty$ is continuous function, so $y'' + xy = 0$ is normal ODE, and

$$W(y_1, y_2)(0) = \begin{vmatrix} y_1 & y_2 \\ y_1' & y_2' \end{vmatrix}(0) = \begin{vmatrix} 1 - \frac{1}{6}x^3 + \cdots & x - \frac{1}{12}x^4 + \cdots \\ -\frac{1}{2}x^2 + \cdots & 1 - \frac{1}{3}x^3 + \cdots \end{vmatrix}(0) = \begin{vmatrix} 1 & 0 \\ 0 & 1 \end{vmatrix} = 1 \neq 0$$

therefore by definition 2-1-1, we know $\{y_1, y_2\}$ is linear dependent.   Q.E.D.

As mentioned at the beginning of this chapter, homogeneous equations are the foundation for solving non-homogeneous equations. In fact, you have already encountered this concept in Chapter 1, when considering first-order ODEs

$$(\frac{d}{dx} + 1)y(x) = \frac{d}{dx}y(x) + y(x) = r(x)$$

Imagine that when you don't know about the integrating factor yet, the most direct way to simplify this problem is to assume $r(x) = 0$ (homogeneous equation), which transforms the original equation into a problem that can be answered using the fundamental theorem of calculus. If now $r(x) \neq 0$, it is natural to first shift the homogeneous solution $u(x)$ (satisfies $\frac{d}{dx}u + u = 0$): $u(x) + p(x)$, then substitute it into the original equation $\frac{d}{dx}p(x) + p(x) = r(x)$. In other words, as long as we can find a particular solution $p(x)$ of the original equation, adding it to the homogeneous solution will give us a large class of solutions to the original equation. Next, we will implement this idea in the case of higher-order equations, starting with the homogeneous solution.

▶ **Theorem 2-1-2**

Solution Space of Homogeneous ODE

If $a_n(x)y^{(n)} + a_{n-1}y^{(n-1)} + \cdots + a_1(x)y' + a_0(x)y = 0$ is a normal ODE, and $y_1(x)$, $y_2(x)$, $\cdots$, $y_n(x)$ is a set of n linearly independent solutions, then the general solution is a linear combination of these homogeneous solutions

$$y(x) = c_1y_1(x) + c_2y_2(x) + \cdots + c_ny_n(x).$$

The set formed by this general solution is also called the solution space of the homogeneous ODE.   ◀

Example 6

Please check if e^{-x} and e^{-2x} are solutions of the ODE $y'' + 3y' + 2y = 0$, and determine if they are linear independent solutions. If they are, please write down the general solution to the ODE.

Solution   By substituting the two exponential functions into the ODE, we can verify that both are solutions. Now through the Wronskian determinant

$$W(y_1, y_2) = \begin{vmatrix} y_1 & y_2 \\ y_1' & y_2' \end{vmatrix} = \begin{vmatrix} e^{-x} & e^{-2x} \\ -e^{-x} & -2e^{-2x} \end{vmatrix} = -e^{-3x} \neq 0 \quad \text{(for all } x \in \mathrm{R}),$$

we know $\{e^{-x}, e^{-2x}\}$ is linear independent. So by definition 2-1-2, the general solution is $y(x) = c_1e^{-x} + c_2e^{-2x}$.   Q.E.D.

Exampe 7

Please check e^{-x} and xe^{-x} are solutions of the ODE $y'' + 2y' + y = 0$, and determine if they are linear independent solutions. If they are, please write down the general solution to the ODE.

Solution   By substituting the two exponential functions into the ODE, we can verify that both are solutions.
Now through the Wronskian determinant

$$W(y_1, y_2) = \begin{vmatrix} y_1 & y_2 \\ y_1' & y_2' \end{vmatrix} = \begin{vmatrix} e^{-x} & xe^{-x} \\ -e^{-x} & e^{-x} - xe^{-x} \end{vmatrix} = e^{-2x} \neq 0 \quad \text{(for all } x \in \mathrm{R}),$$

we know $\{e^{-x}, xe^{-x}\}$ is linear independent. So by definition 2-1-2,

we obtain the homogeneous solution $y(x) = c_1e^{-x} + c_2xe^{-x}$.   Q.E.D.

▶ **Theorem 2-1-3**

The General Solution of Non-Homogeneous ODE

For a normal n-th order linear non-homogeneous equation on $x \in [a, b]$,

$$a_n(x)y^{(n)} + a_{n-1}(x)y^{(n-1)} + \cdots + a_1(x)y' + a_0(x)y = R(x)$$

If $y_p(x)$ is one of the particular solutions, and $y_1(x)$, $y_2(x), \cdots, y_n(x)$ are n linear independent solutions of the homogeneous equation, then the general solution of the non-homogeneous ODE is given by

$$y(x) = c_1 y_1(x) + c_2 y_2(x) + \cdots + c_n y_n(x) + y_p(x).$$ ◀

Example 8

Given ODE $y'' + y = x$.

(1) Please check x is whether a particular solution of this ODE.

(2) Please check $c_1 \cos x + c_2 \sin x$ is a homogeneous solution of ODE.

(3) Please write the general solution of this ODE.

Solution

(1) Substituting x into the original ODE satisfies it, so x is a particular solution of the non-homogeneous ODE.

(2) Substituting $\cos x$, $\sin x$ into the homogeneous ODE $y'' + y = 0$ satisfies it, so $\cos x$, $\sin x$ are two homogeneous solutions. Now through the Wronskian determinant,

we know $W(y_1, y_2) = \begin{vmatrix} y_1 & y_2 \\ y_1' & y_2' \end{vmatrix} = \begin{vmatrix} \cos x & \sin x \\ -\sin x & \cos x \end{vmatrix} = 1 \neq 0$

(for all $x \in R$) and, $\{\cos x, \sin x\}$ is linear independent, so by definition 2-1-2 we obtain the homogeneous solution $y(x) = c_1 \cos x + c_2 \sin x$.

(3) Through definition 2-1-3 know that $y = y_h + y_p = c_1 \cos x + c_2 \sin x + x$ is general solution.

Q.E.D.

At this point, the reader may notice that the solvability of a higher-order linear ODE and the existence of particular solutions are closely related. This topic is systematically discussed in Sections 2-4 to 2-6 of this chapter. However, here we can make some attempts to explore it.

Superposition Principe

Continuing with the notation of differential operators, if in $L(D)y = R(x)$, $R(x) = R_1(x) + R_2(x)$, and y_{p_1} is one of the particular solutions of $L(D)y = R_1(x)$, y_{p_2} is one of the particular solutions of $L(D)y = R_2(x)$. Then the particular solution of $L(D)y = R(x)$ is $y_p = y_{p_1} + y_{p_2}$.

Example 9

If $y_{p_1} = -1$ is the particular solution of $y'' - y = 1$ and $y_{p_2} = \dfrac{1}{3}e^{2x}$ is the particular solution of $y'' - y = e^{2x}$.

Please verify $y_p = y_{p_1} + y_{p_2} = -1 + \dfrac{1}{3}e^{2x}$ is the particular solution of $y'' - y = 1 + e^{2x}$.

Solution   By substituting respectively y_{p_1}, y_{p_2} into the equations and verifying, we can see that they are both particular solutions to their corresponding equations.

Therefore, by the principle of superposition, $y_{p_1} + y_{p_2} = -1 + \dfrac{1}{3}e^{2x}$ is a particular solution to the original equation.

Q.E.D.

2-1 Exercises

Basic questions

1. In the following questions, determine whether $y_1(x)$ and $y_2(x)$ are linearly independent of each other.
   (1) $y_1(x) = x$, $y_2(x) = 2x$.
   (2) $y_1(x) = \sin x$, $y_2(x) = 3\sin x$.
   (3) $y_1(x) = e^x$, $y_2(x) = e^{2x}$.

In the following questions, please verify if $y(x)$ is a solution to the ODE.

2. $y'' - y = 0$, $y(x) = c_1 e^x + c_2 e^{-x}$.

3. $y'' + 3y' + 2y = 0$, $y(x) = c_1 e^{-x} + c_2 e^{-2x}$.

4. $\dfrac{d^2 y}{dx^2} + 9y = 0$, $y(x) = c_1 \cos 3x + c_2 \sin 3x$.

5. $x^2 \dfrac{d^2 y}{dx^2} - 2x \dfrac{dy}{dx} + 2y = 0$,
   $y(x) = c_1 x + c_2 x^2$.

Advanced questions

1. Assuming the function set $\{y_1(x), y_2(x)\}$ is as follows, determine under what conditions it is a linearly independent set of functions.
   (1) e^x, x (2) x, x^2 (3) $\sin x, \cos x$ (4) $x, \sin x$
   (5) $\cos x, \cos 3x$.

2. Please prove that the set of functions $\{1, 2x, 3x^2\}$ is linearly independent.

For the following questions 3 to 12, please verify that y(x) is the general solution of ODE:

3. $\dfrac{d^2 y}{dx^2} - 10 \dfrac{dy}{dx} + 25y = 0$,
   $y(x) = c_1 e^{5x} + c_2 x e^{5x}$.

4. $y'' - 2y' + 2y = 0$,
   $y(x) = e^x (c_1 \cos x + c_2 \sin x)$.

5. $x^2 \dfrac{d^2 y}{dx^2} + x \dfrac{dy}{dx} - y = 0$,
   $y(x) = c_1 x + c_2 \dfrac{1}{x}$.

6. $\dfrac{d^2 y}{dx^2} - 2 \dfrac{dy}{dx} + y = 4e^x$,
   $y(x) = c_1 e^x + c_2 x e^x + 2x^2 e^x$.

7. $\dfrac{d^2 y}{dx^2} + 4y = -12 \sin 2x$,
   $y(x) = c_1 \cos 2x + c_2 \sin 2x + 3x \cos 2x$.

8. $y'' - 2y' + y = 1 + x + e^x$,
   $y(x) = c_1 e^x + c_2 x e^x + \dfrac{1}{2} x^2 e^x + x + 3$.

9. $y'' - 3y' = 8e^{3x} + 4 \sin x$,
   $y(x) = c_1 + c_2 e^{3x} + \dfrac{8}{3} x e^{3x}$
   $+ \dfrac{2}{5} (3 \cos x - \sin x)$.

10. $x^2 y'' - 2xy' + 2y = 2x^3 \cos x$,
    $y(x) = c_1 x + c_2 x^2 - 2x \cos x$.

11. $x^2 y'' + 5xy' - 12y = 12 \ln x$,
    $y(x) = c_1 x^{-6} + c_2 x^2 - \ln x - \dfrac{1}{3}$.

12. $xy'' - y' = (3 + x)x^2 e^x$,
    $y(x) = c_1 + c_2 x^2 + x^2 e^x$.

2-2 Solving Higher-Order ODE with the Reduction of Order Method

In addition to the method of reducing the original ODE to "homogeneous solution and particular solution," we can also use the method of reduction in order to solve second-order homogeneous ODEs. This method is attributed to the great French mathematician Joseph Lagrange (1736-1813) and involves the important concept of reducing higher-order ODEs. Here, we provide an example of reducing a $a_2(x)y'' + a_1(x)y' + a_0(x)y = 0$ (second-order) ODE to a $y' + p(x)y = 0$ (first-order) ODE.

▶ **Theorem 2-2-1**

Given One Homogeneous Solution, Find Another Homogeneous

Solution

For the second order ODE $a_2(x)y'' + a_1(x)y' + a_0(x)y = 0$, if one homogeneous solution is $y_1(x)$, then another linearly independent homogeneous solution is

$$y_2(x) = y_1 \int \frac{e^{-\int p(x)dx}}{y_1^2} dx.$$

Proof

In the context of the Riccati equation discussed in sections 1-4, the approach is essentially to shift the known solution and then substitute it into the original equation for testing. Here, we once again utilize this method, but this time we modify the known solution through "stretching." If $y_2(x)$ is another linear independent homogeneous solution, then $\dfrac{y_2(x)}{y_1(x)}$ cannot be a constant. In other words, there exists a non-zero function $u(x)$ such that $\dfrac{y_2(x)}{y_1(x)} = u(x)$ satisfies the equation.

u can be regarded as the scaling factor for adjusting the magnitude of y_1. Substituting $y_2(x) = y_1(x)u(x)$ into the original equation and simplifying, we obtain

$$y_1 u'' + (2y_1' + p(x)y_1)u' = 0 .$$

Here $p(x) = \dfrac{a_1(x)}{a_2(x)}$. Now the original equation has been transformed into a linear equation, althogh is a second order. Therefore, by letting $z = u'(x)$ be reduced to the first order $y_1 z' + (2y_1' + p(x)y_1)z = 0$. This equation can be solved using the linear equation solving method that we learned in Chapter 1.

(integration factor $I = y_1^2 e^{\int p(x)dx}$) we obtain $u(x) = c \int \dfrac{e^{-\int p(x)dx}}{y_1^2} dx$, that is

$$y_2(x) = y_1 \int \frac{e^{-\int p(x)dx}}{y_1^2} dx$$

(where c is arbitrary constant, here we take $c = 1$), then through Wronskian determinant we know

$$W(y_1(x), y_2(x)) = \begin{vmatrix} y_1 & y_2 \\ y_1' & y_2' \end{vmatrix} \neq 0.$$

So $y_1(x)$, $y_2(x)$ are linear independent, thus the proof is established. ◄

Example 1

For $y'' - y = 0$, if e^x is known to be a homogeneous solution, find the general solution of this ODE.

Solution　In the formula of theorem 2-2-1, $p_1(x) = 0$, thus

$$y_2(x) = y_1 \int \frac{e^{-\int p(x)dx}}{y_1^2} dx = e^x \int \frac{e^{-\int 0 dx}}{(e^x)^2} dx = e^x \int ce^{-2x} dx = \frac{-k}{2} e^{-x}.$$

We take $y_2(x) = e^{-x}$ (that is taking $k = -2$), so the general solution is
$$y(x) = c_1 e^x + c_2 e^{-x}.$$ Q.E.D.

Example 2

For $y'' - 6y' + 9y = 0$, if e^{3x} is known to be a homogeneous solution, find the general solution of this ODE.

Solution　In the formula of definition 2-2-1, $p(x) = -6$, so

$$y_2(x) = y_1 \int \frac{e^{-\int p(x)dx}}{y_1^2} dx = e^{3x} \int \frac{e^{-\int (-6)dx}}{(e^{3x})^2} dx = e^{3x} \int 1 dx = xe^{3x}.$$

We take $y_2(x) = xe^{3x}$, so the general solution is $y(x) = c_1 e^{3x} + c_2 xe^{3x}$. Q.E.D.

Example 3

For $x^2 y'' - 4xy' + 6y = 0$, $\forall x > 0$, if x^2 is known to be a homogeneous solution, find the general solution of this ODE.

Solution   In the formula of definition 2-2-1, $p(x) = \dfrac{-4x}{x^2} = -\dfrac{4}{x}$, so

$$y_2(x) = y_1 \int \frac{e^{-\int p(x)dx}}{y_1^2} dx = x^2 \int \frac{e^{-\int (-\frac{4}{x})dx}}{(x^2)^2} dx = x^2 \int 1 dx = x^3 .$$

We take $y_2(x) = x^3$, so the general solution is $y(x) = c_1 x^2 + c_2 x^3$.   Q.E.D.

2-2 Exercises

Basic questions

In the following ODE, if $y_1(x)$ is known to be a homogeneous solution, find the general solution of the ODE.

1. $y'' + y' - 2y = 0, y_1(x) = e^x$.

2. $y'' + 3y' + 2y = 0, y_1(x) = e^{-x}$.

3. $y'' - 4y' + 3y = 0, y_1(x) = e^x$.

4. $y'' + y = 0, y_1(x) = \cos x$.

5. $x^2 y'' - 2xy' + 2y = 0, y_1(x) = x$.

Advanced questions

In the following ODE, if $y_1(x)$ is known to be a homogeneous solution, find the general solution of the ODE.

1. $y'' - 4y' + 4y = 0, y_1(x) = e^{2x}$.

2. $y'' + 6y' + 9y = 0, y_1(x) = e^{-3x}$.

3. $y'' + 4y = 0, y_1(x) = \sin 2x$.

4. $x^2 y'' - 3xy' + 4y = 0, y_1(x) = x^2$.

5. $xy'' + y' = 0, y_1(x) = \ln |x|$.

2-3 Homogeneous Solutions of Higher-Order ODEs

Linear ODEs can be classified into two main categories: those with constant coefficients and those with variable coefficients. Among them, ODEs with constant coefficients are the most common, mainly because most parameter values in physical systems do not change with time. Building upon the ideas presented at the beginning of this chapter, this section introduces a systematic method for solving homogeneous constant coefficient equations. Sections 2-4 to 2-6 further discuss important techniques for finding particular solutions. By doing so, we obtain a systematic approach for solving the general solution.

We start from the first order, rearrange $a_1 y' + a_0 y = 0$ and we obtain $\int \dfrac{dy}{y} = \dfrac{a_0^*}{a_1} \int dx$,

therefore $e^{\frac{a_0^*}{a_1}x + \frac{a_0^*}{a_1}c} = y$, in other words, the solution to a first order constant coefficient ODE has the following form:

$$y = k e^{mx}$$

Does the solution also have this form in higher-order cases? Let's take the second order as an example to explain below.

2-3-1 Homogeneous Solution

Substituting $y(x) = e^{mx}$ into the second-order equation $a_2 y'' + a_1 y' + a_0 y = 0$, we can obtain $(a_2 m^2 + a_1 m + a_0)e^{mx} = 0$, because $e^{mx} \neq 0$, we have

$$f(m) := a_2 m^2 + a_1 m + a_0 = 0 \text{, and } \Delta = a_1^2 - 4 a_2 a_0$$

This quadratic equation is called the auxiliary equation or characteristic equation. We know that the discriminant of a quadratic equation determines the existence of its solutions. Below, we will discuss it separately.

1. $\Delta > 0$

   In this case, if $f(m)$ has two distinct real roots, we denote them as m_1 and m_2, consider the Wronskian determinant on $S = \{e^{m_1 x}, e^{m_2 x}\}$,

   then $\det \begin{bmatrix} e^{m_1 x} & e^{m_2 x} \\ m_1 e^{m_1 x} & m_2 e^{m_2 x} \end{bmatrix} = (m_2 - m_1)e^{m_1 x}e^{m_2 x} \neq 0$ so S is linear independent set,

   then through definition 2-1-1, we know the general solution is

   $$y(x) = c_1 e^{m_1 x} + c_2 e^{m_2 x}.$$

2. $\Delta = 0$

In this case, where $f(m)$ has a repeated real root m_0, we first have one solution $y_1 = e^{m_0 x}$, as for the other solution, we can refer to the method used in Example 7 of Section 2-1, we let $y_2 = xe^{m_0 x}$, and by substituting it into the original equation, we can verify that it is indeed a solution. Then through the Wronskian determinant, we

can obtain $\det\begin{bmatrix} e^{m_0 x} & xe^{m_0 x} \\ m_0 e^{m_0 x} & e^{m_0 x} + m_0 xe^{m_0 x} \end{bmatrix} = e^{2m_0 x} \neq 0$, so $e^{m_0 x}$ and $xe^{m_0 x}$ is linear

independent, so the general solution is

$$y(x) = c_1 e^{m_0 x} + c_2 xe^{m_0 x}.$$

3. $\Delta < 0$

In this case, where it has a pair of complex conjugate roots $m = \alpha \pm i\beta$, then we substitute these roots into the equation $y = e^{mx}$ and rearrange it using Euler's formula: $y = e^{\alpha x}[a\cos\beta x + b\sin\beta x]$, now the Wronskian determinant of $\{e^{\alpha x}\cos\beta x, e^{\alpha x}\sin\beta x\}$ tell us

$$\det\begin{bmatrix} e^{\alpha x}\cos\beta x & e^{\alpha x}\sin\beta x \\ \alpha e^{\alpha x}\cos\beta x - \beta e^{\alpha x}\sin\beta x & \alpha e^{\alpha x}\sin\beta x + \beta e^{\alpha x}\cos\beta x \end{bmatrix} = 2\beta e^{2\alpha x} \neq 0,$$

therefore the general solution is

$$y(x) = c_1 e^{\alpha x}\cos\beta x + c_2 e^{\alpha x}\sin\beta x.$$

For higher-order differential equations (3rd order and above), both real and complex roots can appear simultaneously, resulting in a mixed-type general solution that is not seen in second-order equations. Specifically, the general solution for an nth order homogeneous equation is given by:

▶ **Theorem 2-3-1**

General Form of Homogeneous Solutions for Higher-Order ODEs.

From the solutions of the characteristic equation $f(x) = a_n x^n + a_{n-1}x^{n-1} + \cdots + a_1 x + a_0$, homogeneous solutions can be classified into the following four types:

(1) $f(x)$ has n distinct real roots $m_1, m_2, \cdots, m_n$. Then the general solution is

$$y(x) = c_1 e^{m_1 x} + c_2 e^{m_2 x} + \cdots + c_n e^{m_n x}.$$

(2) $f(x)$ has k identical real roots m. Then the general solution is

$$y(x) = c_1 \cdot e^{mx} + c_2 xe^{mx} + c_3 x^2 e^{mx} + c_4 x^3 e^{mx} + \cdots + c_k x^{k-1} e^{mx}.$$

(3) $f(x)$ has n roots that are all complex conjugate with multiplicity $\alpha \pm i\beta$. Then the general solution is

$$y(x) = e^{\alpha x}[(c_1 + c_2 x + c_3 x^2 + \cdots + c_n x^{n-1})\cos \beta x + (d_1 + d_2 x + d_3 x^2 + \cdots + d_n x^{n-1})\sin \beta x].$$

(4) If $f(x)$ has both complex conjugate roots and real roots, the general solution in this case is the sum of the three forms (1), (2), and (3). ◀

Example 1

$y'' - 3y' - 4y = 0$.

Solution Let the characteristic equation

$$f(m) = m^2 - 3m - 4 = (m - 4)(m + 1) = 0,$$

$\therefore m = 4, -1$, therefore, based on Theorem 2-3-1, we can conclude the following regarding case 1. The general solution is $y(x) = c_1 e^{4x} + c_2 e^{-x}$. Q.E.D.

Example 2

$4y'' + 4y' + y = 0$, $y(0) = -2$, $y'(0) = 1$.

Solution

(1) Let $f(m) = 4m^2 + 4m + 1 = (2m + 1)^2 = 0$. We obtain

$$m = -\frac{1}{2}, -\frac{1}{2}.$$

Therefore, based on Theorem 2-3-1, we can conclude the following regarding case (2)

$$y(x) = c_1 e^{-\frac{1}{2}x} + c_2 x e^{-\frac{1}{2}x}.$$

(2) Now through $y(0) = -2$, we obtain $c_2 = -2$. Therefore

$$y(x) = -2e^{-\frac{1}{2}x} + c_2 x e^{-\frac{1}{2}x}.$$

Then through

$$y'(x) = e^{-\frac{1}{2}x} + c_2 e^{-\frac{1}{2}x} - \frac{1}{2}c_2 x e^{-\frac{1}{2}x},$$

we obtain

$$y'(0) = 1 = 1 + c_2 + 0 \Rightarrow c_2 = 0,$$

$$\therefore y(x) = -2e^{-\frac{1}{2}x}.$$ Q.E.D.

Example 3

$$y'' - 2y' + 2y = 0,\ y(0) = -3,\ y(\tfrac{1}{2}\pi) = 0.$$

Solution

(1) Let the characteristic equation $f(m) = m^2 - 2m + 2 = 0$, we obtain

$$m = \frac{2 \pm \sqrt{4-8}}{2} = 1 \pm i.$$

Therefore, based on Theorem 2-3-1, we can conclude the following regarding case (3). We know

$$y(x) = e^x (c_1 \cos x + c_2 \sin x).$$

(2) $y(0) = -3 \Rightarrow c_1 = -3$,

$$y(\frac{\pi}{2}) = 0 \Rightarrow y(\frac{\pi}{2}) = e^{\frac{\pi}{2}} \cdot c_2 = 0 \Rightarrow c_2 = 0,$$

$$\therefore y(x) = -3e^x \cos x.$$

Q.E.D.

Example 4

$$y''' + 3y'' + y' + 3y = 0.$$

Solution   Let the characteristic equation $f(m) = (m^3 + m^2 + m + 3) = (m^2 + 1)(m + 3) = 0$, we obtain

$$m = -3, \pm i.$$

Therefore, based on Theorem 2-3-1, we can conclude the following regarding case (4) The general solution is

$$y(x) = c_1 e^{-3x} + c_2 \cos x + c_3 \sin x.$$

Q.E.D.

2-3 Exercises

Basic questions

1. Find the general solution of the following homogeneous ODE.

    (1) $y'' + y' - 2y = 0.$

    (2) $y' + 6y' + 9y = 0.$

    (3) $y'' - 36y = 0.$

    (4) $y'' - y' - 6y = 0.$

    (5) $y'' - 3y' + 2y = 0.$

    (6) $y'' + 8y' + 16y = 0.$

Solve the initial value problems for the following ODEs in questions 2 to 4.

2. $y'' + 3y' + 2y = 0, y(0) = 1, y'(0) = 0.$

3. $y'' - 4y' + 3y = 0, y(0) = 4, y'(0) = 0.$

4. $y'' - 4y' + 3y = 0, y(0) = -1, y'(0) = 3.$

Advanced questions

1. Find the general solution of the following homogeneous ODE

    (1) $y'' - y' + 10y = 0.$

    (2) $4y'' + y' = 0.$

    (3) $y'' - 10y' + 25y = 0.$

    (4) $y'' + 9y = 0.$

    (5) $y'' - 4y' + 5y = 0.$

    (6) $12y'' - 5y' - 2y = 0.$

    (7) $2y'' + 2y' + y = 0.$

    (8) $y''' + 3y'' + y' + 3y = 0.$

    (9) $\dfrac{d^4 y}{dx^4} - y = 0$

Solve the initial value problems for the following ODEs in questions 2 to 3.

2. $y'' + 2y' + 17y = 0, y(0) = 1, y'(0) = 0.$

3. $y'' + 8y' + 16y = 0, y(0) = 3, y'(0) = 3.$

4. Find the second-order linear homogeneous ODE with $e^x \cos x$ and $e^x \sin x$ as two linear independent solutions.

5. If the characteristic equation of a ninth-order linear with constant coefficient homogeneous ODE has roots 2, 2, 2, $3 \pm 4i$, $3 \pm 4i$, $3 \pm 4i$, please write down the general solution of this ODE.

6. Find the minimum order linear differential operator L with respect to y such that $L(D)y = 0$, where $y = \cos 2x + x^3 e^{7x} + x + 1 + \sin 4x + \sin 9x.$

2-4 Finding Particular Solution Using the Method of Undetermined Coefficients

From this section to Section 2-6, we will introduce how to find the particular solution of the following ODE.

$$a_n y^{(n)}(x) + a_{n-1} y^{(n-1)}(x) + \cdots + a_1 y'(x) + a_0 y(x) = R(x)$$

First, we have the method of undetermined coefficients. This method, in simple terms, involves observing the non-homogeneous term (source function) $R(x)$ on the right-hand side of the ODE. We then consider its composition and introduce unknown parameters in a similar form. By substituting these parameters into the original equation, we can solve for their values.

For instance: equation

$$y'' - 2y' - y = e^x.$$

The composition of the right-hand side $R(x)$ is in the form of an exponential function. Therefore, we assume a particular solution to be of the form:

$$y_p(x) = Ae^x.$$

Substituting to the original equation we obtain $Ae^x - 2Ae^x - 3Ae^x = e^x$, that is $A = -\dfrac{1}{4}$, the particular solution is $y_p(x) = -\dfrac{1}{4}e^x$.

Following this idea, we can make different particular solution assumptions for different types of non-homogeneous terms (source functions).

2-4-1 Assumption Principles of Undetermined Coefficient Method

Table 1 lists the most common applicable cases for the method of undetermined coefficients, noting that $R(x)$ should be c (constant), e^{ax}, $\sin bx$, $\cos bx$, x^n (where n is a positive integer), or their linear combinations. If the assumed term $y_p(x)$ contains the same term as $y_h(x)$, then when making the assumption, it must be corrected by multiplying the same term by x^m, where m is the smallest positive integer that ensures the particular solution $y_p(x)$ is not a duplicate of the homogeneous solution $y_h(x)$.

Table 1

Sequence number	$R(x)$ (source function)	$y_p(x)$ (particular solution)
①	c	k
②	e^{ax}	ke^{ax}
③	$\sin bx,\ \cos bx$	$A\cos bx + B\sin bx$
④	x^n	$A_n x^n + A_{n-1}x^{n-1} + \cdots + A_1 x + A_0$
⑤	$e^{ax}\sin bx,\ e^{ax}\cos bx$	$e^{ax}(c_1 \cos bx + c_2 \sin bx)$
⑥	$x^n e^{ax}$	$e^{ax}\cdot(c_n x^n + c_{n-1}x^{n-1} + \cdots + c_1 x + c_0)$
⑦	$x^n\sin bx,\ x^n\cos bx$	$(c_n x^n + c_{n-1}x^{n-1} + \cdots + c_1 x + c_0)\cos bx$ $+(d_n x^n + d_{n-1}x^{n-1} + \cdots + d_1 x + d_0)\sin bx$

2-4-2 Example Description

Next, we will provide examples to illustrate the method of undetermined coefficients for cases involving trigonometric functions, polynomials, and mixed terms of polynomials and exponential functions.

Example 1

Solve the following ODE, and find the particular solution using the method of undetermined coefficients.
$y'' - 2y' - 3y = e^{2x}$.

Solution

Homogeneous Solution
By solving the characteristic equation $f(m) = m^2 - 2m - 3 = 0$, we find $m = 3, -1$, thus the homogeneous solution is $\quad y_h(x) = c_1 e^{3x} + c_2 e^{-x}$.

Particular Solution
Using the method of undetermined coefficients (as given in Table 1 ②), we let

$y_p(x) = Ae^x$ and substitute to the original equation, and we obtain $\quad A = -\dfrac{1}{3}$,

so the particular solution is $\quad y_p(x) = -\dfrac{1}{3}e^{2x}$.

General Solution

$$y(x) = y_h(x) + y_p(x) = c_1 e^{3x} + c_2 e^{-x} - \frac{1}{3}e^{2x}.$$

Q.E.D.

Example 2

Solve the following ODE, and find the particular solution using the method of undetermined coefficients.

$y'' - 2y' - 3y = \cos 2x$.

Solution

Homogeneous Solution

Same as Example1, $y_h(x) = c_1 e^{3x} + c_2 e^{-x}$.

Particular Solution

Using the method of undetermined coefficients (as given in Table 1 ③),
we let $y_p(x) = A \cos 2x + B \sin 2x$ then we substitute to ODE and obtain

$$\begin{cases} -7A - 4B = 1 \\ -7B + 4A = 0 \end{cases}.$$

The solution is

$$A = -\frac{7}{65}, B = -\frac{4}{65}.$$

Thus

$$y_p(x) = -\frac{7}{65} \cos 2x - \frac{4}{65} \sin 2x.$$

General Solution

$$y(x) = y_h(x) + y_p(x) = c_1 e^{3x} + c_2 e^{-x} - \frac{7}{65} \cos 2x - \frac{4}{65} \sin 2x. \quad \boxed{\text{Q.E.D.}}$$

Example 3

Solve the following ODE, and find the particular solution using the method of undetermined coefficients.

$y'' - 2y' - 3y = x + 1$.

Solution

Homogeneous Solution

Same as Example1, $y_h(x) = c_1 e^{3x} + c_2 e^{-x}$.

Particular Solution

Using the method of undetermined coefficients (as given in Table 1 ④),
we let $y_p(x) = Ax + B$ substituting to the original equation and
obtained

$$\begin{cases} -2A - 3B = 1 \\ -3A = 1 \end{cases}.$$

The solution is

$$A = -\frac{1}{3}, \; B = -\frac{1}{9},$$

so

$$y_p(x) = -\frac{1}{3}x - \frac{1}{9}.$$

General Solution

$$y(x) = y_h(x) + y_p(x) = c_1 e^{3x} + c_2 e^{-x} - \frac{1}{3}x - \frac{1}{9}. \qquad \boxed{\text{Q.E.D.}}$$

Example 4

Solve the following ODE, and find the particular solution using the method of undetermined coefficients.

$$y'' - 2y' + y = x^2 e^x.$$

Solution

Homogeneous Solution

Solving the characteristic equation $f(m) = m^2 - 2m + 1 = 0$, $m = 1, 1$, therefore, the homogeneous solution is

$$y_h(x) = c_1 e^x + c_2 x e^x.$$

Particular Solution

Consider the case of Table 1 ⑥, we let $y_p(x) = x^2(A + Bx + Cx^2)e^x$, where multiplying x^2 to avoid overlap with $y_h(x)$, substituting into the original equation we obtain

$$\begin{cases} A = 0 \\ B = 0 \\ C = \dfrac{1}{12} \end{cases}, \text{ thus } y_p(x) = \frac{1}{12}x^4 e^x.$$

General Solution

$$y(x) = y_h(x) + y_p(x) = c_1 e^x + c_2 x e^x + \frac{1}{12}x^4 e^x. \qquad \boxed{\text{Q.E.D.}}$$

2-4　Exercises

Basic questions

Using the method of undetermined coefficients, find the particular solutions for the following questions.

1. $y'' - 2y' - 3y = 1$.

2. $y'' - 2y' - 3y = e^x$.

3. $y'' - 2y' - 3y = x + 1$.

4. $y'' - 3y' + 2y = \cos 3x$.

5. $y'' + 3y' + 2y = 6$.

6. $y'' + y' - 6y = 2x$.

Advanced questions

1. For $y'' - 2y' + y = 4e^x$, please verify the solution of this ODE is $y(x) = c_1 e^x + c_2 x e^x + 2x^2 e^x$.

2. Using the method of undetermined coefficients to solve the differential equation:
   $y'' - 6y' + 9y = 6x^2 + 2 - 12e^{3x}$,
   What form should the particular solution y_p be set to be more appropriate?
   (1) $Ax^3 + Bx^2 + Cx + E + Fe^{3x}$
   (2) $Ax^2 + Bx + C + Ee^{3x}$
   (3) $Ax^2 + Bx + C + Exe^{3x}$
   (4) $Ax^2 + Bx + C + Ex^2 e^{3x}$
   (5) $Ax^2 + Bx + C + Ex^{3x} + Fxe^{3x}$.

3. Find the general solution of the following ODE using the method of undetermined coefficients.
   (1) $y'' - 5y' + 4y = 8e^x$.
   (2) $y'' - 2y' + y = e^x$.
   (3) $y'' + 3y' + 2y = x^3 + x$
   (4) $y'' - 2y' + 10y = 20x^2 + 2x - 8$.
   (5) $y'' - 16y = 2e^{4x}$.
   (6) $y'' + 4y = 3\sin 2x$.
   (7) $y'' + 4y' + 5y = 35e^{-4x}$
   　　$y(0) = -3, y'(0) = 1$.
   (8) $y'' - 4y' + 4y = e^{3x} - 1$.

4. Find the solution using the method of undetermined coefficients.
   $y'' - 2y' + 5y + 4\cos t - 8\sin t = 0$
   $y(0) = 1, y'(0) = 3$.

5. $y'' - 4y' + 4y = 3t^2 + 5te^{2t} + t\cos t$, using the method of undetermined coefficients, just assume the form of $y_p(t)$ without the need for solving.

2-5 Finding Particular Solution Using the Method of Variation of Parameters

Although the method of undetermined coefficients is straightforward to compute, its applicability is limited. The reason is that in order to use this method, the source function $R(x)$ must possess some degree of differential periodicity (such as $\dfrac{d}{dx}e^x = e^x$ differentiation of trigonometric functions, trigonometric functions, etc.). If one wants to avoid relying on the differential behavior of $R(x)$, a direct approach is to "piece together" a particular solution from the homogeneous solutions (in the case where $R(x) = 0$).

▶ Theorem 2-5-1

General Solution of Second Order ODE by Variation of Parameters

Given a second-order ordinary differential equation $a_2(x)y'' + a_1(x)y' + a_0(x)y = R(x)$. If $y_1(x)$, $y_2(x)$ are homogeneous solutions of this ODE (where $y_1(x)$, $y_2(x)$ are linearly independent) and $W(y_1, y_2)$ represents the Wronskian determinant of these two solutions, then the general solution is

$$y(x) = y_h(x) + y_1 \times \int \frac{-[R(x)/a_2(x)]y_2}{W(y_1, y_2)}\,dx + y_2 \times \int \frac{[R(x)/a_2(x)]y_1}{W(y_1, y_2)}\,dx\ .$$

Proof

A linear combination of homogeneous solutions is still a homogeneous solution. Therefore, to obtain a particular solution, it is convenient to replace the coefficients in the linear combination with variable coefficients. For a second-order equation,

$$a_2(x)y'' + a_1(x)y' + a_0(x)y = R(x)\ ,$$

for instnce, we let $y_p(x) = y_1(x)\phi_1(x) + y_2(x)\phi_2(x)$, then

$$y_p'(x) = \phi_1 y_1' + \phi_2 y_2' + y_1 \phi_1' + y_2 \phi_2'\ .$$

Here, we specially assume $y_1\phi_1' + y_2\phi_2' = 0$ (called formula ①). Next, we differentiate twice obtained

$$y_p''(x) = \phi_1 y_1'' + \phi_2 y_2'' + y_1' \phi_1' + y_2' \phi_2'\ ,$$

We obtain after substituting to the original equation

$$y_1'\phi_1' + y_2'\phi_2' = \frac{R(x)}{a_2(x)}\ .$$

So, through the assumption in equation ①, we have transformed the original problem into a linear algebra problem, which is to solve:

$$\begin{bmatrix} y_1 & y_2 \\ y_1' & y_2' \end{bmatrix}\begin{bmatrix} \phi_1' \\ \phi_2' \end{bmatrix} = \begin{bmatrix} 0 \\ \dfrac{R(x)}{a_2(x)} \end{bmatrix}$$, note at here, because $\{y_1, y_2\}$ is linear independent, so

$W(y_1, y_2) = \det\begin{bmatrix} y_1 & y_2 \\ y_1' & y_2' \end{bmatrix} \neq 0$. Therefore, using Cramer's rule we obtain

$$\phi_1' = \frac{\begin{vmatrix} 0 & y_2 \\ R(x)/a_2(x) & y_2' \end{vmatrix}}{W(y_1, y_2)} = \frac{-[R(x)/a_2(x)]y_2}{W(y_1, y_2)}.$$

Same reason we have $\phi_2' = \dfrac{\begin{vmatrix} y_1 & 0 \\ y_1' & R(x)/a_2(x) \end{vmatrix}}{W(y_1, y_2)} = \dfrac{[R(x)/a_2(x)]y_1}{W(y_1, y_2)}$, for ϕ_1', ϕ_2'

integrate separately, then we obtain

$$\phi_1 = \int \frac{-[R(x)/a_2(x)]y_2}{W(y_1, y_2)}\,dx, \quad \phi_2 = \int \frac{[R(x)/a_2(x)]y_1}{W(y_1, y_2)}\,dx.$$

Finally, substituting to the original equation and get the general solution.   ◀

In fact, in general cases of third order or higher, we can also proceed step by step using assumptions similar to equation ①. This allows us to simplify each order of differentiation while constructing a system of linear differential equations. Finally, we can use Cramer's rule to solve the system and obtain the particular solution. The detailed explanation of this method is left as an exercise for the reader.

Example 1

$y'' - y = xe^x$.

Solution   First, we use the characteristic equation to find the homogeneous solution $y_h = k_1 e^x + k_2 e^{-x}$, therefore through definition 2-5-1, we know $y_1 = e^x$, $y_2 = e^{-x}$, $R(x) = xe^x$, $a_2 = 1$, $a_1 = 0$, $a_0 = -1$, calculating then we obtain

$$\phi_1(x) = \int \frac{-\dfrac{xe^x}{1}\cdot e^{-x}}{W(e^x, e^{-x})}\,dx = \frac{1}{4}x^2, \quad \phi_2(x) = \int \frac{\dfrac{xe^x}{1}\cdot e^x}{W(e^x, e^{-x})}\,dx = -\frac{1}{4}xe^{2x} + \frac{1}{8}e^{2x}.$$

So

$$y_p = \phi_1 y_1 + \phi_2 y_2 = \frac{1}{4}x^2 e^x - \frac{1}{4}xe^x + \frac{1}{8}e^x,$$

the genetal solution is

$$y = y_h + y_p = c_1 e^x + c_2 e^{-x} + \frac{1}{4} x^2 e^x - \frac{1}{4} x e^x$$

where $(k_1 + \frac{1}{8}) = c_1$, $k_2 = c_2$.

Q.E.D.

Example 2

$y'' + y = \sec x$.

Solution   First, we use the characteristic equation to find the homogeneous solution $y_h(x) = c_1 \cos x + c_2 \cos x$, therefore through definition 2-5-1, we know $y_1 = \cos x$, $y_2 = \sin x$, $R(x) = \sec x$, $a_2 = 1$, $a_1 = 0$, $a_0 = 1$. Calculating then we obtain

$$\phi_1(x) = \int \frac{-\frac{\sec x}{1} \cdot \sin x}{W(\cos x, \sin x)} dx = \ln|\cos x|, \quad \phi_2(x) = \int \frac{\frac{\sec x}{1} \cdot \cos x}{W(\cos x, \sin x)} dx = x,$$

thus

$$y_p = \phi_1 y_1 + \phi_2 y_2 = (\ln|\cos x|)\cos x + x \sin x,$$

so the general solution is

$$y = y_h + y_p = c_1 \cos x + c_2 \sin x + \cos x (\ln|\cos x|) + x \sin x.$$

Q.E.D.

2-5 Exercises

Basic questions

Find the solution to the following ODE, and use the method of variation of parameters to find the particular solution:

1. $y'' - 3y' + 2y = e^{3x}$.

2. $y'' - 3y' + 2y = e^x$, $y(0) = 0$, $y'(0) = 1$.

3. $y'' + y = \cos x$.

4. $y'' - 6y' + 9y = e^{3x}$.

5. $y'' + 9y = \sec 3x$.

Advanced questions

Find the solution to the following ODE, and use the method of variation of parameters to find the particular solution:

1. $y'' + 2y' + 5y = e^{-x} \sin 2x$.

2. $4y'' + 36y = \csc 3x$.

3. $y'' + y = \sin x$.

4. $y'' + y = \cos^2 x$.

5. $y'' - 9y = \frac{9x}{e^{3x}}$.

6. $y'' + 3y' + 2y = \sin e^x$.

7. $y'' + 4y = \sin 2t$.

8. $y'' + y = \tan x$.

2-6 Finding Particular Solution Using the Method of Inverse Differential Operators

We now come to the final method for finding particular solutions, known as the method of inverse differential operators. Although some mathematicians do not quite agree with this method, and some other textbooks do not include this method. However, this method is more efficient and worthy of use in engineering from an engineering perspective. If some readers find its rigor lacking or harbor reservations, you can choose to use other previous methods and skip this section. Let's review the previous methods briefly. In Sections 2-4 and 2-5, we made assumptions about the form of the particular solution based on the analysis of $R(x)$ or the avoidance of relying on $R(x)$ by using homogeneous solutions. In practice, these methods involved setting up systems of equations and reducing the problem to linear algebra. However, these methods can become inefficient when the order of the ODE exceeds 2.

In contrast, the method of inverse differential operators treats differentiation as an algebraic operation on the space of functions. It defines the inverse operator of a differential operator $L(D)$, allowing us to apply abstract algebraic methods such as polynomial long division directly to find particular solutions. The advantage of this method is that we can easily find particular solutions for higher-order linear ODEs.

Naturally, let's start by making some observations about $y' - \lambda y = R(x)$. By the integrating factor $I = e^{-\lambda x}$, we obtain

$$y = c \cdot e^{\lambda x} + e^{\lambda x} \int e^{-\lambda x} R(x) dx,$$

where particular solution $y_p(x) = e^{\lambda x} \int e^{-\lambda x} R(x) dx$. and $y_p' - \lambda y_p = R(x)$ can be rewritten in operator form $(D - \lambda) y_p = R(x)$, we have

$$y_p = \frac{R(x)}{D - \lambda} = e^{\lambda x} \int e^{-\lambda x} R(x) dx .$$

And when $\lambda = 0$, $\dfrac{R(x)}{D} = \int R(x) dx$, therefore $\dfrac{1}{D}$ just like integral.

2-6-1   Inverse Differential Operators

▶ **Definition 2-6-1**

Inverse Differential Operators

For all continuous functions $R(x)$, the (differential) inverse operator is defined as

$$(D - \lambda)^{-1} R(x) \equiv \frac{R(x)}{D - \lambda} \equiv e^{\lambda x} \int e^{-\lambda x} R(x) dx .$$ ◀

▶ **Theorem 2-6-1**

Decomposition of Inverse Differential Operators

If the differential operator $L(D)$ can be decomposed of $(D-\lambda_1)(D-\lambda_2)\cdots(D-\lambda_n)$, then

(1) $\dfrac{1}{L(D)}R(x) = \left[\dfrac{A_1}{D-\lambda_1} + \dfrac{A_2}{D-\lambda_2} + \cdots + \dfrac{A_n}{D-\lambda_n}\right]R(x)$ [1].

(2) $\dfrac{1}{L(D)}R(x) = \dfrac{1}{(D-\lambda_1)(D-\lambda_2)\cdots(D-\lambda_n)}R(x)$

$= \dfrac{1}{(D-\lambda_1)}[\dfrac{1}{(D-\lambda_2)}[\cdots[\dfrac{1}{(D-\lambda_n)}R(x)]\cdots]]$ [2]. ◀

Example 1

$y'' - y' - 2y = e^{-x}$, use the method of inverse differential operators to find its particular solution.

Solution   Let $L(D) = D^2 - D - 2$, then

$$y_p = \frac{1}{D^2-D-2}e^{-x} = \frac{1}{(D-2)(D+1)}e^{-x} = [\frac{\frac{1}{3}}{(D-2)} + \frac{-\frac{1}{3}}{(D+1)}]e^{-x}.$$

Now according to definition 2-6-1, we can obtain

$$\frac{1}{D-2}e^{-x} = e^{2x}\int e^{-2x}e^{-x}dx = \frac{-1}{3}e^{-x},$$

$$\frac{1}{D+1}e^{-x} = e^{-x}\int e^{x}e^{-x}dx = xe^{-x},$$

substituting to the original equation then we obtain $y_p = -\dfrac{1}{9}e^{-x} - \dfrac{1}{3}xe^{-x}.$ Q.E.D.

---

[1] The method is called partial fraction decomposition.

[2] The method is called integration by parts.

Example 2

$y'' - y' - 2y = \sin(e^{-x})$, use the method of inverse differential operators to find its particular solution.

Solution Let $L(D) = D^2 - 3D + 2$, then through definition 2-6-1 (2)

$$y_p = \frac{1}{D^2 - 3D + 2}\sin(e^{-x}) = \frac{1}{(D-2)(D-1)}\sin(e^{-x})$$

$$= \frac{1}{(D-2)} \cdot \frac{1}{(D-1)} \cdot \sin(e^{-x}) = \frac{1}{D-2} \cdot e^x \int e^{-x}\sin(e^{-x})dx$$

$$= \frac{1}{D-2}e^x \cdot \int -\sin(e^{-x})d(e^{-x}) = \frac{1}{D-2}e^x\cos(e^{-x})$$

$$= e^{2x} \cdot \int e^{-2x} \cdot e^x \cdot \cos(e^{-x})dx = -e^{2x}\sin(e^{-x}).$$

Q.E.D.

2-6-2 Commonly Used Inverse Operator Formulas

In 2-6-1 Inverse Operator Definition, by substituting different functions into $R(x)$, we obtain different inverse operator formulas. We can summarize these formulas in Table 2 (where not otherwise specified, $L(D)$ represents a differential operator that can be decomposed into the form of $(D-\lambda_1)(D-\lambda_2)\cdots(D-\lambda_n)$).

Table 2

Number	$R(x)$	$L(D)$	$L(D)^{-1}[R(x)] = \dfrac{R(x)}{L(D)}$
①	e^{ax}	$L(a) \neq 0$	$\dfrac{1}{L(a)}e^{ax}$
②	$e^{ax}Q(x)$		$e^{ax}\dfrac{1}{L(D+a)}Q(x)$
③	e^{ax}	$(D-a)^m$	$e^{ax}\dfrac{1}{D^m}(1) = e^{ax}\dfrac{x^m}{m!}$
④	e^{ax}	$(D-a)^m F(D),$ where $F(a) \neq 0$	$\dfrac{1}{(D-a)^m}\dfrac{1}{F(D)}e^{ax} = \dfrac{1}{F(a)}e^{ax} \cdot \dfrac{x^m}{m!}$
⑤a	$\cos(ax + b)$	$L(D)$ change into $L(D^2)$, where $L(-a^2) \neq 0$	$\dfrac{1}{L(-a^2)}\cos(ax + b)$
⑤b	$\sin(ax + b)$		$\dfrac{1}{L(-a^2)}\sin(ax + b)$

Number	$R(x)$	$L(D)$	$L(D)^{-1}[R(x)] = \dfrac{R(x)}{L(D)}$
⑥a	$\cos ax$	$D^2 + a^2$	$\dfrac{x}{2a}\sin ax$
⑥b	$\sin ax$		$\dfrac{-x}{2a}\cos ax$
⑦a	$\cosh(ax+b)$	$L(D)$ change into $L(D^2)$ where $L(a^2) \neq 0$	$\dfrac{1}{L(a^2)}\cosh(ax+b)$
⑦b	$\sinh(ax+b)$		$\dfrac{1}{L(a^2)}\sinh(ax+b)$
⑧a	$\cosh ax$	$D^2 - a^2$	$\dfrac{x}{2a}\sinh ax$
⑧b	$\sinh ax$		$\dfrac{x}{2a}\cosh ax$
⑨	$a_0 + a_1 x + a_2 x^2$ $+ \cdots + a_n x^n$	$b_0 + b_1 D + \cdots + b_n D^n$	$(c_0 + c_1 D + c_2 D^2 + \cdots + b_n D^n + \cdots) \times$ $(a_0 + a_1 x + \cdots + a_n x^n)$, where by using long division we obtain $\dfrac{1}{b_0 + b_1 D + \cdots + b_n D^n} = c_0 + c_1 D + c_2 D^2 + \cdots$

In table 2, the differential operator $D^2 + a^2$ in Formula ⑥ is irreducible over the real numbers. In order to utilize other formulas to comprehend this differential (inverse) operator, we can consider its factorization in the complex domain $D^2 + a^2 = (D + ai)$ $(D - ai)$. When the source function $R(x)$ takes one of the following forms: ①, ②, or ③, we can employ alternative formulas to understand the differential (inverse) operator.

$$\begin{cases} e^{ax}\sin bx, \ e^{ax}\cos bx & \cdots ① \\ x^n e^{ax} & \cdots ② \\ x^n \sin bx, \ x^n \cos bx & \cdots ③ \end{cases}$$

We can simplify the problem step by step using the formulas in the table as follows:

1. $\dfrac{1}{L(D)}e^{ax}\sin bx = e^{ax} \cdot \dfrac{1}{L(D+a)}\sin bx$, then utilize formula ⑤ or ⑥ to simplify.

2. $\dfrac{1}{L(D)}x^n e^{ax} = e^{ax} \cdot \dfrac{1}{L(D+a)}x^n$, then utilize formula ⑨ to simplify.

3. Through Euler's theorem ($e^{ibx} = \cos bx + i\sin bx$), also we can obtain:
$\dfrac{1}{L(D)}x^n \sin bx = \mathrm{Im}[\dfrac{1}{L(D)}(x^n e^{ibx})]$, $\dfrac{1}{L(D)}x^n \cos bx = \mathrm{Re}[\dfrac{1}{L(D)}(x^n e^{ibx})]$.

The following example demonstrates how to obtain a particular solution using the formulas listed in the table, and then combining it with the homogeneous solution obtained from Section 2-3 to obtain the general solution of the original equation. The general solution is the sum of the homogeneous solution and the particular solution. Detailed explanations will not be provided here, and readers are encouraged to practice on their own.

Example 3

$y'' - 4y' + 4y = e^{3x} - 1$, find the general solution of this ODE.

Solution

Homogeneous Solution

From the characteristic equation $m^2 - 4m + 4 = 0$, we obtain $m = 2, 2$ (multiple root), so

$$y_h(x) = c_1 e^{2x} + c_2 x e^{2x}.$$

Particular Solution

Let $L(D) = D^2 - 4D + 4$, through formula ① of Table 6, the particular solution is

$$y_p(x) = \frac{1}{D^2 - 4D + 4}(e^{3x} - 1)$$

$$= \frac{1}{D^2 - 4D + 4}e^{3x} - \frac{1}{D^2 - 4D + 4}e^{0x}$$

$$= \frac{1}{3^2 - 4 \cdot 3 + 4}e^{3x} - \frac{1}{0^2 - 4 \cdot 0 + 4}e^{0x}$$

$$= e^{3x} - \frac{1}{4}.$$

General Solution

$$y(x) = y_h(x) + y_p(x) = c_1 e^{2x} + c_2 e^{2x} + e^{3x} - \frac{1}{4}.$$

Q.E.D.

Example 4

$y'' - 2y' + y = xe^x$, find the general solution of this ODE.

Solution

Homogeneous Solution

From the characteristic equation $m^2 - 2m + 1 = 0$, we obtain $m = 1, 1$, thus

$$y_h(x) = (c_1 + c_2 x)e^x.$$

Particular Solution

Let $L(D) = D^2 - 2D + 1$, through formula ② of Table 2 ($Q(x) = x$), then we obtain

$$y_p(x) = \frac{1}{D^2 - 2D + 1}(xe^x) = \frac{1}{(D-1)^2}xe^x = e^x\frac{1}{(D+1-1)^2}x$$

$$= e^x\frac{1}{(D)^2}x = e^x\iint[\int xdx]dx = e^x(\frac{1}{6}x^3) = \frac{1}{6}x^3e^x.$$

General Solution

$$y(x) = y_h(x) + y_p(x) = c_1e^x + c_2xe^x + \frac{1}{6}x^3e^x.$$

Q.E.D.

Example 5

$y''' - y'' - 8y' + 12y = 7e^{2x}$, find the general solution of this ODE.

Solution

Homogeneous Solution

Considering the characteristic equation, $m^3 - m^2 - 8m + 12 = 0$, using Newton's method for testing roots, we can observe that there is a root at $m = 2$,
then, using long division, we have

$$m^3 - m^2 - 8m + 12 = (m-2)(m+3)(m-2) = 0, m = 2, 2, -3,$$

so

$$y_h(x) = (c_1 + c_2x)e^{2x} + c_3e^{-3x}.$$

Particular Solution

Let $L(D) = D^2 - D^2 - 8D + 12$, through formula ④ of Table2 ($F(D) = D + 3$). The particular solution is

$$y_p(x) = \frac{1}{(D-2)^2(D+3)}7e^{2x} = 7\times\frac{1}{(D-2)^2}\times\frac{1}{5}e^{2x} = \frac{7}{5}e^{2x}\cdot\frac{1}{D^2}(1)$$

$$= \frac{7}{10}x^2e^{2x} \quad \text{(formula ③).}$$

General Solution

$$y(x) = y_h(x) + y_p(x) = c_1e^{2x} + c_2xe^{2x} + c_3e^{-3x} + \frac{7}{10}x^2e^{2x}.$$

Q.E.D.

Example 6

$y'' - 5y' + 6y = -3 \sin 2x$, find the general solution of this ODE.

Solution

Homogeneous Solution

Through the characteristic equation $m^2 - 5m + 6 = 0$, we obtain $m = 2, 3$, therefore

$$y_h(x) = c_1 e^{2x} + c_2 e^{3x}.$$

Particular Solution

Let $L(D) = D^2 - 5D + 6$, through formula ⑤ of Table 2. We obtain

$$y_p = \frac{1}{(D^2 - 5D + 6)}(-3\sin 2x) = -3\frac{1}{-2^2 - 5D + 6}\sin 2x \quad \text{(formula ⑤)}$$

$$= -3\frac{(2 + 5D)}{(4 - 25D^2)}\sin 2x = -3\frac{2 + 5D}{4 - 25 \times (-2^2)}\sin 2x$$
$$\qquad\qquad\qquad\qquad\qquad\qquad \text{(formula ⑤)}$$

$$= -3\frac{2 + 5D}{104}\sin 2x = -\frac{3}{104}(2\sin 2x + 5 \times 2\cos 2x)$$

$$= -\frac{3}{52}(\sin 2x + 5\cos 2x).$$

General Solution

$$y(x) = y_h(x) + y_p(x) = c_1 e^{2x} + c_2 e^{3x} - \frac{3}{52}(\sin 2x + 5\cos 2x). \qquad \boxed{\text{Q.E.D.}}$$

Example 7

$y'' + 4y = \cos 2x + \cos 4x$, find the general solution of this ODE.

Solution

Homogeneous Solution

Through the characteristic equation $m^2 + 4 = 0$, we have $m = \pm 2i$, so

$$y_h(x) = c_1 \cos 2x + c_2 \cos 2x.$$

Particular Solution

Let $L(D) = D^2 + 2^2$, through formula ⑥ and ⑤ of Table2. We obtain

$$y_p = \frac{1}{(D^2 + 4)}(\cos 2x + \cos 4x) = \frac{1}{(D^2 + 2^2)}\cos 2x + \frac{1}{(D^2 + 4)}\cos 4x$$

$$= \frac{x}{2 \times 2}\sin 2x + \frac{1}{-4^2 + 4}\cos 4x = \frac{x}{4}\sin 2x - \frac{1}{12}\cos 4x.$$

General Solution

$$y(x) = y_h(x) + y_p(x) = c_1 \cos 2x + c_2 \sin 2x + \frac{x}{4}\sin 2x - \frac{1}{12}\cos 4x. \qquad \boxed{\text{Q.E.D.}}$$

Example 8

$y'' - 6y' + 9y = 6x^2 + 2 - 12e^{3x}$, find the general solution of this ODE.

Solution

Homogeneous Solution

Through the characteristic equation $m^2 - 6m + 9 = 0$, we obtain $m = 3, 3$, so

$$y_h(x) = c_1 e^{3x} + c_2 x e^{3x}.$$

Particular Solution

Let $L(D) = D^2 - 6D + 9$, then

$$y_p = \frac{1}{(D^2 - 6D + 9)} 6x^2 + \frac{2}{(D^2 - 6D + 9)} e^{0x} - 12 \frac{1}{(D^2 - 6D + 9)} e^{3x}.$$

Among the three inverse operators on the right side, the last two terms correspond to the applications of Formula ① and Formula ③ from Table 2. As for the remaining inverse operator, we can refer to Formula ⑨. Using long division, we can convert the inverse operator into a differential operator, as shown in the attached figure.

Therefore

$$\frac{1}{(D^2 - 6D + 9)} = \frac{1}{9} + \frac{2}{27} D + \frac{1}{27} D^2 + \cdots .$$

$$
\begin{array}{r}
\frac{1}{9} + \frac{2}{27} D + \frac{1}{27} D^2 + \cdots \\
9 - 6D + D^2 \overline{\smash{\big)}\ 1 } \\
1 - \frac{2}{3} D + \frac{1}{9} D^2 \\
\hline
\frac{2}{3} D + \frac{1}{9} D^2 \\
\frac{2}{3} D + \frac{4}{9} D^2 + \frac{2}{27} D^3 \\
\hline
\frac{1}{3} D^2 - \frac{2}{27} D^3 \\
\frac{1}{3} D^2 - \frac{6}{27} D^3 + \frac{1}{27} D^4 \\
\hline
\cdots\cdots
\end{array}
$$

We substitute all the derived formulas into y_p as follows, we obtain

$$y_p = \frac{2}{3} x^2 + \frac{8}{9} x + \frac{6}{9} - 6x^2 e^{3x}.$$

General Solution

$$y(x) = y_h(x) + y_p(x) = c_1 e^{3x} + c_2 x e^{3x} + \frac{2}{3} x^2 + \frac{8}{9} x + \frac{6}{9} - 6x^2 e^{3x}. \qquad \text{Q.E.D.}$$

Example 9

$y'' + 9y = x \cos x$, find the general solution of this ODE.

Solution

Homogeneous Solution

Through the characteristic equation $m^2 + 9 = 0$, we have $m = \pm 3i$, so

$$y_h(x) = c_1 \cos 3x + c_2 \sin 3x.$$

Particular Solution

Let $L(D) = D^2 + 9$, then we obtain

$$y_p = \frac{1}{(D^2+9)} x \cos x$$

$$= \mathrm{Re}\{\frac{1}{(D^2+9)} x \cdot e^{ix}\} \quad \text{(through \textbf{Euler's formula})}$$

$$= \mathrm{Re}\{e^{ix} \frac{1}{((D+i)^2+9)} x\} \quad \text{(formula \textcircled{2})}$$

$$= \mathrm{Re}\{e^{ix} \frac{1}{8+i2D+D^2} x\}$$

$$= \mathrm{Re}\{e^{ix}(\frac{1}{8} - \frac{1}{32}iD + \cdots)x\} \quad \text{(Long division, as shown in the attached image)}$$

$$= \mathrm{Re}\{(\cos x + i\sin x)(\frac{1}{8}x - \frac{1}{32}i)\}$$

$$= \frac{1}{8}x\cos x + \frac{1}{32}\sin x .$$

The long division shown on the right:

$$
\begin{array}{r}
\frac{1}{8} - i\frac{D}{32} + \cdots \\[4pt]
8+i2D+D^2\,\overline{\big)\ 1} \\
1 + i\frac{D}{4} + \frac{D^2}{4} \\ \hline
-i\frac{D}{4} - \frac{D^2}{8} \\
-i\frac{D}{4} + \frac{D^2}{16} - i\frac{D^3}{32} \\ \hline
-\frac{3}{16}D^2 + i\frac{D^3}{32}
\end{array}
$$

General Solution

$$y(x) = y_h(x) + y_p(x) = c_1\cos 3x + c_2\sin 3x + \frac{1}{8}x\cos 2x + \frac{1}{32}\sin x .$$

Q.E.D.

2-6 Exercises

Basic questions

Solve the following ODE, where the particular solution is to be obtained using the method of inverse differential operators.

1. $y'' - 3y' + 2y = e^{3x}$.

2. $y'' - 3y' + 2y = e^{2x}$.

3. $y'' - 4y' = \cos 3x$.

4. Solve the following three small questions
   (1) $y'' + 3y' + 2y = e^x$.
   (2) $y'' - 2y' + y = e^x \cdot x^{-3}$.
   (3) $y'' - y' - 2y = 2e^{-x}$.

5. Solve the following two small questions
   (1) $y' + y = \sin x$.
   (2) $y'' - 4y' + 4y = e^{3x} - 1$.

6. $y'' - 3y' + 2y = x + 2$.

Advanced questions

Solve the following ODE, where the particular solution is to be obtained using the method of inverse differential operators.

1. $y'' + 4y' + 13y = 26e^{-4t}$,
   $y(0) = 5, y'(0) = -29$.

2. $y'' + 2y' + 5y = e^{-x} \sin 2x$.

3. $y'' + 4y' + 4y = 7x - 3\cos 2x + 5e^{-2x}$.

4. $y'' + 4y = \cos^2 x$.

5. $y'''' + 2y''' + 2y'' = x + 1$.

6. $y'' + 5y' = e^{-x} \sin 3x$.

7. $y'' + 4y = 6x \sin 2x$.

8. $(D^2 - D - 2)y = \sin x$.

9. $y'' + \lambda^2 y = \cos \lambda x, \lambda > 0$.

10. $y''' - 2y'' + y' = x^3 + 2e^x$.

11. $y'' - 2y' - 8y = 40 \sin 2x$.

12. $y'' - 2y' + 2y = 2e^x \cos x$.

13. $\dfrac{d^2x}{dt^2} + 2\dfrac{dx}{dt} + 2x = 10\cos 2t$,
    $x(0) = 1, x'(0) = 0$.

14. $y'' + w^2 y = r(t), r(t) = \cos \alpha t + \cos \beta t$,
    $w^2 \neq \alpha^2$ or β^2 and $w > 0$.

15. $y' - y = e^{2x} + x^2$.

16. $y'' + 2y' + y = 3e^{-x} + x$.

17. $y'' + y = x \cos x - \cos x$.

18. $y'''' + 2y'' + y = \cos x$.

19. $y'' - 4y = e^{2x} + 2x$.

20. $y'' - 3y' - 4y = 4x^2 + 2 \sin x$.

21. $y'' + 9y = x \cos x$.

22. $y'' + 5y' + 7y = \cos 3x$.

23. $y''' - 6y'' + 11y' - 6y = e^{2x} \cos x$.

24. $y'' - 6y' + 9y = e^{3x}$.

25. $y'' + 3y' = 28\cosh 4x$.

26. $y'' - y' - 12y = 2\sinh^2 x$.

27. $y'' + 4y' + 13y = \dfrac{1}{3}e^{-2t} \sin 3t$.

2-7 **Equidimensional Linear ODEs**

We have already discussed the solution of *n*th-order linear ODEs with constant coefficients. Now, we will introduce the solution of *n*th-order linear ODEs with variable coefficients. Solving variable coefficents ODEs can be quite challenging, and there are various methods available, including (1) equidimensional ODEs, (2) high-order exact ODEs, (3) variable transformation reduction of order (finding the general solution of the ODE given a homogeneous solution), (4) variable transformation, (5) factorization method, and (6) power series solution. Among these methods, equidimensional ODEs are the most common and relatively easier to solve. Therefore, in this section, we will focus on solving ODEs of this type.

2-7-1   Euler-Cauchy Equidimensional ODEs

We refer to equations of the following form as the Euler-Cauchy standard form.

$$a_n x^n y^{(n)} + a_{n-1} x^{n-1} y^{(n-1)} + \cdots + a_1 xy' + a_0 y = R(x)$$

To solve, let's first consider the case when $n = 1$ and then generalize it to the case of *n*-th order equations. In the equation

$$a_1 xy'(x) + a_0 y(x) = R(x),$$

if we make the variable transformation $x = e^t$, $t = \ln x$, then we have $\dfrac{dt}{dx} = \dfrac{1}{x}$, applying the chain rule, we can rewrite the equation as follows:

$$y' = \frac{dy}{dx} = \frac{dy}{dt} \cdot \frac{dt}{dx} = \frac{1}{x}\frac{dy}{dt} \Rightarrow xy' = \frac{d}{dt}y.$$

Substituting the expression back into the original equation, we obtain a linear constant coefficient ODE $a_1 \dfrac{dy}{dt} + a_1 y = R(t)$.

Similarly, by applying the chain rule, we can derive formulas for higher-order differentials step by step.

$$x^2 y'' = \frac{d}{dt}(\frac{d}{dt} - 1)y$$

$$x^3 y''' = \frac{d}{dt}(\frac{d}{dt} - 1)(\frac{d}{dt} - 2)y$$

$$x^4 y'''' = \frac{d}{dt}(\frac{d}{dt} - 1)(\frac{d}{dt} - 2)(\frac{d}{dt} - 3)y$$

$$\vdots$$

After substituting the given differentials into the original equation, we can convert the variable coefficients into constant coefficients term by term. At this point, we can use the knowledge from previous sections to first find particular solutions. If we assume $R(x) = 0$, we can assume $y_h(x) = e^{mt} = x^m$ then substitute (here $e^t = x$)   and solve for the value of m, which gives us the homogeneous solution. In the case of repeated roots, we can take another homogeneous solution as $y_h(x) = te^{mt} = \ln x \cdot x^m$. Thus, we obtain the general solution $y = y_h + y_p$. In the following examples, we will denote the differential equation with constant coefficients $\dfrac{d}{dt}$ as D, to facilitate the use of knowledge learned in previous sections.

Example 1

$x^2 y'' - 4xy' + 6y = 2x^4 + x^2$, find the general solution of this ODE.

Solution   Coefficient transformation to constant coefficient (assume $x = e^t$), substitute the above first-order and second-order differential equations into the original equation, then we obtain

$$(D^2 - 5D + 6)y = 2e^{4t} + e^{2t}.$$

Homogeneous Solution
From the characteristic equation $m^2 - 5m + 6 = 0$, we obtain $m = 2, 3$, therefore

$$y_h(t) = c_1 e^{2t} + c_2 e^{3t} = c_1 x^2 + c_2 x^3,$$

alternatively, let $y_h(x) = x^m$ and substitute, we obtain

$$m(m - 1) - 4m + 6 = 0 \Rightarrow m = 2, 3,$$

so

$$y_h(x) = c_1 x^2 + c_2 x^3.$$

Particular Solution
Let $L(D) = D^2 - 5D + 6$, then through formula ② and ③ of 2-6 Table 2. We obtain

$$y_p = \frac{1}{D^2 - 5D + 6}(2e^{4t} + e^{2t})$$

$$= \frac{1}{(D-2)(D-3)}2e^{4t} + \frac{1}{(D-2)(D-3)}e^{2t}$$

$$= e^{4t} - e^{2t} \cdot t$$

$$= e^{4t} - te^{2t}$$

$$= x^4 - x^2 \ln x.$$

General Solution

$$y = y_h + y_p = c_1 x^2 + c_2 x^3 + x^4 - x^2 \ln x.$$

Q.E.D.

Example 2

$x^3 y''' - 3x^2 y'' + 6xy' - 6y = x^4 \ln x$; $x > 0$, find the general solution of this ODE.

Solution   Coefficient transformation to constant coefficient (let $x = e^t$), substitute the first-order, second-order, and third-order differential equations into the original equation, then we obtain

$$(D - 1)(D - 2)(D - 3)y = te^{4t}.$$

Homogeneous Solution

Based on the previous step, we have determined that the characteristic equation has solutions $m = 1, 2, 3$, therefore

$$y_h(t) = c_1 e^t + c_2 e^{2t} + c_3 e^{3t} = c_1 x + c_2 x^2 + c_3 x^3.$$

Particular Solution

Let $L(D) = (D - 1)(D - 2)(D - 3)$, then through formula ② and ⑨ of 2-6 Table 2, we obatin

$$y_p(t) = \frac{1}{(D-1)(D-2)(D-3)} te^{4t}$$

$$= e^{4t} \frac{1}{(D+3)(D+2)(D+1)} t \quad \text{(formula ②)}$$

$$= e^{4t} \times \frac{1}{D^3 + 6D^2 + 11D + 6} t$$

$$= e^{4t} (\frac{1}{6} - \frac{11}{36} D + \cdots) t \quad \text{(formula ⑨)}$$

$$= e^{4t} (\frac{1}{6} t - \frac{11}{36}) = x^4 (\frac{1}{6} \ln x - \frac{11}{36}).$$

General Solution

$$y(t) = y_h(t) + y_p(t) = c_1 e^t + c_2 e^{2t} + c_3 e^{3t} + e^{4t} (\frac{1}{6} t - \frac{11}{36})$$

$$= c_1 x + c_2 x^2 + c_3 x^3 + x^4 (\frac{1}{6} \ln x - \frac{11}{36}).$$ Q.E.D.

2-7-2   Legendre Equidimensional ODEs

An ODE of the following form is referred to as the Legendre equation

$$a_n (bx + c)^n y^{(n)} + a_{n-1} (bx + c)^{n-1} y^{(n-1)} + \cdots + a_1 (bx + c) y' + a_0 y = R(x).$$

Find the Solution

Comparing the Legendre equation and the Euler-Cauchy equation, we observe that the coefficient term in front of y is a similar linear function. Therefore, by making the variable transformation $bx + c = e^t$ (referred to as equation ①), we can mimic the situation of the Euler-Cauchy equation.

$$(bx+c)y' = b\frac{dy}{dt} = b\frac{d}{dt}y$$

$$(bx+c)^2 y'' = b^2 \frac{d}{dt}(\frac{d}{dt}-1)y$$

$$(bx+c)^3 y''' = b^3 \frac{d}{dt}(\frac{d}{dt}-1)(\frac{d}{dt}-2)y$$

$$(bx+c)^4 y^{(4)} = b^4 \frac{d}{dt}(\frac{d}{dt}-1)(\frac{d}{dt}-2)(\frac{d}{dt}-3)y$$

$$\vdots$$

Similarly, if we denote the variable transformation $\dfrac{d}{dt}$ as D, we can transform the original variable coefficient ODE into a constant coefficient ODE.

Example 3

$$(x^2 +2x+1)\frac{d^2y}{dx^2}+(5x+5)\frac{dy}{dx}+5y = 0.$$

Solution　First, we observe that the original equationis equal to $(x + 1)^2 y'' + 5(x + 1) y' + 5y = 0$, therefore we let $b = 1$ and $c = 1$ in the variable transformation of equation ①, that is $x + 1 = e^t$, Applying the differential formula $[D(D - 1) + 5D + 5]y = 0$ mentioned earlier, we revert back to the homogeneous equation. Therefore, We can utilize the methods discussed in Section 2-2. We obtain

$$y_h(t) = e^{-2t}(c_1 \cos t + c_2 \sin t).$$

The general solution is

$$y(x) = \frac{1}{(x+1)^2}[c_1 \cos(\ln|x+1|)+c_2 \sin(\ln|x+1|)].$$
　　　　　　　　　　　　　　　　　　　　　　　　　　　　Q.E.D.

Example 4

$(3x+2)^2 y'' + 3(3x+2)y' - 36y = 3x^2 + 4x + 1$.

Solution　By observing the original equation, we can simplify it as

$$(3x+2)^2 y'' + 3(3x+2)y' - 36y = \frac{1}{3}(3x+2)^2 - \frac{1}{3}.$$

Following the approach in Example 3 (let $b = 3$, $c = 2$, $3x + 2 = e^t$), we can transform the original equation into

$$(D^2 - 4)y = \frac{1}{27}(e^{2t} - 1).$$

Now, from the characteristic equation $m^2 - 4 = 0$, so through Table 2 of 2-6, the homogeneous solution is

$$y_h(t) = c_1 e^{2t} + c_2 e^{-2t} = c_1(3x+2)^2 + c_2(3x+2)^{-2}.$$

Then through formula ② of 2-4 Table1, the particular solution is

$$y_p(t) = \frac{1}{108}(te^{2t} + 1) = \frac{1}{108}\left[1 + (3x+2)^2 \ln(3x+2)\right].$$

Therefore, the general solution is

$$y(x) = c_1(3x+2)^2 + c_2(3x+2)^{-2} + \frac{1}{108}\cdot\left[1 + (3x+2)^2 \ln|3x+2|\right]. \quad \boxed{\text{Q.E.D.}}$$

2-7　Exercises

Basic questions

Find the solution to the following homogeneous ODE.

1. $x^2 y'' + xy' - 4y = x^{-2}$, $x > 0$.

2. $x^2 y'' - 2xy' + 2y = \frac{4}{x^2}$.

3. $x^2 y'' - 3xy' + 4y = x^6 + 1$.

4. $x^2 y'' - 3xy' - 5y = 6x^5$.

5. $x^2 y'' + 2xy' - 2y = 6x$.

Advanced questions

Find the solution to the following homogeneous ODE.

1. $xy'' - y' = 2x^2 e^x$.

2. $x^2 y'' + xy' + 4y = \cos(2\ln x)$.

3. $x^3 y''' - 3x^2 y'' + 6xy' - 6y = 0$.

4. $x^2 y'' - 3xy' + 3y = 2x^4 e^x$.

5. $y'' - \frac{5}{x}y' + \frac{8}{x^2}y = \frac{2\ln x}{x^2}$.

6. $x^2 y'' - 2xy' + 2y = x^2$.

7. $x^3 y''' + 2xy' - 2y = x^2 \ln x + 3x$.

8. $x^3 y''' - 5x^2 y'' + 18xy' - 26y = 0$, $x > 0$.

9. $x^2 y'' - 7xy' + 15y = 15\ln x$.

10. $t^2 y'' + 10ty' + (t + 8) = 0$.

11. $x^2 y'' - 4xy' + 4y = x^4 + x^2$.

12. $x^2 y'' - 4xy' + 6y = 6x + 12$.

13. $(4x^2 + 12x + 9)y'' + (12x + 18)y' + 4y = 0$.

2-8 The Applications of Higher-Order ODEs in Engineering

In Chapter 1, we learned how to model and solve engineering problems using first-order ordinary differential equations (ODEs). However, many physical problems in engineering require higher-order ODEs to fully capture their behavior. Therefore, in this chapter, we will utilize higher-order ODEs to model common engineering problems and apply the solution methods introduced earlier to gain a deeper understanding of their physical significance.

2-8-1 Mass-Damping-Spring Vibration System

From Newton's second law $\vec{F} = m\vec{a}$, we can derive the governing equations for systems, which can be categorized into two types:

1. Single-Degree-of-Freedom Vibrational Systems

$$m\frac{d^2 y}{dt^2} + c\frac{dy}{dt} + ky = F(t)$$

Figure 2-1   Single-degree-of-freedom mck vibrational system schematic

2. Multi-Degree-of-Freedom Vibrational Systems

$$\begin{cases} m_1\dfrac{d^2 y_1}{dt^2} = -c_1\dfrac{dy_1}{dt} - k_1 y_1 + c_2(\dfrac{dy_2}{dt} - \dfrac{dy_1}{dt}) + k_2(y_2 - y_1) \\ m_2\dfrac{d^2 y_2}{dt^2} = -c_2(\dfrac{dy_2}{dt} - \dfrac{dy_1}{dt}) - k_2(y_2 - y_1) + F(t) \end{cases}$$

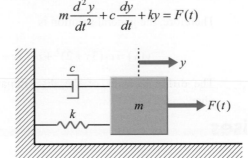

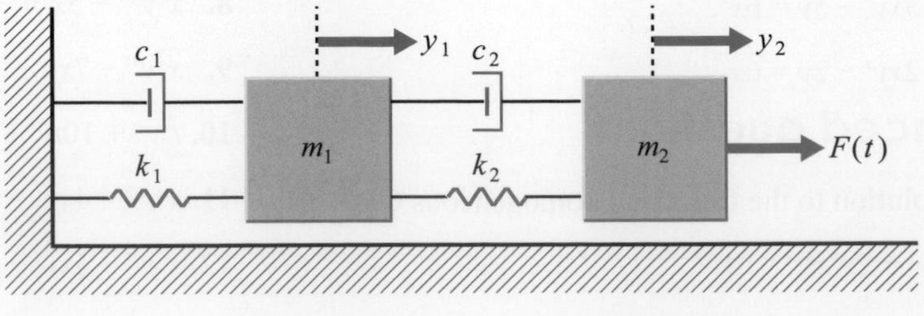

Figure 2-2   Multi-degree-of-freedom mck vibrational system schematic

Example 1

There is a vibrational system depicted in the diagram above, consisting of a mass, damping, and a spring. Using a differential equation, please establish its dynamic equation, where the system parameters m, c, and k are positive constants. and discuss the solutions of the system when the external force $f(t) = 0$ and when there is an external force $f(t) = A \cos \omega t$.

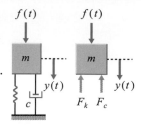

Solution   From Newton's second law, we know $f(t) - F_k - F_c = m \dfrac{d^2 y}{dt^2}$, where

$F_k = ky$, $F_c = c \dfrac{dy}{dt}$, thus, the dynamic equation for this system can be expressed as

$m \dfrac{d^2 y}{dt^2} + c \dfrac{dy}{dt} + ky = f(t)$, also

【**No external force ($f(t) = 0$)**】

At this point, the system is also referred to as "unforced motion", the original equation can be transformed to $m \dfrac{d^2 y}{dt^2} + c \dfrac{dy}{dt} + ky = 0$,

where the characteristic equation $m \lambda^2 + c \lambda + k = 0$ tell us

$$\lambda_1 = \frac{-c + \sqrt{c^2 - 4mk}}{2m}, \quad \lambda_2 = \frac{-c - \sqrt{c^2 - 4mk}}{2m}.$$

The different cases of the discriminant $c^2 - 4mk$ correspond to different physical phenomena as follows:

(1) $c^2 - 4mk > 0$, here we called **overdamping motion**, at this point

$$\lambda_1 = \frac{-c + \sqrt{c^2 - 4mk}}{2m}, \quad \lambda_2 = \frac{-c - \sqrt{c^2 - 4mk}}{2m},$$

and λ_1, λ_2 are all are negative real numbers. The solution is

$$y_h(t) = c_1 e^{\lambda_1 t} + c_2 e^{\lambda_2 t} \quad \text{(Transient response)},$$

and $\lim\limits_{t \to \infty} y_h(t) = 0$, the system response will exponentially decay to zero. (Illustrated in the following graphical simulation.)

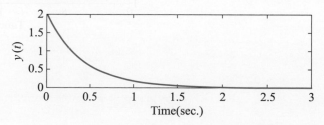

(2) $c^2 - 4mk = 0$, here we called **critical damping motion**, at this point $\lambda_1 = \dfrac{-c}{2m}$,

$\lambda_2 = \dfrac{-c}{2m}$, and λ_1, λ_2 are equal negative real roots. The solution is

$$y_h(t) = (c_1 + c_2 t)e^{-\frac{c}{2m}t} \quad \text{(Transient response)}.$$

Because $\lim\limits_{t \to \infty} y_h(t) = 0$, the system response will still decay to zero, but its decay form is different from that of overdamped motion. (Illustrated in the following graphical simulation.)

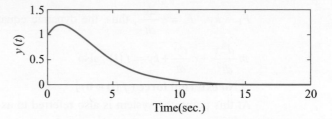

(3) $c^2 - 4mk < 0$, here we called **underdamping motion**, at this point

$$\lambda_1 = \frac{-c + i\sqrt{-(c^2 - 4mk)}}{2m}, \quad \lambda_2 = \frac{-c - i\sqrt{-(c^2 - 4mk)}}{2m},$$

and x_1, λ_2 are complex conjugate negative roots with negative real parts. Let

$$\beta = \frac{\sqrt{-(c^2 - 4mk)}}{2m} > 0 . \text{ We obtain}$$

$$y_h(t) = e^{-\frac{c}{2m}t}(c_1 \cos \beta t + c_2 \sin \beta t) \quad \text{(Transient response)}.$$

Because $\lim\limits_{t \to \infty} y_h(t) = 0$, the system response will exhibit a steady decay to zero with oscillatory waveform in the form of trigonometric functions. (Illustrated in the following graphical simulation.)

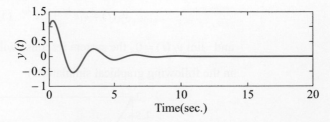

【With an external force ($f(t) = A\cos\omega t$)】

At this point, the system is referred to as "forced motion." The original equation becomes $m\dfrac{d^2y}{dt^2} + c\dfrac{dy}{dt} + ky = A\cos\omega t$. From the previous discussion, we know that the homogeneous solution $y_h(t)$ satisfies $\lim\limits_{t\to\infty} y_h(t) = 0$, so only the particular solution $y_p(t)$ remains for the homogeneous ODE at $t \to \infty$. In this case, the particular solution $y_p(t)$ is referred to as the steady-state response of the system. Now, let's discuss the solution for $y_p(t)$.

The original equation can be transformed into $(mD^2 + cD + k)y = A\cos\omega t$.

$$y_p(t) = \frac{1}{mD^2 + cD + k}(A\cos\omega t) = \frac{1}{m\cdot(-\omega^2) + cD + k}(A\cos\omega t)$$

$$= \frac{1}{cD + k - m\omega^2}(A\cos\omega t) = \frac{cD + (m\omega^2 - k)}{c^2D^2 - (m\omega^2 - k)^2}(A\cos\omega t)$$

$$= \frac{cD}{c^2(-\omega^2) - (m\omega^2 - k)^2}(A\cos\omega t) + \frac{m\omega^2 - k}{c^2(-\omega^2) - (m\omega^2 - k)^2}(A\cos\omega t)$$

$$= \frac{mA(\omega^2 - k/m)}{c^2(-\omega^2) - (m\omega^2 - k)^2}(\cos\omega t) + \frac{-cA\omega}{c^2(-\omega^2) - (m\omega^2 - k)^2}(\sin\omega t)$$

$$= \frac{mA(k/m - \omega^2)}{c^2(\omega^2) + m^2(\omega^2 - k/m)^2}(\cos\omega t) + \frac{cA\omega}{c^2(\omega^2) + m^2(\omega^2 - k/m)^2}(\sin\omega t)$$

$$= \frac{mA(\omega_0^2 - \omega^2)}{c^2\omega^2 + m^2(\omega^2 - \omega_0^2)^2}(\cos\omega t) + \frac{cA\omega}{c^2\omega^2 + m^2(\omega^2 - \omega_0^2)^2}(\sin\omega t),$$

where $\omega_0 - \sqrt{\dfrac{k}{m}}$ is called natural frequency, we obtain

$$\lim_{t\to\infty} y(t) = y_p(t)$$

$$= \frac{mA(\omega_0^2 - \omega^2)}{c^2\omega^2 + m^2(\omega^2 - \omega_0^2)^2}(\cos\omega t) + \frac{cA\omega}{c^2\omega^2 + m^2(\omega^2 - \omega_0^2)^2}(\sin\omega t).$$

(Illustrated in the following graphical simulation.)

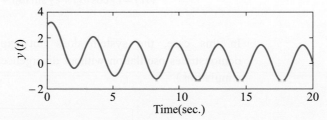

Example 2

In the given mass-spring-damper system, if $c = 0$, please analyze the solutions of the system when an external force $f(t) = A \cos \omega t$ is applied.

Solution The dynamic equation for this system is $m\dfrac{d^2 y}{dt^2} + ky = A\cos \omega t$. Solving the characteristic equation $m\lambda^2 + k = 0$, we have $\lambda_{1,2} = \pm i\sqrt{\dfrac{k}{m}} = \pm i\omega_0$. The homogeneous solution is

$$y_h(t) = c_1 \cos \omega_0 t + c_2 \sin \omega_0 t.$$

To find the particular solution, we apply the method of the inverse operator, because

$$(mD^2 + k)y = A \cos \omega t,$$

therefore

$$y_p(t) = \frac{1}{m \cdot D^2 + k}(A\cos \omega t) = \frac{A}{m}\frac{1}{D^2 + \omega_0^2}\cos \omega t.$$

The different cases of ω in the equation correspond to different phenomena:

(1) $\omega \neq \omega_0$

Here excitation frequency ω is not equal to natural frequency ω_0, then

$$y_p(t) = \frac{A}{m}\frac{1}{\omega_0^2 - \omega^2}\cos \omega t.$$

The general solution is

$$y(t) = c_1 \cos \omega_0 t + c_2 \sin \omega_0 t + \frac{A}{m}\frac{1}{\omega_0^2 - \omega^2}\cos \omega t.$$

In this case, the system does not experience resonance, and the solution demonstrates oscillation without divergence. (Illustrated in the following graphical simulation.)

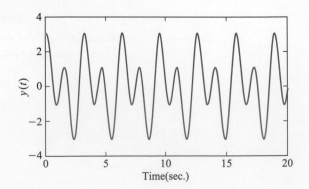

(2) $\omega = \omega_0$

The excitation frequency ω is equal to natural frequency ω_0, then the particular solution is $y_p(t) = \dfrac{A}{m}\dfrac{t}{2\omega_0}\sin\omega t$. The general solution is

$$y(t) = c_1\cos\omega_0 t + c_2\sin\omega_0 t + \frac{A}{m}\frac{t}{2\omega_0}\sin\omega t.$$

In this case, the system undergoes resonance, and the solution demonstrates oscillations that diverge, that is $\lim\limits_{t\to\infty} y(t) = \infty$. (Illustrated in the following graphical simulation.)

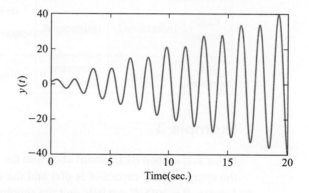

Q.E.D.

2-8-2 RLC Electric Circuit System

An RLC circuit system can be categorized into "single-loop" and "multi-loop" configurations, which are explained as follows:

1. Single-Loop System

The system's control equations can be derived using Kirchhoff's voltage law. Let's consider a loop with current $I(t)$ flowing through it and a capacitor with charge $Q(t)$, then

$$L\frac{d^2Q(t)}{dt^2} + R\frac{dQ(t)}{dt} + \frac{1}{C}Q(t) = E(t) \quad \text{or} \quad L\frac{d^2I(t)}{dt^2} + R\frac{dI(t)}{dt} + \frac{1}{C}I(t) = \frac{dE(t)}{dt}.$$

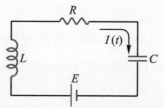

Figure 2-3   RLC circuit system

2. Multi-Loop System

(1) Kirchhoff's Voltage Law, KVL

Kirchhoff's voltage law states that the algebraic sum of all voltage drops along any closed loop in a circuit is equal to zero.

(2) Kirchhoff's Current Law, KCL

Kirchhoff's current law states that the algebraic sum of currents flowing into a node in a network is zero. This law can be utilized to derive the control equations for the system. A comparison between RLC systems and mck systems is presented in Table 3.

Table 3

RLC system	Inductance L	resistance R	reciprocal of capacitance $\dfrac{1}{C}$	the derivative of electromotive force $E'(t)$	electric current $I(t)$
mck system	mass m	damping constant c	spring constant k	External force $F(t)$	displacement $y(t)$

Example 3

In the single-loop RLC circuit shown in the diagram, assuming the charge on the capacitor is $q(t)$ and the current is $i(t)$, with $L = 1\text{H}$, $R = 20\Omega$, $C = 0.01\text{F}$, and the supply voltage $E(t) = 120 \sin 10t$ volts, with initial conditions $q(0) = 0$ and $i(0) = 0$, determine the steady-state current in this circuit.

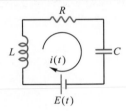

Solution　The dynamic equation for the original circuit is

$L\dfrac{d^2q}{dt^2} + R\dfrac{dq}{dt} + \dfrac{1}{C}q = E(t)$. Substituting the given parameter values, we obtain

$\dfrac{d^2q}{dt^2} + 20\dfrac{dq}{dt} + 100q = 120\sin 10t$, then the solution of ODE is

$$q(t) = c_1 e^{-10t} + c_2 t e^{-10t} - \frac{3}{5}\cos 10t \text{ , also}$$

$$i(t) = \frac{dq}{dt} = -10c_1 e^{-10t} + c_2 e^{-10t} - 10c_2 t e^{-10t} + 6\sin 10t \text{ .}$$

By $q(0) = 0$, $i(0) = 0$, we obtain $c_1 - \dfrac{3}{5} = 0 \Rightarrow c_1 = \dfrac{3}{5}$, $-10c_1 + c_2 = 0 \Rightarrow c_2 = 6$,

$i(t) = -60t e^{-10t} + 6\sin 10t$, the steady-state current is $\lim\limits_{t \to \infty} i(t) = 6\sin 10t$.　 Q.E.D.

2-8-3 Mixing System with Feedback

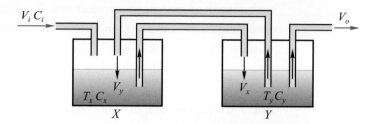

Figure 2-4   Diagram of a mixing system

Let's consider a system where water tank X contains T_x liters (ℓ) of saltwater with concentration $C_x(g/\ell)$, and water tank Y contains T_y liters of saltwater with concentration $C_y(g/\ell)$. If saltwater with concentration $C_i(g/\ell)$ and flow rate flows $V_i(\ell/\min)$ into tank X, and the mixture in tank X flows into tank Y with a flow rate of $V_x(\ell/\min)$, while tank Y is pumped back into tank X with a flow rate of $V_y(\ell/\min)$ (establishing feedback), and simultaneously, water is pumped out of tank Y at a rate of $V_o(\ell/\min)$ (as shown in Figure 2-4), we need to determine the salt content in both tanks at time t. By applying the principle of mass conservation, we can obtain the control equations for the system.

Modeling

Let $x(t)$, $y(t)$ demonstrate the salt content in water tanks X and Y at time t. By $\dfrac{dx}{dt} =$ input rate $-$ output rate $=$ the rate of salt inflow minus the rate of salt outflow per unit time, we obtain

$$\begin{cases} \dfrac{dx}{dt} = V_i \cdot C_i + V_y \cdot \dfrac{y}{T_y + (V_x - V_y - V_o)t} - V_x \cdot \dfrac{x}{T_x + (V_i + V_y - V_x)t} \\ \dfrac{dy}{dt} = V_x \cdot \dfrac{x}{T_x + (V_i + V_y - V_x)t} - (V_y + V_o) \cdot \dfrac{y}{T_y + (V_x - V_y - V_o)t} \end{cases},$$

and

$$\begin{cases} x(0) = T_x \cdot C_x \\ y(0) = T_y \cdot C_y \end{cases}.$$

By solving the coupled ODEs derived from the equation above, we can obtain $x(t)$, $y(t)$.

Example 4

Let's consider a system where water tank X contains 10 liters (ℓ) of saltwater with a concentration of $1.0(g/\ell)$, and water tank Y contains 20 liters of pure water. If pure water is added to tank X with a flow rate of $2.0(\ell/\min)$, and the mixture in tank X flows into tank Y with a flow rate of $4.0(\ell/\min)$, while tank Y is pumped back into tank X with a flow rate of $2.0(\ell/\min)$, and simultaneously, water is pumped out of tank Y with a flow rate of $2.0(\ell/\min)$, we need to determine the salt content in both tanks at time t.

Solution　By modeling, we have
$$\begin{cases} \dot{x} = -\dfrac{2}{5}x + \dfrac{1}{10}y & \cdots ① \\[2mm] \dot{y} = \dfrac{2}{5}x - \dfrac{1}{5}y & \cdots ② \end{cases}.$$

Through ① we know $y = 10\dot{x} + 4x$, then substitute into ② we have $\dot{y} = -\dfrac{2}{5}x - 2\dot{x}$,

also $\dot{x} = -\dfrac{2}{5}x + \dfrac{1}{10}\dot{y} \Rightarrow \ddot{x} + \dfrac{3}{5}\dot{x} - \dfrac{1}{25}x = 0$.

The general solution is
$$x(t) = c_1 e^{\frac{-3+\sqrt{5}}{10}t} + c_2 e^{\frac{-3-\sqrt{5}}{10}t}.$$

By $y = 10\dot{x} + 4x$, we obtain $y(t) = (1+\sqrt{5})c_1 e^{\frac{-3+\sqrt{5}}{10}t} + (1-\sqrt{5})c_2 e^{\frac{-3-\sqrt{5}}{10}t}$;
by $x(0) = 10$, $y(0) = 0$, we have $c_1 = 5-\sqrt{5}$, $c_2 = 5+\sqrt{5}$.
The system solution is
$$x(t) = (5-\sqrt{5})e^{\frac{-3+\sqrt{5}}{10}t} + (5+\sqrt{5})e^{\frac{-3-\sqrt{5}}{10}t}, \quad y(t) = 4\sqrt{5}e^{\frac{-3+\sqrt{5}}{10}t} - 4\sqrt{5}e^{\frac{-3-\sqrt{5}}{10}t}. \quad \boxed{\text{Q.E.D.}}$$

2-8   Exercises

Basic questions

1. In the mck vibration system, $m = 1$, $c = 5$, $k = 4$, the external force term $F(t) = 0$, and the initial position $y(0) = 1$, with an initial velocity $y'(0) = 1$ please determine the system response.

2. In the RLC circuit, $R = 2\Omega$, $L = 1H$, $C = 0.1F$, and $E(t) = 0$, the initial charge $Q(0) = -2$, and the initial current $I(0) = \dfrac{dQ}{dt}\bigg|_{t=0} = 0$. Find the charge $Q(t)$ of the circuit.

Advanced questions

1. Consider an mck vibration system with $m = 1$, $c = 2$, $k = 6$, and an external force $F = \sin 2t + 2\cos 2t$. The system has an initial displacement of 1.0 and an initial velocity of 0. Please determine the response of the system.

2. Consider a single-loop RLC circuit with $R = 50\Omega$, $L - 30H$, $C = 0.025F$, and $E(t) = 200\sin 4t$ V. Try to determine the steady-state current of the circuit.

3. Consider a tank X contains 10 liters (ℓ) of saltwater with a concentration of $1.0(g/\ell)$, while tank Y contains 20 liters of pure water. Pure water is being added to tank X at a rate of $2.5(\ell/\min)$, and the mixture from tank X is flowing into tank Y at a rate of $4.0(\ell/\min)$. At the same time, tank Y is being pumped back into tank X at a rate of $1.5(\ell/\min)$, and water is flowing out of tank Y at a rate of $2.5(\ell/\min)$. When will tank Y have the maximum salt content? What is the salt content at that time?

3

Laplace Transform

Laplace
(1749~1827,
France)

Pierre-Simon Laplace (1749-1827) introduced the Laplace transformation in the study of probability theory. Later, the Laplace transformation became widely used in physics and engineering for solving differential and integral equations, especially in linear time-invariant systems. Laplace made significant contributions to the advancement of modern mathematical and engineering sciences.

Learning Objectives

3-1

The Definition of
Laplace Transform

3-1-1 Understanding exponential function

3-1-2 Understanding and determining the existence of Laplace transform

3-1-3 Understanding the existence of the Laplace transform through
decomposing improper integrals

3-2

Basic Characteristics
and Theorems

3-2-1 Learning to derive the Laplace transform of common functions

3-2-2 Mastering the common properties of Laplace transforms

3-3

Laplace Transform of
Special Functions

3-3-1 Mastering the Laplace transform of the unit impulse function

3-3-2 Mastering the Laplace transform of the periodic function

3-3-3 Mastering the convolution theorem and its application

3-4

Laplace Inverse
Transform

3-4-1 Understanding Laplace inverse transform and its properties

3-4-2 Mastering the common formula and theorem of Laplace inverse
transform

3-4-3 Mastering the partial fraction decomposition of Laplace transforms

3-5

The Application of
Laplace Transform

3-5-1 Learn to use Laplace transforms to solve constant coefficients ODE

3-5-2 Learn to use Laplace transforms to solve variable coefficients ODE

3-5-3 Mastering the solution of convolution-type differential equations

ExampleVideo

In solving general engineering problems, it is common to use differential equations or integral equations for modeling, followed by applying established methods for solving these differential equations. However, solving differential equations is not as straightforward as solving algebraic equations. Therefore, the transformation of differential or integral equations into another space where they can be represented as algebraic equations and solved becomes an important topic.

In 1744, the Swiss mathematician Leonhard Euler began studying solutions of differential equations in the form of $z = \int X(x)e^{ax}dx$, $z = \int X(x)x^{A}dx$, etc. However, he did not pursue this line of research extensively. Later, the French mathematician Joseph-Louis Lagrange, while studying probability theory, considered probability density functions in integral form: $\int X(x)e^{-ax}dx$.

This is widely regarded as the origin of the modern Laplace transform. The complete development of the Laplace transform was accomplished by Pierre-Simon Laplace.

Subsequently, in physics and engineering, the Laplace transform has been extensively used to solve differential and integral equations, particularly in linear time-invariant systems. This chapter systematically organizes the formulas and results of the Laplace transform and illustrates how it is applied to solve physical problems. The Laplace transform finds wide application in engineering problems across various fields, which is a highly important chapter.

3-1   The Definition of Laplace Transform

In order to compute the integral form defined by Euler, the integral itself must exist. Therefore, we first distinguish this class of functions.

3-1-1   Exponential Order Function

▶ **Definition 3-1-1**

Exponential Order Function and Convergent Abscissa

If a function $f(t)$ satisfies $|f(t)| \leq Me^{\alpha t}$, $\forall t$, where M and α are constants, then $f(t)$ is called an exponential order function. If α_0 is a lower bound for $|f(t)| \leq Me^{\alpha t}$, then α_0 is referred to as the convergent abscissa.

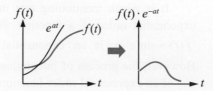

Figure 3-1   Diagrams of exponential order function ◀

According to the definition 3-1-1 of exponential order functions, they exhibit behavior similar to exponential functions. Therefore, in general, it is common to calculate the limit of the function

$$\lim_{t\to\infty}\frac{|f(t)|}{e^{\alpha t}}.$$

To determine whether $f(t)$ is an exponential order function, the convergence of the limit mentioned in the definition is considered. The abscissa mentioned in the definition is not difficult to imagine. For example, if the limit of the function as t approaches infinity is given by $f(t) = e^{2t}$, then the value of α must be at least 2 for the limit to converge. Therefore, $\alpha_0 = 2$, the convergence abscissa is 2.

Similarly, if the limit as t approaches negative infinity is given by $f(t) = e^{-5t}$, then the value of a must be at least -5 for the limit to converge. Hence, $\alpha_0 = -5$, the convergence abscissa is -5.

Example 1

Which of the following functions is an exponential order function:
(1) t^n (2) e^{t^2} (3) t^t.

Solution   We calculate the limits for each sub-questions $\lim_{t\to\infty}\dfrac{|f(t)|}{e^{\alpha t}}$.

(1) $\lim_{t\to\infty}\dfrac{t^n}{e^{\alpha t}} = \lim_{t\to\infty}t^n\cdot e^{-\alpha t} = \lim_{t\to\infty}\exp[\ln t^n]\cdot e^{-\alpha t} = \lim_{t\to\infty}e^{n\cdot\ln t}\cdot e^{-\alpha t}$

$= \lim_{t\to\infty}e^{n\ln t-\alpha t} = 0 = M$, taking $\alpha > 0$,

$\therefore t^n$ is an exponential order function.

(2) $\lim_{t\to\infty}\dfrac{e^{t^2}}{e^{\alpha t}} = \lim_{t\to\infty}e^{t^2}\cdot e^{-\alpha t} = \lim_{t\to\infty}e^{t^2-\alpha t} = \lim_{t\to\infty}e^{t\cdot(t-\alpha)} \to \infty$,

$\therefore e^{t^2}$ is not an exponential order function.

(3) $\lim_{t\to\infty}\dfrac{t^t}{e^{\alpha t}} = \lim_{t\to\infty}t^t\cdot e^{-\alpha t} = \lim_{t\to\infty}\exp[t\ln t]\cdot e^{-\alpha t} = \lim_{t\to\infty}e^{t\ln t-\alpha t}$

$= \lim_{t\to\infty}e^{t\cdot(\ln t-\alpha)} \to \infty$,

$\therefore t^t$ is not an exponential order function.   Q.E.D.

It is worth mentioning that the process of differentiation does not preserve the exponential order of a function. For example, because the sin function is bounded, $f(t) = \sin(e^{t^2})$ is an exponential order function, but $f'(t) = 2te^{t^2}\cos e^{t^2}$ is not. However, the process of integration can preserve the piecewise continuity (denoted by C_P) of an exponential order function. Piecewise continuity means that the function has a finite number of discontinuity points and its function values are bounded, as shown in Figure 3-2.

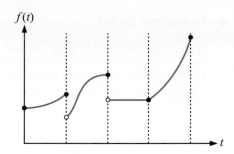

Figure 3-2   Diagram of a piecewise continuous function

Furthermore, it can be verified using the L'Hôpital's rule that if a piecewise continuous function $f(x)$ is an exponential order at the same time, then its integral $\int_0^T f(x)dx$ is also exponential order. With the concept of exponential order functions, we can now proceed to define what is known as the Laplace transform.

3-1-2   Laplace Transform

▶ **Definition 3-1-2**

Laplace Transform and Inverse Transform

The Laplace transform and Inverse transform of a function $f(t)$ is defined as follows :

(1) $\mathcal{L}\{f(t)\} = \int_0^{\infty} f(t)e^{-st}dt - F(s)$.

(2) $\mathcal{L}^{-1}\{F(s)\} = \dfrac{1}{2\pi i}\int_{a-i\,\infty}^{a+i\,\infty} f(s)e^{st}ds\,^1$.   ◀

From a physical perspective, if $f(t)$ represents a signal, the Laplace transform will map the original signal in the "time-domain" to a one-to-one corresponding signal in another space called the "s-domain," without any confusion. Whether this transformation is possible or not, which is described by Theorem 3-1-1, as follows.

---

[1] The process of Laplace inverse transform, which involves integrating, is discussed in the context of complex variables for solving it.

► **Theorem 3-1-1**

The Conditions for the Existence of Laplace Transform

If $f(t)$ satisfies that $f(t)$ is piecewise continuous on $[0, T]$ and is an exponential order function at $t > T$, then the Laplace transform of $f(t)$ exists and there exist positive real numbers M and α_0 causes

$$\mathscr{L}\{f(t)\} = F(s) \le \frac{M}{s - \alpha_0}, \text{ where } s > \alpha_0 \text{ and } |f(t)| \le Me^{\alpha_0 t}.$$

Proof

Because $f(t)$ is exponential order, the convergence abscissa can be set as α_0, and according to Definition 3-1-1, there exists a positive real number M causes

$$|f(t)| \le Me^{\alpha_0 t},$$

thus $\mathscr{L}\{f(t)\} = \int_0^\infty f(t)e^{-st}dt \le \int_0^\infty |f(t)|e^{-st}dt \le M\int_0^\infty e^{\alpha_0 t}e^{-st}dt$

$$= M\int_0^\infty e^{-(s-\alpha_0)t}dt = M\times\frac{-1}{(s-\alpha_0)}\times e^{-(s-\alpha_0)t}\Big|_0^\infty = \frac{M}{s-\alpha_0}(s>\alpha_0),$$

so $\mathscr{L}\{f(t)\} = F(s) \le \frac{M}{s-\alpha_0}$ (where $s > \alpha_0$). ◄

Through Definition 3-1-1, we obtain $\begin{cases} (1) \lim\limits_{s\to\infty} F(s) = 0 \\ (2) \lim\limits_{s\to\infty} sF(s) = M = \text{constant} \end{cases}$.

3-1-3   Existence Analysis of Laplace Transform (Selected Reading)

Let's now discuss how to analyze the existence of the Laplace transform. First, the Laplace transform $\int_0^\infty f(t)e^{-st}dt$ itself is an improper integral. We can split the integral as follows:

$$\int_0^T f(t)e^{-st}dt + \int_T^\infty f(t)e^{-st}dt.$$

For the integral on the right-hand side, its convergence depends on the existence of an upper bound function for f on $[T, \infty]$. On the other hand, for the integral on the left-hand side, its existence depends on whether the function f has any singular points on the interval $[0, T]$. Therefore, based on the proof of Theorem 3-1-1, we obtain the following result.

1. If $f(t)$ is C_p, then there must exist a T such that the discontinuity points of $f(t)$ lie within the interval $[0, T]$ and $\int_0^T f(t) e^{-st} dt$ exists.

2. If $f(t)$ is an exponential order function, then the integral of the second part of the Laplace transform $\int_T^\infty f(t) e^{-st} dt$ must exist.

When both conditions mentioned above (1-2) are satisfied, the existence of the Laplace transform is ensured. Therefore, when determining the existence of a Laplace transform, our main task is to find a suitable value of T to split the improper integral and then discuss each part separately.

Example 2

Try to determine the existence of the Laplace transform for the following functions.

$(1) f(t) = t^{-3}$ $(2) f(t) = t^{-\frac{1}{2}}$ $(3) f(t) = e^{at}$ $(4) f(t) = e^t$ $(5) f(t) = \dfrac{1}{1+t}$.

Solution

(1) $f(t) = t^{-3}$ is exponential order, but $\int_0^T t^{-3} dt$ doesn't exist, $\therefore \mathcal{L}\{f(t)\}$ doesn't exist.

(2) $f(t) = t^{-\frac{1}{2}}$ is exponential order, although $f(t)$ is not C_p, but $\int_0^T t^{-\frac{1}{2}} dt = 2\sqrt{t} \Big|_0^T$ exists, $\therefore \mathcal{L}\{f(t)\}$ exists.

(3) $f(t) = e^{at}$ is both exponential order and continuous., $\therefore \mathcal{L}\{f(t)\}$ exists.

(4) $f(t) = t^t$ is not exponential order, $\therefore \mathcal{L}\{f(t)\}$ doesn't exist.

(5) $f(t) = \dfrac{1}{1+t}$ is both exponential order and piecewise continuous at $[0, \infty)$, $\therefore \mathcal{L}\{f(t)\}$ exists.

Q.E.D.

3-1 Exercises

Basic questions

1. Try to determine which functions are exponential order functions.

$(1) e^{at}$. $(2) \sin wt$. $(3) e^{t^3}$. $(4) t^2$. $(5) \cosh wt$. $(6) t^{\frac{1}{2}t}$.

Advanced questions

1. Try to determine which functions can undergo a Laplace transform.

$(1) t^{-2}$. $(2) t^3$. $(3) t^{-\frac{1}{3}}$. $(4) e^{3t}$. $(5) \sin 2t$. $(6) \cos 3t$.
$(7) \sinh 5t$. $(8) e^{t^3}$. $(9) t^{\frac{1}{2}t}$.

3-2 Basic Characteristics and Theorems

In this chapter, we will focus on finding the Laplace transform of common functions. For more complex functions, you can refer to Appendix 2 for their Laplace transforms.

3-2-1 The Common Functions of L-T

▶ **Theorem 3-2-1**

Unit Step Function L-T

Define $u(t) = \begin{cases} 1 \ , \ t \geq 0 \\ 0 \ , \ t < 0 \end{cases}$,

then $\mathcal{L}\{u(t)\} = \dfrac{1}{s}$

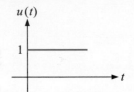

Figure 3-3　Unit step function

Proof

$\mathcal{L}\{u(t)\} = \displaystyle\int_0^\infty 1 \cdot e^{-st}dt = -\dfrac{1}{s}e^{-st}\Big|_0^\infty = \dfrac{1}{s}$, taking $s > 0$,

$\therefore \mathcal{L}\{u(t)\} = \dfrac{1}{s}$　$(\mathcal{L}^{-1}\{\dfrac{1}{s}\} = u(t) \text{ or } 1)$.　[2]　◀

▶ **Theorem 3-2-2**

Exponential Function L-T

$\mathcal{L}\{e^{at}\} = \dfrac{1}{s-a}$.

Proof

$\mathcal{L}\{e^{at}\} = \displaystyle\int_0^\infty e^{at} \cdot e^{-st}dt = \int_0^\infty e^{-(s-a)t}dt = \dfrac{-1}{(s-a)}e^{-(s-a)t}\Big|_0^\infty = \dfrac{1}{s-a}$, where $s > a$,

$\therefore \mathcal{L}\{e^{at}\} = \dfrac{1}{s-a}$　$(\mathcal{L}^{-1}\{\dfrac{1}{s-a}\} = e^{at})$.　◀

---

[2] In Laplace transform problems, we typically consider the range $t \geq 0$. Therefore, the unit step function $u(t)$ can be directly expressed as 1 in these cases, that is $\mathcal{L}\{1\} = \dfrac{1}{s}$, $\mathcal{L}^{-1}\{\dfrac{1}{s}\} = 1 = u(t)$.

► **Theorem 3-2-3**

Trigonometric Function L-T

$$\mathcal{L}\{\sin wt\} = \frac{w}{s^2 + w^2},$$

$$\mathcal{L}\{\cos wt\} = \frac{s}{s^2 + w^2}.$$

Proof

First, we recall Euler's formula: $e^{iwt} = \cos wt + i\sin wt$,

therefore $\mathcal{L}\{e^{iwt}\} = \mathcal{L}\{\cos wt + i\sin wt\} = \mathcal{L}\{\cos wt\} + i\mathcal{L}\{\sin wt\}$.

Now through the Laplace transform of the exponential function, according to Theorem 3-2-2, we know

$$\mathcal{L}\{e^{iwt}\} = \frac{1}{s - iw} = \frac{s + iw}{(s - iw)(s + iw)} = \frac{s + iw}{s^2 + w^2} = \frac{s}{s^2 + w^2} + i\frac{w}{s^2 + w^2},$$

after comparison coefficient

we obtain $\quad \mathcal{L}\{\cos wt\} = \dfrac{s}{s^2 + w^2} \quad (\mathcal{L}^{-1}\{\dfrac{s}{s^2 + w^2}\} = \cos wt)$,

$$\mathcal{L}\{\sin wt\} = \frac{w}{s^2 + w^2} \quad (\mathcal{L}^{-1}\{\frac{w}{s^2 + w^2}\} = \sin wt). \quad ◄$$

► **Theorem 3-2-4**

Hyperbolic Function L-T

$$\mathcal{L}\{\cosh wt\} = \frac{s}{s^2 - w^2},$$

$$\mathcal{L}\{\sinh wt\} = \frac{w}{s^2 - w^2}.$$

Proof

Because $\cosh wt = \dfrac{e^{wt} + e^{-wt}}{2}$, through the Laplace transform formula for the exponential function,

we know $\quad \mathcal{L}\{\cosh wt\} = \dfrac{1}{2}[\dfrac{1}{s - w} + \dfrac{1}{s + w}] = \dfrac{s}{s^2 - w^2} \quad (\mathcal{L}^{-1}\{\dfrac{s}{s^2 - w^2}\} = \cosh wt)$,

similarly, $\mathcal{L}\{\sinh wt\} = \dfrac{w}{s^2 - w^2} \quad (\mathcal{L}^{-1}\{\dfrac{w}{s^2 - w^2}\} = \sinh wt). \quad ◄$

Before discussing the Laplace transform of power functions, let's introduce the Gamma function because we will need to utilize it for the final calculations. The Gamma function was originally introduced by Euler as an extension of the factorial to non-integer values. Its significance was further established through the research of mathematicians such as Gauss and Legendre. Euler initially defined it as follows: $\Gamma(x) = \int_0^1 (-\log t)^{x-1} dt$ $\Gamma(x) = \int_0^1 (-\log t)^{x-1} dt$. Later, a commonly used modern form was obtained through the variable transformation $u = -\log t$.

▶ **Definition 3-2-1**

Gamma Function

For $x > 0$, Gamma function can be defined as $\Gamma(x) = \int_0^\infty e^{-t} t^{x-1} dt$.

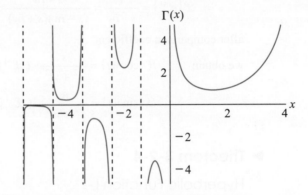

Figure 3-4　The graph of the Gamma function plotted by a computer ◀

The next Theorem 3-2-5, can be proved using partial integration (integration by parts), except for case (4) where the result requires a double integral and a change of coordinates to polar coordinates. The details of this particular case can be found in standard calculus textbooks.

▶ **Theorem 3-2-5**

The Characteristics of Gamma Function

(1) $\Gamma(x) = \dfrac{\Gamma(x+n)}{x(x+1)(x+2)\cdots(x+n-1)}$, where $x < 0$; $x + n > 0$, x is not integer.

(2) $\Gamma(1) = 1$.

(3) $\Gamma(x + 1) = x\Gamma(x)$, $x > 0$.

(4) $\Gamma(\dfrac{1}{2}) = \sqrt{\pi}$.

(5) $\Gamma(k + 1) = k!$, k is positive integers. ◀

Example 1

Please calculate (1) $\Gamma(10)$ (2) $\Gamma(\frac{5}{2})$.

Solution

(1) By Theorem 3-2-5 formula (5), we know $\Gamma(10) = 9!$.

(2) By Theorem 3-2-5 formula (3), (4), we know

$$\Gamma(\frac{5}{2}) = \Gamma(1+\frac{3}{2}) = \frac{3}{2}\Gamma(\frac{3}{2}) = \frac{3}{2}\times\frac{1}{2}\Gamma(\frac{1}{2}) = \frac{3}{4}\sqrt{\pi} .$$

Q.E.D.

When considering the Laplace transform of power functions, it is natural to think about the Gamma function due to its close connection. Through the process of transformation, additional coefficients arise, which can be simplified by using variable substitutions. This simplification is a result of the transformation process.

▶ Theorem 3-2-6

Power Function L-T

For $a > -1$, we have

$$\mathscr{L}\{t^a\} = \frac{\Gamma(a+1)}{s^{a+1}} = \frac{a!}{s^{a+1}} .$$

Proof

Through L'Hôpital's rule, we know $\lim_{t\to\infty}\frac{t^a}{e^t} = 0$, therefore the function t^a is exponential order. Then by Theorem 3-1-1, its Laplace transform exists. Now according to definition, what we want to calculate is $\mathscr{L}\{t^a\} = \int_0^\infty t^a e^{-st}dt$, which is approximately in the form of the Gamma function. However, the variable needs to be slightly simplified, let $st = \xi$, $t = \frac{\xi}{s}$, $dt = \frac{1}{s}d\xi$,

therefore $\mathscr{L}\{t^a\} = \int_0^\infty t^a e^{-st}dt = \int_0^\infty (\frac{\xi}{s})^a e^{-\xi}\frac{1}{s}d\xi = \int_0^\infty \frac{\xi^a}{s^a} e^{-\xi}\frac{1}{s}d\xi$

$$= \frac{1}{s^{a+1}}\int_0^\infty e^{-\xi}\cdot\xi^{(a+1)-1}d\xi = \frac{\Gamma(a+1)}{s^{a+1}} = \frac{a!}{s^{a+1}} \quad (\mathscr{L}^{-1}\{\frac{1}{s^{a+1}}\} = \frac{t^a}{a!}). \blacktriangleleft$$

Example 2

Find the Laplace transform of function $t^{-\frac{1}{2}}$.

Solution

$$\mathcal{L}\{t^{-\frac{1}{2}}\} = \frac{\Gamma(-\frac{1}{2}+1)}{s^{-\frac{1}{2}+1}} = \frac{\Gamma(\frac{1}{2})}{s^{\frac{1}{2}}} = \sqrt{\frac{\pi}{s}} \ .$$

Q.E.D.

In general, we have the following results, and readers can try to refer to the derivation process of the Laplace transform of power functions to write down the proof.

▶ **Theorem 3-2-7**

Polynomial Function L-T

If $a = n$ is positive integer 1, 2, 3, $\cdots\cdots$,

then
$$\begin{cases} \mathcal{L}\{t^n\} = \dfrac{n!}{s^{n+1}} \\ \mathcal{L}^{-1}\{\dfrac{1}{s^{n+1}}\} = \dfrac{t^n}{n!} \end{cases} .$$

◀

3-2-2 The Common Theorem

From section 3-2-1, we observe that when we apply the Laplace transform to common functions, the results are rational functions. This provides us with an important insight for solving ordinary differential equations (ODEs). Even for linear ODEs, the source function can be a challenging factor. However, the Laplace transform allows us to transform these functions into relatively easier-to-handle rational functions. This motivates us to examine the overall behavior of the ODE under the Laplace transform.

In this section, we derive the Laplace transform for functions involving differentiation, integration, and other operations. In section 3-5, we discuss the application of these results in solving ODEs.

▶ **Theorem 3-2-8**

The Laplace Transform Is Linear

$$\mathcal{L}\{c_1 f(t) \pm c_2 g(t)\} = \mathcal{L}\{c_1 f(t)\} \pm \mathcal{L}\{c_2 g(t)\} = c_1 \mathcal{L}\{f(t)\} \pm c_2 \mathcal{L}\{g(t)\} \ .$$

◀

Example 3

Find the Laplace transform of function $3t - 5 \sin 2t$.

Solution   From the Laplace transform of trigonometric functions and power functions, we can deduce that $\mathcal{L}\{3t - 5\sin 2t\} = 3\mathcal{L}\{t\} - 5\mathcal{L}\{\sin 2t\} = \dfrac{3}{s^2} - 5 \times \dfrac{2}{s^2 + 2^2}$.   **Q.E.D.**

▶ Theorem 3-2-9

Differential Function L-T

If $f(t)$ and its differential $f'(t)$ both are exponentially bounded continuous functions, then $\mathcal{L}\{f'(t)\} = s\mathcal{L}\{f(t)\} - f(0)$.

Proof

From the integration by parts (as shown in Figure 3-5), we have

$$\mathcal{L}\{f'(t)\} = \int_0^\infty f'(t)\, e^{-st}\, dt = f(t)e^{-st}\Big|_0^\infty + s\int_0^\infty f(t)e^{-st}\, dt$$

$$= -f(0) + s\mathcal{L}\{f(t)\} = s\mathcal{L}\{f(t)\} - f(0).$$

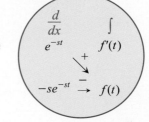

Figure 3-5 ◀

▶ Theorem 3-2-10

Higher-Order Derivative Functions L-T

$$\mathcal{L}\{f^{(n)}(t)\} = s^n F(s) - s^{n-1} f(0) - s^{n-2} f'(0) - \cdots - f^{(n-1)}(0).$$

Proof

In Theorem 3-2-9, we substitute $f'(t)$ for $f(t)$ then obtain
$$\mathcal{L}\{f''(t)\} = s\mathcal{L}\{f'(t)\} - f(0) = s^2\mathcal{L}\{f(t)\} - sf'(0) - f(0),$$

therefore, by the principle of mathematical induction, the proof is established.   ◀

Example 4

Solve $\mathcal{L}\{\sin^2 t\}$.

Solution

Method 1. Using Theorem 3-2-10

$f(t) = \sin^2 t$, then $f'(t) = 2\sin t\cos t = \sin 2t$ and $f(0) = 0$, then

$$\mathcal{L}\{f'(t)\} = s\mathcal{L}\{f(t)\} - f(0),$$

$$\mathcal{L}\{\sin 2t\} = s\mathcal{L}\{f(t)\} - f(0) = s\mathcal{L}\{\sin^2 t\}.$$

Therefore

$$\mathscr{L}\{\sin^2 t\} = \frac{1}{s}\mathscr{L}\{\sin 2t\} = \frac{1}{s}\frac{2}{s^2 + 2^2} = \frac{2}{s(s^2 + 4)}.$$

Method 2. Using Half-Angle Formulas

$$\mathscr{L}\{\sin^2 t\} = \mathscr{L}\{\frac{1 - \cos 2t}{2}\} = \frac{1}{2}(\frac{1}{s} - \frac{s}{s^2 + 2^2}) = \frac{2}{s(s^2 + 4)}. \qquad \text{Q.E.D.}$$

For convenience, from now on, the Laplace transform $\mathscr{L}\{f(t)\}$ of any function $f(t)$ will be denoted as $F(s)$.

▶ **Theorem 3-2-11**

The Laplace Transform of Integral

$$\mathscr{L}\{\int_0^t f(\tau)d\tau\} = \frac{F(s)}{s}.$$

Proof

The Laplace transform itself is a type of integral transformation, so the theorem involves the computation of a double integral. In such cases, drawing the integration region and appropriately interchanging the order of integration can be a useful technique to facilitate the calculation.

$$\mathscr{L}\{\int_0^t f(\tau)d\tau\} = \int_0^\infty (\int_0^t f(\tau)d\tau)e^{-st}dt \quad \text{(The integration region is depicted as shown in Figure 3-6.)}$$

$$= \int_{\tau=0}^{\tau=\infty} (\int_{t=\tau}^{t=\infty} e^{-st}dt)f(\tau)d\tau \quad \text{(Interchanging the order of integration)}$$

$$= \int_0^\infty -\frac{1}{s}e^{-st}\Big|_\tau^\infty f(\tau)d\tau$$

$$= \int_0^\infty \frac{1}{s}e^{-s\tau}f(\tau)d\tau$$

$$= \frac{1}{s}\int_0^\infty f(\tau)e^{-s\tau}d\tau$$

$$= \frac{F(s)}{s}.$$

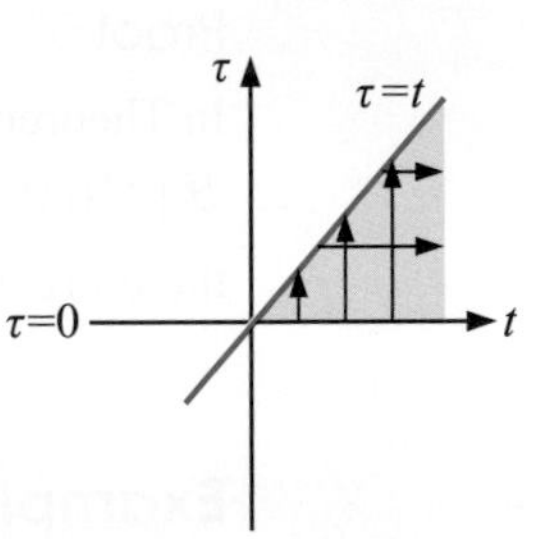

Figure 3-6　The schematic of integration region ◀

▶ **Theorem 3-2-12**

Laplace Transform of Multiple Integrals

$$\mathscr{L}\{\int_0^t \int_0^{t_1} \cdots \int_0^{t_{n-1}} f(\tau)d\tau dt_{n-1} \cdots dt_1\} = \frac{F(s)}{s^n}.$$

Proof

Roughly speaking, Theorem 3-2-11 can be understood as "the Laplace transform of an integral is equal to the Laplace transform of the integrand multiplied by a modifying coefficient $\frac{1}{s}$." Therefore, the proof can be established by utilizing the linearity property of the Laplace transform and the principle of mathematical induction.   ◀

Example 5

Solve $\mathscr{L}\{\int_0^t \int_0^{\tau}(1-\cos wu)dud\tau\}$.

Solution   By theorem 3-2-12($n = 2$), we obtain

$$\mathscr{L}\{\int_0^t \int_0^{\tau}(1-\cos wu)dud\tau\} = \frac{1}{s^2}\mathscr{L}\{1-\cos wt\}$$

$$= \frac{1}{s^2}(\frac{1}{s} - \frac{s}{s^2+w^2})$$

$$= \frac{w^2}{s^3(s^2+w^2)}$$

Q.F.D.

▶ **Theorem 3-2-13**

Scale Transformation

$$\mathscr{L}\{f(at)\} = \frac{1}{a}F(\frac{s}{a}) = \frac{1}{a}\mathscr{L}\{f(t)\}\Big|_{s\to\frac{s}{a}}.$$

Proof

According to definition $\mathscr{L}\{f(at)\} = \int_0^{\infty} f(at)\,e^{-st}dt$, now let $at - \xi$, then $dt = \frac{1}{a}d\xi$,

so $\mathscr{L}\{f(at)\} = \int_0^{\infty} f(at)\,e^{-st}dt = \int_0^{\infty} f(\xi)\,e^{-s\frac{\xi}{a}}\frac{1}{a}d\xi$

$$= \frac{1}{a}\int_0^{\infty} f(\xi)\,e^{-\frac{s}{a}\xi}d\xi = \frac{1}{a}F(\frac{s}{a}).$$ ◀

Example 6

The mth order Bessel function is defined as $J_m(t) = \sum_{n=0}^{\infty} \frac{(-1)^n}{\Gamma(n+1+m)n!}(\frac{t}{2})^{2n+m}$,

and $\mathcal{L}\{J_0(t)\} = \dfrac{1}{\sqrt{s^2+1}}$, solve $\mathcal{L}\{J_0(at)\}$.

Solution　By Theorem 3-2-13, we know

$$\mathcal{L}\{J_0(at)\} = \frac{1}{a}\mathcal{L}\{J_0(t)\}\Big|_{s\to\frac{s}{a}} = \frac{1}{a}\frac{1}{\sqrt{(\frac{s}{a})^2+1}} = \frac{1}{\sqrt{s^2+a^2}} .$$ Q.E.D.

▶ Theorem 3-2-14

Initial Value Theorem

If $f(t)$ and its derivatives $f'(t)$ are both exponentially bounded continuous functions, then $\lim_{s\to\infty} sF(s) = f(0^+)$.

Proof

First, from the given premises, we know that there exists a non-zero positive number M, α causes $|f'(t)| \le Me^{\alpha t}$,

therefore $\lim_{s\to\infty}[\mathcal{L}\{f'(t)\}] = \lim_{s\to\infty}[\int_0^{\infty} f'(t)\,e^{-st}\,dt]$

$$\le M\lim_{s\to\infty}\int_0^{\infty} e^{\alpha t}\,e^{-st}\,dt = M\lim_{s\to\infty}\frac{e^{\alpha-s}}{\alpha-s} = 0 ,$$

but at the same time, from the Laplace transform formula for derivatives we know
$\lim_{s\to\infty}[\mathcal{L}\{f'(t)\}] = \lim_{s\to\infty}[sF(s) - f(0^+)] = 0$

so $\quad \lim_{s\to\infty} sF(s) = f(0^+)$.　◀

▶ Theorem 3-2-15

The Generalized Initial Value Theorem

If $\mathcal{L}\{f(t)\} = F(s)$, $\mathcal{L}\{g(t)\} = G(s)$, then $\lim_{s\to\infty}\dfrac{F(s)}{G(s)} = \dfrac{f(0^+)}{g(0^+)}$.

Proof

Through Theorem3-2-14

we know $\quad \lim_{s\to\infty}\dfrac{F(s)}{G(s)} = \dfrac{\lim_{s\to\infty} sF(s)}{\lim_{s\to\infty} sG(s)} = \dfrac{f(0^+)}{g(0^+)}$.　◀

▶ **Theorem 3-2-16**

Final Value Theorem

If $f(t)$ and its derivatives $f'(t)$ are both exponentially bounded continuous functions and converge as the abscissa approaches negative infinity (i.e., the final value exists), then $\lim_{s \to 0^+} sF(s) = f(\infty) = \lim_{t \to \infty} f(t)$.

Proof

Because $\mathcal{L}\{f'(t)\} = sF(s) - f(0)$, therefore $\lim_{s \to 0^+}[\mathcal{L}\{f'(t)\}] = \lim_{s \to 0^+}[sF(s) - f(0^+)]$,

$$\lim_{s \to 0^+}[\mathcal{L}\{f'(t)\}] = \lim_{s \to 0^+}(\int_0^\infty f'(t) \cdot e^{-st}dt)$$

$$= \int_0^\infty f'(t)dt$$

$$= f(t)\big|_0^\infty$$

$$= \lim_{t \to \infty} f(t) - f(0^+),$$

so $\lim_{s \to 0^+}[sF(s) - f(0^+)] = \lim_{t \to \infty} f(t) - f(0^+)$,

as a result $\lim_{s \to 0^+} sF(s) = f(\infty) = \lim_{t \to \infty} f(t)$.   ◀

▶ **Theorem 3-2-17**

The Generalized Final Value Theorem

If $\mathcal{L}\{f(t)\} = F(s)$, $\mathcal{L}\{g(t)\} = G(s)$, then $\lim_{s \to 0}\dfrac{F(s)}{G(s)} = \lim_{t \to \infty}\dfrac{f(t)}{g(t)}$.

Proof

By utilizing the result of the final value theorem and following a similar argument as in the generalized initial value theorem, we can get a conclusion.   ◀

Example 7

If $Y(s) = \mathcal{L}\{y(t)\} = \dfrac{s+2}{s(s^2 + 9s + 14)}$,

solve (1) $y(0)$ (2) $\lim_{t \to \infty} y(t)$.

Solution

(1) By initial value theorem, we know $\lim_{s \to \infty} sY(s) = y(0)$,

$$\therefore y(0) = \lim_{s \to \infty} s \times \frac{s+2}{s(s^2 + 9s + 14)} = 0 .$$

(2) By final value theorem, we know [3] $\lim\limits_{s \to 0} sY(s) = \lim\limits_{t \to \infty} y(t)$,

$$\therefore \lim_{t \to \infty} y(t) = \lim_{s \to 0} s \times \frac{s+2}{s(s^2+9s+14)} = \frac{1}{7}.$$

 Q.E.D.

► **Theorem 3-2-18**

The First Shifting Theorem

$$\mathcal{L}\{f(t)e^{at}\} = F(s-a) = \mathcal{L}\{f(t)\}\big|_{s \to s-a}.$$

Proof

$$\mathcal{L}\{f(t)e^{at}\} = \int_0^\infty f(t)\,e^{at}e^{-st}\,dt = \int_0^\infty f(t)e^{-(s-a)t}\,dt = F(s-a) = \mathcal{L}\{f(t)\}\big|_{s \to s-a}. \quad \blacktriangleleft$$

Example 8

Find the Laplace transform of the following functions.
(1) $f(t) = e^{-2t}\cos 6t$ (2) $f(t) = t^3 e^{4t}$.

Solution

(1) $\mathcal{L}\{e^{-2t}\cos 6t\} = \mathcal{L}\{\cos 6t\}_{s \to s+2} = \dfrac{s+2}{(s+2)^2 + 6^2}.$

(2) $\mathcal{L}\{t^3 e^{4t}\} = \mathcal{L}\{t^3\}_{s \to s-4} = \dfrac{3!}{(s-4)^4} = \dfrac{6}{(s-4)^4}.$ Q.E.D.

Next, we will derive the second shifting theorem. Recalling the step function defined earlier as $u(t) = \begin{cases} 1, t \ge 0 \\ 0, t < 0 \end{cases}$, when we shift this function to the right by a units, it becomes $u(t-a) = \begin{cases} 1, t \ge a \\ 0, t < a \end{cases}$, as shown in Figure 3-7. When we consider a function $f(t)$ multiplied by $u(t-a)$, intuitively, it means extracting the function values (or graph) corresponding to $x \ge a$, as can be seen from Figure 3-8.

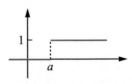

Figure 3-7   The unit step function t shifted.

---

[3] To confirm the applicability of the final value theorem, we need to check if the real parts of the roots of $s \cdot F(s)$ denominators are all negative.

Taking this example: $sY(s) = \dfrac{s+2}{(s+2)(s+7)}$, the roots are -2 and -7, and both have negative real parts. Therefore, the final value theorem is applicable.

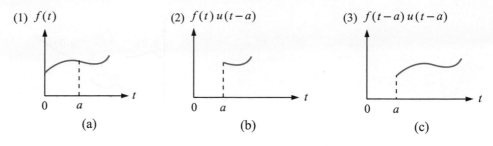

Figure 3-8 The schematic of signal delay

In Figure 3-8(a), we have the graph of a signal function $f(t)$. In Figure 3-8(b), $f(t) \cdot u(t-a)$ represents the signal function starting from $t = a$, while $f(t-a) \cdot u(t-a)$ indicates that the signal $f(t)$ is delayed and starts at $t = a$. How do we compute the Laplace transform of such a function with a delay effect? Let's derive it below.

▶ Theorem 3-2-19

The Second Shifting Theorem

$$\mathscr{L}\{f(t-a)u(t-a)\} = e^{-as}F(s) = e^{-as}\mathscr{L}\{f(t)\}.$$

Proof

According to the definition of the Laplace transform,

we have $\quad \mathscr{L}\{f(t-a)u(t-a)\} = \int_0^\infty f(t-a)\, u(t-a)e^{-st}dt = \int_{t=a}^{t=\infty} f(t-a)e^{-st}dt$.

Let $t - a = \tau$,

we obtain $\quad \int_{\tau=0}^{\tau=\infty} f(\tau)\, e^{-s(a+\tau)}d\tau = e^{-as} \int_0^\infty f(\tau)\, e^{-s\tau}d\tau$

$$= e^{-as}\mathscr{L}\{f(t)\}$$

$$= e^{-as}F(s). \qquad \blacktriangleleft$$

In the proof of the above theorem, it can be observed that if we replace $f(t - a)$ with $f(t)$, we obtain the following formula:

$$\mathscr{L}\{f(t)u(t-a)\} = e^{-as}\mathscr{L}\{f(t+a)\}$$

Before applying the example, let's introduce a commonly used technique in mathematical analysis: expressing piecewise continuous functions using unit step functions. In simple terms, suppose we have a function $f(t) = \begin{cases} g_1(t), 0 \le t < a \\ g_2(t), a \le t < b \\ g_3(t), b \le t \end{cases}$, as

shown in Figure 3-9. Then, based on the definition of the step function, we know that $f(t) = g_1(t)[u(t-0)-u(t-a)] + g_2(t)[u(t-a)-u(t-b)] + g_3(t)u(t-b)$.

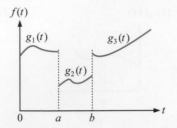

Figure 3-9   The schematic of function interval

Example 9

Let $f(t) = \begin{cases} 2t, t < 3 \\ 1, t \geq 3 \end{cases}$, as shown in the figure.

Please find $\mathscr{L}\{f(t)\}$.

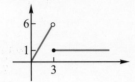

Solution   Following the approach of function decomposition as mentioned earlier we obtain $f(t) = 2t[u(t) - u(t-3)] + 1 \times u(t-3)$, then using the linearity property of the Laplace transform and the second shifting theorem, we obtain

$$\mathscr{L}\{f(t)\} = \mathscr{L}\{2t \times u(t) - 2t \times u(t-3) + 1 \times u(t-3)\}$$

$$= \mathscr{L}\{2t\} - e^{-3s}\,\mathscr{L}\{2(t+3)\} + e^{-3s}\,\mathscr{L}\{1\}$$

$$= \frac{2}{s^2} - e^{-3s}(\frac{2}{s^2} + \frac{6}{s}) + e^{-3s}(\frac{1}{s})$$

$$= \frac{2}{s^2} - e^{-3s}(\frac{2}{s^2} + \frac{5}{s}).$$ `Q.E.D.`

Example 10

If function $f(t) = \begin{cases} 1, 0 \leq t < 3 \\ -5, 3 \leq t < 7 \\ 1 + 2t, t \geq 7 \end{cases}$, then $\mathscr{L}\{f(t)\} = ?$

Solution   First, we decompose the function
$f(t) = [u(t) - u(t-3)] - 5[u(t-3) - u(t-7)] + (1 + 2t)u(t-7)$.
Then, using the second shifting theorem we obtain
$$\mathscr{L}\{f(t)\} = \mathscr{L}\{[u(t) - u(t-3)] - 5[u(t-3) - u(t-7)] + (1 + 2t)u(t-7)\}$$

$$= \mathscr{L}\{[u(t) - u(t-3)]\} - 5\,\mathscr{L}\{[u(t-3) - u(t-7)]\} + \mathscr{L}\{(1 + 2t)u(t-7)\}$$

$$= [\frac{1}{s} - \frac{1}{s}e^{-3s}] - [\frac{5}{s}e^{-3s} - \frac{5}{s}e^{-7s}] + e^{-7s}\,\mathscr{L}\{1 + 2(t+7)\}$$

$$= \frac{1}{s} - \frac{6}{s}e^{-3s} + \frac{5}{s}e^{-7s} + e^{-7s}(\frac{15}{s} + \frac{2}{s^2})$$

$$= \frac{1}{s} - \frac{6}{s}e^{-3s} + \frac{20}{s}e^{-7s} + \frac{2}{s^2}e^{-7s}.$$ `Q.E.D.`

Example 11

$$f(t) = \begin{cases} 0, & t < 8 \\ t^2 - 4, & t \geq 8 \end{cases}, \text{ then } \mathcal{L}\{e^{-3t} f(t)\} = ?$$

Solution   First, through step function demonstrates $f(t) = (t^2 - 4)u(t - 8)$, then by the second shifting theorem, we obtain

$$\mathcal{L}\{f(t)\} = \mathcal{L}\{(t^2 - 4)u(t - 8)\} = e^{-8s} \mathcal{L}\{(t + 8)^2 - 4\}$$

$$= e^{-8s} \mathcal{L}\{t^2 + 16t + 60\}$$

$$= e^{-8s}[\frac{2!}{s^3} + \frac{16}{s^2} + 60 \times \frac{1}{s}]$$

$$= e^{-8s}[\frac{2}{s^3} + \frac{16}{s^2} + \frac{60}{s}].$$

Using the first shifting theorem again, we get

$$\mathcal{L}\{e^{-3t} f(t)\} = \mathcal{L}\{f(t)\}_{s \to s+3}$$

$$= e^{-8(s+3)}[\frac{2}{(s+3)^3} + \frac{16}{(s+3)^2} + \frac{60}{(s+3)}]. \quad \boxed{\text{Q.E.D.}}$$

▶ Theorem 3-2-20

The Laplace Transform of the First Derivative

$$\mathcal{L}\{t f(t)\} = -\frac{dF(s)}{ds}.$$

Proof

According to definition, $\mathcal{L}\{f(t)\} = \int_0^\infty f(t) e^{-st} dt$,

$$\frac{dF(s)}{ds} = \int_0^\infty f(t) \frac{\partial e^{-st}}{\partial s} dt = \int_0^\infty f(t)(-te^{-st}) dt$$

$$= -\int_0^\infty [tf(t)] e^{-st} dt = -\mathcal{L}\{t f(t)\}. \qquad \blacktriangleleft$$

Readers can recursively use the second shifting theorem to obtain the generalized version of the following.

▶ Theorem 3-2-21

The Laplace Transform of the nth Derivative

$$\mathcal{L}\{t^n f(t)\} = (-1)^n \frac{d^n F(s)}{ds^n}. \qquad \blacktriangleleft$$

Example 12

$\mathscr{L}\{te^{-3t}\sin 2wt\} = ?$

Solution   In Theorem 3-2-20, we let $f(t) = e^{-3t}\sin 2wt$, then obtain

$$\mathscr{L}\{te^{-3t}\sin 2wt\} = -\frac{d}{ds}\mathscr{L}\{e^{-3t}\sin 2wt\}$$

$$= -\frac{d}{ds}[\frac{2w}{(s+3)^2 + (2w)^2}] \qquad \text{(The Laplace transform of trigonometric functions combined with the first shifting theorem)}$$

$$= \frac{4w(s+3)}{[(s+3)^2 + 4w^2]^2} \,. \qquad \text{Q.E.D.}$$

▶ **Theorem 3-2-22**

The Laplace Transform of the First Integral

Suppose that $\lim\limits_{t\to 0^+}\dfrac{f(t)}{t}$ exists, then $\mathscr{L}\{\dfrac{f(t)}{t}\} = \displaystyle\int_s^\infty F(u)\,du$.

Proof

According to definition $\mathscr{L}\{f(t)\} = \displaystyle\int_0^\infty f(t)e^{-st}dt$, then

$$\int_s^\infty F(u)du = \int_s^\infty [\int_0^\infty f(t)e^{-ut}dt]du = \int_0^\infty f(t)(\int_s^\infty e^{-ut}du)du$$

$$= \int_0^\infty f(t)(-\frac{1}{t}e^{-ut}\Big|_s^\infty)dt = \int_0^\infty \frac{f(t)}{t}e^{-st}dt = \mathscr{L}\{\frac{f(t)}{t}\}\,,$$

so $\mathscr{L}\{\dfrac{f(t)}{t}\} = \displaystyle\int_s^\infty F(u)du$.   ◀

▶ **Theorem 3-2-23**

The Laplace Transform of the nth Integral

$$\mathscr{L}\{\frac{f(t)}{t^n}\} = \int_s^\infty \int_{s_1}^\infty \cdots \int_{s_{n-1}}^\infty F(u)du\,ds_{n-1}\cdots ds_1 \,.\qquad ◀$$

Example 13

(1) $\mathcal{L}\{\dfrac{\sin kt}{t}\} = ?$ (2) $\displaystyle\int_0^\infty \dfrac{\sin t}{t}dt = ?$

Solution

(1) $\mathcal{L}\{\dfrac{\sin kt}{t}\} = \displaystyle\int_s^\infty \mathcal{L}\{\sin kt\}\,du = \int_s^\infty \dfrac{k}{u^2+k^2}\,du = \tan^{-1}\dfrac{u}{k}\Big|_s^\infty$

$\qquad = \dfrac{\pi}{2} - \tan^{-1}(\dfrac{s}{k}) = \tan^{-1}(\dfrac{k}{s}).$

(2) $\displaystyle\int_0^\infty \dfrac{\sin t}{t}dt = \int_0^\infty \dfrac{\sin t}{t}\cdot e^{-st}dt\Big|_{s=0} = \mathcal{L}\{\dfrac{\sin t}{t}\}_{s=0}$

$\qquad = (\dfrac{\pi}{2} - \tan^{-1}\dfrac{s}{1})\Big|_{s=0} = \dfrac{\pi}{2} - 0 = \dfrac{\pi}{2}.$ Q.E.D.

In Example 13, part (2), we learn that when faced with difficult improper integrals, it is worth trying the Laplace transform. However, readers may perceive this as a coincidence because the form of the integrated function happens to fit the conclusion of Theorem 3-2-23. In the chapter on complex variables, we will encounter a more general method for calculating improper integrals, the Cauchy principal value theorem. Interested readers can directly refer to Section 7 of Chapter 11.

3-2   Exercises

Basic questions

1. Find the Laplace transform of following functions.
   (1) $f(t) = 5$ (2) $f(t) = t$ (3) $f(t) = e^{2t}$
   (4) $f(t) = \cos 3t$ (5) $f(t) = \sinh(2t)$
   (6) $f(t) = t - \sin t$.

2. Find the Laplace transform of following functions.
   (1) $f(t) = t\cdot e^{4t}$
   (2) $f(t) = e^{-t}\sin t$
   (3) $f(t) = 3 - 2t + 4t^2$
   (4) $f(t) = -5e^{4t} - 6e^{-5t}$
   (5) $f(t) = 5\sin 2t + 3\cos 4t$.

3. $\mathcal{L}\{\displaystyle\int_0^t (4 - e^{-3\tau} + 2\tau^4)d\tau\} = ?$

4. Find $\mathcal{L}[f(t)]$,
   where $f(t) = e^{-2t}\displaystyle\int_0^t e^{2\tau}\cos(3\tau)d\tau$.

5. Find $\mathcal{L}\{g(t)\}$, where $g(t) = \begin{cases} 0, & t<3 \\ t^2, & t\geq 3 \end{cases}$.

Advanced questions

1. Find the Laplace transform of following functions:
   (1) $f(t) = (t+1)^3$.
   (2) $f(t) = \cos^2 t$.
   (3) $f(t) = \sin^3 t$.
   (4) $f(t) = e^{2t}(t+2)^2$.
   (5) $f(t) = e^{-2t}(\cos 2t - \sin 2t)$.

2. Find the Laplace transform of following functions:
   (1) $\mathscr{L}\{\int_0^t \int_0^\tau (ue^{2u})dud\tau\} = ?$

   (2) $\mathscr{L}\{\int_0^t \int_0^\tau (e^{-3u}\sin^2 u)dud\tau\} = ?$

3. If $Y(s) = \dfrac{s^2 + 2}{(s^3 + 6s^2 + 11s + 6)}$,
   and $\mathscr{L}\{y(t)\} = Y(s)$, find:
   (1) $y(0)$. (2) $\lim\limits_{t\to\infty} y(t)$.

4. Find the Laplace transform of following functions:
   (1) $f(t) = \int_0^t \dfrac{\cos a\tau - \cosh a\tau}{\tau} d\tau$.

   (2) $f(t) = \begin{cases} \dfrac{1}{\sqrt{t}}, & t > 0 \\ 0, & t \le 0 \end{cases}$.

   (3) $f(t) = \begin{cases} \sin t, & 0 \le t < 2\pi \\ \sin t + \cos t, & t \ge 2\pi \end{cases}$.

   (4) $f(t) = \begin{cases} 0, & t < 1 \\ t^2 - 2t + 2, & t \ge 1 \end{cases}$.

   (5) $f(t) = \begin{cases} 5t, & 0 \le t \le 1 \\ t, & t > 1 \end{cases}$.

   (6) $f(t) = \int_2^t u^2 e^{3u} du$.

5. Find the Laplace transform of following functions:
   (1) $f(t) = t \cos wt$.
   (2) $f(t) = t \sin wt$.
   (3) $f(t) = t^2 \cos wt$.

6. $f(t) = \begin{cases} 2, & 0 < t < \pi \\ 0, & \pi < t < 2\pi, \text{ find } \mathscr{L}\{f(t)\} = ? \\ \sin t, & t > 2\pi \end{cases}$

7. $f(t) = \cos(t-2) \cdot u(t-2) - 2u(t-4) \cdot t$,
   then $\mathscr{L}\{f(t)\} = ?$

8. The graph of the linear function $f(t)$ is shown on the right. then $\mathscr{L}\{f(t)\} = ?$

9. Find $\mathscr{L}\{f(t)\}$,
   where $f(t) = \begin{cases} 1 - e^{-t}, & 0 < t < \pi \\ 0, & t > \pi \end{cases}$.

10. Find following the Laplace transform.
    (1) $f(t) = \begin{cases} 2t+1, & 0 \le t < 1 \\ 0, & t \ge 1 \end{cases}$

    (2) $f(t) = \begin{cases} \sin t, & 0 \le t < \dfrac{\pi}{2} \\ 0, & t \ge \dfrac{\pi}{2} \end{cases}$

    (3) $f(t) = t \cos t$.
    (4) $f(t) = t^2 + 6t - 3$.
    (5) $f(t) = \cos 5t + \sin 2t$.
    (6) $f(t) = e^t \sinh t$.
    (7) $f(t) = \sin 2t \cos 2t$.

11. (1) Find $\mathscr{L}\{\dfrac{\sin^2 t}{t}\}$.

    (2) Calculate $\int_0^\infty \dfrac{e^{-t}\sin^2 t}{t} dt$.

3-3 Laplace Transform of Special Functions

In this section, we mainly focus on the Laplace transform of commonly used special functions in engineering, including the impulse function, periodic function, and convolution function. We discuss the methods for calculating their Laplace transforms.

3-3-1 Unit Impulse Function

The unit impulse function, also known as Dirac's Delta function, is an essential function in engineering and even in the solution of partial differential equations. Strictly speaking, it is a generalized function defined through the limit of step functions. First, let's define this function and then explore how its Laplace transform is calculated.

Consider function $P_\varepsilon(t) = \begin{cases} \dfrac{1}{\varepsilon}, & 0 \le t < \varepsilon \\ 0, & t \ge \varepsilon \end{cases}$, where ε is an arbitrary positive real

number. In other words, by adjusting ε, we obtain a series of step functions in the first quadrant with an area of 1, as shown in Figure 3-10, these are referred to as energy

functions. Taking the limit as $\lim\limits_{\varepsilon \to 0} P_\varepsilon(t) = \begin{cases} \infty, & t = 0 \\ 0, & t \ne 0 \end{cases}$, we obtain a generalized function

$\delta(t)$, known as the unit impulse function, which can be represented as

$\delta(t-a) = \begin{cases} \infty, & t = a \\ 0, & t \ne a \end{cases}$, as shown in Figure 3-11.

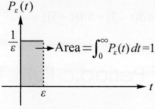

Figure 3-10   The graph
of the energy function.

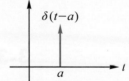

Figure 3-11   The graph
of the impulse function.

Naturally, from the process of constructing the impulse function, we know that

$$\int_0^\infty \delta(t)\,dt = \lim_{a \to 0^+} \int_0^a \delta(t)\,dt = \lim_{a \to 0^+} \int_0^a \lim_{\varepsilon \to 0} P_\varepsilon(t)\,dt = \lim_{a \to 0^+} \lim_{\varepsilon \to 0} \int_0^a P_\varepsilon(t)\,dt = \lim_{a \to 0^+} \lim_{\varepsilon \to 0} 1 = 1$$

An example of the physical manifestation of the impulse function is when a ball is pressed against a tabletop. The pressure exerted at the contact point between the ball and the tabletop can be represented by an impulse function.

▶ **Theorem 3-3-1**

The Laplace Transform of the Impulse Function

$$\mathcal{L}\{\delta(t)\} = 1, \quad \mathcal{L}\{\delta(t-a)\} = e^{-as}.$$

Proof

Through the definition of the step function,

$$P_\varepsilon(t) = \frac{1}{\varepsilon}[u(t) - u(t-\varepsilon)], \text{ so } \delta(t) = \lim_{\varepsilon \to 0} \frac{[u(t) - u(t-\varepsilon)]}{\varepsilon},$$

$$\mathcal{L}\{\delta(t)\} = \mathcal{L}\{\lim_{\varepsilon \to 0} \frac{[u(t) - u(t-\varepsilon)]}{\varepsilon}\}$$

$$= \lim_{\varepsilon \to 0} \frac{\dfrac{1}{s} - \dfrac{e^{-\varepsilon s}}{s}}{\varepsilon} = \lim_{\varepsilon \to 0} \frac{1 - e^{-\varepsilon s}}{s\varepsilon}$$

$$= \lim_{\varepsilon \to 0} \frac{s \cdot e^{-\varepsilon s}}{s} \quad \text{(L'Hôpital's Rule)}$$

$$= 1,$$

therefore $\mathcal{L}\{\delta(t)\} = 1$ and $\mathcal{L}\{\delta(t-a)\} = e^{-as}$. ◀

Example 1

Find $\mathcal{L}\{2u(t-1) - 4\delta(t-2) - 5\delta(t-3)\}$.

Solution By Theorem 3-3-1 we know

$$\mathcal{L}\{2u(t-1) - 4\delta(t-2) - 5\delta(t-3)\} = \frac{2}{s}e^{-s} - 4e^{-2s} - 5e^{-3s}.$$ Q.E.D.

3-3-2 L-T of Periodic Function

Periodic functions are common in engineering systems. For example, the 110V alternating current (AC) used in our homes is a periodic waveform function. Another example is the cutting force generated by the contact between the cutting tool and the workpiece during machining, which is periodic due to the rotation of the tool. If a function $f(t)$ is periodic with a period T, it satisfies $f(t) = f(t+T)$, as shown in Figure 3-12. Here, we specifically restrict $t \geq 0$, as it is necessary for calculating the Laplace transform.

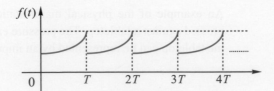

Figure 3-12 Periodic function diagram

▶ **Theorem 3-3-2**

The Laplace Transform of Periodic Function

If $f(t)$ is a piecewise continuous exponential order function with period T, then the

Laplace transform of $f(t)$ is given by $\mathcal{L}\{f(t)\} = \dfrac{\int_0^T f(t)\cdot e^{-st}\,dt}{1-e^{-Ts}}$.

Proof

Based on the definition of the Laplace transform and the periodic function, we know

$$\mathcal{L}\{f(t)\} = \int_0^\infty f(t)\,e^{-st}\,dt = \int_0^T f(t)\,e^{-st}\,dt + \int_T^{2T} f(t)\,e^{-st}\,dt + \cdots = \sum_{n=0}^\infty \int_{nT}^{(n+1)T} f(t)e^{-st}\,dt.$$

Now let $\xi(t) = t - nT$, the transformation of the integration limits corresponds to

$\begin{array}{c|c|c} \xi & 0 & T \\ \hline t & nT & (n+1)T \end{array}$. In other words, through the aforementioned variable transformation,

we "incorporate" all the integrals within one period into the first segment of the periodic variation. Then, from the variable substitution method in integration techniques, we know that

$$\int_0^\infty f(t)\,e^{-st}\,dt = \int_0^\infty f(\xi+nT)\,e^{-s(\xi+nT)}\,d\xi$$

$$= \sum_{n=0}^\infty \int_0^T f(\xi+nT)e^{-s\cdot(\xi+nT)}\,d\xi$$

$$= \sum_{n=0}^\infty \int_0^T f(\xi)\,e^{-s\xi}e^{-snT}\,d\xi$$

$$= \sum_{n=0}^\infty e^{-nTs} \int_0^T f(\xi)\,e^{-s\xi}\,d\xi$$

$$= [\int_0^T f(\xi)e^{-s\xi}\,d\xi]\sum_{n=0}^\infty e^{-nTs}$$

$$= \frac{\int_0^T f(\xi)e^{-s\xi}\,d\xi}{1-e^{-Ts}}.$$

∴The Laplace transform of periodic function $f(t)$ is $\mathcal{L}\{f(t)\} = \dfrac{\int_0^T f(t)e^{-st}\,dt}{1-e^{-Ts}}$. ◀

If we define $f(t) = \begin{cases} g(t), & t\in[0,T] \\ 0, & T<t \end{cases}$, then the formula of Theorem 3-3-2 can

rewrite as

$$\mathcal{L}\{f(t)\} = \frac{\mathcal{L}\{g(t)\}}{1-e^{-Ts}}$$

This also facilitates our calculations. Let's look at the following examples.

Example 2

If on $[0, 1]$, $f(t) = t$ and the period of f is 1, as shown in the figure. Try to calculate $\mathscr{L}\{f(t)\}$.

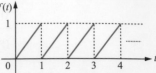

Solution Now $g(t) = t[u(t) - u(t-1)]$, and in formula

$$\mathscr{L}\{f(t)\} = \frac{\mathscr{L}\{g(t)\}}{1 - e^{-Ts}},$$

$$\mathscr{L}\{g(t)\} = \mathscr{L}\{tu(t) - tu(t-1)\}$$

$$= \frac{1}{s^2} - e^{-s} \mathscr{L}\{t+1\}$$

$$= \frac{1}{s^2} - e^{-s}[\frac{1}{s^2} + \frac{1}{s}],$$

therefore

$$\mathscr{L}\{f(t)\} = \frac{\dfrac{1}{s^2} - e^{-s}[\dfrac{1}{s^2} + \dfrac{1}{s}]}{1 - e^{-s}}.$$

Q.E.D.

Example 3

If on $[0, 2c]$, $f(t) = \begin{cases} 1, & t \in [0,c) \\ -1, & t \in [c, 2c] \end{cases}$ and the

period of f is $2c$, as shown in the figure. Try to calculate $\mathscr{L}\{f(t)\}$.

Solution Because $g(t) = u(t) - 2u(t-c) + u(t-2c)$, therefore in the formula

$$\mathscr{L}\{f(t)\} = \frac{\mathscr{L}\{g(t)\}}{1 - e^{-2cs}},$$

$$\mathscr{L}\{g(t)\} = \mathscr{L}\{u(t) - 2u(t-c) + u(t-2c)\}$$

$$= \frac{1}{s} - e^{-cs} \times \frac{2}{s} + e^{-2cs} \times \frac{1}{s},$$

so

$$\mathscr{L}\{f(t)\} = \frac{\dfrac{1}{s} - \dfrac{2}{s}e^{-cs} + \dfrac{1}{s}e^{-2cs}}{1 - e^{-2cs}} = \frac{(1 - e^{-cs})^2}{s(1 - e^{-cs})(1 + e^{-cs})} = \frac{1 - e^{-cs}}{s(1 + e^{-cs})}.$$

Q.E.D.

3-3-3 Convolution Theorem

In recent years, in the field of Artificial Intelligence (AI), convolution has been widely used for feature extraction from images or signals. The most famous application of convolution is the Convolutional Neural Network (CNN). Let's delve into what convolution means.

▶ Definition 3-3-1

Convolution

The Convolution of function $f(t)$ and $g(t)$ defined as

$$f(t)*g(t) \triangleq \int_0^t f(\tau)g(t-\tau)d\tau = \int_0^t g(\tau)f(t-\tau)d\tau .$$ ◀

In the definition of convolution 3-3-1 mentioned above, if we consider $t = 100$, the integral can be approximated using Thomas Simpson's rule (Thomas Simpson, 1710~1761). This rule involves dividing the integration interval into equal parts, in this case, 100 equal parts. Then, we would have terms like $f(0) \cdot g(100)$, $f(1) \cdot g(99)$, $f(2) \cdot g(98)$, $\cdots$, $f(100) \cdot g(0)$. This process resembles folding and multiplying along the coordinate axis, hence the term "convolution."

Example 4

Find the following convolution functions.
(1) $1*1$.　(2) $t*e^{2t}$.

Solution

(1) $1*1 = \int_0^t 1 \times 1\, d\tau = \int_0^t 1\, d\tau = \tau \big|_0^t = t$.

(2) $t*e^{2t} = \int_0^t \tau e^{2(t-\tau)}d\tau = e^{2t}\int_0^t \tau e^{-2\tau}d\tau$

$$= e^{2t}[-\frac{1}{2}\tau e^{-2\tau} - \frac{1}{4}e^{-2\tau}]\Big|_0^t \quad \text{(Integration by parts)}$$

$$= -\frac{1}{2}t - \frac{1}{4} + \frac{1}{4}e^{2t} .$$

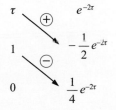

Q.E.D.

The principle of converting the complex convolution integral into algebraic multiplication is known as the convolution theorem. It provides a convenient way to calculate convolutions. Let's introduce the theorem below.

▶ **Theorem 3-3-3**

Convolution Theorem

Suppose that the Laplace transform of function $f(t)$, $g(t)$ separately are $\mathscr{L}\{f(t)\} = F(s)$, $\mathscr{L}\{g(t)\} = G(s)$,

then $\mathscr{L}\{f(t) * g(t)\} = F(s) \cdot G(s)$.

Proof

$\mathscr{L}\{f(t) * g(t)\}$

$= \int_0^\infty [\int_0^t f(\tau)g(t-\tau)d\tau]e^{-st}dt$ (the region of integration as shown in Figure3-13)

$= \int_{\tau=0}^{\tau=\infty} \int_{t=\tau}^{t=\infty} f(\tau)g(t-\tau)e^{-st}dtd\tau$

(changing the order of integration)

$= \int_0^\infty \int_{x=0}^\infty f(\tau)g(x)e^{-s\cdot(x+\tau)}dxd\tau$

(variable transformations $t - \tau = x$)

$= \int_0^\infty f(\tau)\,e^{-s\tau}d\tau \times \int_0^\infty g(x)\,e^{-sx}dx$

$= \mathscr{L}\{f(t)\}\,\mathscr{L}\{g(t)\} = F(s)G(s)$.

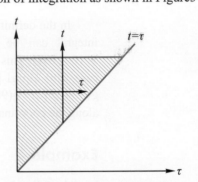

Figure 3-13   Convolution integral region diagram ◀

In physical systems, if we want to determine the response of a system $G(s)$, we can use an input function $f(t) = \delta(t)$ that is a unit impulse function, denoted as $\mathscr{L}\{f(t) * g(t)\} = \mathscr{L}\{f(t)\} \cdot \mathscr{L}\{g(t)\} = F(s) \cdot G(s)$. The convolution of the unit impulse function with the system $G(s)$ yields the response of the system, denoted as $g(t)$, the output result.

Example 5

Find the Laplace transform of following functions.

(1) $1*1$. (2) $t*e^{2t}$. (3) $e^{-2t}*e^{2t}$. (4) $\cos wt*\sin wt$.

Solution

(1) $\mathscr{L}\{1*1\} = \mathscr{L}\{1\}\,\mathscr{L}\{1\} = \dfrac{1}{s} \times \dfrac{1}{s} = \dfrac{1}{s^2}$.

(2) $\mathscr{L}\{t*e^{2t}\} = \mathscr{L}\{1t\}\,\mathscr{L}\{e^{2t}\} = \dfrac{1}{s^2} \times \dfrac{1}{s-2} = \dfrac{1}{s^2(s-2)}$.

(3) $\mathscr{L}\{e^{-2t}*e^{2t}\} = \mathscr{L}\{e^{-2t}\}\,\mathscr{L}\{e^{2t}\} = \dfrac{1}{s+2} \times \dfrac{1}{s-2} = \dfrac{1}{s^2-4}$.

(4) $\mathscr{L}\{\cos wt*\sin wt\} = \mathscr{L}\{\cos wt\}\,\mathscr{L}\{\sin wt\} = \dfrac{s}{s^2+w^2} \times \dfrac{w}{s^2+w^2} = \dfrac{ws}{(s^2+w^2)^2}$.[4] Q.E.D.

Example 6

Find the Laplace transform of following functions $e^{-2t}\displaystyle\int_0^t e^{2\tau}\cos 3\tau\,d\tau$.

Solution   In the first shifting theorem, let $f(t) = \displaystyle\int_0^t e^{2\tau}\cos 3\tau\,d\tau$, then

$\mathscr{L}\{e^{-2t}\displaystyle\int_0^t e^{2\tau}\cos 3\tau\,d\tau\}$

$= \mathscr{L}\{\displaystyle\int_0^t e^{-2(t-\tau)}\cos 3\tau\,d\tau\}$

$= \mathscr{L}\{e^{-2t}*\cos 3t\} = \mathscr{L}\{e^{-2t}\}\,\mathscr{L}\{\cos 3t\}$ (the convolution theorem)

$= \dfrac{1}{s+2} \times \dfrac{s}{s^2+9} = \dfrac{s}{(s+2)(s^2+9)}$.   Q.E.D.

---

[4] Comparing the first two questions in Examples 4 and 5, we can observe that the process of first calculating the convolution integral and then taking the Laplace transform can be more complicated compared to directly using the convolution theorem. However, despite the difference in the calculation process, the results obtained are the same.

3-3 Exercises

Basic questions

1. Find the Laplace transform of following functions.
   (1) $1*t$.
   (2) $t*\cos 2t$.
   (3) $e^{-2t}*\cos 2t$.
   (4) $t*t^2*t^3$.
   (5) $\delta(t)*t$.

2. Find the Laplace transform of function
   $\int_0^t \sin 2\tau \cdot \sinh 2(t-\tau)d\tau$.

3. Find the Laplace transform of function
   $e^{-3t}\int_0^t e^{3\tau}\tau^3 d\tau$.

4. Find the Laplace transform of periodic function
   $$f(t) = \begin{cases} t, & 0 < t < 2 \\ t-2, & 2 < t < 4 \end{cases},$$
   $f(t+4) = f(t)$.

Advanced questions

1. The half-wave rectification of sin t can be expressed as a periodic function.
   $$f(t) = \begin{cases} \sin t, & 0 < t < \pi \\ 0, & \pi < t < 2\pi \end{cases},$$
   $f(t+2\pi) = f(t)$, find the Laplace transform.

2. Find the Laplace transform of the following periodic functions.

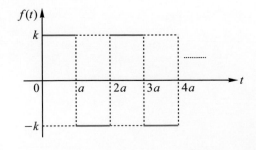

3. Find the Laplace transform of the following periodic functions.

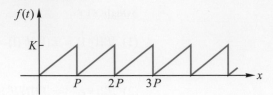

4. The full-wave rectification of sin t can be expressed as a periodic function $f(t) = |\sin t|$, find the Laplace transform.

5. Find the Laplace transform of the following periodic functions.

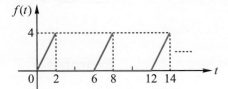

6. Find the Laplace transform of the following periodic functions.

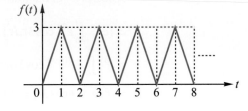

3-4 Laplace Inverse Transform

In practical applications, the Laplace transform is used to transform complex signals in the time domain (t domain) into frequency domain (s domain) functions that are easier to analyze. In order to apply the results obtained in the frequency domain, it is often necessary to translate them back to the time domain. This process is known as the Laplace inverse transform or inverse Laplace transform. In the next section, we will explore the techniques and methods used for the Laplace inverse transform.

3-4-1 Laplace Inverse Transform

The Laplace transform involves multiplying a function by an exponential function e^{-st} and then integrating it. Naturally, the inverse transform is defined as multiplying the function by e^{-st} and then integrating it. Therefore, we have the following definition:

▶ **Definition 3-4-1**

Laplace Inverse Transform

The Laplace inverse transform of a function $f(t)$ is defined as:

$$\mathscr{L}^{-1}\{F(s)\} = \frac{1}{2\pi i} \int_{a-i\infty}^{a+i\infty} F(s)e^{st}\,ds \,,$$

where a is real number, and constant $s > 0$. ◀

Therefore, $\mathscr{L}^{-1}\{F(s)\} = \dfrac{1}{2\pi i} \int_{a-i\infty}^{a+i\infty} [\int_{0}^{\infty} f(t)e^{-st}\,dt]e^{st}\,ds = f(t)$, the question is, how do we determine whether the inverse Laplace transform of a function exists? Just like the origin of exponential-order functions, this naturally relates to the behavior of the function. Please refer to the following theorem.

▶ **Theorem 3-4-1**

The Existence of Laplace Inverse Transform

If $\begin{cases} \lim\limits_{s\to\infty} F(s) = 0 \\ \lim\limits_{s\to\infty} sF(s) \text{ is bounded} \end{cases}$, then the inverse of $F(s)$ exists. ◀

3-4-2 The Basic Formula and Theorem for Laplace Inverse Transform

Since the Laplace inverse transform is the inverse operation of the Laplace transform, by referring to the results of the Laplace transform, we can naturally determine the original function from which it was transformed. This is what the Laplace inverse transform requires. Therefore, referring to the formulas in the beginning sections of this chapter, we can construct Table 1. We will use this as a basis to calculate the Laplace inverse transform of various types of functions. For the Laplace inverse transform of less common functions, you can refer to Appendix II. Additionally, after understanding the content of Chapter 11 on complex functions, readers can try to directly compute the inverse transform using the definition, skipping this part.

Table 1

function	Laplace transform	Laplace inverse transform
$u(t)$	$\mathscr{L}\{u(t)\} = \dfrac{1}{s}$	$\mathscr{L}^{-1}\{\dfrac{1}{s}\} = u(t)$ or 1
e^{at}	$\mathscr{L}\{e^{at}\} = \dfrac{1}{s-a}$	$\mathscr{L}^{-1}\{\dfrac{1}{s-a}\} = e^{at}$
$\sin wt$	$\mathscr{L}\{\sin wt\} = \dfrac{w}{s^2+w^2}$	$\mathscr{L}^{-1}\{\dfrac{1}{s^2+w^2}\} = \dfrac{1}{w}\sin wt$
$\cos wt$	$\mathscr{L}\{\cos wt\} = \dfrac{s}{s^2+w^2}$	$\mathscr{L}^{-1}\{\dfrac{s}{s^2+w^2}\} = \cos wt$
$\sinh wt$	$\mathscr{L}\{\sinh wt\} = \dfrac{w}{s^2-w^2}$	$\mathscr{L}^{-1}\{\dfrac{1}{s^2-w^2}\} = \dfrac{1}{w}\sinh wt$
$\cosh wt$	$\mathscr{L}\{\cosh wt\} = \dfrac{s}{s^2-w^2}$	$\mathscr{L}^{-1}\{\dfrac{s}{s^2-w^2}\} = \cosh wt$
t^n	$\mathscr{L}\{t^n\} = \dfrac{n!}{s^{n+1}}$	$\mathscr{L}^{-1}\{\dfrac{1}{s^{n+1}}\} = \dfrac{t^n}{n!}$; n is positive integer

Many of the theorems in Laplace transform also hold true in Laplace inverse transform. The introduction is as follows.

▶ **Theorem 3-4-2**

The Laplace Inverse Transform Is Linear .

If $\mathscr{L}\{f(t)\} = F(s)$, $\mathscr{L}\{g(t)\} = G(s)$, then

$$\mathscr{L}^{-1}\{c_1 F(s) \pm c_2 G(s)\} = c_1 \mathscr{L}^{-1}\{F(s)\} \pm c_2 \mathscr{L}^{-1}\{G(s)\} .$$ ◀

Example 1

Find the Laplace inverse transform of the following functions:

(1) $F(s) = \dfrac{1}{s} + \dfrac{5}{s^2} - \dfrac{3}{s^4}$.   (2) $F(s) = \dfrac{1}{s+3} + \dfrac{s+1}{s^2+9}$.

Solution

(1) $\mathscr{L}^{-1}\{\dfrac{1}{s} + \dfrac{5}{s^2} - \dfrac{3}{s^4}\} = \mathscr{L}^{-1}\{\dfrac{1}{s}\} + \mathscr{L}^{-1}\{\dfrac{5}{s^2}\} + \mathscr{L}^{-1}\{\dfrac{-3}{s^4}\}$

$$= 1 + 5\dfrac{t}{1!} - 3\dfrac{t^3}{3!} = 1 + 5t - \dfrac{1}{2}t^3 .$$

(2) $\mathscr{L}^{-1}\{\dfrac{1}{s+3} + \dfrac{s}{s^2+9} + \dfrac{1}{s^2+9}\} = \mathscr{L}^{-1}\{\dfrac{1}{s+3}\} + \mathscr{L}^{-1}\{\dfrac{s}{s^2+3^2}\} + \mathscr{L}^{-1}\{\dfrac{1}{s^2+3^2}\}$

$$= e^{-3t} + \cos 3t + \dfrac{1}{3}\sin 3t .$$ Q.E.D.

Recalling the beginning of this chapter, we mentioned the importance of examining the overall result of the ODE under Laplace transformation, which led to the derivation of Theorem 3-2-10, namely $\mathscr{L}\{f^{(n)}(t)\} = s^n F(s) - s^{n-1}f(0) - s^{n-2}f'(0) - \cdots - f^{(n-1)}(0)$. However, after dealing with the result of the Laplace transform, it is crucial to know how to restore it. By setting $f(0) = f'(0) = \cdots\cdots = f^{(n-1)}(0) = 0$ in the above formula, we obtain

$$\mathscr{L}\{f^{(n)}(t)\} = s^n F(s)$$

▶ Theorem 3-4-3

The Inverse Transform of Multiplying s^n

$\mathscr{L}^{-1}\{s^n F(s)\} = f^{(n)}(t)$.   ◀

From the above theorem, we can observe that during the Laplace inverse transform, an s corresponds to a first-order derivative of $f(t)$. In the following Example 2, we will make an initial attempt to solve an ODE using Laplace transform. However, a systematic discussion of the theory will be presented in the next section.

Example 2

Find the solutions of following ODE:
$y'' + y = 0$, $y(0) = 1$, $y'(0) = 1$.

Solution   Let $\mathcal{L}\{y(t)\} = Y(s)$, taking the Laplace transform of the original ODE,
then $\mathcal{L}(y'') + \mathcal{L}(y) = [s^2 Y(s) - sy(0) - y'(0)] + Y(s) = \mathcal{L}(0) = 0$ after   arrangement

$(s^2 + 1)Y(s) = s + 1$, that is $Y(s) = \dfrac{s}{s^2 + 1} + \dfrac{1}{s^2 + 1}$, applying the Laplace inverse

transform to both sides of the equation, we obtain

$$y(t) = \mathcal{L}^{-1}\{Y(s)\} = \mathcal{L}^{-1}\{\frac{s}{s^2+1}\} + \mathcal{L}^{-1}\{\frac{1}{s^2+1}\} = \cos t + \sin t .$$ Q.E.D.

Recalling Theorem 3-2-12, we obtained the relationship between the Laplace transform of the n-fold integral and the original function

$$\mathcal{L}\{\int_0^t \int_0^{t_1} \cdots \int_0^{t_{n-1}} f(\tau)d\tau dt_{n-1} \cdots dt_1\} = \frac{F(s)}{s^n}$$

In other words, we have

▶ Theorem 3-4-4

The Inverse Transform of Dividing s^n

$$\mathcal{L}^{-1}\{\frac{F(s)}{s^n}\} = \int_0^\tau \int_0^{t_1} \cdots \int_0^{t_{n-1}} f(\tau)d\tau dt_{n-1} \cdots dt_1 .$$ ◀

From this theorem, we can conclude that during the Laplace inverse transform, an $'\dfrac{1}{s}'$ corresponds to a first-order integral of $f(t)$.

Example 3

Find the Laplace inverse transform:

$$\mathcal{L}^{-1}\{\frac{1}{s(s^2+1)}\} .$$

Solution

$$\mathcal{L}^{-1}\{\frac{1}{s(s^2+1)}\} = \mathcal{L}^{-1}\{\frac{\frac{1}{s^2+1}}{s}\}$$

$$= \int_0^t \mathcal{L}^{-1}\{\frac{1}{s^2+1}\}d\tau$$

$$= \int_0^t \sin\tau\, d\tau = -\cos\tau \Big|_0^t$$

$$= 1-\cos t .$$

Q.E.D.

Through first shifting theorem, we know $\mathcal{L}\{e^{at}f(t)\} = \mathcal{L}\{f(t)\}_{s\to s-a} = F(s-a)$, therefore

▶ Theorem 3-4-5

s-shift Inverse Transform

$$\mathcal{L}^{-1}\{F(s-a)\} = e^{at}f(t) = e^{at}\mathcal{L}^{-1}\{F(s)\} .$$

◀

Example 4

Find the Laplace inverse transform:

$$\mathcal{L}^{-1}\{\frac{1}{s^2+2s+5}\} .$$

Solution

$$\mathcal{L}^{-1}\{\frac{1}{s^2+2s+5}\} = \mathcal{L}^{-1}\{\frac{1}{(s+1)^2+2^2}\}$$

$$= e^{-t}\mathcal{L}^{-1}\{\frac{1}{(s)^2+2^2}\}$$

$$= \frac{1}{2}e^{-t}\sin(2t) .$$

Q.E.D.

Example 5

$$\mathscr{L}^{-1}\{\frac{6s-4}{s^2-4s+20}\} = ?$$

Solution

$$\mathscr{L}^{-1}\{\frac{6s-4}{s^2-4s+20}\} = \mathscr{L}^{-1}\{\frac{6\cdot(s-2)+8}{(s-2)^2+4^2}\} = \mathscr{L}^{-1}\{\frac{6(s-2)}{(s-2)^2+4^2} + 2\times\frac{4}{(s-2)^2+4^2}\}$$

$$= 6e^{2t}\cos 4t + 2e^{2t}\sin 4t .$$ Q.E.D.

Through second shifting theorem, we know $\mathscr{L}\{f(t-a)u(t-a)\} = e^{-as}F(s)$, therefore we obtain

▶ Theorem 3-4-6

t-shift Inverse Transform

$$\mathscr{L}^{-1}\{e^{-as}F(s)\} = \mathscr{L}^{-1}\{F(s)\}_{t\to t-a}\cdot u(t-a) .$$ ◀

Examle 6

Find the Laplace inverse transform:

$$\mathscr{L}^{-1}\{\frac{2}{s} - \frac{3e^{-s}}{s^2} + \frac{5e^{-2s}}{s^2}\} .$$

Solution

$$\mathscr{L}^{-1}\{\frac{2}{s} - \frac{3e^{-s}}{s^2} + \frac{5e^{-2s}}{s^2}\} = \mathscr{L}^{-1}\{\frac{2}{s}\} - \mathscr{L}^{-1}\{\frac{3e^{-s}}{s^2}\} + \mathscr{L}^{-1}\{\frac{5e^{-2s}}{s^2}\}$$

$$= 2 - 3(t-1)u(t-1) + 5(t-2)u(t-2) .$$ Q.E.D.

From Theorem 3-2-13, we know that when the domain of a function is scaled, the corresponding formula for the Laplace transform also changes accordingly satisfies $\mathscr{L}\{f(at)\} = \frac{1}{a}F(\frac{s}{a})$, therefore we obtain

▶ Theorem 3-4-7

Scale Transformation Inverse Transform

$$\mathscr{L}^{-1}\{F(as)\} = \frac{1}{a}f(\frac{t}{a}) .$$ ◀

Example 7

Find the Laplace inverse transform:

$\mathcal{L}^{-1}\{\dfrac{1}{4s^2+1}\}$.

Solution

$$\mathcal{L}^{-1}\{\frac{1}{4s^2+1}\}=\mathcal{L}^{-1}\{\frac{1}{(2s)^2+1}\}=\frac{1}{2}\,\mathcal{L}^{-1}\{\frac{1}{s^2+1}\}\Big|_{t\to\frac{t}{2}}=\frac{1}{2}\sin(\frac{t}{2})\ .$$

Q.E.D.

If we multiply a function $f(t)$ by a polynomial t^n to modify it, Theorem 3-2-21 tells us that there exists a differential relationship between the modified Laplace transform and the original transform.

$$\mathcal{L}\{t^n f(t)\}=(-1)^n\,\frac{d^n F(s)}{ds^n}$$

Therefore, we obtain following Theorem 3-4-8.

▶ **Theorem 3-4-8**

The Inverse Transform of Multiplying t^n

$$t^n f(t)=(-1)^n\,\mathcal{L}^{-1}\{\frac{d^n F(s)}{ds^n}\}\ .$$

◀

It is worth mentioning that when dealing with functions that are not easily inverted by Laplace inverse transform, a common simplification technique is to differentiate the function before performing the inverse transform. This is particularly useful for cases involving logarithmic, inverse trigonometric functions, and others. Let's look at the following example. The commonly used formula for this technique is

$$f(t)=\frac{(-1)^n}{t^n}\,\mathcal{L}^{-1}\{\frac{d^n F(s)}{ds^n}\}\ ,\ \text{when } n=1\ \Rightarrow f(t)=-\frac{1}{t}\,\mathcal{L}^{-1}\{\frac{dF(s)}{ds}\}$$

Example 8

Find the Laplace inverse transform:
$\mathcal{L}^{-1}\{\ln(s^2+1)\}$.

Solution　In Theorem 3-4-8, we let $n=1$,

$$\mathcal{L}^{-1}\{\ln(s^2+1)\}=-\frac{1}{t}\,\mathcal{L}^{-1}\{\frac{d}{ds}[\ln(s^2+1)]\}=-\frac{1}{t}\,\mathcal{L}^{-1}\{\frac{2s}{s^2+1}\}=-\frac{2}{t}\cos t\ .$$

Q.E.D.

In a situation similar to Theorem 3-4-8, if we multiply a function $\dfrac{1}{t^n}$, Theorem 3-2-23 tells us that there exists an integral relationship between the modified Laplace transform and the original function

$$\mathcal{L}\{\frac{f(t)}{t^n}\} = \int_s^\infty \int_{s_1}^\infty \cdots \int_{s_{n-1}}^\infty F(u)\,du\,ds_{n-1}\cdots ds_1$$

Therefore, we obtain the following Theorem 3-4-9.

▶ **Theorem 3-4-9**

The Inverse Transform of Dividing t^n

$$\frac{f(t)}{t^n} = \mathcal{L}^{-1}\{\int_s^\infty \int_{s_1}^\infty \cdots \int_{s_{n-1}}^\infty F(u)\,du\,ds_{n-1}\cdots ds_1\}\,.$$ ◀

Similarly, when encountering a situation where it is difficult to directly invert $F(s)$ in Laplace transform, one can attempt to integrate $F(s)$ and then perform the Laplace inverse transform on the result.

Example 9

Find the Laplace inverse transform:

$$\mathcal{L}^{-1}\{\frac{s}{(s^2+1)^2}\}\,.$$

Solution   In Theorem 3-4-9, we let $n = 1$,

$$\mathcal{L}^{-1}\{\frac{s}{(s^2+1)^2}\} = t\,\mathcal{L}^{-1}\{\int_s^\infty \frac{u}{(u^2+1)^2}\,du\}\,,\ \text{(variable transformation } t = u^2 + 1\text{)}$$

$$= t\,\mathcal{L}^{-1}\{\frac{1}{2}\frac{1}{s^2+1}\} = \frac{t}{2}\sin t\,. \qquad \boxed{\text{Q.E.D.}}$$

Theorem 3-3-3 states that taking the convolution of two functions, $f(t) * g(t) = \int_0^t f(\tau)g(t-\tau)\,d\tau$ followed by the Laplace transform, is equivalent to first taking the Laplace transform of each function separately and then multiplying the resulting transforms. Mathematically, it can be expressed as:

$$\mathcal{L}\{f(t) * g(t)\} = F(s)\cdot G(s)$$

Therefore, the Laplace inverse transform will map the product of functions back to the convolution of the original functions.

▶ **Theorem 3-4-10**

The Convolution of Inverse Transform

$$\mathcal{L}^{-1}\{F(s)\cdot G(s)\} = f(t)*g(t).$$ ◀

Example 10

Find the Laplace inverse transform:

$$\mathcal{L}^{-1}\{\frac{1}{s(s^2+1)}\}.$$

Solution

$$\mathcal{L}^{-1}\{\frac{1}{s(s^2+1)}\} = \mathcal{L}^{-1}\{\frac{1}{s}\times\frac{1}{s^2+1}\}$$
$$= 1*\sin t$$
$$= \int_0^t \sin\tau\,d\tau = 1-\cos t.$$ Q.E.D.

3-4-3   Partial Fractions

When performing the Laplace inverse transform on a fractional function $F(s)$, if $F(s)$ is complex and cannot be directly applied to basic formulas, it is necessary to use partial fraction decomposition to express the original function as a product of simpler functions. Then, each component can be separately inverse transformed. Here are several representative cases of partial fraction decomposition. In general, the coefficients of the partial fractions are determined using methods such as differentiation and substitution.

1. **Proper Rational Function with Distinct Linear Factors in the Denominator**

$$F(s) = \frac{h(s)}{(s-a)(s-b)(s-c)(s-d)} = \frac{A_1}{s-a}+\frac{A_2}{s-b}+\frac{A_3}{s-c}+\frac{A_4}{s-d}.$$

2. **Disassemble**

By multiplying both sides of the equation by $(s-a)$, and substituting $s=a$, we

obtain $A_1 = \left.\frac{h(s)}{(s-b)(s-c)(s-d)}\right|_{s\to a}$,

we can determine A_2, A_3, A_4 and so on:

$$A_2 = \left.\frac{h(s)}{(s-a)(s-c)(s-d)}\right|_{s\to b}, \quad A_3 = \left.\frac{h(s)}{(s-a)(s-b)(s-d)}\right|_{s\to c},$$

$$A_4 = \left.\frac{h(s)}{(s-a)(s-b)(s-c)}\right|_{s\to d}.$$

Example 11

Find $\mathscr{L}^{-1}\{\dfrac{2s^2-5s-7}{(s-1)(s-3)(s-2)(s-5)}\}$.

Solution

$$\mathscr{L}^{-1}\{\frac{2s^2-5s-7}{(s-1)(s-3)(s-2)(s-5)}\}=\mathscr{L}^{-1}\{\frac{\frac{5}{4}}{s-1}+\frac{1}{s-3}+\frac{-3}{s-2}+\frac{\frac{3}{4}}{s-5}\}$$

$$=\frac{5}{4}e^t+e^{3t}-3e^{2t}+\frac{3}{4}e^{5t}.$$

Q.E.D.

3. **Proper Rational Function with a Repeated Factor in the Denominator**

$$F(s)=\frac{h(s)}{(s-a)^n Q(s)}=\frac{A_1}{s-a}+\frac{A_2}{(s-a)^2}+\cdots+\frac{A_n}{(s-a)^n}+R(s),$$

and $(s-a)$ does not divide $Q(s)$.

4. **Disassemble**

By multiplying both sides of the equation by $(s-a)^n$ and differentiating, we obtain A_k, as shown in the following formula:

$$A_{n-k}=\lim_{s\to a}\frac{1}{k!}\frac{d^k}{ds^k}\{\frac{h(s)}{Q(s)}\},\quad k=0,1,2...,n-1,\text{ and }A_n=\lim_{s\to a}\{\frac{h(s)}{Q(s)}\}.$$

Example 12

$$\mathscr{L}^{-1}\{\frac{2s+1}{(s+1)(s-2)^2}\}=?$$

Solution

$$\mathscr{L}^{-1}\{\frac{2s+1}{(s+1)(s-2)^2}\}=\mathscr{L}^{-1}\{\frac{B}{(s+1)}+\frac{A_1}{s-2}+\frac{A_2}{(s-2)^2}\},$$

$$B=\frac{2s+1}{(s-2)^2}\bigg|_{s\to-1}=-\frac{1}{9},\quad A_2=\frac{2s+1}{(s+1)}\bigg|_{s\to2}=\frac{5}{3},\quad A_1=\frac{d}{ds}[\frac{2s+1}{(s+1)}]\bigg|_{s\to2}=\frac{1}{9}.$$

Therefore,

$$\mathscr{L}^{-1}\{\frac{2s+1}{(s+1)(s-2)^2}\}=\mathscr{L}^{-1}\{\frac{\frac{-1}{9}}{(s+1)}+\frac{\frac{1}{9}}{s-2}+\frac{\frac{5}{3}}{(s-2)^2}\}$$

$$=-\frac{1}{9}e^{-t}+\frac{1}{9}e^{2t}+\frac{5}{3}te^{2t}.$$

Q.E.D.

In Example 12, the coefficient A_1 can also be determined by taking a limit. In the process of partial fraction decomposition $\dfrac{2s+1}{(s+1)(s-2)^2} = \dfrac{B}{(s+1)} + \dfrac{A_1}{s-2} + \dfrac{A_2}{(s-2)^2}$, by multiplying both sides by s and taking the limit as s approaches infinity,

we obtain $\displaystyle\lim_{s\to\infty} s \times \dfrac{2s+1}{(s+1)(s-2)^2} = \lim_{s\to\infty} s \times \dfrac{B}{(s+1)} + \lim_{s\to\infty} s \times \dfrac{A_1}{s-2} + \lim_{s\to\infty} s \times \dfrac{A_2}{(s-2)^2}$,

then $\quad 0 = B + A_1$, that is $A_1 = -B = \dfrac{1}{9}$.

5. Proper Rational Function with Distinct Quadratic Factors in the Denominator

$$F(s) = \frac{h(s)}{[(s+a)^2 + b^2]Q(s)} = \frac{As+B}{(s+a)^2 + b^2} + R(s)$$

6. Disassemble

By substituting $s = 0$, we can determine B. By multiplying both sides by s and taking the limit as s approaches infinity, we can determine A.

Example 13

$$\mathcal{L}^{-1}\{\frac{18+11s-s^2}{(s^2-1)(s^2+3s+3)}\} = ?$$

Solution

$$\mathcal{L}^{-1}\{\frac{18+11s-s^2}{(s^2-1)(s^2+3s+3)}\} = \mathcal{L}^{-1}\{\frac{18+11s-s^2}{(s+1)(s-1)(s^2+3s+3)}\}$$

$$= \mathcal{L}^{-1}\{\frac{C}{s+1} + \frac{D}{s-1} + \frac{As+B}{s^2+3s+3}\}$$

where

$$C = \frac{18+11s-s^2}{(s-1)(s^2+3s+3)}\Big|_{s\to-1} = -3,$$

$$D = \frac{18+11s-s^2}{(s+1)(s^2+3s+3)}\Big|_{s\to1} = 2.$$

Substitute $s = 0$ into original equation and obtain $B = -3$, multiplying both sides by s and taking $\displaystyle\lim_{s\to\infty}$, we know $0 = -3 + 2 + A \Rightarrow A = 1$, so

$$\mathcal{L}^{-1}\{\frac{18+11s-s^2}{(s^2-1)(s^2+3s+3)}\} = \mathcal{L}^{-1}\{\frac{-3}{s+1} + \frac{2}{s-1} + \frac{s-3}{s^2+3s+3}\}$$

$$= \mathcal{L}^{-1}\{\frac{-3}{s+1} + \frac{2}{s-1} + \frac{(s+\frac{3}{2})-\frac{9}{2}}{(s+\frac{3}{2})^2 + (\frac{\sqrt{3}}{2})^2}\}$$

$$= -3e^{-t} + 2e^t + e^{-\frac{3}{2}t} \cdot [\cos\frac{\sqrt{3}}{2}t - 3\sqrt{3}\sin\frac{\sqrt{3}}{2}t].$$ Q.E.D.

In addition to the techniques mentioned above, partial fraction decomposition can also be solved using the method of substitution. We can substitute relatively simple values such as $s = 0, \pm 1, \pm 2$, and so on, into both sides of the partial fraction equation. By solving the resulting system of equations, we can determine the coefficients of the partial fractions.

Now that we have gained experience in decomposing partial fractions, let's return to the topic of Laplace inverse transform and explore how these partial fraction techniques can simplify problems and facilitate calculations.

Example 14

$$\mathscr{L}^{-1}\{\frac{3}{s^2+3s-10}\} = ?$$

Solution   Consider the decomposition method for partial fractions with a product of distinct linear factors in the denominator, we obtain

$$\mathscr{L}^{-1}\{\frac{3}{s^2+3s-10}\} = \mathscr{L}^{-1}\{\frac{-\dfrac{3}{7}}{s+5}+\frac{\dfrac{3}{7}}{s-2}\} = \frac{3}{7}e^{2t}-\frac{3}{7}e^{-5t}.$$ Q.E.D.

Example 15

$$\mathscr{L}^{-1}\{\frac{1}{s^2\cdot(s-3)}\} = ?$$

Solution   Consider the decomposition method for partial fractions with a product of a repeated linear factor in the denominator, we obtain

$$\mathscr{L}^{-1}\{\frac{1}{s^2\cdot(s-3)}\} = \mathscr{L}^{-1}\{\frac{-\dfrac{1}{9}}{s}+\frac{-\dfrac{1}{3}}{s^2}+\frac{\dfrac{1}{9}}{s-3}\}$$

$$= -\frac{1}{9}-\frac{1}{3}t+\frac{1}{9}e^{3t}.$$ Q.E.D.

3-4   Laplace Inverse Transform   **3-45**

Example 16

$$\mathscr{L}^{-1}\{\frac{e^{-2s}}{s^3+4s^2+5s+2}\}=?$$

Solution   Since the numerator involves an exponential function, we can refer to the Second Shifting Theorem. Let $F(s)=\dfrac{1}{s^3+4s^2+5s+2}$. We obtain

$$\mathscr{L}^{-1}\{e^{-as}\cdot F(s)\}=\mathscr{L}^{-1}\{F(s)\}_{t\to t-a}u(t-a).$$

Now consider decomposition method for partial fractions with a product of a repeated linear factor in the denominator, we obtain

$$\mathscr{L}^{-1}\{\frac{1}{s^3+4s^2+5s+2}\cdot e^{-2s}\}=\mathscr{L}^{-1}\{\frac{1}{(s+1)^2\cdot(s+2)}e^{-2s}\}$$

$$=\mathscr{L}^{-1}\{[\frac{1}{(s+2)}+\frac{-1}{(s+1)}+\frac{1}{(s+1)^2}]\cdot e^{-2s}\}$$

$$=(e^{-2t}-e^{-t}+te^{-t})|_{t\to t-2}\cdot u(t-2)$$

$$=[e^{-2(t-2)}-e^{-(t-2)}+(t-2)e^{-(t-2)}]u(t-2).\qquad \text{Q.E.D.}$$

3-4 Exercises

Basic questions

1. Find the Laplace inverse transform of following functions:

(1) $F(s) = \dfrac{1}{s+2}$ (2) $F(s) = \dfrac{1}{s^4}$

(3) $F(s) = \dfrac{s}{s^2+4}$ (4) $F(s) = \dfrac{1}{s^2+4}$

(5) $F(s) = \dfrac{1}{s^2-4}$.

2. Find the Laplace inverse transform of following functions:

(1) $F(s) = \dfrac{s+3}{(s-2)(s+1)}$ (2) $F(s) = \dfrac{1}{s^3-s}$

(3) $F(s) = \dfrac{e^{-2s}}{s^2+s-2}$ (4) $F(s) = \dfrac{e^{-4s}}{s^2}$

(5) $F(s) = \dfrac{1-e^{-2s}}{s^2}$ (6) $F(s) = \dfrac{s}{s^2+2s-3}$

(7) $F(s) = \dfrac{1}{s(s-1)}$.

Advanced questions

Find the Laplace inverse transform of following functions:

1. $F(s) = \dfrac{1}{(s^2+4)(s+12)}$.

2. $F(s) = \dfrac{1}{s^2(s+1)^2}$.

3. $F(s) = \dfrac{1}{s^2(s-a)}$.

4. $F(s) = \dfrac{ab}{(s^2+a^2)(s^2+b^2)}$.

5. $F(s) = \dfrac{2s^2+3s+3}{(s+1)(s+3)^3}$.

6. $F(s) = \dfrac{6s-4}{s^2-4s+20}$.

7. $F(s) = \dfrac{1}{s^2(s-3)}$.

8. $F(s) = \ln(1+\dfrac{1}{s^2})$.

9. $F(s) = \dfrac{3e^{-2s}}{(s+1)^2(s^2+2s+10)}$.

10. $F(s) = \dfrac{e^{-3s}}{(s-1)^3}$.

11. $F(s) = \dfrac{1}{s^3+4s^2+5s+2}$.

12. $F(s) = \dfrac{(3s+5)e^{-3s}}{s(s^2+2s+5)}$.

13. $F(s) = \ln(1-\dfrac{a^2}{s^2})$.

14. $F(s) = \dfrac{\pi}{2} - \tan^{-1}\dfrac{s}{2}$.

15. $F(s) = \dfrac{e^{-s}}{s(s+1)(s+2)}$.

16. $F(s) = \dfrac{1}{s^2} - \dfrac{48}{s^5}$.

17. $F(s) = \dfrac{(s+1)^3}{s^4}$.

18. $F(s) = \dfrac{s+1}{s^2+2}$.

19. $F(s) = \dfrac{s^2+1}{s(s-1)(s+1)(s-2)}$.

20. $F(s) = \dfrac{1}{s^2-6s+10}$.

21. $F(s) = \dfrac{2s+5}{s^2+6s+34}$.

22. $F(s) = \dfrac{1}{s^2(s-1)}$.

23. $F(s) = \dfrac{1}{s^2(s-2)}e^{-2s}$.

24. $F(s) = \ln(\dfrac{s^2+1}{s^2+s})$.

3-5 **The Application of Laplace Transform**

With the preparation from Sections 3-1 to 3-4, we have gained an understanding of the result of a linear ordinary differential equation (ODE) under the Laplace transform and how to obtain the original solution through the Laplace inverse transform (as demonstrated in Example 2 of Section 3-4). The complete process is depicted in Figure 3-14.

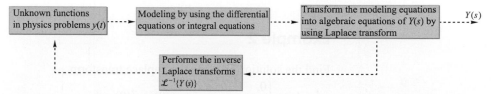

Figure 3-14   Conceptual diagram for solving physics problems using the Laplace transform

3-5-1 Solving Constant Coefficients ODE Using the Laplace Transform

Recalling Theorem 3-2-10, we have the Laplace transform of a higher-order derivative as follows:

$$\mathscr{L}\{y(t)\} = Y(s)$$
$$\mathscr{L}\{y'(t)\} = sY(s) - y(0)$$
$$\mathscr{L}\{y''(t)\} = s^2Y(s) - sy(0) - y'(0)$$
$$\vdots$$
$$\mathscr{L}\{y^{(n)}(t)\} = s^nY(s) - s^{n-1}y(0) - s^{n-2}y'(0)\cdots\cdots - y^{(n-1)}(0)$$

Next, let's consider the solution of higher-order linear ODEs using the Laplace transform.

Example 1

Find the solution by using the Laplace transform:
$y'' - 4y' + 4y = \delta(t-1)$, $y(0) = 0$, $y'(0) = 1$.

Solution   Let $\mathscr{L}\{y(t)\} = Y(s)$, now by the Laplace transform of first, second derivative and Laplace transform of an impulse function (Theorem 3-3-1), we know

$$[s^2Y(s) - sy(0) - y'(0)] - 4[sY(s) - y(0)] + 4Y(s) = e^{-s}.$$

Adding original conditions $y(0) = 0, y'(0) = 1$, we obtain after rearrangement

$$Y(s) = \frac{1+e^{-s}}{s^2 - 4s + 4} = \frac{1}{(s-2)^2} + \frac{e^{-s}}{(s-2)^2}$$

Taking the Laplace inverse transform we obtain

$$y(t) = \mathscr{L}^{-1}\{Y(s)\} = \mathscr{L}^{-1}[\frac{1}{(s-2)^2} + \frac{1}{(s-2)^2}e^{-s}]$$

$$= te^{2t} + (t-1)e^{2(t-1)}u(t-1). \text{ (Theorem 3-2-18, Theorem 3-2-19)} \qquad \boxed{\text{Q.E.D.}}$$

Example 2

Find the solution by using the Laplace transform

$$y'' + 4y = \begin{cases} 0, & 0 \le t < \pi \\ 3\cos t, & t \ge \pi \end{cases}, y(0) = y'(0) = 1.$$

Solution Express the source function using a step function, then we obtain

$$f(t) = \begin{cases} 0, & 0 \le t < \pi \\ 3\cos t, & t \ge \pi \end{cases} = 3\cos t \cdot u(t-\pi),$$

by the Laplace transforms of the first and second derivatives, as well as the Laplace transform of trigonometric functions, we obtain

$$[s^2 Y(s) - sy(0) - y'(0)] + 4Y(s) = 3e^{-\pi s} \cdot \mathscr{L}\{\cos(t+\pi)\}.$$

After rearrangement, we obtain $Y(s) = \frac{1}{s^2 + 2^2} + \frac{s}{s^2 + 2^2} - [\frac{s}{s^2 + 1} - \frac{s}{s^2 + 4}]e^{-\pi s}.$

Now taking the Laplace inverse transform then we obtain

$$y(t) = \mathscr{L}^{-1}\{Y(s)\} = \mathscr{L}^{-1}\{\frac{1}{s^2 + 2^2} + \frac{s}{s^2 + 2^2} - \frac{s}{s^2 + 1}e^{-\pi s} + \frac{s}{s^2 + 4}e^{-\pi s}\}$$

$$= \frac{1}{2}\sin 2t + \cos 2t - \cos(t-\pi)u(t-\pi) + \cos 2(t-\pi)u(t-\pi) \quad \text{(Theorem 3-2-19)}$$

$$= \frac{1}{2}\sin 2t + \cos 2t + \cos t \cdot u(t-\pi) + \cos 2t \cdot u(t-\pi). \qquad \boxed{\text{Q.E.D.}}$$

3-5-2 Solve Variable Coefficients ODE

Combing Theorem3-2-10 and 3-2-21, we obtain following common formulas

$$\mathscr{L}\{t^m y^{(n)}(t)\} = (-1)^m \frac{d^m}{ds^m}[\mathscr{L}\{y^{(n)}(t)\}] \quad \text{(Theorem 3-2-21)}$$

$$= (-1)^m \frac{d^m}{ds^m}[s^n Y(s) - s^{n-1}y(0) \cdots\cdots - y^{(n-1)}(0)] \quad \text{(Theorem 3-2-10)}.$$

Therefore, let $\mathscr{L}\{y(x)\} = Y(s)$, based on the general formulas mentioned above, the translation of the listed formulas for functions involving different orders of derivatives is as follows.

$$\mathscr{L}\{xy\} = -\frac{d}{ds}Y(s)$$

$$\mathscr{L}\{y'\} = sY(s)$$

$$\mathscr{L}\{(xy'')\} = -\frac{d}{ds}[s^2 Y(s) - sy(0) - y'(0)]$$

$$= -(2sY(s) + s^2 \frac{dY(s)}{ds}), \text{ assume where } y(0) = 0, y'(0) = c$$

$\vdots$

please see the practical application below.

Example 3

Please using the Laplace transform to solve following ODE, $xy'' + 2y' + (2 - x)y = 2e^x$, $y(0) = 0$.

Solution   Let $\mathscr{L}\{y(x)\} = Y(s)$, substituting the results mentioned in the previous discussion, we obtain

$$2sY(s) - s^2 \frac{dY(s)}{ds} + 2sY(s) + 2Y(s) + \frac{d}{ds}Y(s) = \frac{2}{s-1},$$

rearranging the equation, we obtain $\dfrac{dY(s)}{ds} - \dfrac{2}{s^2 - 1}Y(s) = \dfrac{-2}{(s^2 - 1)(s - 1)}.$

The above equation represents a first-order linear ODE in terms of the Laplace transform $Y(s)$ as the unknown variable. Referring back to Chapter 1, to solve a first-order linear ODE with variable coefficients, we use the integrating factor method,

$$I = \exp[\int -\frac{2}{s^2 - 1} ds] = \frac{s+1}{s-1},$$

the general solution is $Y(s) = \dfrac{1}{s^2 - 1} + c \times \dfrac{s-1}{s+1}.$

Now, we want to determine the constant c determined by the initial conditions. This relates to the relationship between $y(0)$ and $Y(s)$. Referring to the initial value theorem (Theorem 3-2-14), we can determine $\lim\limits_{s \to \infty} sY(s) = \lim\limits_{s \to \infty} s[\dfrac{1}{s^2 - 1} + c\dfrac{s-1}{s+1}] = y(0) = 0$,

that is $c = 0$, so $Y(s) = \dfrac{1}{(s+1)(s-1)} = \dfrac{1}{s^2 - 1^2}$, now by the inverse Laplace transform of hyperbolic functions (reference to Table 1). We obtain

$$y(x) = \mathscr{L}^{-1}\{Y(s)\} = \sinh x.$$ Q.E.D.

3-5-3 Solve an Integral Equation of Convolution Using the Laplace Transform

The common integral equation is

$$y(t) = f(t) + \int_0^t y(\tau)k(t-\tau)d\tau = f(t) + y(t) * k(t) \cdots (*)$$

Taking the Laplace transform (combind using the convolution theorem), we obtain $Y(s) = F(s) + Y(s)K(s)$, then simplifying and we get $Y(s) = \dfrac{F(s)}{1-K(s)}$, next applying the inverse Laplace transform, we can obtain the general solution.

$$y(t) = \mathscr{L}^{-1}\{Y(s)\} = \mathscr{L}^{-1}\{\frac{F(s)}{1-K(s)}\}$$

Example 4

Solve by using the Laplace transform $y(t) = u(t) + \int_0^t e^{-(t-\tau)}y(\tau)d\tau$, where

$$u(t) = \begin{cases} 1, & t \geq 0 \\ 0, & t < 0 \end{cases}.$$

Solution   In formula (*) we let $f(t) = u(t)$, $k(t) = e^{-t}$. Then by Chapter 3-1, the Laplace transform formulas, we obtain

$$\mathscr{L}\{f(t)\} = \mathscr{L}\{u(t)\} = \frac{1}{s}, \quad \mathscr{L}\{k(t)\} = \mathscr{L}\{e^{-t}\} = \frac{1}{s+1}.$$

Substituting into general soluton formula, we obtain

$$y(t) = \mathscr{L}^{-1}\{Y(s)\} = \mathscr{L}^{-1}\{\frac{1}{s} + \frac{1}{s^2}\} = u(t) + t = 1 + t .$$ Q.E.D.

Example 5

"Integral-differential" equation $x'(t)+3x(t)+2\int_0^t x(\tau)\,d\tau = 5u(t)$, where $x(0) = 1$. Using the Laplace transform to solve $x(t)$.

Solution　After rearranging the original equation

$x(t) = \dfrac{1}{3}[-x'(t)+5u(t)] + \int_0^t (-\dfrac{2}{3})x(\tau)\,d\tau$, in formula (A) we let

$f(t) = \dfrac{1}{3}[-x'(t)+5u(t)]$, $k(t) = -\dfrac{2}{3}$. Using the Laplace transform formulas from Chapter 3-1, we obtain

$$\mathscr{L}\{f(t)\} = \frac{1}{3}[-sX(s)+x(0)+\frac{5}{s}],$$

$$\mathscr{L}\{k(t)\} = -\frac{2}{3s}.$$

Substituting these results into the general solution equation, we get

$X(s) = \dfrac{4}{(s+1)} + \dfrac{-3}{(s+2)}$, then taking Laplace inverse transform the general solution is

$$x(t) = \mathscr{L}^{-1}\{\frac{4}{(s+1)} + \frac{-3}{(s+2)}\} = 4e^{-t} - 3e^{-2t}.$$

【Another Method】

Let $\mathscr{L}\{x(t)\} = \hat{x}(s)$, apply $L - T$ to the original integral-differential equation

$$\Rightarrow s\hat{x}(s) - x(0) + 3\hat{x}(s) + 2 \cdot \frac{1}{s}\hat{x}(s) = \frac{5}{s}$$

$$\Rightarrow (s + \frac{2}{s} + 3)\hat{x}(s) = 1 + \frac{5}{s} \Rightarrow (\frac{s^2+3s+2}{s})\hat{x}(s) = \frac{s+5}{s},$$

$$\therefore \hat{x}(s) = \frac{s+5}{(s+1)(s+2)} = \frac{4}{s+1} + \frac{-3}{s+2},$$

$$\therefore x(t) - \mathscr{L}^{-1}\{\frac{4}{s+1} + \frac{-3}{s+2}\} = 4e^{-t} - 3e^{-2t}.$$

Q.E.D.

3-5   Exercises

Basic questions

Solve following initial value problem by using Laplace transform.

1. $y' + 4y = 0, y(0) = 2.$

2. $y'' + 5y' + 4y = 2e^{-2t}, y(0) = 0, y'(0) = 0.$

3. $y'' - y' - 6y = 0, y(0) = 2, y'(0) = 3.$

4. $y'' + 9y = 10e^{-t}, y(0) = 0, y'(0) = 0.$

5. $y'' + 3y' + 2y = f(t)$, where

$$f(t) = \begin{cases} 1, & 0 < t < 1 \\ 0, & t > 1 \end{cases}, y(0) = y'(0) = 0.$$

6. $y'' - 2y' + 10y = 0, y(0) = 6, y'(0) = 0.$

7. $y'' - 4y' + 3y = 4e^{3x}, y(0) = -1, y'(0) = 3.$

8. $y'' - 4y' + 4y = \delta(t-1), y(0) = 0, y'(0) = 1.$

9. $f(t) = ?$

$$f(t) = \int_0^t \sin 2\lambda \cdot \sinh 2(t-\lambda) d\lambda.$$

10. $y'(t) = 1 - \int_0^t y(t-\tau)e^{-2\tau} d\tau, y(0) = 1.$

Advanced questions

Solve the following equation or system using the Laplace transform.

1. $y'' + 2y' + 5y = \delta(t-1) + \delta(t-3), y(0) = y'(0) = 0.$

2. $y'' + 9y = f(t), y'(0) = y(0) = 1,$

$$f(t) = \begin{cases} 0, & 0 \le t < \pi \\ \cos t, & t \ge \pi \end{cases}.$$

3. $y'' - 3y' = 2e^{2x} \sin x, y(0) = 1, y'(0) = 2.$

4. $y'' - 3y' + 2y = 4t + e^{3t}, y(0) = 0, y'(0) = -1.$

5. $y'' + 4y' + 4y = 1 + \delta(t-1), y(0) = 0, y'(0) = 0.5.$

6. Considering the LC circuit as follows, if its initial current $i(0) = i'(0) = 0$, and a voltage source is

$$E(t) = \begin{cases} 25t, 0 \le t \le 4 \\ 100, t > 4 \end{cases},$$

$$C = 0.04, L = 1,$$

find $i(t)$, where $t > 0$.

7. $y'' - y = 2\sin t + \delta(t-1),$
    $y(0) = 0, y'(0) = 2.$

8. $y'' + 4y' + 13y = 26e^{-4t},$
    $y(0) = 5, y'(0) = -29.$

9. $y'' + 4y = f(t), y(0) = 0, y'(0) = 1,$

$$f(t) = \begin{cases} 0, 0 < t < 3 \\ 1, t > 3 \end{cases}.$$

10. $y'' + y = r(t), y(0) = y'(0) = 0,$

$$r(t) = \begin{cases} t, 0 < t < 1 \\ 0, t > 1 \end{cases}.$$

11. $y''' - y'' - y' + y = 0, y(0) = 2,$
    $y'(0) = 1, y''(0) = 0.$

12. $ty'' - ty' - y = 0, y(0) = 0, y'(0) = 3.$

13. Consider the RLC circuit shown in the diagram. At time $t = 0$, a square wave with an amplitude of 10 V and a pulse width of 1 second is applied to this circuit. The circuit has no initial current, and the capacitor has no initial charge. Determine the current for $t \ge 0$? $(R = 3\Omega, L = 1H, C = 0.5F)$

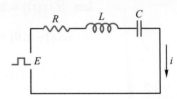

14. $y(t) = \sin 5t - 6\int_0^t y(t-\lambda) \cdot \cos 5\lambda d\lambda.$

15. $y' + \int_0^t y(\alpha) \cdot \cos 2(t-\alpha) d\alpha = \delta(t-3), y(0) = 1.$

4

Power Series Solution of Ordinary Differential Equations

Frobenius
(1849~1917,
Germany)

Mathematician Frobenius (1849-1917) introduced the concept of general power series solutions, overcoming the issue of ODE being unable to consider series solutions at non-analytic points. His work was instrumental in solving many engineering problems involving variable coefficient ODEs and made significant contributions to the field.

Learning Objectives

4-1

Expansion at a Regular Point for Solving ODE

4-1-1 Understanding the existence of power series solutions: regular points and singular points

4-1-2 Mastering the method of expanding power series around regular points to solve ODEs

4-2

Regular Singular Point Expansion for Solving ODE(Selected Reading)

4-2-1 Understanding characteristic equations

4-2-2 Mastering the solutions of different ODEs corresponding to different characteristic roots

ExampleVideo

Most ordinary differential equations (ODEs) do not have simple analytical solutions, so mathematicians turned to power series methods to solve ODEs, particularly in the case of variable coefficient ODEs. By expressing the solution as a Taylor expansion and substituting it into the original ODE, mathematicians were able to determine the coefficients by solving a recursive relationship, transforming the analytical problem into a problem of solving a series of equations. In this chapter, we introduce the "Frobenius series solution," which employs the expansion at regular singular points, to address the challenge of solving ODEs at non-analytic points. This approach opens up another avenue for solving variable coefficient ODEs that are difficult to solve otherwise.

4-1   Expansion at a Regular Point for Solving ODE

First, in this section, we will begin by defining the terms "regular point" and "singular point" in the context of ODEs. We will then introduce how the Taylor series can be used to expand around a regular point for obtaining the power series solution of the ODE.

4-1-1   The Existence of Power Series Solution

The power series solution, as the name suggests, represents the solution of

$$y(x) = \sum_{n=0}^{\infty} a_n x^n$$

an equation in the form of a power series. By substituting this series into the original equation, the problem is transformed into an algebraic problem of solving for the coefficients, denoted as "a_n." However, the issue of convergence arises when dealing with series. It is necessary for the series to converge in order to utilize this method effectively.

The most commonly used form of series for functions is the Taylor series expansion $f(x) = \sum_{n=0}^{\infty} \frac{f^{(n)}(a)}{n!}(x-a)^n$. Hence, we can determine the existence range (convergence radius) of the Taylor series using limit-based methods. Naturally, we consider the series solution of the ODE within the convergence radius. However, the story does not end here. When dealing with variable coefficient ODEs, we also need to expand the coefficient functions as series and substitute them, along with the assumed solution $y(x)$, into the original equation. Only then we can truly transform the problem into a coefficient-solving problem. Let's consider the case of a second-order linear ODE and further classify the situations.

$$y'' + P(x)y' + Q(x)y = 0 \cdots ①$$

1. **Regular Point**

   If $P(x)$ and $Q(x)$ are analytic at $x = a$, meaning that $P(x)$ and $Q(x)$ have existing and continuous derivatives of any order at $x = a$, then $x = a$ is a regular point of ①.

2. **Singular Point**

   (1) Regular singular point

      If $P(x)$ and $Q(x)$ are not analytic at $x = a$, but both $(x - a)P(x)$ and $(x - a)^2 Q(x)$ are analytic at $x = a$, then $x = a$ is referred to as a regular singular point of ①.

   (2) Irregular singular point

      If $x = a$ is not a regular point or a regular singular point of ①, then it is called an irregular singular point of ①.

Example 1

Determine the types of singular points for the following ODE.

$(1) (1+x)y'' + 2xy' + 5xy = x^2$.

$(2) (x+1)^2 y'' + 2xy' + y = x^2$.

$(3) (x+1)^2 y'' + 2(x+1)y' + 5xy = x^2$.

Solution

(1) Let $P(x) = \dfrac{2x}{1+x}$, $Q(x) = \dfrac{5x}{1+x}$,

   because $P(x)$, $Q(x)$ at $x = -1$ are not continuous, derivative not exist but

   $(1+x) \cdot \dfrac{2x}{1+x} = 2x$, $(1+x)^2 \cdot \dfrac{5x}{1+x} = 5x(1+x)$ are analytic functions,

   so $x = -1$ is a regular singular point of the ODE, while the rest are regular points.

(2) Let $P(x) = \dfrac{2x}{(x+1)^2}$, $Q(x) = \dfrac{1}{(x+1)^2}$,

   $P(x)$ and $Q(x)$ at $x = -1$ are not continuous, so they are not analytic,

   also $(x+1) \cdot P(x) = \dfrac{2x}{x+1}$ at $x = -1$ is not analytic,

   so $x = -1$ is a regular singular point of the ODE, while the rest are regular points.

(3) Let $P(x) = \dfrac{2}{x+1}$, $Q(x) = \dfrac{5x}{(x+1)^2}$,

   $P(x)$ and $Q(x)$ are not analytic at $x = -1$

   but $(x+1)P(x) = 2$ and $(x+1)^2 Q(x) = 5x$ at $x = -1$ are analytic,

   so $x = -1$ is a regular singular point, while the rest are regular points. Q.E.D.

4-1-2 Expansion at a Regular Point for Solving ODE

Considering the equation $y'' + P(x)y' + Q(x)y = 0$, if $x = a$ is a regular point, then $y(x)$ at $x = a$ is analytic. In this case, we can expand the solution $y(x)$ as a Taylor series and substitute it into the ODE to solve it.

(1) If we let $y(x) = \sum_{n=0}^{\infty} \dfrac{y^{(n)}(a)}{n!}(x-a)^n$, by using initial conditions to solve for the

coefficients $y^{(n)}(a)$, then substitute into the ODE and find solution, which is called the direct substitution method.

(2) If we let $y(x) = \sum_{n=0}^{\infty} a_n(x-a)^n$ and substitute into ODE, and solve the

coefficients a_n. It is referred to as the method of undetermined coefficients.

1. **The Direct Substitution Method**

Taking a homogeneous ODE $y'' + P(x)y' + Q(x)y = 0$ as an example, let's consider the series solution at the regular point $x = a$. Therefore we assume

$$y(x) = \sum_{n=0}^{\infty} \frac{y^{(n)}(a)}{n!}(x-a)^n$$

the prerequisite for completely determining the analytic solution is to find all the derivative values $y^{(n)}(a)$. However, it is generally impossible to obtain all these derivative values explicitly. Hence, we rely on the original ODE to calculate the derivatives up to a certain order:

(1) $y(a) = c_1$, $y'(a) = c_2$

(2) $y''(a) = -P(a)y'(a) - Q(a)y(a)$

(3) $y'''(a) = -P'(a)y'(a) - P(a)y''(a) - Q'(a)y(a) - Q(a)y'(a)$

(4) $y^{(4)}(a) = \cdots\cdots$ so on and so forth

(5) Substituting $y(a)$, $y'(a)$, $y''(a)$, $y'''(a)\cdots\cdots$ into the Taylor series expansion of the ODE, we can obtain an approximate solution for the ODE.

The advantage of this method is that we do not need to explicitly derive algebraic equations involving the coefficients or obtain the general solution. Instead, we can obtain an approximate solution with a very small error compared to the analytic solution. However, the drawback is that the solution obtained through this method is not an exact solution. If the ODE is non-homogeneous, it becomes challenging to deduce the equations satisfied by the coefficients in the Taylor series expansion. In such cases, it is often easier to use the direct substitution method rather than the method of undetermined coefficients.

Example 2

$y'' + (\sin x)y = e^{x^2}$, solve it by using power series solution.

Solution

Because $x = 0$ is a regular point of ODE, solve by using the direct substitution method,

let $y(x) = \sum_{n=0}^{\infty} \dfrac{y^{(n)}(0)}{n!} x^n$, and $y(0) = c_1$, $y'(0) = c_2$,

then $y'' = -(\sin x)y + e^{x^2}$,

$y''' = -(\cos x)y - (\sin x)y' + 2xe^{x^2}$,

$y^{(4)} = (\sin x)y - (\cos x)y' - (\cos x)y' - (\sin x)y'' + 2e^{x^2} + 4x^2 e^{x^2}$.

Let $x = 0$ substitute and we obtain

$$y''(0) = 1, \quad y'''(0) = -y(0) = -c_1, \quad y^{(4)}(0) = -2y'(0) + 2 = -2c_2 + 2 .$$

Next, substituting the approximate solution obtained from the Taylor series expansion of $y(x)$ into the ODE, we have

$$y(x) = \sum_{n=0}^{\infty} \frac{y^{(n)}(0)}{n!} x^n = y(0) + \frac{y'(0)}{1!}x + \frac{y''(0)}{2!}x^2 + \frac{y'''(0)}{3!}x^3 + \frac{y^{(4)}(0)}{4!}x^4 + \cdots$$

$$= c_1 + c_2 x + \frac{1}{2}x^2 - \frac{c_1}{6}x^3 + \frac{2 - 2c_2}{24}x^4 + \cdots$$

$$= c_1(1 - \frac{x^3}{6} + \cdots) + c_2(x - \frac{1}{12}x^4 + \cdots) + (\frac{1}{2}x^2 + \frac{1}{12}x^4 + \cdots) \qquad \text{Q.E.D.}$$

2. Power Series with Unknown Coefficients

Similarly to the direct substitution method, let's consider a homogeneous ODE as an example. Assuming that $x = a$ is a regular point, we substitute $y(x) = \sum_{n=0}^{\infty} a_n(x-a)^n$ into the original ODE. By combining like terms and comparing coefficients, we can derive the relationships between the coefficients a_n. We then solve these relationship equations to determine the coefficients. It's important to note that the solutions obtained using this method are not approximate solutions but rather analytic solutions. When solving the equations, it is recommended to follow the following principles.

(1) Give priority to terms with x^n.   (2) Merge terms with consistent subscripts and the same exponent.

Example 3

$y'' + xy' + y = 0$; $y(0) = 2$, $y'(0) = 0$, solve it by using power series solution.

Solution

(1) $x = 0$ is a regular point of ODE, let $y(x) = \sum_{n=0}^{\infty} a_n x^n$, then

$$y'(x) = \sum_{n=0}^{\infty} n a_n x^{n-1} = \sum_{n=1}^{\infty} n a_n x^{n-1},$$

$$y''(x) = \sum_{n=1}^{\infty} n \cdot (n-1) a_n x^{n-2} = \sum_{n=2}^{\infty} n \cdot (n-1) a_n x^{n-2}.$$

(2) Substituting the given expressions into the original ODE and simplifying, we have $\sum_{n=2}^{\infty} n(n-1) a_n x^{n-2} + \sum_{n=1}^{\infty} n \cdot a_n x^n + \sum_{n=0}^{\infty} a_n x^n = 0$.

To unify the upper and lower indices of the summation Σ, we perform the following adjustments.

① Adjusting the Exponent:
   Completing the missing term in the middle

   $$\sum_{n=2}^{\infty} n(n-1) a_n x^{n-2} + \sum_{n=1}^{\infty} n a_n x^n + \sum_{n=0}^{\infty} a_n x^n = 0,$$

   we have $\sum_{n=0}^{\infty} (n+2)(n+1) a_{n+2} x^n + \sum_{n=1}^{\infty} n a_n x^n + \sum_{n=0}^{\infty} a_n x^n = 0$.

② Adjusting the Lower Index:
   Add the intermediate term $\sum_{n=0}^{\infty} (n+2)(n+1) a_{n+2} x^n + \sum_{n=0}^{\infty} n a_n x^n + \sum_{n=0}^{\infty} a_n x^n = 0$.

③ Combining: (Combining the terms with the same exponent and subscript)
   We obtain $\sum_{n=0}^{\infty} [(n+2)(n+1) a_{n+2} + (n+1) a_n] x^n = 0$,

   therefore $(n+2)(n+1) a_{n+2} + (n+1) a_n = 0$, organizing the terms, we obtain the recurrence relation $a_{n+2} = -\dfrac{1}{n+2} a_n$, where $n = 0, 1, 2, \cdots$.

(3) Now, from the recurrence relation, we can express all the coefficients in terms of a_0:

$n = 0 \Rightarrow a_2 = -\dfrac{1}{2} a_0$, $n = 1 \Rightarrow a_3 = -\dfrac{1}{3} a_1$,

$n = 2 \Rightarrow a_4 = -\dfrac{1}{4} a_2 = -\dfrac{1}{4} \cdot (-\dfrac{1}{2} a_0) = (-1)^2 \dfrac{1}{2 \cdot 4} a_0 = \dfrac{1}{8} a_0$,

$n = 3 \Rightarrow a_5 = -\dfrac{1}{5} a_3 = -\dfrac{1}{5} \cdot (-\dfrac{1}{3} a_1) = (-1)^2 \dfrac{1}{3 \cdot 5} a_1 = \dfrac{1}{15} a_1$,

similarly, $a_6 = -\dfrac{1}{48} a_0$, $a_7 = -\dfrac{1}{105} a_1$, $\cdots$.

(4) Then $y(x) = a_0 + a_1 x + a_2 x^2 + a_3 x^3 + a_4 x^4 + a_5 x^5 + \cdots$

$$= a_0 + a_1 x - \frac{1}{2} a_0 x^2 - \frac{1}{3} a_1 x^3 + \frac{1}{8} a_0 x^4 + \frac{1}{15} a_1 x^5 + \cdots$$

$$= a_0 (1 - \frac{1}{2} x^2 + \cdots) + a_1 (x - \frac{1}{3} x^3 + \cdots).$$

(5) Now, by substituting the initial condition $y(0) = 2$ into the series representation, we obtain $a_0 = 2$ and through $y'(0) = 0$, we obtain $a_1 = 0$, so

$$y(x) = 2(1 - \frac{1}{2} x^2 + \cdots).$$

`Q.E.D.`

Example 4

Solve $y'' - xy' + y = 0$ by using power series solution.

Solution

(1) $x = 0$ is a regular point of original ODE, let $y(x) = \sum_{n=0}^{\infty} a_n x^n$, then

$$y'(x) = \sum_{n=1}^{\infty} n a_n x^{n-1}, \quad y''(x) = \sum_{n=2}^{\infty} n \cdot (n-1) a_n x^{n-2}.$$

(2) Substitute above equations into original ODE

$$\Rightarrow \sum_{n=2}^{\infty} n \cdot (n-1) a_n x^{n-2} - x \sum_{n=1}^{\infty} n \cdot a_n x^{n-1} + \sum_{n=0}^{\infty} a_n x^n = 0$$

$$\Rightarrow \sum_{n=2}^{\infty} n \cdot (n-1) a_n x^{n-2} - \sum_{n=1}^{\infty} n \cdot a_n x^n + \sum_{n=0}^{\infty} a_n x^n = 0$$

$$\Rightarrow \sum_{n=0}^{\infty} (n+2)(n+1) a_{n+2} x^n - \sum_{n=0}^{\infty} n a_n x^n + \sum_{n=0}^{\infty} a_n x^n = 0$$

$$\Rightarrow \sum_{n=0}^{\infty} [(n+2)(n+1) a_{n+2} - (n-1) a_n] x^n = 0$$

$$\Rightarrow (n+2)(n+1) a_{n+2} - (n-1) a_n = 0,$$

therefore $a_{n+2} = \dfrac{(n-1)}{(n+2)(n+1)} a_n$; $n = 0, 1, 2, \cdots$.

(3) Therefore, the odd-indexed terms can be expressed as a_0, while the even-indexed terms can be expressed as a_1. Here are a few examples:

$$n = 0 \Rightarrow a_2 = -\frac{1}{2} a_0; \quad n = 1 \Rightarrow a_3 = \frac{0}{3 \cdot 2} a_1 = 0;$$

$$n = 2 \Rightarrow a_4 = \frac{1}{4 \cdot 3} a_2 = \frac{1}{12} a_2 = \frac{1}{12} \cdot (-\frac{1}{2} a_0) = -\frac{1}{24} a_0; \quad n = 3 \Rightarrow a_5 = \frac{2}{5 \cdot 4} a_3 = 0;$$

$$n = 4 \Rightarrow a_6 = \frac{3}{6 \cdot 5} a_4 = \frac{3}{6 \cdot 5} \cdot (-\frac{1}{24} a_0) = -\frac{1}{240} a_0;$$

$$\vdots$$

(4) Therefore, we obtain the analytic solution (partial general solution) as follows:

$$y(x) = \sum_{n=0}^{\infty} a_n x^n = a_0 + a_1 x + a_2 x^2 + a_3 x^3 + a_4 x^4 + a_5 x^5 + \cdots$$

$$= a_0 + a_1 x - \frac{1}{2} a_0 x^2 + 0 \cdot x^3 - \frac{1}{24} a_0 x^4 + 0 \cdot x^5 - \frac{1}{240} a_0 x^6 + \cdots$$

$$= a_1 x + a_0 \cdot [1 - \frac{1}{2} x^2 - \frac{1}{24} x^4 - \frac{1}{240} x^6 \cdots] .$$ Q.E.D.

Example 5

Please utilize the power series method for solving by expanding at a regular point $(1 - x^2) y' = 2xy$.

Solution

(1) $x = 0$ is a regular point of ODE, let $y(x) = \sum_{n=0}^{\infty} a_n x^n$, $y'(x) = \sum_{n=1}^{\infty} n a_n x^{n-1}$, substitute into original ODE.

(2) $(1 - x^2) \sum_{n=1}^{\infty} n \cdot a_n x^{n-1} = 2x \sum_{n=0}^{\infty} a_n x^n \Rightarrow \sum_{n=1}^{\infty} n a_n x^{n-1} - \sum_{n=0}^{\infty} n \cdot a_n x^{n+1} = 2 \sum_{n=0}^{\infty} a_n x^{n+1}$

$$\Rightarrow \sum_{n=0}^{\infty} (n+1) a_{n+1} x^n - \sum_{n=1}^{\infty} (n-1) a_{n-1} x^n = 2 \sum_{n=1}^{\infty} a_{n-1} x^n$$

$$\Rightarrow \sum_{n=0}^{\infty} (n+1) a_{n+1} x^n \quad \sum_{n=1}^{\infty} (n+1) a_{n-1} x^n = 0 \Rightarrow a_1 + \sum_{n=1}^{\infty} (n+1) a_{n+1} x^n - \sum_{n=1}^{\infty} (n+1) a_{n-1} x^n = 0$$

$$\Rightarrow a_1 + \sum_{n=1}^{\infty} [(n+1) a_{n+1} - (n+1) a_{n-1}] x^n = 0 \text{ , we obtain the relationship between the}$$

coefficients a_n $\begin{cases} a_1 = 0 \\ a_{n+1} = a_{n-1} \end{cases}$, where $n = 1, 2, \cdots$, therefore, the even-indexed terms can be expressed as a_0, the odd-indexed terms can be expressed as a_1, that is

$$a_{2n} = a_0, \ a_{2n+1} = 0, \ n = 1, 2, \cdots .$$

(4) Substituting all the coefficients into $y(x) = \sum_{n=0}^{\infty} a_n x^n$, we obtain the partial general

solution $y(x) = \sum_{n=0}^{\infty} a_{2n} x^{2n} + \sum_{n=0}^{\infty} a_{2n+1} x^{2n+1} = \sum_{n=0}^{\infty} a_0 x^{2n} = a_0 \sum_{n=0}^{\infty} (x^2)^n = \frac{a_0}{1 - x^2}$.   Q.E.D.

4-1 Exercises

Basic questions

Please solve following ODE by using power series expansion at a regular point.

1. $y'' + 12y' + x^3y = 0$
   (at least to the first five nonzero terms of the solution).

2. $y'' + xy = 0$ (at least to the first five nonzero terms of the solution).

3. $y'' - xy = 1$, if its solution can be written as
   $y = c_0(1 + Px^3 + \cdots) + c_1(Qx + Rx^4 + \cdots)$
   $+ Sx^2 + Tx^5 + \cdots$,
   find P, Q, R, S, T.

4. $y'' + e^x y' + y = 0$.

5. $y'' - 2y' + x^3y = 0$ (at least to the first five non-zero solutions).

Advanced questions

Please solve following ODE by using power series expansion at a regular point.

1. $y''(x) - 2xy'(x) + 2y'(x) + 8y(x) = 0$,
   $y(1) = 3$, $y'(1) = 0$.

2. $y'' + (1 + x^2)y = 0$.

3. $y'' + y' + 2xy = 0$ at
   (1) $y(0) = 0$, $y'(0) = 1$.
   (2) $y(0) = 1$, $y'(0) = 0$.

4. $y'' + xy = 0$.

5. $y'' - xy' - x^2y = 0$.

4-2   Regular Singular Point Expansion for Solving ODE (Selected Reading)

Many differential equations arising from physical systems are non-analytic at certain crucial points or positions, making it impossible to solve them using Taylor series expansion. In this section, we introduce the concept of Frobenius series solution to address this class of variable coefficient ODEs and to compensate for the limitations of power series expansion at regular points.

4-2-1   Introduction

Considering a second-order linear ODE $y'' + P(x)y' + Q(x)y = 0 \cdots ①$, Frobenius (1849~1917) discovered that if $x = 0$ is a regular singular point of the ODE, then there exists a solution of the form $y(x) = \sum_{n=0}^{\infty} a_n x^{n+r} = x^r \cdot [a_0 + a_1 x + a_2 x^2 + \cdots]$. Now, let's use this solution to derive the general solution. Based on the definition of a regular point, $xP(x)$, $x^2 Q(x)$ are analytic, so their Taylor series expansions exist, that is,

$$xP(x) = b_0 + b_1 x + b_2 x^2 + \cdots; \quad x^2 Q(x) = c_0 + c_1 x + c_2 x^2 + \cdots.$$

Now by applying the Frobenius series solution, we have
$$y'(x) = a_0 r x^{r-1} + a_1 (r+1)x^r + \cdots;$$
$$y''(x) = a_0 r(r-1)x^{r-2} + a_1 r(r+1)x^{r-1} + \cdots$$
substituting above information into the original ODE ① (in the form of $x^2 y'' + x \cdot xP(x)y' + x^2 Q(x)y = 0$), we have

$$x^2 [a_0 r(r-1)x^{r-2} + a_1 r(r+1)x^{r-1} + \cdots]$$
$$+ x(b_0 + b_1 x + b_2 x^2 + \cdots)[a_0 r x^{r-1} + a_1 (r+1)x^r + \cdots]$$
$$+ [c_0 + c_1 x + c_2 x^2 + \cdots] \times x^r \times [a_0 + a_1 x + a_2 x^2 + \cdots] = 0,$$

then rearranged as $a_0 [r(r-1) + b_0 r + c_0]x^r + [\quad]x^{r+1} + \cdots = 0$, that is $a_0 [r(r-1) + b_0 r + c_0] = 0$, because $a_0 \neq 0$, therefore

$$r(r-1) + b_0 r + c_0 = 0.$$

The above equation is called the indicial equation of the Frobenius series solution. The roots of this equation are known as the indicial roots or indices, where $b_0 = \lim_{x \to 0} xP(x)$, $c_0 = \lim_{x \to 0} x^2 Q(x)$.

Example 1

Solve ODE, and find indicial roots of $4xy'' + 2y' + y = 0$.

Solution

Rearranging the original equation, we have $y'' + \dfrac{1}{2x}y' + \dfrac{1}{4x}y = 0$, since $x = 0$ is a

regular singular point of the ODE, so $\quad b_0 = \lim\limits_{x \to 0} x \cdot \dfrac{1}{2x} = \dfrac{1}{2}; \quad c_0 = \lim\limits_{x \to 0} x^2 \cdot \dfrac{1}{4x} = 0$,

we obtain the indicial equation : $r \cdot (r-1) + \dfrac{1}{2}r = 0$, we find the indicial roots $r = 0, \ \dfrac{1}{2}$.

Q.E.D.

b_0, c_0 can be easily determined, and to complete the Frobenius series solution, we need to find the roots r by solving the indicial equation and then substitute them back into the original equation. The remaining portion of this chapter will explore the influence of the roots of the indicial equation on the solution given by equation ①. For example, under what conditions do the indicial roots directly provide us with a set of linearly independent solutions? And under what circumstances do the indicial roots only give us one solution, and how can we determine the remaining solution using known theoretical methods? These are crucial questions that will be addressed in the upcoming sections.

4-2-2　The Relationship Between Indicial Roots and Series Solution

Assuming $x = a$ is a regular singular point of $y'' + P(x)y' + Q(x)y = 0$, in the discussion of the 4-2-1 Introduction, let x be a new variable $x - a$, then by using the same reasoning, we can derive the same indicial equation. From this point onwards, all indicial roots will be denoted as r_1, r_2.

▶ Theorem 4-2-1

Solutions Where the Two Indicial Roots Differ by a Non-Integer Value.

If $x = 0$ is a regular singular point and where the two indicial roots r_1 and r_2 have a non-integer difference $(r_1 - r_2)$, by substituting r_1 and r_2 we find the solution $\{y_1(x) = y(x)\big|_{r=r_1}, \ y_2(x) = y(x)\big|_{r=r_2}\}$, which gives us two linearly independent solutions. That is, the general solution of the original ODE is

$$y(x) = c_1 y_1(x) + c_2 y_2(x).$$

◀

Example 2

Solve $4xy'' + 2y' + y = 0$ by using the Frobenius series solution.

Solution   From Example 1, we found that the indicial roots are $r = 0$, $\dfrac{1}{2}$. Therefore,

we obtain the series solution in the form: $y(x) = \displaystyle\sum_{n=0}^{\infty} a_n x^n$ or $y(x) = \displaystyle\sum_{n=0}^{\infty} a_n x^{n+\frac{1}{2}}$.

Substituting these forms into the original equation and following the organization as demonstrated in Example 3 of Section 4-1,

we have
$$\begin{cases} a_n = \dfrac{-1}{(2n)(2n-1)} a_{n-1}, & r = 0 \\[3mm] a_n = \dfrac{-1}{(2n)(2n+1)} a_{n-1}, & r = \dfrac{1}{2} \end{cases}$$

where $n = 1, 2, \cdots$, we observe that all the terms a_n can be expressed in terms of a_0. Therefore, we have two cases to consider:

① $r = 0$, $a_1 = -\dfrac{1}{2} a_0$, $a_2 = \dfrac{1}{24} a_0$, $a_3 = -\dfrac{1}{720} a_0$, $\cdots\cdots$,

$$\therefore y_1(x) = \sum_{n=0}^{\infty} a_n x^{n+r} \Big|_{r=0} = a_0 + a_1 x + a_2 x^2 + a_3 x^3 + \cdots$$

$$= a_0 - \frac{1}{2} x a_0 + \frac{1}{24} x^2 a_0 - \frac{1}{720} x^3 a_0 + \cdots$$

$$= a_0 \cdot (1 - \frac{1}{2} x + \frac{1}{24} x^2 - \frac{1}{720} x^3 + \cdots).$$

② $r = \dfrac{1}{2}$, $a_1 = -\dfrac{1}{6} a_0$, $a_2 = \dfrac{1}{120} a_0$, $a_3 = -\dfrac{1}{5040} a_0$, $\cdots\cdots$,

$$\therefore y_2(x) = \sum_{n=0}^{\infty} a_n x^{n+r} \Big|_{r=\frac{1}{2}} = \sum_{n=0}^{\infty} a_n x^{n+\frac{1}{2}} = x^{\frac{1}{2}} \sum_{n=0}^{\infty} a_n x^n$$

$$= x^{\frac{1}{2}} \cdot (a_0 + a_1 x + a_2 x^2 + a_3 x^3 + \cdots)$$

$$= x^{\frac{1}{2}} \cdot a_0 (1 - \frac{1}{6} x + \frac{1}{120} x^2 - \frac{1}{5040} x^3 + \cdots).$$

So the general solution is
$$y(x) = c_1 y_1(x) + c_2 y_2(x)$$

$$= c_1 \cdot (1 - \frac{1}{2} x + \frac{1}{24} x^2 - \frac{1}{720} x^3 + \cdots) + c_2 x^{\frac{1}{2}} (1 - \frac{1}{6} x + \frac{1}{120} x^2 - \frac{1}{5040} x^3 + \cdots).$$

Q.E.D.

Actually, we can directly substitute $y(x) = \sum_{n=0}^{\infty} a_n x^{n+r}$ into the ODE and use the approach described in the introduction to derive the indicial equation and the substitution relationship for the coefficients a_n. Here is an example that demonstrates this approach:

Example 3

Solve $4xy'' + 2y' + y = 0$ by using the Frobenius series solution.

Solution

(1) $x = 0$ is a regular singular point of ODE, let $y(x) = \sum_{n=0}^{\infty} a_n x^{n+r}$,

then
$$\begin{cases} y'(x) = \sum_{n=0}^{\infty} a_n (n+r) x^{n+r-1} \\ y''(x) = \sum_{n=0}^{\infty} a_n (n+r)(n+r-1) x^{n+r-2} \end{cases},$$

substituting into original ODE.

(2) $4x \cdot \sum_{n=0}^{\infty} a_n (n+r)(n+r-1) x^{n+r-2} + 2 \cdot \sum_{n=0}^{\infty} a_n (n+r) x^{n+r-1} + \sum_{n=0}^{\infty} a_n x^{n+r} = 0$

① Truncating the leading term:

$$\Rightarrow \sum_{n=0}^{\infty} 4a_n (n+r)(n+r-1) x^{n+r-1} + \sum_{n=0}^{\infty} 2a_n (n+r) x^{n+r-1} + \sum_{n=0}^{\infty} a_n x^{n+r} = 0$$

$$\Rightarrow \sum_{n=0}^{\infty} [4(n+r)(n+r-1) + 2(n+r)] a_n x^{n+r-1} + \sum_{n=0}^{\infty} a_n x^{n+r} = 0$$

$$\Rightarrow \sum_{n=0}^{\infty} [2(n+r) \cdot (2(n+r-1) + 1)] a_n x^{n+r-1} + \sum_{n=0}^{\infty} a_n x^{n+r} = 0$$

② Adjusting the exponents: prioritize complex terms

$$\Rightarrow \sum_{n=0}^{\infty} [(2n+2r)(2n+2r-1) a_n] x^{n+r-1} + \sum_{n=0}^{\infty} a_n x^{n+r} = 0$$

$$\Rightarrow \sum_{n=0}^{\infty} [(2n+2r)(2n+2r-1) a_n] x^{n+r-1} + \sum_{n=1}^{\infty} a_{n-1} x^{n+r-1} = 0$$

③ Adjusting the subscripts to be consistent: releasing any surplus terms

$$\Rightarrow 2r \cdot (2r-1) a_0 x^{r-1} + \sum_{n=1}^{\infty} [(2n+2r)(2n+2r-1) a_n] x^{n+r-1} + \sum_{n=1}^{\infty} a_{n-1} x^{n+r-1} = 0$$

$$\Rightarrow 2r \cdot (2r-1) a_0 x^{r-1} + \sum_{n=1}^{\infty} [(2n+2r)(2n+2r-1) a_n + a_{n-1}] x^{n+r-1} = 0$$

$$\Rightarrow \begin{cases} 2r \cdot (2r-1)a_0 = 0 \text{ , } a_0 \neq 0 \Rightarrow 2r \cdot (2r-1) = 0 \quad \therefore r = 0, \frac{1}{2} \\ (2n+2r)(2n+2r-1)a_n + a_{n-1} = 0 \cdots \text{①} \end{cases}$$

From formula ① $\Rightarrow a_n = \dfrac{-1}{(2n+2r)(2n+2r-1)} a_{n-1}; \quad n = 1, 2, \cdots$

$r = 0, \dfrac{1}{2}$ (The two indicial roots differ by a non-integer value).

(3) $n = 1 \Rightarrow a_1 = \dfrac{-1}{(2+2r)(2+2r-1)} a_0 = \dfrac{-1}{(2r+2)(2r+1)} a_0$

$n = 2 \Rightarrow a_2 = \dfrac{-1}{(4+2r)(4+2r-1)} a_1 = \dfrac{-1}{(2r+4)(2r+3)} a_1$

$\qquad = \dfrac{-1}{(2r+4)(2r+3)} \cdot \dfrac{-1}{(2r+2)(2r+1)} a_0$

$\qquad = \dfrac{(-1)^2}{(2r+4)(2r+3)(2r+2)(2r+1)} a_0$

$n = 3 \Rightarrow a_3 = \dfrac{-1}{(6+2r)(2r+5)} a_2 = \dfrac{(-1)^3}{(2r+1)(2r+2)(2r+3)+\cdots+(2r+6)} a_0$

$\vdots$

① $r = 0 \Rightarrow a_1 = -\dfrac{1}{2}a_0, \quad a_2 = \dfrac{1}{24}a_0, \quad a_3 = -\dfrac{1}{720}a_0, \quad \cdots$

$\therefore y_1(x) = \sum_{n=0}^{\infty} a_n x^{n+r}\Big|_{r=0} = a_0 + a_1 x + a_2 x^2 + a_3 x^3 + \cdots$

$\qquad = a_0 - \dfrac{1}{2}xa_0 + \dfrac{1}{24}x^2 a_0 - \dfrac{1}{720}x^3 a_0 + \cdots$

$\qquad = a_0 \cdot (1 - \dfrac{1}{2}x + \dfrac{1}{24}x^2 - \dfrac{1}{720}x^3 + \cdots)$

② $r = \dfrac{1}{2} \Rightarrow a_1 = -\dfrac{1}{6}a_0, \quad a_2 = \dfrac{1}{120}a_0, \quad a_3 = -\dfrac{1}{5040}a_0, \quad \cdots$

$\therefore y_2(x) = \sum_{n=0}^{\infty} a_n x^{n+r}\Big|_{r=\frac{1}{2}} = \sum_{n=0}^{\infty} a_n x^{n+\frac{1}{2}} = x^{\frac{1}{2}} \sum_{n=0}^{\infty} a_n x^n$

$\qquad = x^{\frac{1}{2}} \cdot (a_0 + a_1 x + a_2 x^2 + a_3 x^3 + \cdots)$

$\qquad = x^{\frac{1}{2}} \cdot a_0 (1 - \dfrac{1}{6}x + \dfrac{1}{120}x^2 - \dfrac{1}{5040}x^3 + \cdots)$

(4) So the general solution is $y(x) = c_1 y_1(x) + c_2 y_2(x)$,

that is $c_1 \cdot (1 - \dfrac{1}{2}x + \dfrac{1}{24}x^2 - \dfrac{1}{720}x^3 + \cdots) + c_2 x^{\frac{1}{2}}(1 - \dfrac{1}{6}x + \dfrac{1}{120}x^2 - \dfrac{1}{5040}x^3 + \cdots)$. **Q.E.D.**

▶ **Theorem 4-2-2**

The Solution When the Indicial Roots Are Repeated

If $x = 0$ is a regular singular point and the indicial roots r_1 and r_2 are repeated

$(r_1 = r_2 = r_0)$, then substituting r_1 and r_2, the solutions $y_1(x) = y(x)\big|_{r=r_1} = x^{r_0} \sum_{n=0}^{\infty} a_n(r_0)x^n$,

$y_2 = (\ln x) \cdot y_1(x) + \sum_{n=1}^{\infty} \frac{\partial a_n}{\partial r}\bigg|_{r=r_0} x^{n+r_0}$ are two linearly independent solutions. In this case,

the general solution of the ODE is $y(x) = c_1 y_1(x) + c_2 y_2(x)$.

Proof

Because r_1 and r_2 are repeated roots, through one of solutions y_1, we obtain another solution:

$$y_2(x) = \frac{\partial y}{\partial r}\bigg|_{r=r_0} = \sum_{n=0}^{\infty} \frac{\partial a_n}{\partial r}\bigg|_{r=r_0} x^{n+r_0} + \sum_{n=0}^{\infty} a_n x^{n+r_0} \cdot \frac{\partial}{\partial r}[(n+r)\ln x]_{r=r_0}$$

$$= \sum_{n=0}^{\infty} \frac{\partial a_n}{\partial r}\bigg|_{r=r_0} x^{n+r_0} + \ln(x)y_1(x)^{\,1}.$$ ◀

▶ **Theorem 4-2-3**

Generalization of the Solution for Repeated Indicial Roots

According to Theorem 4-2-2, if regular singular point is $x = a$, two linearly independent solutions are

$$y_1(x) = \sum_{n=0}^{\infty} a_n(r_0)(x-a)^{n+r_0} ; \quad y_2(x) = \ln(x-a) \cdot y_1(x) + \sum_{n=1}^{\infty} \frac{\partial a_n}{\partial r}\bigg|_{r=r_0} (x-a)^{n+r_0}{}^{\,2}.$$ ◀

---

[1] In fact, y_2 can also be obtained using the method of reducing order mentioned in Chapter 1, we have $y_2(x) = y_1 \cdot \int \dfrac{e^{-\int P(x)dx}}{y_1^2} dx$.

[2] When dealing with repeated indicial roots, the calculation of y_2 often involves the evaluation of logarithmic terms $\dfrac{\partial a_n}{\partial r}$. In such cases, the

following formula is frequently used: $\left(\dfrac{f_1^{a_1} \cdot f_2^{a_2}}{g_1^{b_1} \cdot g_2^{b_2}}\right)' = \left(\dfrac{f_1^{a_1} \cdot f_2^{a_2}}{g_1^{b_1} \cdot g_2^{b_2}}\right) \cdot [a_1 \dfrac{f_1'}{f_1} + a_2 \dfrac{f_2'}{f_2} - b_1 \dfrac{g_1'}{g_1} - b_2 \dfrac{g_2'}{g_2}]$.

Example 4

Use the method of expanding at a regular singular point to solve $4x^2 y'' + (4x+1)y = 0$.

Solution

Because $b_0 = \lim\limits_{x \to 0} x \times 0 = 0$, $c_0 = \lim\limits_{x \to 0} x^2 \times \dfrac{4x+1}{4x^2} = \dfrac{1}{4}$, the indicial roots equation is

$(r - \dfrac{1}{2})^2 = 0$, the indicial roots are found to be a repeated root $\dfrac{1}{2}$, therefore, we still need to determine the pattern of the coefficients a_n. Now by substituting the form

$y(x) = \sum\limits_{n=0}^{\infty} a_n x^{n+\frac{1}{2}}$ of the solution into the original equation and simplifying, we obtain

$a_n = \dfrac{-1}{n^2} a_{n-1}$, $n = 1, 2, 3, \cdots$, so we have the first solution

$y_1(x) = a_0 x^{\frac{1}{2}} \cdot (1 - x + \dfrac{1}{4}x^2 - \dfrac{1}{36}x^3 + \cdots)$, now according to Theorem 4-2-2,

we obtain $\quad y_2(x) = \ln x \cdot y_1(x) + \sum\limits_{n=1}^{\infty} \dfrac{\partial a_n}{\partial r} \cdot x^{n+r}\Big|_{r=\frac{1}{2}} = \ln x \cdot y_1(x) + \sum\limits_{n=1}^{\infty} \dfrac{\partial a_n}{\partial r}\Big|_{r=\frac{1}{2}} \cdot x^{n+\frac{1}{2}}$

$= \ln x \cdot y_1(x) + x^{\frac{1}{2}} \cdot \sum\limits_{n=1}^{\infty} \dfrac{\partial a_n}{\partial r}\Big|_{r=\frac{1}{2}} x^n$

where $\dfrac{\partial a_0}{\partial r} = 0$,

therefore $\quad \dfrac{\partial a_1}{\partial r}\Big|_{r=\frac{1}{2}} = -\dfrac{4a_0}{(2r+1)^2} \cdot (-2 \cdot \dfrac{2}{(2r+1)})\Big|_{r=\frac{1}{2}} = \dfrac{16a_0}{(2r+1)^3}\Big|_{r=\frac{1}{2}} = 2a_0$

$\dfrac{\partial a_2}{\partial r}\Big|_{r=\frac{1}{2}} = \dfrac{16a_0}{(2r+1)^2(2r+3)^2} \cdot [-2 \cdot \dfrac{2}{2r+1} - 2 \cdot \dfrac{2}{2r+3}]_{r=\frac{1}{2}} = -\dfrac{3}{4}a_0$

we obtain the second solution

$y_2(x) = \ln x \cdot y_1(x) + x^{\frac{1}{2}} \cdot (\dfrac{\partial a_1}{\partial r}\Big|_{r=\frac{1}{2}} x + \dfrac{\partial a_2}{\partial r}\Big|_{r=\frac{1}{2}} x^2 + \cdots)$

$= \ln x \cdot y_1(x) + x^{\frac{1}{2}} \cdot (2a_0 x - \dfrac{3}{4}a_0 x^2 + \cdots)$

$= \ln x \cdot y_1(x) + a_0 x^{\frac{1}{2}} \cdot (x - \dfrac{3}{4}x^2 + \cdots)$.

The general solution is $y(x) = c_1 y_1(x) + c_2 y_2(x)$. **Q.E.D.**

If the difference between the indicial roots is an integer, that is $r_2 - r_1 \in \mathbf{Z}$, then the

solution $\left. y_1(x) = y(x) \right|_{r=r_2} = (x-a)^{r_2} \sum_{n=0}^{\infty} a_n(t_2)(x-a)^n$ corresponding to the larger

indicial root r_2 must exist. However, the solution corresponding to the smaller indicial root r_1 may not necessarily exist. In such cases, further subdivisions are required to determine the behavior of the solution.

(1) Neither r_2 nor r_1 is an integer.

(2) Both r_2 and r_1 are integers, these two conditions.

▶ **Theorem 4-2-4**

The Solution of the Smaller Indicial Root Exists

Assuming the two indicial roots r_1 and r_2 differ by an integer $(r_2 > r_1)$,
if $\left. y_2(x) = y(x) \right|_{r=r_1}$ exists, then now

$\left. y_1(x) = y(x) \right|_{r=r_2}$, $\left. y_2(x) = y(x) \right|_{r=r_1}$ are linearly independent, and the general solution of
ODE is $y(x) = c_1 y_1(x) + c_2 y_2(x)$.

Example 5

Use the method of expanding at a regular singular point to solve

$$x^2 y'' + xy' + (x^2 - \frac{1}{4})y = 0 .$$

Solution

In this example, we will practice deriving the answer using the approach outlined at the beginning of this section. During the process, the indicial equation and the relationship

between the coefficients a_n will exist together. Let $y(x) = \sum_{n=0}^{\infty} a_n x^{n+r}$ substituting into

original ODE, then by the organization techniques shown in Example 3 of Section 4-1,

we obtain $\quad (r+\frac{1}{2})(r-\frac{1}{2})a_0 x^r + (1+r+\frac{1}{2})(1+r-\frac{1}{2})a_1 x^{r+1}$

$$+ \sum_{n=2}^{\infty} [(n+r+\frac{1}{2})(n+r-\frac{1}{2})a_n + a_{n-2}]x^{n+r} = 0$$

therefore $\begin{cases} (r+\frac{1}{2})(r-\frac{1}{2}) = 0 \\ a_1 = 0 \\ a_n = \dfrac{-a_{n-2}}{(n+r+\frac{1}{2})(n+r-\frac{1}{2})}, n = 2, 3, \cdots \end{cases}$, the first equation inside the curly

brackets represents the indicial equation, and by solving it, we find the indicial roots to

be $r_2 = \frac{1}{2}$, $r_1 = \frac{-1}{2}$, differ by an integer, and neither of them is an integer, Therefore, according to Theorem 4-2-4, when $r = -\dfrac{1}{2}$, $a_n = -\dfrac{a_{n-2}}{n(n-1)}$, $n = 2, 3, \cdots$, the solutions corresponding to are

$$y_2(x) = y(x)\Big|_{r=-\frac{1}{2}} = \sum_{n=0}^{\infty} a_n x^{n+r}\Big|_{r=-\frac{1}{2}} = x^{-\frac{1}{2}}(a_0 + a_1 x + a_2 x^2 + a_3 x^3 + \cdots)$$

$$= a_0 x^{-\frac{1}{2}} \cdot (1 - \frac{1}{2!}x^2 + \frac{1}{4!}x^4 - \frac{1}{6!}x^6 + \cdots).$$

When $r = \dfrac{1}{2}$, $a_n = \dfrac{-a_{n-2}}{n(n+1)}$, $n = 2, 3, \cdots$, the solutions corresponding to are

$$y_1(x) = y(x)\Big|_{r=r_2=\frac{1}{2}} = \sum_{n=0}^{\infty} a_n x^{n+r}\Big|_{r=\frac{1}{2}}$$

$$= x^{\frac{1}{2}} \cdot (a_0 - \frac{1}{3!}a_0 x^2 + \frac{1}{5!}a_0 x^4 - \frac{1}{7!}a_0 x^6 + \cdots)$$

$$= a_0 x^{\frac{1}{2}}(1 - \frac{1}{3!}x^2 + \frac{1}{5!}x^4 - \frac{1}{7!}x^6 + \cdots).$$

So the general solution is $y(x) = c_1 y_1(x) + c_2 y_2(x)$,

that is $c_1 x^{\frac{1}{2}}(1 - \frac{1}{3!}x^2 + \frac{1}{5!}x^4 - \frac{1}{7!}x^6 + \cdots) + c_2 x^{-\frac{1}{2}}(1 - \frac{1}{2!}x^2 + \frac{1}{4!}x^4 - \frac{1}{6!}x^6 + \cdots).$

Q.E.D.

When both indicial roots are integers, there is a possibility that the coefficient $a_n(r_1)$ corresponding to the smaller indicial root r_1 may not exist. If $a_n(r_1)$ does not exist, we cannot determine $y_2(x)$. In such cases, we need to modify the term a_n, in general, we let $a_0^* = c(r - r_1)$ as new a_n to modify a_n as a_0^*, we obtain $a_n^*(r)$ after substituting, that is

$$y^* = \sum_{n=0}^{\infty} a_n^*(x-a)^{n+r} \quad \text{where } y_1, y_2 \text{ are linearly independent solutions given by}$$

$$y_1 = y^*\Big|_{r=r_1} = \sum_{n=0}^{\infty} a_n^*(r_1)(x-a)^{n+r_1},$$

$$y_2 = \frac{\partial y^*}{\partial r}\Big|_{r=r_1} = \ln(x-a) \cdot y_1 + \sum_{n=0}^{\infty} \frac{\partial a_n^*}{\partial r}\Big|_{r=r_1} (x-a)^{n+r_1}.$$

The following is illustrated with a case.

Example 6

Please use Frobenius series to solve $xy'' + xy' - y = 0$.

Solution

(1) By $b_0 = \lim\limits_{x\to 0} xP(x) = \lim\limits_{x\to 0} x \times 1 = 0$, $c_0 = \lim\limits_{x\to 0} x^2 Q(x) = \lim\limits_{x\to 0} x^2 \times (-\dfrac{1}{x}) = 0$, we obtain

the indicial equation $r(r - 1) = 0$, thus the characteristic roots are $r_1 = 1$, $r_2 = 0$, this corresponds to the situation described in Theorem 4-2-4,

we obtain $\begin{cases} a_0 \neq 0, \quad r \cdot (r-1) = 0 \Rightarrow r = 0,1, \text{ the difference is an integer} \\ a_n(n+r)(n+r-1) + a_{n-1}(n+r-2) = 0 \end{cases}$,

$\Rightarrow a_n = \dfrac{-(n+r-2)a_{n-1}}{(n+r)(n+r-1)}, n = 1, 2, 3, \cdots$.

(2) $a_n = -\dfrac{(n+r-2)a_{n-1}}{(n+r)(n+r-1)}, n = 1, 2, \cdots$,

$n = 1 \Rightarrow a_1 = -\dfrac{(r-1)}{r \cdot (r+1)} a_0$ ($a_1(r = 0)$ does not exist),

$n = 2 \Rightarrow a_2 = -\dfrac{r}{(r+2)(r+1)} a_1 = (-1)^2 \cdot \dfrac{(r-1)}{(r+1)^2(r+2)} a_0$,

$n = 3 \Rightarrow a_3 = -\dfrac{(r+1)}{(r+3)(r+2)} a_2 = (-1)^3 \cdot \dfrac{(r-1)}{(r+1)(r+2)^2(r+3)} a_0$,

$\vdots$

let $a_0^* = a_0 = k(r-0) = kr$ substitute into $\Rightarrow a_0^* = kr$,

$\therefore a_1^* = -kr\dfrac{(r-1)}{r \cdot (r+1)} = -k\dfrac{r-1}{r+1}$, $a_2^* = k \cdot \dfrac{(r-1)}{(r+2)(r+1)^2}$,

$a_3^* = -k \cdot \dfrac{r(r-1)}{(r+1)(r+2)^2(r+3)}$, $\cdots$, $\therefore y^*(x) = \sum\limits_{n=0}^{\infty} a_n^*(r)x^{n+r}$,

we obtain $y_1(x) = y^*\Big|_{r=0} = \sum\limits_{n=0}^{\infty} a_n^*\Big|_{r=0} x^n = a_0^*\Big|_{r=0} + a_1^*\Big|_{r=0} x + a_2^*\Big|_{r=0} x^2 + \cdots = kx$.

(3) $y_2(x) = \dfrac{\partial y^*}{\partial r}\bigg|_{r=0} = \ln x \cdot y_1 + \sum\limits_{n=0}^{\infty} \dfrac{\partial a_n^*}{\partial r}\bigg|_{r=0} x^{n+0}$

$= \ln x \cdot y_1(x) + [\dfrac{\partial a_0^*}{\partial r}(r=0) + \dfrac{\partial a_1^*}{\partial r}(r=0)x + \dfrac{\partial a_2^*}{\partial r}(r=0)x^2 + \cdots]$,

where $\dfrac{\partial a_0^*}{\partial r}\bigg|_{r=0} = k$;

$\dfrac{\partial a_1^*}{\partial r}\bigg|_{r=0} = -k \cdot \dfrac{r-1}{r+1} \cdot [\dfrac{-1}{r+1} + \dfrac{+1}{r-1}]\bigg|_{r=0} = -k \cdot (-1) \cdot (-1-1) = -2k$;

$$\left.\frac{\partial a_2^*}{\partial r}\right|_{r=0} = k \cdot \frac{r \cdot (r-1)}{(r+2)(r+1)^2} \cdot \left\{\frac{1}{r} + \frac{1}{r-1} - \frac{1}{r+2} - 2\frac{1}{r+1}\right\}\Big|_{r=0} = k \cdot \frac{-1}{2 \times 1} = -\frac{1}{2}k ,$$

then we obtain $y_2(x) = \ln(x) \cdot y_1(x) + [k - 2kx - \frac{1}{2}kx^2 + \cdots]$

$$= \ln(x) \cdot y_1(x) + k \cdot (1 - 2x - \frac{1}{2}x^2 + \cdots)^3 .$$

(4) $\therefore y(x) = c_1^* y_1(x) + c_2^* y_2(x) = c_1 y_1 + c_2[y_1 \ln(x) + (1 - \frac{1}{2}x^2 + \cdots)]$,

where the term $(-2x)$ can be merged into y_1.

`Q.E.D.`

---

[3] $y_1(x) = x$, then $y_2(x)$ can also be obtained using the method of reducing order: $y_2(x) = y_1 \cdot \int \frac{e^{-\int P(x)dx}}{y_1^2} dx$; $P(x) = 1$,

$$\therefore y_2 = x \cdot \int \frac{e^{-x}}{x^2} dx = x \cdot \int \frac{1}{x^2}[1 - \frac{x}{1} + \frac{x^2}{2} - \frac{x^3}{6} + \cdots]dx = x \cdot \int [\frac{1}{x^2} - \frac{1}{x} + \frac{1}{2} - \frac{1}{6}x + \cdots]dx$$

$$= x \cdot [-\frac{1}{x} - \ln|x| + \frac{1}{2}x - \frac{1}{12}x^2 + \cdots] = -1 - x\ln(x) + \frac{1}{2}x^2 - \frac{1}{12}x^3 + \cdots$$

$$= -[x\ln(x) + (1 - \frac{1}{2}x^2 + \frac{1}{12}x^3 + \cdots)] = -[y_1 \ln(x) + (1 - \frac{1}{2}x^2 + \cdots)],$$

$\therefore y(x) = d_1 y_1 + d_2 y_2$. If $d_1 = c_1$, $d_2 = -c_2$, then it is the same as the original series solution.

4-2 Exercises

Basic questions

1. Find the indicial roots of the following ODE when expanded around $x = 0$ as a regular singular point.

   (1) $x^2 y'' - 2xy' - (x^2 - 2)y = 0$.

   (2) $x(x-1)y'' + 3y' - 2y = 0$.

   (3) $2xy'' + (x+1)y' + y = 0$.

Use Frobenius series to solve following ODE.

2. $2xy'' - y' + 2y = 0$.

3. $2xy'' + (x+1)y' + y = 0$.

4. $16x^2 y'' + 3y = 0$.

Advanced questions

Use Frobenius series to solve following ODE.

1. $x^2 y'' - xy' + (x+1)y = 0$.

2. $x^2 y'' + xy' + (x^2 - \frac{1}{9})y = 0$.

3. $xy'' + 6y' + 2x^3 y = 0$.

5

Vector Operations and Vector Spaces

Descartes
$(1596\sim1650,$
France)

In engineering problems, there are two common types of physical quantities: scalars and vectors. Scalars only have magnitude and do not have a direction, such as mass, temperature, and height. On the other hand, vectors have both magnitude and direction. Common examples of vectors include velocity, acceleration, and force. With the invention of Cartesian coordinates by Descartes, the description of these physical quantities underwent a revolutionary advancement, enabling a significant leap forward in the symbolic and abstract thinking in the scientific community.

Learning Objectives

5-1
The Basic Operations of Vector

5-1-1 Proficient in basic vector operations

5-1-2 Proficient in vector dot product

5-1-3 Proficient in vector cross product

5-1-4 Proficient in vector triple product and its applications

5-2
Vector Geometry

5-2-1 Proficient in deriving equations of lines in R^3

5-2-2 Proficient in deriving equations of planes in R^3

5-2-3 Proficient in calculating distance from a point to a line in R^3

5-2-4 Proficient in calculating distance from a point to a skew line in R^3

5-2-5 Proficient in calculating distance from a point to a plane in R^3

5-2-6 Proficient in calculating angle between two planes in R^3

5-2-7 Proficient in deriving equations of the intersection line between two planes in R^3

5-3
Vector Spaces R^n

5-3-1 Understanding R^n and the conditions for determining its subspaces

5-3-2 Understanding linear dependence and linear independence, along with their determination methods

5-3-3 Understanding bases, dimensions, and related properties, including calculations

5-3-4 Proficient in Gram-Schmidt orthogonalization

ExampleVideo

This chapter primarily aims to review the fundamental concepts of vectors that you have previously learned, such as dot product and cross product. However, in the last section, we will delve into the more abstract concept of vector spaces. We will discuss the concepts of vector space basis, dimension, and how to use a basis to represent any vector in the vector space. These are all crucial concepts that are important to understand.

5-1   The Basic Operations of Vector

This section introduces coordinate vectors in three-dimensional space known as Cartesian coordinates. It was established by the French mathematician René Descartes (1596-1650) and is widely used to describe various physical and engineering problems.

5-1-1   The Basic Characteristics of Vector

► **Definition 5-1-1**

Vector

Any quantity that has both magnitude and direction is called a vector. It is typically represented by a symbol $\vec{A}$, as shown in Figure 5-1.

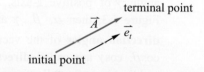

Figure 5-1   The schematic diagrams of vector   ◄

If $|\vec{A}|$ represents the magnitude of $\vec{A}$, $\vec{e_t}$ represents the direction of $\vec{A}$ and $|\vec{e_t}|=1$, then $\vec{A}=|\vec{A}|\vec{e_t}$. For example, in the real number plane, we have constant vectors like $\vec{A}=2\vec{i}+3\vec{j}$, and function vectors like $\vec{A}(t)=t\vec{i}+2t\vec{k}$.

1. **Cartesian Coordinate**

In three-dimensional real space, the x-axis, y-axis, and z axis are mutually perpendicular, and the unit vectors on three axes are represented by $\vec{i}$, $\vec{j}$, and $\vec{k}$ respectively, as shown in Figure 5-2. The position vector of point P relative to the origin, is given by $\overrightarrow{OP}=(x, y, z)-(0, 0, 0)=x\vec{i}+y\vec{j}+z\vec{k}$, is called the Cartesian coordinates of point P (position vector), that is $\overrightarrow{OP}=P(x, y, z)$. If we have another point $Q(a, b, c)$, then the vector formed by P and Q is given by $\overrightarrow{PQ}=(a-x, b-y, c-z)$, as shown in Figure 5-2.

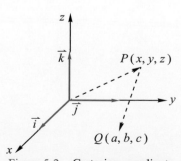

Figure 5-2   Cartesian coordinate

▶ **Definition 5-1-2**

The Common Definition of Vectors

(1) If $\vec{A} = (a_1, a_2, a_3)$, defining the magnitude of $\vec{A}$ is $|\vec{A}| = \sqrt{a_1^2 + a_2^2 + a_3^2}$.

(2) If $|\vec{A}| = 0$, then $\vec{A}$ is called **zero vector**.

(3) For $\vec{A}$, **the unit vector** $\dfrac{\vec{A}}{|\vec{A}|}$ is the direction on $\vec{A}$, commonly used to represent the

direction of $\vec{A}$.

(4) If $\vec{A} = (a_1, a_2, a_3)$, then the real number scalar multiplication of $\vec{A}$ is defined as

$m\vec{A} = (ma_1, ma_2, ma_3)$, where m is a real number.   ◀

2. **The Properties**

(1) If $\vec{A} = (a_1, a_2, a_3)$, $\vec{B} = (b_1, b_2, b_3)$, then $\vec{A} = \vec{B}$ if and only if $a_1 = b_1$, $a_2 = b_2$, $a_3 = b_3$.

(2) If the included angle α of $\vec{A} = (a_1, a_2, a_3)$ and positive x-axis and the included angle β of positive y-axis, and the included angle γ of positive z-axis, as shown in Figure 5-3, then α, β, γ are called **the direction angles** of the vector $\vec{A}$, $\cos\alpha$, $\cos\beta$, $\cos\gamma$ are called **direction cosines**, and they have the following properties:

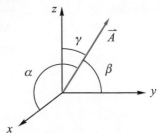

Figure 5-3   The azimuth angle of vector

① $\dfrac{\vec{A}}{|\vec{A}|} = \cos\alpha\,\vec{i} + \cos\beta\,\vec{j} + \cos\gamma\,\vec{k} = (\dfrac{a_1}{|\vec{A}|}\vec{i} + \dfrac{a_2}{|\vec{A}|}\vec{j} + \dfrac{a_3}{|\vec{A}|}\vec{k})$.

② $\cos^2\alpha + \cos^2\beta + \cos^2\gamma = 1$.

(3) Assuming $\vec{A} = a_1\vec{i} + a_2\vec{j} + a_3\vec{k} = (a_1, a_2, a_3)$, $\vec{B} = b_1\vec{i} + b_2\vec{j} + b_3\vec{k} = (b_1, b_2, b_3)$.

Defined operation: $\vec{A} + \vec{B} = (a_1 + b_1, a_2 + b_2, a_3 + b_3)$,

$\vec{A} - \vec{B} = (a_1 - b_1, a_2 - b_2, a_3 - b_3)$,

these operations have the following geometric interpretations as shown in Figure 5-4.

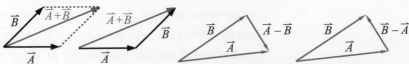

Figure 5-4   Geometric Illustration of Vector Addition and Subtraction

(4) For a vector $\vec{A} = a_1\vec{i} + a_2\vec{j} + a_3\vec{k}$, $m\vec{A} = ma_1\vec{i} + ma_2\vec{j} + ma_3\vec{k}$. The geometric interpretation of scalar multiplication is scaling the vector $\vec{A}$ by a factor of m. If $\vec{A} // \vec{B}$, the classifications of scalar multiplication are as follows:

$$\vec{B} = m\vec{A} \Rightarrow \begin{cases} m > 1 & \Rightarrow \text{enlarge} \\ m = 1 & \Rightarrow \vec{A} = \vec{B} \\ 0 < m < 1 & \Rightarrow \text{shrink in the positive direction} \\ -1 < m < 0 \Rightarrow \text{shrink in the negative direction} \\ m < -1 & \Rightarrow \text{enlarge in the negative direction} \end{cases}$$

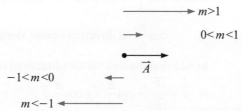

$$m > 1$$
$$0 < m < 1$$
$$\vec{A}$$
$$-1 < m < 0$$
$$m < -1$$

Figure 5-5   Vector scalar multiplication $\vec{B} = m\vec{A}$

The corresponding geometric interpretation is shown in Figure 5-5.

(5) Under operations in the context of (3)~(4), vectors in R^3 satisfies the following properties

① $\vec{A} + \vec{B} = \vec{B} + \vec{A}$ (**commutative law**)

② $(\vec{A} + \vec{B}) + \vec{C} = \vec{A} + (\vec{B} + \vec{C})$ (**associative law**)

③ $m(\vec{A} + \vec{B}) = m\vec{A} + m\vec{B}$ (**coefficient distributive law**)

④ $\vec{A} \pm \vec{0} = \vec{A}$ (**has an additive identity element**)

Let's go through some examples to familiarize with the definitions and properties mentioned above.

Example 1

There are two points in the Cartesian coordinate space : $p_1(-2, 1, 3)$, $p_2(-3, -1, 5)$, then
(1) find $\overrightarrow{p_1p_2}$ (2) find $|\overrightarrow{p_1p_2}|$ (3) find the unit vector on $\overrightarrow{p_1p_2}$.

Solution

(1) $\overrightarrow{p_1p_2} = (-1, -2, 2)$.

(2) $|\overrightarrow{p_1p_2}| = \sqrt{(-1)^2 + (-2)^2 + (2)^2} = 3$.

(3) the unit vector on $\overrightarrow{p_1p_2}$ is $\vec{e} = \dfrac{\overrightarrow{p_1p_2}}{|\overrightarrow{p_1p_2}|} = (\dfrac{-1}{3}, \dfrac{-2}{3}, \dfrac{2}{3})$.   Q.E.D.

Example 2

In the Cartesian coordinate plane, if the angle between $\vec{P}$ and the x-axis is $\dfrac{\pi}{3}$, and the length of this vector is 10, what is the vector?

Solution

$$\vec{P}=|\vec{P}|\cdot\cos(\frac{\pi}{3})\vec{i}+|\vec{P}|\cdot\cos(\frac{\pi}{6})\vec{j}=(10\cdot\frac{1}{2})\vec{i}+(10\cdot\frac{\sqrt{3}}{2})\vec{j}=5\vec{i}+5\sqrt{3}\,\vec{j}\;,$$

where $\cos\dfrac{\pi}{3}$ is direction cosine along the x-axis,

$\cos\dfrac{\pi}{6}$ is direction cosine along the y-axis.

In addition the unit vector (direction) on $\vec{P}$

is $\vec{e}=\dfrac{\vec{P}}{|\vec{P}|}=\cos\dfrac{\pi}{3}\vec{i}+\cos\dfrac{\pi}{6}\vec{j}$.

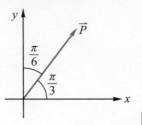

Q.E.D.

5-1-2 Inner Product (Dot Product)

The inner product space plays a significant role in mathematics and engineering fields. For example, it is involved in vector orthogonalization processes, and when applied to sets of functions, it becomes the foundation of the important Fourier series in signal analysis. Now, let's review the concept of the inner product and its related properties.

▶ **Definition 5-1-3**

Inner Product

Assuming $\vec{A}$, $\vec{B} \in \mathrm{R}^3$ be vectors in three-dimensional space, then the inner product between $\vec{A}$ and $\vec{B}$ is defined as $\vec{A}\cdot\vec{B}=|\vec{A}\|\vec{B}|\cos\theta$, as shown in Figure 5-6.

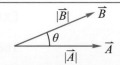

Figure 5-6 Schematic diagram of vector inner product

◀

If assuming $\vec{A} = (a_1, a_2, a_3)$, $\vec{B} = (b_1, b_2, b_3)$, then above definition is equal to $\vec{A}\cdot\vec{B} = a_1b_1 + a_2b_2 + a_3b_3$, moreover, the above definition can be extended to R^n: if $\vec{A} = (a_1, a_2, \cdots, a_n)$, $\vec{B} = (b_1, b_2, \cdots, b_n)$, then

$$\vec{A}\cdot\vec{B} = |\vec{A}\|\vec{B}|\cos\theta = a_1b_1 + a_2b_2 + a_3b_3 + \cdots\cdots + a_nb_n.$$

1. **Orthogonal (Perpendicular) of Vector**

   Let $\vec{A}$, $\vec{B} \in \mathrm{R}^3$, if $|\vec{A}| \neq 0$, $|\vec{B}| \neq 0$, and $\vec{A} \cdot \vec{B} = 0$, then $\vec{A}$ and $\vec{B}$ are called orthogonal, therefore, if two vectors are orthogonal, according to the equivalent definition, the angle between them is $\dfrac{\pi}{2}$; simultaneously, $\vec{A} \cdot \vec{B} = 0$ implies

   (1) at least one of the vectors in $\vec{A}$, $\vec{B}$ is $\vec{0}$.　　(2) $\vec{A} \perp \vec{B}$.

Example 3

Assuming $\vec{A} = -4\vec{i} + \vec{j} + 2\vec{k}$, $\vec{B} = 2\vec{i} + 4\vec{k}$, $\vec{C} = \vec{i} + 2\vec{j} + 3\vec{k}$,

please verify $\vec{A}$ is perpendicular to $\vec{B}$, while $\vec{A}$ is not perpendicular to $\vec{C}$.

Solution

(1) $\vec{A} \cdot \vec{B} = -4 \cdot 2 + 1 \cdot 0 + 2 \cdot 4 = 0$, therefore $\vec{A}$ is perpendicular to $\vec{B}$.

(2) $\vec{A} \cdot \vec{C} = -4 \cdot 1 + 1 \cdot 2 + 2 \cdot 3 = 4 \neq 0$, therefore $\vec{A}$ is not perpendicular to $\vec{C}$.　　**Q.E.D.**

2. **Properties of Vector**

   (1) $\vec{A} \cdot \vec{A} = |\vec{A}|^2 \Rightarrow |\vec{A}| = \sqrt{\vec{A} \cdot \vec{A}} = \sqrt{a_1^2 + a_2^2 + a_3^2}$, is called the norm of the vector $\vec{A}$.

   (2) $\cos\theta = \dfrac{\vec{A} \cdot \vec{B}}{|\vec{A}||\vec{B}|} \Rightarrow \theta = \cos^{-1}(\dfrac{\vec{A} \cdot \vec{B}}{|\vec{A}||\vec{B}|})$ represents the included angle of two vectors.

   (3) $\vec{i} = (1, 0, 0)$, $\vec{j} = (0, 1, 0)$, $\vec{k} = (0, 0, 1)$ are called the standard vectors in R^3, according to the definition of the inner product in the Euclidean space, we have
   $$\begin{cases} \vec{i} \cdot \vec{i} = \vec{j} \cdot \vec{j} = \vec{k} \cdot \vec{k} = 1 \\ \vec{i} \cdot \vec{j} = \vec{j} \cdot \vec{k} = \vec{k} \cdot \vec{i} = 0 \end{cases}.$$

   (4) $\vec{A} \cdot \vec{B} = \vec{B} \cdot \vec{A}$ (commutative law) ; $\vec{A} \cdot (\vec{B} + \vec{C}) = \vec{A} \cdot \vec{B} + \vec{A} \cdot \vec{C}$ (distributive law) $\alpha(\vec{A} \cdot \vec{B}) = (\alpha\vec{A}) \cdot \vec{B} = \vec{A} \cdot (\alpha\vec{B})$ (associativity of scalar multiplication). Note that vector dot product does not have division operation, that is $\vec{A} \cdot \vec{B} = \vec{A} \cdot \vec{C}$ cannot infer $\vec{B} = \vec{C}$. As a counterexample: $\vec{i} \cdot \vec{j} = \vec{i} \cdot \vec{k} = 0$, but $\vec{j} \neq \vec{k}$.

(5) Projection of vector

If $\vec{A}$, $\vec{B} \in \mathbb{R}^3$, then we define $\text{Proj}_{\vec{B}}(\vec{A}) = \vec{A} \cdot e_B$ is the projection

of $\vec{A}$ on $\vec{B}$, where $e_B = \dfrac{\vec{B}}{|\vec{B}|}$ is the unit vector on the direction $\vec{B}$.

in other words, $\text{Proj}_{\vec{B}}(\vec{A}) = |\vec{A}| \cdot \cos\theta = \vec{A} \cdot \dfrac{\vec{B}}{|\vec{B}|}$, as shown in figure 5-7,

while the projection vector of $\vec{A}$ on $\vec{B}$ is $\vec{P} = (\vec{A} \cdot e_B) \dfrac{\vec{B}}{|\vec{B}|} = (\dfrac{\vec{A} \cdot \vec{B}}{|\vec{B}|^2})\vec{B}$.

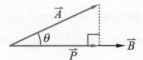

Figure 5-7   Schematic diagram of projection vector

Example 4

Let $\vec{A} = -2\vec{i} + \vec{j} + 2\vec{k}$, $\vec{B} = 3\vec{i} + 4\vec{k}$, please find

(1) the dot product and included angle between $\vec{A}$ and $\vec{B}$

(2) the projection vector of $\vec{A}$ onto $\vec{B}$.

Solution

(1) $\vec{A} \cdot \vec{B} = -2 \cdot 3 + 1 \cdot 0 + 2 \cdot 4 = 2$. $|\vec{A}| = \sqrt{(-2)^2 + (1)^2 + (2)^2} = 3$,

$|\vec{B}| = \sqrt{(3)^2 + (0)^2 + (4)^2} = 5$, $\cos\theta = \dfrac{\vec{A} \cdot \vec{B}}{|\vec{A}||\vec{B}|} = \dfrac{2}{3 \cdot 5} = \dfrac{2}{15}$, so $\theta = \cos^{-1}(\dfrac{2}{15})$.

(2) The projection vector of $\vec{A}$ onto $\vec{B}$ is $\vec{P} = (\dfrac{\vec{A} \cdot \vec{B}}{|\vec{B}|^2})\vec{B} = \dfrac{2}{25}(3\vec{i} + 4\vec{k})$.   Q.E.D.

Example 5

Let $\vec{u}$, $\vec{v}$ be two mutually perpendicular unit vectors, then $|\vec{u} + \vec{v}| = ?$

Solution

Because $\vec{u}$, $\vec{v}$ are two mutually perpendicular unit vectors, therefore

$|\vec{u}| = 1$, $|\vec{v}| = 1$ and $\vec{u} \cdot \vec{v} = 0$, then

$|\vec{u} + \vec{v}| = \sqrt{(\vec{u} + \vec{v}) \cdot (\vec{u} + \vec{v})} = \sqrt{\vec{u} \cdot \vec{u} + \vec{u} \cdot \vec{v} + \vec{v} \cdot \vec{u} + \vec{v} \cdot \vec{v}} = \sqrt{1 + 0 + 0 + 1} = \sqrt{2}$ Q.E.D.

5-1-3 Vector Product, Cross Product

To find the common perpendicular vector of two linearly independent vectors in space, the cross product is used. Here is an introduction to the cross product.

▶ Definition 5-1-4

The Cross Product of Vector

Let $\vec{A} = A_1\vec{i} + A_2\vec{j} + A_3\vec{k}$, $\vec{B} = B_1\vec{i} + B_2\vec{j} + B_3\vec{k} \in \mathbf{R}^3$,

define $\vec{A} \times \vec{B} = \begin{vmatrix} \vec{i} & \vec{j} & \vec{k} \\ A_1 & A_2 & A_3 \\ B_1 & B_2 & B_3 \end{vmatrix} = \begin{vmatrix} A_2 & A_3 \\ B_2 & B_3 \end{vmatrix}\vec{i} - \begin{vmatrix} A_1 & A_3 \\ B_1 & B_3 \end{vmatrix}\vec{j} + \begin{vmatrix} A_1 & A_2 \\ B_1 & B_2 \end{vmatrix}\vec{k}$

is called the cross product of $\vec{A}$, $\vec{B}$.   ◀

It can be shown that the magnitude of $\vec{A} \times \vec{B}$ is equal to the area of the parallelogram formed by $\vec{A}$, $\vec{B}$, therefore $\vec{A} \times \vec{B} = |\vec{A}||\vec{B}|\sin\theta$ also $\vec{A} \times \vec{B}$ is perpendicular to both $\vec{A}$, $\vec{B}$. The direction of the cross product can be determined using the right-hand rule, as illustrated in Figure 5-8.

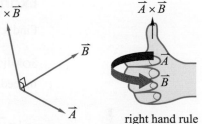

right hand rule

Figure 5-8   Schematic diagram of cross product

Example 6

Let $\vec{A} = -2\vec{i} + \vec{j} + 2\vec{k}$, $\vec{B} = 3\vec{i} + 4\vec{k}$, find the cross product of $\vec{A}$ and $\vec{B}$.

Solution

the cross product of $\vec{A}$ and $\vec{B}$:

$\vec{A} \times \vec{B} = \begin{vmatrix} 1 & 2 \\ 0 & 4 \end{vmatrix}\vec{i} - \begin{vmatrix} -2 & 2 \\ 3 & 4 \end{vmatrix}\vec{j} + \begin{vmatrix} -2 & 1 \\ 3 & 0 \end{vmatrix}\vec{k}$

$-4\vec{i} + 14\vec{j} - 3\vec{k}$.   Q.E.D.

1. **The Properties of Cross Product:**

    (1) the area of the parallelogram formed by $\vec{A}$, $\vec{B}$ so $|\vec{A} \times \vec{B}| = |\vec{A}||\vec{B}||\sin\theta| = |\vec{A}| \times h$, as shown in Figure 5-9.

    (2) If $\vec{A} \times \vec{B} = 0$, one of the following statements is true.

    ① $\vec{A}$, $\vec{B}$ at least one of them is $\vec{0}$.   ② $\vec{A}//\vec{B}$.

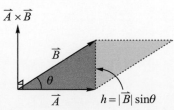

$h = |\vec{B}|\sin\theta$

Figure 5-9   The cross product of two vectors forms a parallelogram

(3) $\vec{A}\times(\vec{B}+\vec{C})=\vec{A}\times\vec{B}+\vec{A}\times\vec{C}$ (left distributive law)

$\vec{A}\times\vec{B}=-\vec{B}\times\vec{A}$ (anticommutative property)

$(\vec{A}+\vec{B})\times\vec{C}=\vec{A}\times\vec{C}+\vec{B}\times\vec{C}$ (right distributive law)

$\alpha\vec{A}\times\vec{B}=\vec{A}\times(\alpha\vec{B})$ (commutativity of real numbers)

(4) $\vec{i}\times\vec{j}=\vec{k}$, $\vec{j}\times\vec{i}=-\vec{k}$, $\vec{j}\times\vec{k}=\vec{i}$, $\vec{k}\times\vec{j}=-\vec{i}$,

$\vec{k}\times\vec{i}=\vec{j}$, $\vec{i}\times\vec{k}=-\vec{j}$, as shown in figure5-10.

Figure 5-10

(5) Lagrange's equality $|\vec{A}\times\vec{B}|=\sqrt{|\vec{A}|^2|\vec{B}|^2-(\vec{A}\cdot\vec{B})^2}$

Calculate the inner product and cross product involve trigonometric functions that are complementary to each other (with a 90-degree difference in angles). Therefore, by directly writing down the sum of squares

$$|\vec{A}\times\vec{B}|^2+(\vec{A}\cdot\vec{B})^2=|\vec{A}|^2|\vec{B}|^2\sin^2\theta+|\vec{A}|^2|\vec{B}|^2\cos^2\theta=|\vec{A}|^2|\vec{B}|^2$$

then simplifying the equation through rearrangement, we can prove Lagrange's equality.

Example 7

Find a unit vector that is simultaneously perpendicular to $2\vec{j}-3\vec{k}$ and $2\vec{i}$?

Solution

Indeed, the cross product of the two vectors satisfies the requirement,

so let $\vec{A}=2\vec{j}-3\vec{k}$, $\vec{B}=2\vec{i}$, $\vec{N}=\vec{A}\times\vec{B}=\begin{vmatrix} \vec{i} & \vec{j} & \vec{k} \\ 0 & 2 & -3 \\ 2 & 0 & 0 \end{vmatrix}=-6\vec{j}-4\vec{k}$,

taking $\vec{u}=\pm\dfrac{\vec{N}}{|\vec{N}|}=\pm\dfrac{(-6\vec{j}-4\vec{k})}{2\sqrt{13}}=\pm(0,\dfrac{3}{\sqrt{13}},\dfrac{2}{\sqrt{13}})$. **Q.E.D.**

Example 8

$A(2, 2, 2)$, $B(3, 0, 4)$, $C(5, 2, -2)$ are three points in space, as shown in figure, calculate the area of the triangle formed by these three points.

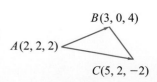

$B(3, 0, 4)$

$A(2, 2, 2)$

$C(5, 2, -2)$

Solution

According to basic properties (1), the area of triangle ABC $=\dfrac{1}{2}|\overrightarrow{AB}\times\overrightarrow{AC}|$,

and by $\overrightarrow{AB}=(1, -2, 2)$,

$\overrightarrow{AC}=(3, 0, -4)$, we have $\overrightarrow{AB}\times\overrightarrow{AC}=\begin{vmatrix} \vec{i} & \vec{j} & \vec{k} \\ 1 & -2 & 2 \\ 3 & 0 & -4 \end{vmatrix}=(8, 10, 6)$,

therefore the area of triangle ABC $=\dfrac{1}{2}|\overrightarrow{AB}\times\overrightarrow{AC}|=\dfrac{1}{2}\sqrt{(8^2+10^2+6^2)}=5\sqrt{2}$. **Q.E.D.**

5-1-4   Scalar Triple Product of Vector

In the context of vector geometry, it is often necessary to calculate the volume of a parallelepiped or tetrahedron spanned by three linearly independent vectors. This becomes particularly relevant when computing the lattice of metallic materials, and it involves the concept of scalar triple product. Let's delve into a detailed explanation.

▶ **Definition 5-1-5**

Scalar Triple Product

Let $\vec{A}$, $\vec{B}$, $\vec{C} \in R^3$, then $\vec{A} \cdot (\vec{B} \times \vec{C})$ or $(\vec{A} \times \vec{B}) \cdot \vec{C}$ is called scalar triple product of $\vec{A}$, $\vec{B}$, $\vec{C}$.   ◀

1.   Critical Properties

(1) As shown in figure 5-11, the volume of the parallelepiped with sides $\vec{A}$, $\vec{B}$,

$\vec{C}$ is $|\vec{A} \cdot \vec{B} \times \vec{C}| = |\vec{A}| |\vec{B} \times \vec{C}| |\cos\theta| = |\vec{B} \times \vec{C}| |\vec{A}| |\cos\theta|$. The six-sided solid is divided into three equal parts from its vertices.

① Tetrahedron formed by $\vec{A}$, $\vec{B}$, $\vec{C}$,

the volume $= \dfrac{1}{6} |\vec{A} \cdot (\vec{B} \times \vec{C})|$.

② Triangular prism formed by $\vec{A}$,

$\vec{B}$, $\vec{C}$, the volume $= \dfrac{1}{2} |\vec{A} \cdot (\vec{B} \times \vec{C})|$.

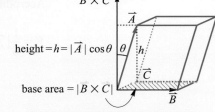

height $= h = |\vec{A}| \cos\theta$

base area $= |B \times C|$

Figure 5-11   Schematic diagram of Scalar Triple Product

(2) If $\vec{A} = a_1 \vec{i} + b_1 \vec{j} + c_1 \vec{k}$, $\vec{B} = a_2 \vec{i} + b_2 \vec{j} + c_2 \vec{k}$,

$\vec{C} = a_3 \vec{i} + b_3 \vec{j} + c_3 \vec{k}$, then it can be calculated from the definition of inner and

cross products $\vec{A} \cdot (\vec{B} \times \vec{C}) = \begin{vmatrix} a_1 & b_1 & c_1 \\ a_2 & b_2 & c_2 \\ a_3 & b_3 & c_3 \end{vmatrix}$.

(3) When $\vec{A} \cdot (\vec{B} \times \vec{C}) = 0$, one of following conditions will establish:

① $\vec{A}$, $\vec{B}$, $\vec{C}$ at least one of them is $\vec{0}$.

② $\vec{A}$, $\vec{B}$, $\vec{C}$ three vectors in the same plane (The volume is 0).

(4) $\vec{A} \cdot (\vec{B} \times \vec{C}) = (\vec{A} \times \vec{B}) \cdot \vec{C}$ (The commutative of $\times$ and $\cdot$).

Example 9

Let $\vec{a} = 3\vec{i} + 2\vec{j} + \vec{k}$, $\vec{b} = 2\vec{i} - \vec{j} + \vec{k}$, $\vec{c} = \vec{j} + \vec{k}$, find the volume of the parallelepiped formed by $\vec{a}$, $\vec{b}$, $\vec{c}$.

Solution

Accroding ro critical properties (1), the volume equals to

$$|\vec{a} \cdot (\vec{b} \times \vec{c})| = \left| \begin{matrix} 3 & 2 & 1 \\ 2 & -1 & 1 \\ 0 & 1 & 1 \end{matrix} \right| = |-8| = 8.$$ Q.E.D.

Example 10

Let $\vec{a} = \vec{i} + 2\vec{k}$, $\vec{b} = 4\vec{i} + 6\vec{j} + 2\vec{k}$, $\vec{c} = 3\vec{i} + 3\vec{j} - 6\vec{k}$, as shown in figure, find the volume of tetrahedron formed by $\vec{a}$, $\vec{b}$, $\vec{c}$.

Solution

Accroding ro critical properties1., the volume equals to

$$V = \frac{1}{6}\left| \vec{a} \cdot (\vec{b} \times \vec{c}) \right| = \frac{1}{6}\left| \begin{matrix} 1 & 0 & 2 \\ 4 & 6 & 2 \\ 3 & 3 & -6 \end{matrix} \right| = 9.$$ Q.E.D.

5-1   Exercises

Basic questions

1. Known $\vec{A} = (1, 1, -2)$, $\vec{B} = (2, -3, 5)$, try to calculate following questions.
   (1) $\vec{A} + \vec{B}$.
   (2) $\vec{A} - \vec{B}$.
   (3) $-\vec{A}$.
   (4) $4\vec{A}$.
   (5) $4\vec{A} + 3\vec{B}$.
   (6) $4\vec{A} - 3\vec{B}$.

2. Find the inner product and cross product of $\vec{A}$ and $\vec{B}$ on Cartesian coordinate system.
   (1) $\vec{A} = (-3, 6, 1)$, $\vec{B} = (-1, -2, 1)$.
   (2) $\vec{A} = (2, -3, 4)$, $\vec{B} = (-3, 2, 0)$.
   (3) $\vec{A} = (5, 3, 4)$, $\vec{B} = (20, 0, 6)$.
   (4) $\vec{A} = (18, -3, 4)$, $\vec{B} = (0, 22, -1)$.
   (5) $\vec{A} = (-4, 0, 6)$, $\vec{B} = (1, -2, 7)$.

Advanced questions

1. What's the angle between $\vec{A}$ and $\vec{B}$? And find the projection vector $\vec{A}$ on $\vec{B}$?
   (1) $\vec{A} = (3, 4, 5)$, $\vec{B} = (-1, -2, 2)$.
   (2) $\vec{A} = (2, -3, 4)$, $\vec{B} = (3, 0, 4)$.
   (3) $\vec{A} = (2, 2, 1)$, $\vec{B} = (0, 5, -12)$.

2. A, B, C are three points in space, what's the volume of triangle formed by these three points?
   (1) $A(6, 1, 1)$, $B(7, -2, 4)$, $C(8, -4, 3)$.
   (2) $A(4, 2, -3)$, $B(6, 2, -1)$, $C(2, -6, 4)$.

3. What's the volume of tetrahedron formed by $\vec{a}$, $\vec{b}$, $\vec{c}$?
   (1) $\vec{a} = (-5, 1, 6)$, $\vec{b} = (2, 4, 6)$, $\vec{c} = (-1, 0, 5)$.
   (2) $\vec{a} = (-3, -2, 1)$, $\vec{b} = (2, -6, -1)$, $\vec{c} = (1, -4, -5)$.
   (3) $\vec{a} = (1, 1, 1)$, $\vec{b} = (5, 0, 2)$, $\vec{c} = (-3, 3, 5)$.

4. Find the volume of parallelepiped surrounded by $\vec{a} = (2, 0, 3)$, $\vec{b} = (0, 6, 2)$, $\vec{c} = (3, 3, 0)$.

5. Find the volume of tetrahedron with the vertex of $A(1, 0, 1)$, $B(0, 1, -1)$, $C(2, 1, 0)$, $D(3, 5, 2)$.

6. Find the volume of tetrahedron with the vertex of $A(0, 1, 2)$, $B(5, 5, 6)$, $C(1, 2, 1)$, $D(3, 3, 1)$.

5-2 Vector Geometry

This section introduces how to derive the straight line and plane equations in $\mathbb{R}^3$, and explains how to use these equations to describe the relationship between points, lines and planes.

5-2-1 Equation of Line in $\mathbb{R}^3$

1. Two-Point Form

Given two points $P(x_1, y_1, z_1)$, $Q(x_2, y_2, z_2)$ in the known space, assuming any point on the straight line passing through the two points P, Q is $A(x, y, z)$, then $\overrightarrow{PA} /\!/ \overrightarrow{PQ}$, which means $\overrightarrow{PA} = t\overrightarrow{PQ}$, therefore

$$(x - x_1, y - y_1, z - z_1) = t(x_2 - x_1, y_2 - y_1, z_2 - z_1)$$

comparing coordinate, rearrange as
$$\begin{cases} x = x_1 + (x_2 - x_1)t \\ y = y_1 + (y_2 - y_1)t \, , t \in \mathbb{R} \\ z = z_1 + (z_2 - z_1)t \end{cases}$$

is called parametric equation of line pass through two points P, Q, as shown in Figure 5-12. The parametric equation of a line can also be rearranged as
$$\frac{x - x_1}{x_2 - x_1} = \frac{y - y_1}{y_2 - y_1} = \frac{z - z_1}{z_2 - z_1} = t$$ and referred to as the symmetric form.

Figure 5-12   The two-point form of a line in space

2. Point-Direction Form

Given a point $P(x_1, y_1, z_1)$ and a vector $\vec{u} = (a, b, c)$, let L be the line passing through the point P and parallel to $\vec{u}$. Then $\overrightarrow{PA} /\!/ \vec{u}$, that is $(x - x_1, y - y_1, z - z_1) = t(a, b, c)$, the parameter equation of line L can be obtained by comparing the coordinates as
$$\begin{cases} x = x_1 + a \cdot t \\ y = y_1 + b \cdot t \, , t \in \mathbb{R} \\ z = z_1 + c \cdot t \end{cases}$$ or written as the symmetric form:
$$\frac{x - x_1}{a} = \frac{y - y_1}{b} = \frac{z - z_1}{c} = t$$, as shown in Figure5-13.

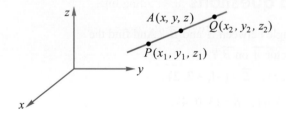

Figure 5-13   point-direction form

Example 1

Find the parameter equation and symmetric form of the line L passing through the points $P(1, 0, 4)$ and $Q(2, 1, 1)$.

Solution

Both the point-direction form and the two-point form can be used to obtain the

parameter equation of the line L $\begin{cases} x = 1 + (2-1) \cdot t \\ y = 0 + (1-0) \cdot t \\ z = 4 + (1-4) \cdot t \end{cases}$, that is $\begin{cases} x = 1 + t \\ y = t \\ z = 4 - 3t \end{cases}$,

so any point on L can be expressed as $(1 + t, t, 4 - 3t)$, $t \in \mathbb{R}$.

The symmetric form can be derived from the parameter equation as follows

$$\frac{x-1}{2-1} = \frac{y-0}{1-0} = \frac{z-4}{1-4},$$

thus we obtain $\dfrac{x-1}{1} = \dfrac{y-0}{1} = \dfrac{z-4}{-3}$.   Q.E.D.

5-2-2   Equation of Plane

In space, the distribution of points and lines forms a plane if it satisfies one of the following four conditions, as shown in Figure 5-14:

(1) three non-collinear points,

(2) two intersecting lines,

(3) two parallel lines,

(4) one line and a point outside the line.

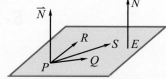

Figure 5-14   Diagram of three points determine a plane in space

The following theorem demonstrates how to obtain the equation of a plane from known points.

▶ Theorem 5-2-1

Equation of Plane

If $P(x_1, y_1, z_1)$, $Q(x_2, y_2, z_2)$, $R(x_3, y_3, z_3)$ are non-collinear points, then the plane formed by these three points satisfies $a(x - x_1) + b(y - y_1) + c(z - z_1) = 0$, where $\overrightarrow{N} = (a, b, c)$ is the normal vector of this plane.

Proof

Assuming that the plane formed by P, Q, R is denoted as E, let $S(x, y, z)$ be a point on E take $\overrightarrow{N} = \overrightarrow{PQ} \times \overrightarrow{PR} = a\vec{i} + b\vec{j} + c\vec{k} = (a, b, c)$ be the normal vector of E, then $\overrightarrow{PS} \cdot \overrightarrow{N} = 0$,

now $\overrightarrow{PS} = (x - x_1, y - y_1, z - z_1)$, substituting into above equation, we have $a(x - x_1) + b(y - y_1) + c(z - z_1) = 0$.   ◀

The equation of a plane in Theorem 5-2-1 can also be derived using determinants. Since $\overrightarrow{PS}$, $\overrightarrow{PQ}$ and $\overrightarrow{PR}$ are three vectors lying on the same plane, that is the volume of the parallelepiped formed by them, $\overrightarrow{PS} \cdot (\overrightarrow{PQ} \times \overrightarrow{PR}) = 0$ is zero. We have

$$\begin{vmatrix} x - x_1 & y - y_1 & z - z_1 \\ x_2 - x_1 & y_2 - y_1 & z_2 - z_1 \\ x_3 - x_1 & y_3 - y_1 & z_3 - z_1 \end{vmatrix} = 0.$$

Example 2

Find the equation of plane that pass through three points $(1, 2, 3)$, $(3, 2, 2)$, $(-2, -1, 3)$.

Solution

By determinants we obtain $\begin{vmatrix} x-1 & y-2 & z-3 \\ 2 & 0 & -1 \\ -3 & -3 & 0 \end{vmatrix} = 0$, expanded and we have $x - y + 2z = 5$.

Q.E.D.

5-2-3 The Shortest Distance from a Point to a Line

The equation of the line going along $\overrightarrow{A} = (a, b, c)$ and passing through $Q(x_1, y_1, z_1)$ is L: $\dfrac{x - x_1}{a} = \dfrac{y - y_1}{b} = \dfrac{z - z_1}{c}$. If $P(x_0, y_0, z_0)$ is a point outside the line L (as shown in Figure 5-15), then the shortest distance from P to L can be expressed using the cross product as:

$$\overrightarrow{PH} = d(P, L) = |\overrightarrow{QP}| \cdot \sin\theta = |\overrightarrow{QP}| \cdot |\vec{e}| \cdot \sin\theta = |\overrightarrow{QP} \times \vec{e}| = |\overrightarrow{QP} \times \frac{\overrightarrow{A}}{|\overrightarrow{A}|}|$$

where $\vec{e} = \dfrac{\overrightarrow{A}}{|\overrightarrow{A}|} = \dfrac{a\vec{i} + b\vec{j} + c\vec{k}}{\sqrt{a^2 + b^2 + c^2}}$, as shown in Figure 5-15.

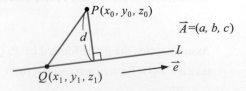

Figure 5-15   The distance from a point to a line

Example 3

Find the distance of pont $P(3, 2, 4)$ to the line L: $x = 1 + t, y = 3 - 2t, z = 6 + 3t$.

Solution

Let $t = 0$, we obtain a pont $Q(1, 3, 6)$ on the line,

then $\overrightarrow{PQ} = (-2, 1, 2)$. Since $\overrightarrow{A} = (1, -2, 3)$,

$\overrightarrow{PQ} \times \overrightarrow{A} = (7, 8, 3)$ can be calculated by cross product, therefore

$$d(P, L) = |\overrightarrow{PQ} \times \frac{\overrightarrow{A}}{|\overrightarrow{A}|}| = \sqrt{\frac{61}{7}}.$$

Q.E.D.

5-2-4   The Shortest Distance Between Skew Lines

Let L and M be two skew lines that do not lie in the same plane, let $\overrightarrow{u_1}$ and $\overrightarrow{u_2}$ be vectors parallel to L and M, and $\overrightarrow{u_1} \times \overrightarrow{u_2} \neq 0$. If P, Q are points on M, L respectively, and $\overrightarrow{N} = \overrightarrow{u_1} \times \overrightarrow{u_2}$, then the projection of $\overrightarrow{PQ}$ on $\overrightarrow{N}$ represents the distance between the two skew lines, in other words,

$$d = |\overrightarrow{PQ} \cdot \frac{\overrightarrow{N}}{|\overrightarrow{N}|}| = |\overrightarrow{PQ} \cdot \frac{\overrightarrow{u_1} \times \overrightarrow{u_2}}{|\overrightarrow{u_1} \times \overrightarrow{u_2}|}|, \text{ as shown in}$$

Figure5-16.

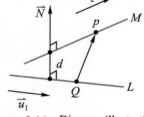

Figure 5-16   Diagram illustrating the distance between two skew lines

Example 4

Find the distance between two skew lines $\begin{cases} L_1 : x = 1 + t \text{ , } y = 2t \text{ , } z = 3 - 3t \\ L_2 : x = 2 - s \text{ , } y = 2 + s \text{ , } z = 3 - s \end{cases}$.

Solution

Take a pont $P(1, 0, 3)$ on L_1, direction vector $\overrightarrow{u_1} = (1, 2, -3)$;

take a point $Q(2, 2, 3)$ on L_2, direction vector $\overrightarrow{u_2} = (-1, 1, -1)$,

then $\overrightarrow{u_1} \times \overrightarrow{u_2} = \overrightarrow{N} = (1, 4, 3)$, $\overrightarrow{PQ} = (1, 2, 0)$,

then $d = |(1, 2, 0) \cdot \frac{(1, 4, 3)}{\sqrt{1^2 + 4^2 + 3^2}}| = \frac{9}{\sqrt{26}}$.

Q.E.D.

5-2-5 The Shortest Distance from a Point to a Plane

Given a plane $E: ax + by + cz + d = 0$, and $P(x_0, y_0, z_0)$ is a point outside E. Now $Q(x, y, z)$ be an arbitrary point on E, and let $\vec{N} = (a, b, c)$ be the normal vector of E, then the projection of $\vec{PQ}$ on $\vec{N}$ is the shortest distance, in other words

$$d(P, E) = |\vec{PQ} \cdot \frac{\vec{N}}{|\vec{N}|}| = |(x - x_0, y - y_0, z - z_0) \cdot \frac{(a, b, c)}{\sqrt{a^2 + b^2 + c^2}}|$$

$$= \frac{|ax + by + cz - ax_0 - by_0 - cz_0)|}{\sqrt{a^2 + b^2 + c^2}}$$

$$= \frac{|-d - ax_0 - by_0 - cz_0|}{\sqrt{a^2 + b^2 + c^2}} = \frac{|ax_0 + by_0 + cz_0 + d|}{\sqrt{a^2 + b^2 + c^2}}$$

as shown in Figure 5-17.

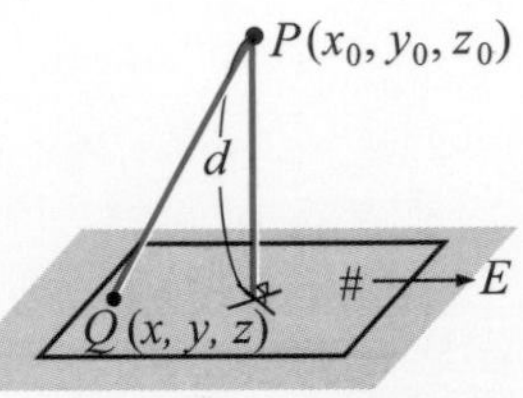

Figure 5-17   Diagram illustrating the distance from a point to a plane

Example 5

Find the shortest distance of the point $P(3, 2, 4)$ to the plane $E: 2x - y + 2z = 5$.

Solution

$$d(P, E) = \frac{|2 \cdot 3 - 2 + 2 \cdot 4 - 5|}{\sqrt{2^2 + (-1)^2 + 2^2}} = \frac{7}{3}.$$

Q.E.D.

5-2-6 Angle Between Two Planes

Consider two planes $E_1: a_1x + b_1y + c_1z + d_1 = 0$, $E_2: a_2x + b_2y + c_2z + d_2 = 0$, then the angle between E_1 and E_2 can be obtained by $\cos\theta = \dfrac{\vec{N}_1 \cdot \vec{N}_2}{|\vec{N}_1||\vec{N}_2|}$. That is, if let $\vec{N}_1 = (a_1, b_1, c_1)$,

$\vec{N}_2 = (a_2, b_2, c_2)$, then $\cos\theta = \pm \dfrac{(a_1a_2 + b_1b_2 + c_1c_2)}{\sqrt{a_1^2 + b_1^2 + c_1^2}\sqrt{a_2^2 + b_2^2 + c_2^2}}$.

So the angle between the two planes $\theta = \cos^{-1}(\dfrac{\vec{N}_1 \cdot \vec{N}_2}{|\vec{N}_1||\vec{N}_2|})$

and $\pi - \theta$, as shown in Figure 5-18.

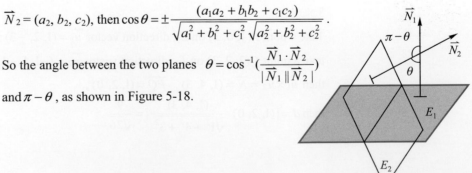

Figure 5-18   Angle between two planes

Example 6

Find the angle between two planes $x + 2y + 3z = 3$ and $x - 2y - 3z = 3$.

Solution

$$\cos\theta = \pm\frac{\vec{N}_1 \cdot \vec{N}_2}{|\vec{N}_1||\vec{N}_2|} = \pm\frac{(1,2,3)\cdot(1,-2,-3)}{\sqrt{1^2+2^2+3^2}\cdot\sqrt{1^2+2^2+3^2}} = \pm\frac{-12}{14} = \pm\frac{6}{7}$$

$$\Rightarrow \theta = \cos^{-1}(\frac{6}{7}) \text{ or } \pi-\theta .$$

Q.E.D.

5-2-7 Line of Two Intersection Planes

Consider $E_1 : a_1x + b_1y + c_1z + d_1 = 0$, and let $\vec{N}_1 = (a_1, b_1, c_1)$ be the normal vectors; $E_2 : a_2x + b_2y + c_2z + d_2 = 0$, the normal vector $\vec{N}_2 = (a_2, b_2, c_2)$. Then $\vec{u} = \vec{N}_1 \times \vec{N}_2 = (l, m, n)$ is the direction of the line L formed by the intersection of the two planes. Let $P(x_0, y_0, z_0)$ be any point on L, solving the system of equations of E_1, E_2,

we have L: $\dfrac{x - x_0}{l} = \dfrac{y - y_0}{m} = \dfrac{z - z_0}{n} = t$

(let one of the variables be 0) , in other words, the parameter equation of L is

$$\begin{cases} x = x_0 + lt \\ y = y_0 + mt \ , \ t \in \mathbb{R}, \\ z = z_0 + nt \end{cases}$$

as shown in Figure5-19.

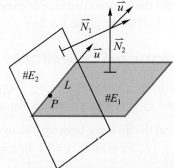

Figure 5-19   Diagram illustrating the intersection line of two planes in space

Example 7

Find the equation of the intersection line between the two planes $2x + y - z = 0$ and $3x + 2y + z = 3$.

Solution

The normal vectors of the two planes are $\vec{N}_1 = (2, 1, -1)$, $\vec{N}_2 = (3, 2, 1)$,then the direction vector of the intersection line L is $\vec{u} = \vec{N}_1 \times \vec{N}_2 = (3, -5, 1)$. Solving the system of equations formed by the two planes: $\begin{cases} 2x + y - z = 0 \\ 3x + 2y + z = 3 \end{cases}$,

and let $x = 0 \Rightarrow y = 1$, $z = 1$, that is $(0, 1, 1)$ is a point on L,

therefore L: $\dfrac{x-0}{3} = \dfrac{y-1}{-5} = \dfrac{z-1}{1} \Rightarrow \begin{cases} x = 3t \\ y = 1-5t \ ; \ t \in \mathbb{R}. \\ z = 1+t \end{cases}$

Q.E.D.

5-2   Exercises

Basic questions

1. Find the parameter equation of the line passing
   through the following two points.
   (1) $(1, 0, 5), (2, 1, -1)$.
   (2) $(4, 0, 0), (-3, 1, 0)$.
   (3) $(2, 1, 1), (2, 1, -4)$.
   (4) $(0, 1, 3), (0, -1, 2)$.
   (5) $(1, 0, 4), (-2, -3, 5)$.

2. Find the plane in space containing the following
   three points.
   (1) $(12, 5, 0), (0, 4, 0), (12, 0, 6)$.
   (2) $(1, 2, 1), (-1, 1, 3), (-2, -2, -2)$.
   (3) $(1, 1, 2), (-1, 1, -26), (0, 2, 1)$.

3. Find the shortest distance of the point
   $(1, 3, 2)$ to the plane $x + 2y + z = 4$.

Advanced questions

1. If the two planes $x + y + z = 1$ and
   $2x + cy + 7z = 0$ are orthogonal, then $c =$?

2. Find the distance between the two skew lines.
   (1) $\begin{cases} L_1 : x = 1 + 2t, \ y = 1 + 3t, \ z = 2 - t \\ L_2 : x = s, \ y = 2s, \ z = 3s \end{cases}$

   (2) $\begin{cases} L_1 : x = 1 + t, \ y = 2 + 2t, \ z = 3 + 3t \\ L_2 : x = 4 + 4s, \ y = 5 + 5s, \ z = 6 + 6s \end{cases}$

   (3) $\begin{cases} L_1 : \dfrac{x-2}{3} = \dfrac{y-5}{2} = \dfrac{z-1}{-1} \\ L_2 : \dfrac{x-4}{-4} = \dfrac{y-5}{4} = \dfrac{z+2}{1} \end{cases}$

3. Find the angle between two planes
   $x + 2y - 2z + 3 = 0$ and
   $2x - y + 2z - 3 = 0$.

5-3   Vector Spaces R^n

In this section, the properties of vectors in R^3 are extended to the n-dimensional vector space R^n. When $n \geq 4$, the vector space R^n can no longer be represented by geometric figures. However, most properties remain the same as in R^3. So, when encountering concepts that are difficult to imagine or understand, one can use R^3 as a tool for exploration and simulation.

5-3-1   The Vector of n-Dimensional Space

▶ **Definition 5-3-1**

n-Dimensional Vector

We represent elements in the n-dimensional vector space R^n using $\vec{A} = (a_1, a_2, \cdots, a_n)$

or $\vec{A} = \begin{bmatrix} a_1 \\ a_2 \\ a_3 \\ \vdots \\ a_n \end{bmatrix}$, which $a_1, a_2, \cdots, a_n$ are all real numbers.   ◀

For convenience, starting from this point, unless specifically needed, all n-dimensional vectors will be referred to simply as "vectors." Unless specified otherwise, all statements and discussions are referring to R^n and its subsets.

▶ **Definition 5-3-2**

The Common Definition of n-Dimensional Vector

$\vec{A} = (a_1, a_2, \cdots, a_n)$, $\vec{B} = (b_1, b_2, \cdots, b_n)$ are two vectors, α is real number,

(1) $\vec{A} \pm \vec{B} = (a_1 \pm b_1, a_2 \pm b_2, \cdots, a_n \pm b_n)$ is **vector addition and subtraction.**

(2) $\alpha\vec{A} = (\alpha a_1, \alpha a_2, \cdots, \alpha a_n)$ is **scalar multiplication of a vector.**

(3) $|\vec{A}| = \|\vec{A}\| = \sqrt{a_1^2 + a_2^2 + \cdots + a_n^2}$ is called the length or magnitude of $\vec{A}$, which is often referred to as the **norm.**

(4) **The unit vector** $\vec{u} = \dfrac{\vec{A}}{\|\vec{A}\|}$ on $\vec{A}$, the action of dividing norms is called

   normalizing $\vec{A}$. This vector can represent the direction of $\vec{A}$.

(5) $\vec{A} \cdot \vec{B} = a_1 b_1 + a_2 b_2 + \cdots + a_n b_n$ is the operation of two vectors doing inner product. If

$\vec{A} \cdot \vec{B} = 0$, called

$\vec{A} = (a_1, a_2, \cdots, a_n)$ and $\vec{B} = (b_1, b_2, \cdots, b_n)$ are **orthogonal**, in R^3 it is also referred

to as being perpendicular. ◀

Example 1

Consider two vectors $\vec{A} = (-1, 1, 1, 2)$, $\vec{B} = (1, -1, 0, 1)$.

(1) Find the norm of $\vec{A}$. (2) Verify $\vec{A}$ and $\vec{B}$ are orthogonal. (3) Normalize $\vec{A}$.

Solution

(1) the norm of $\vec{A}$ is $\|\vec{A}\| = \sqrt{(-1)^2 + 1^2 + 1^2 + 2^2} = \sqrt{7}$.

(2) $\vec{A} \cdot \vec{B} = -1 \cdot 1 + 1 \cdot (-1) + 1 \cdot 0 + 2 \cdot 1 = 0$, so $\vec{A}$ and $\vec{B}$ are orthogonal.

(3) The normalized vector of $\vec{A}$ is $\dfrac{\vec{A}}{\|\vec{A}\|} = \dfrac{1}{\sqrt{7}}(-1, 1, 1, 2)$. Q.E.D.

The Subspace of R^n

Indeed, a subspace is a subset of the original vector space that preserves all the operations of the vector space. Therefore, it is more appropriate to use the symbol S to represent a subspace.

▶ Definition 5-3-3

The Subspace of R^n

Assuming that the set S is a non-empty subset of R^n, if the following conditions hold for any vectors $\vec{a}$, $\vec{b}$, b in S and any scalar k in R:

(1) S contains the zero vector $\vec{0}$ (2) $\vec{a} + \vec{b}$ is still in S (3) $k\vec{a}$ is still in S.

Then we refer to S as a subspace of R^n. ◀

Based on the definition, it can be observed that R^n has at least two subspaces, that is R^n and $\{0\}$. Furthermore, according to the definition 5-3-3 and the graphical representation in R^3. The 1-dimensional subspaces in R^3 are all lines passing through the origin, while the 2-dimensional subspaces are all planes passing through the origin. Similarly, if S is a subspace of R^2, then S must fall into one of the following three categories:

(1) $S = \mathrm{R}^2$ (dimension= 2) .

(2) The set S represents a line passing through the origin (dimension= 1) .

(3) The set S only contains the zero vector $\vec{0}$ (dimension= 0) ,

as shown in Figure 5-20.

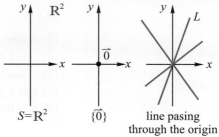

Figure 5-20　The subspace in R^2

It's important to note that for a set to be considered a subspace, the conditions (1) to (3) in definition 5-3-3 must hold simultaneously. The definition 5-3-3 can be simplified to the following statement.

▶ Theorem 5-3-1

The Subspace Theorem

Assuming S is a non-empty subset of R^n then S is a subspace if and only if

(1) zero vector $\vec{0} \in S$　(2) If $\vec{f}$ and $\vec{g} \in S$　also $\alpha \in \mathrm{R}$, then $\vec{f} + \alpha\vec{g} \in S$.　◀

Example 2

Assume that S is a subset of R^4, if all vectors in S are scalar multiples of $\vec{v_1} = (1, 2, 3, 0)$, please prove that S is a subspace of R^4.

Solution　According to the given condition, all vectors in S can be expressed as $c\vec{v_1} = c(1, 2, 3, 0)$, where c is a real number.

(1) Taking $c = 0$, then $0 \cdot \vec{v_1} = 0 \cdot (1, 2, 3, 0) = (0, 0, 0, 0) = \vec{0}$, so $\vec{0}$ is included in S.

(2) Taking two vectors $\vec{f} = \alpha(1, 2, 3, 0)$ and $\vec{g} = \beta(1, 2, 3, 0)$　in S, where $\alpha, \beta \in \mathrm{R}$,

let $k \in \mathrm{R}$ then　$\vec{f} + k\vec{g} = (\alpha + \beta k) \cdot (1, 2, 3, 0)$ is still scalar multiple of

$\vec{v_1} = (1, 2, 3, 0)$.

That is, it remains in S. So by theorem 5-3-1, we know that S is a subspace of R^4.

Q.E.D.

Example 3

Please determine which of the following is a subspace of R^3?

(1) s_1: $\begin{cases} x = 1 + 2 \cdot t \\ y = -3 \cdot t \\ z = 2 - 4 \cdot t \end{cases}$, $t \in \mathrm{R}$ 　(2) s_2: $\begin{cases} x = 2 \cdot t \\ y = -3 \cdot t \\ z = 4 \cdot t \end{cases}$, $t \in \mathrm{R}$ 　(3) s_3: $2x - 3y + 4z = 5$

(4) s_4: $-3x + 5y - 2z = 0$ 　(5) s_5: $x = 0$ 　(6) s_6: y-axis.

Solution

(1) s_1 does not pass through the origin of $\mathbb{R}^3$, it is not a subspace.

(2) s_2 is a line passing through the origin of $\mathbb{R}^3$, it is a subspace.

(3) s_3 does not contain the origin, it is not a subspace.

(4) s_4 is a plane that contains the origin of $\mathbb{R}^3$, it is a subspace.

(5) s_5 is the yz-plane in $\mathbb{R}^3$, contains the origin, it is a subspace.

(6) s_6 is a line passing through the origin, it is a subspace.

Q.E.D.

5-3-2   The Linear Independence and Linear Dependence

When we find particular solutions to ODE, we have discussed the use of the Wronskian determinant to determine the linear independence or dependence of functions. Now, let's formally define what it means for vectors to be linearly independent or dependent.

▶ **Definition 5-3-4**

The Linear Independence and Linear Dependence of Vector

Assuming that $W = \{\vec{v_1}, \vec{v_2}, \cdots, \vec{v_n}\}$ is a subset of $\mathbb{R}^n$, if there exists a vector $\vec{v_k}$ that can be expressed as a linear combination of the other $(n-1)$ vectors, then $\{\vec{v_1}, \vec{v_2}, \cdots, \vec{v_n}\}$ is called linear dependence. Conversely, if no vector in W can be expressed as a linear combination of the other $(n-1)$ vectors, then $\{\vec{v_1}, \vec{v_2}, \cdots, \vec{v_n}\}$ is said to be linearly independent. ◀

Example 4

A set of vectors $\{\vec{A}, \vec{B}, \vec{C}\}$ in $\mathbb{R}^4$, where $\vec{A} = (-1, 1, 1, 0)$, $\vec{B} = (1, -1, 1, 1)$, $\vec{C} = (2, -2, 4, 3)$, please determine whether $\vec{A}$, $\vec{B}$, $\vec{C}$ are linear independent or linear dependent.

Solution

Since $\vec{C} = (2, -2, 4, 3) = (-1, 1, 1, 0) + 3 \cdot (1, -1, 1, 1) = \vec{A} + 3\vec{B}$, we can see that $\vec{C}$ can be expressed as a linear combination of $\vec{A}$ and $\vec{B}$. Therefore, $\{\vec{A}, \vec{B}, \vec{C}\}$ is linearly dependent.

Q.E.D.

Definition 5-3-4 is equivalent to the following theorem, which is also commonly used. This book aims for an intuitive approach, so Definition 5-3-4 is adopted as the primary definition.

▶ **Theorem 5-3-2**

The Linear Independence and Linear Dependence of Vector

Assuming that $\{\vec{v_1}, \vec{v_2}, \cdots, \vec{v_n}\}$ is a subset of R^n, (1) if $\sum_{k=1}^{n} \alpha_k \vec{v_k} = 0$, α_1, α_2, $\cdots$, α_n can

be established only when they are all zero, then $\{\vec{v_1}, \vec{v_2}, \cdots, \vec{v_n}\}$ is linear independence

(2) if $\sum_{k=1}^{n} \alpha_k \vec{v_k} = 0$, α_1, α_2, $\cdots$, α_n can also be established when they are not all zero,

then $\{\vec{v_1}, \vec{v_2}, \cdots, \vec{v_n}\}$ is linear dependence.   ◀

From The Coordinate Axes in R^3, We Can Find Many Examples

(1) For example, in the set $\{(1, 0, 0), (0, 1, 0), (0, 0, 3)\} = s_1$:
   $\alpha_1 \cdot (1, 0, 0) + \alpha_2 \cdot (0, 1, 0) + \alpha_3 \cdot (0, 0, 3) = 0 \Rightarrow \alpha_1 = \alpha_2 = \alpha_3 = 0$,
   therefore, s_1 is linearly independent.

(2) Another example is the set $\{(1, 0, 0), (0, 2, 0), (3. 1, 0)\} = s_2$,

   by $\alpha_1 \cdot (1, 0, 0) + \alpha_2 \cdot (0, 2, 0) + \alpha_3 \cdot (3, 1, 0) = 0$, we have $\begin{cases} \alpha_1 + 3\alpha_3 = 0 \\ 2\alpha_2 + \alpha_3 = 0 \end{cases}$.

   We obtain one solution with $\alpha_3 = 1$, $\alpha_1 = -3$, $\alpha_2 = -\dfrac{1}{2}$, so s_2 is linear dependent.

According to Theorem 5-3-2, determining whether a set is linearly dependent or independent using the original definition is equivalent to solving an $n \times n$ system of equations, which is computationally inefficient. In practice, this computation can be replaced by using determinants.

▶ **Theorem 5-3-3**

The Relationship Between the Determinant Value and The Linear Independence or Dependence of Vectors

Let $W = \{\vec{v_1}, \vec{v_2}, \cdots, \vec{v_n}\}$, where $\vec{v_1} = (a_{11}, a_{12}, \cdots, a_{1n})$, $\vec{v_2} = (a_{21}, a_{22}, \cdots, a_{2n})$, $\cdots$,

$\vec{v_n} = (a_{n1}, a_{n2}, \cdots, a_{nn})$, consider determinants $|A| = \begin{vmatrix} a_{11} & a_{12} & \cdots & a_{1n} \\ a_{21} & a_{22} & & \vdots \\ \vdots & \vdots & & \vdots \\ a_{n1} & \cdots & \cdots & a_{nn} \end{vmatrix}$, the

relationship between the determinant and linear independence is as follows (the proof will be encountered in the linear algebra section):

(1) If $|A| \neq 0 \Rightarrow \{\vec{v_1}, \vec{v_2}, \cdots, \vec{v_n}\}$ is linearly independent.

(2) If $|A| = 0 \Rightarrow \{\vec{v_1}, \vec{v_2}, \cdots, \vec{v_n}\}$ is linearly dependent. ◀

Example 5

A set $\{\vec{A}, \vec{B}, \vec{C}\}$ in R^3, where $\vec{A} = (-1, 1, 1)$, $\vec{B} = (1, -1, 1)$, $\vec{C} = (0, 0, 2)$, please determine if $\{\vec{A}, \vec{B}, \vec{C}\}$ are linearly dependent or independent.

Solution

$$\begin{vmatrix} -1 & 1 & 1 \\ 1 & -1 & 1 \\ 0 & 0 & 2 \end{vmatrix} = 0 \Rightarrow \{\vec{A}, \vec{B}, \vec{C}\} \text{ linear dependence.}$$

Q.E.D.

Example 6

$$\vec{v_1} = \begin{bmatrix} 1 \\ 2 \\ 3 \end{bmatrix}, \quad \vec{v_2} = \begin{bmatrix} 2 \\ -1 \\ 3 \end{bmatrix}, \quad \vec{v_3} = \begin{bmatrix} 0 \\ 1 \\ -1 \end{bmatrix}, \text{ verify } \vec{v_1}, \vec{v_2}, \vec{v_3} \text{ are linearly independent.}$$

Solution

$$\begin{vmatrix} 1 & 2 & 0 \\ 2 & -1 & 1 \\ 3 & 3 & -1 \end{vmatrix} \neq 0, \text{ so } \{\vec{v_1}, \vec{v_2}, \vec{v_3}\} \text{ are linearly independent.}$$

Q.E.D.

5-3-3 Basis and Dimension

A set W is linear independent, when expressing a vector as a linear combination of the vectors in W, the representation is unique. In R^2, $\{(1, 0), (0, 1)\}$ is an example of such a set; Similarly, in R^n, $\{e_1, e_2, \cdots, e_n\}$, is another example. What characteristics do they have in common?

▶ **Definition 5-3-5**

Linear Combination

Let $\vec{v_1}, \vec{v_2}, \cdots, \vec{v_k}$ be vectors in R^n. For any vector in $\alpha_1, \alpha_2, \cdots, \alpha_k \in R$,

$\alpha_1 \vec{v_1} + \alpha_2 \vec{v_2} + \cdots + \alpha_k \vec{v_k}$ is called the **linear combination** of these n vectors. ◀

Example 7

Let $W = \{(1, 0, 0, 0), (1, 2, 0, 0)\}$ be a subset in R^4, please make $\vec{f} = (a, b, 0, 0)$ represent the linear combination of vectors in the set W.

Solution

Let $\vec{f} = (a, b, 0, 0) = \alpha(1, 0, 0, 0) + \beta(1, 2, 0, 0)$, then $(a, b, 0, 0) = (\alpha + \beta, 2\beta, 0, 0)$,

therefore $\begin{cases} a = \alpha + \beta \\ b = 2\beta \end{cases}$, the solution is $\begin{cases} \alpha = a - \dfrac{1}{2}b \\ \beta = \dfrac{1}{2}b \end{cases}$,

then $\vec{f} = (a, b, 0, 0) = (a - \dfrac{1}{2}b) \cdot (1, 0, 0, 0) + \dfrac{b}{2} \cdot (1, 2, 0, 0)$.   Q.E.D.

▶ **Definition 5-3-6**

The Span of Vector

Given a subset $W = \{\vec{v_1}, \vec{v_2}, \cdots, \vec{v_k}\}$ of R^n, then
$\{\vec{f} \mid \vec{f} = \alpha_1 \vec{v_1} + \cdots + \alpha_k \vec{v_k}, \alpha_1, \cdots, \alpha_k \in R\}$ is called the **span** of W, denoted by span(W).

◀

The subset of R^n $\{\vec{e_1} = (1, 0, \cdots, 0), \vec{e_2} = (0, 1, \cdots, 0), \cdots, \vec{e_n} = (0, 0, \cdots, 1)\}$ is indeed a spanning set, because every vector $\vec{f}$ in R^n can be expressed as:
$\vec{f} = a_1(1, 0, \cdots, 0) + a_2(0, 1, \cdots, 0) + \cdots + a_n(0, 0, \cdots, 1)$.

Example 8

If in the subspace V of R^4, every vector can be expressed as $(x + y, z, x - z)$, find one of spans from V.

Solution

$\begin{bmatrix} x + y \\ z \\ x - z \end{bmatrix} = x \begin{bmatrix} 1 \\ 0 \\ 1 \end{bmatrix} + y \begin{bmatrix} 1 \\ 0 \\ 0 \end{bmatrix} + z \begin{bmatrix} 0 \\ 1 \\ -1 \end{bmatrix}$, that is any vector in V can be expressed as

$\left\{ \begin{bmatrix} 1 \\ 0 \\ 1 \end{bmatrix}, \begin{bmatrix} 1 \\ 0 \\ 0 \end{bmatrix}, \begin{bmatrix} 0 \\ 1 \\ -1 \end{bmatrix} \right\}$, therefore $V = \text{span}\left\{ \begin{bmatrix} 1 \\ 0 \\ 1 \end{bmatrix}, \begin{bmatrix} 1 \\ 0 \\ 0 \end{bmatrix}, \begin{bmatrix} 0 \\ 1 \\ -1 \end{bmatrix} \right\}$

or $V = \text{span}\{(1, 0, 1), (1, 0, 0), (0, 1, -1)\}$ Q.E.D.

▶ **Definition 5-3-7**

Basis

Let V be a subset of $\mathbb{R}^n$, $W = \{v_1, \cdots, v_k\}$ is a subset of V. If W satisfies

(1) W is linearly independent,   (2) $\mathrm{span}(W) = V$, then we called W the **basis** of V.   ◀

Example 9

Verify $\{a_1 = \begin{bmatrix} 1 \\ 0 \\ 0 \end{bmatrix}, a_2 = \begin{bmatrix} 0 \\ 1 \\ 0 \end{bmatrix}, a_3 = \begin{bmatrix} 1 \\ 5 \\ 3 \end{bmatrix}\}$ is a basis of $\mathbb{R}^3$.

Solution

(1) First the proof generating:

$$\forall \vec{f} = \begin{bmatrix} x \\ y \\ z \end{bmatrix} \in \mathbb{R}^3, \ \vec{f} = \begin{bmatrix} x \\ y \\ z \end{bmatrix} = p\vec{a_1} + q\vec{a_2} + r\vec{a_3} = p\begin{bmatrix} 1 \\ 0 \\ 0 \end{bmatrix} + q\begin{bmatrix} 0 \\ 1 \\ 0 \end{bmatrix} + r\begin{bmatrix} 1 \\ 5 \\ 3 \end{bmatrix}$$

$$\Rightarrow \begin{cases} x = p + r \\ y = q + 5r \\ z = 3r \end{cases} \Rightarrow \begin{cases} p = x - \dfrac{z}{3} \\ q = y - \dfrac{5}{3}z \\ r = \dfrac{z}{3} \end{cases} \Rightarrow \vec{f} = (x - \dfrac{z}{3})\begin{bmatrix} 1 \\ 0 \\ 0 \end{bmatrix} + (y - \dfrac{5}{3}z)\begin{bmatrix} 0 \\ 1 \\ 0 \end{bmatrix} + \dfrac{z}{3}\begin{bmatrix} 1 \\ 5 \\ 3 \end{bmatrix},$$

∴ Any vector in $\mathbb{R}^3$ can be expressed as a linear combination of $\vec{a_1} \cdot \vec{a_2} \cdot \vec{a_3}$.

(2) Then verify the linear independence:

$\begin{vmatrix} 1 & 0 & 1 \\ 0 & 1 & 5 \\ 0 & 0 & 3 \end{vmatrix} = 3 \neq 0 \Rightarrow \vec{a_1}, \vec{a_2}, \vec{a_3}$ is linear independent, so $\{\vec{a_1}, \vec{a_2}, \vec{a_3}\}$ is a basis of $\mathbb{R}^3$.

Q.E.D.

▶ **Definition 5-3-8**

Dimension

The number of vectors in a basis for a vector space V is called the **dimension** of V, denoted as $\dim(V)$.   ◀

For space $\mathbb{R}^3$, its basis consists of $\vec{i} = \begin{bmatrix} 1 \\ 0 \\ 0 \end{bmatrix}, \vec{j} = \begin{bmatrix} 0 \\ 1 \\ 0 \end{bmatrix}, \vec{k} = \begin{bmatrix} 0 \\ 0 \\ 1 \end{bmatrix}$ three vectors,

therefore $\dim \mathbb{R}^3 = 3$. By extension, we can infer that $\dim \mathbb{R}^n = n$.

► Theorem 5-3-4

The Other Properties of Vector Space

In R^n, we have the following properties.

(1) If a subset contains more than $n + 1$ vectors, then it is linearly dependent.

(2) A set W that contains exactly n linearly independent vectors is a basis.

Proof   (selected reading)

1. If a set is linearly independent, represents $R^{n+1} \subseteq R^n$, in this case, there must exist n

    nonzero real numbers $c_1, \cdots\cdots, c_n$ such that $\underbrace{(0, \cdots, 0, 1)}_{n} = \sum_{i=1}^{n} c_i e_i$, which clearly

    leads to a contradiction.

2. If W is not a spanning set, then there exists a nonzero vector v such that
    $v \in R^n \setminus \text{span}(W)$. In other words, $\text{span}(W) \cup \{v\} = R^{n+1} \subseteq R^n$, following the same
    reasoning as in (1), we arrive at a contradiction.   ◄

Example 10

(1) Verify $V_1 = \begin{bmatrix} 1 \\ 2 \\ 3 \end{bmatrix}$, $V_2 = \begin{bmatrix} 2 \\ -1 \\ 3 \end{bmatrix}$, $V_3 = \begin{bmatrix} 0 \\ 1 \\ -1 \end{bmatrix}$, $V_4 = \begin{bmatrix} 4 \\ -1 \\ 5 \end{bmatrix}$ are linearly dependent each

   other.

(2) Verify V_1, V_2, V_3 are linearly independent each other.

Solution

(1) In Theorem 5-3-4, we know by $n = 3$.

(2) $\begin{vmatrix} 1 & 2 & 0 \\ 2 & -1 & 1 \\ 3 & 3 & -1 \end{vmatrix} \neq 0$, so V_1, V_2, V_3 are linearly independent.   Q.E.D.

Example 11

Let V be the subspace of R^6, and all the vectors in V can be represented as
$(x, y, 2x - y, z, 3x + y - 2z, z)$, where x, y, z are real numbers, find a basis for V, and
solve $\dim(V)$.

Solution

(1) Through
   $(x, y, 2x - y, z, 3x + y - 2z, z)$
   $= x \cdot (1, 0, 2, 0, 3, 0) + y \cdot (0, 1, -1, 0, 1, 0) + z \cdot (0, 0, 0, 1, -2, 1)$
   we know $V = \text{span}\{(1, 0, 2, 0, 3, 0), (0, 1, -1, 0, 1, 0), (0, 0, 0, 1, -2, 1)\}$.

According to Theorem 5-3-2, we know
$\{(1, 0, 2, 0, 3, 0), (0, 1, -1, 0, 1, 0), (0, 0, 0, 1, -2, 1)\}$ is combination of linear independence.
Hence $\{(1, 0, 2, 0, 3, 0), (0, 1, -1, 0, 1, 0), (0, 0, 0, 1, -2, 1)\}$ is one basis in V.

(2) Through the definition of dimension, we know $\dim(V) = 3$.

▶ **Theorem 5-3-5**

Properties of Subspace Dimension

If V is a subspace of a vector space U, then $\dim(V) \leq \dim(U)$.

Proof

If W is a basis for a subspace V, then $\mathrm{span}(W) = V \subseteq U$. Therefore, the number of vectors in s at most $\dim(U)$, or it would lead to a contradiction, as stated in Theorem 5-3-4(1). ◀

If we consider the polynomial x^k as a vector (representing e_k), then a polynomial $f(x) = a_0 + a_1 x + \cdots + a_n x^n$ can be represented as $v = a_0 + a_1 e_1 + \cdots + a_n e_n$. In fact, based on the assumption of polynomials, the set $\{1, x, x^2, \cdots, x^n\}$ is linearly independent, therefore, the set of all polynomials with $\deg \leq n$

$$P_n = \mathrm{span}\{1, x, x^2, \cdots, x^n\}$$

can be treated as $\mathbb{R}^{n+1}$, with the basis $\{1, x, x^2, \cdots, x^n\}$. Similarly, $n \times n$ matrices $M_{n \times n}(\mathbb{R})$, sets of all analytic functions $C^\infty(\mathbb{R})$ and so on, can be viewed as vector spaces. We will discuss these in more detail in Chapter 6.

Example 12

Consider the second-order linear homogeneous ODE: $y'' + 9y = 0$, its solution space is V, find a set of basis in V, and solve $\dim(V)$.

Solution

According to the theory of ODE, the general solution is $y(x) = c_1 \cos 3x + c_2 \sin 3x$, the solution space $V = \mathrm{span}\{\cos 3x, \sin 3x\}$ and $\{\cos 3x, \sin 3x\}$ is one basis of the solution space, and the number of vectors in basis are two, therefore $\dim(V) = 2$. Q.E.D.

In summary, based on the concepts discussed earlier, we actually have the following theorem, which readers can try to prove.

▶ **Theorem 5-3-6**

The Relationship Between Subspace and Basis

A subset W of a vector space V is a basis if and only if it satisfies one of the following three conditions:
(1) W is a minimal span.
(2) W is a maximal linearly independent subset.
(3) The number of elements in W is equal to dim(V). ◀

5-3-4   Orthogonal Set

We are accustomed to representing vectors $v \in \mathrm{R}^3$ using the standard basis $\{\hat{i}, \hat{j}, \hat{k}\}$, which means that $v = a_1\hat{i} + a_2\hat{j} + a_3\hat{k}$. This representation aligns with our intuitive understanding of perpendicular coordinate axes in three-dimensional space. In fact, according to the standard inner product in R^n, the vectors in $\{e_i \mid 1 \le i \le n\}$ are also mutually perpendicular. Consequently, every non-zero vector v can be represented in the form $v = \sum_{i=1}^{n} <v, e_i> e_i$. In other words, using the inner product to transform the vectors in the basis to be mutually perpendicular makes it easier for us to describe the vector space. This section provides a detailed explanation of the relevant methods.

▶ **Definition 5-3-9**

Orthogonal Set and Orthonormal Set

(1) If $\{\vec{v_1}, \vec{v_2}, \cdots, \vec{v_n}\}$ is a subset of R^n, satisfies $<\vec{v_i}, \vec{v_j}> = \begin{cases} 0, & i \ne j \\ |\vec{v_i}|^2, & i = j \end{cases}$,

($i, j = 1, 2, \cdots, n$), then $\{\vec{v_1}, \vec{v_2}, \cdots, \vec{v_n}\}$ is called an **orthogonal set.**

(2) If $<\vec{v_i}, \vec{v_j}> = \begin{cases} 0, & i \ne j \\ 1, & i = j \end{cases}$ ($i, j = 1, 2, \cdots, n$), then we called $\{\vec{v_1}, \vec{v_2}, \cdots, \vec{v_n}\}$ is an

orthonormal set. ◀

Therefore, by normalizing the vectors in an orthogonal set $\vec{u_1} = \dfrac{\vec{v_1}}{|\vec{v_1}|}$, $\vec{u_2} = \dfrac{\vec{v_2}}{|\vec{v_2}|}$,

······, $\vec{u_n} = \dfrac{\vec{v_n}}{|\vec{v_n}|}$ we can obtain a orthonormal set.

Gram-Schmidt Process

First, let's consider the orthogonalization of two vectors, through a process that making $\{\vec{a_1}, \vec{a_2}\}$ became $\{\vec{v_1}, \vec{v_2}\}$, causes $\{\vec{v_1}, \vec{v_2}\}$ is orthogonal set (as shown in Figure 5-21) and $\text{span}\{\vec{a_1}, \vec{a_2}\} = \text{span}\{\vec{v_1}, \vec{v_2}\}$, specifically, we are seeking a coefficient λ causes $\vec{a_2} - \lambda \vec{a_1}$ is perpendicular to $\vec{a_1}$, as shown in Figure 5-21. Therefore, we can solve the equation $<\vec{a_2} - \lambda\vec{a_1}, \vec{a_1}> = 0$ to obtain $\lambda = \dfrac{<\vec{a_2}, \vec{v_1}>}{<\vec{v_1}, \vec{v_1}>}$;

$\text{span}\{\vec{a_1}, \vec{a_2} - \lambda\vec{a_1}\} = \text{span}\{\vec{a_1}, \vec{a_2}\}$, this can be verified from the definition of linear combination. Let's summarize the above process into an algorithm and generalize it for orthogonalizing n vectors set $\{\vec{a_1}, \vec{a_2}, \cdots, \vec{a_n}\}$ as follows:

▶ **Theorem 5-3-7**

Gram-Schmidt Orthogonalization

Assuming that $\{\vec{a_1}, \vec{a_2}, \cdots, \vec{a_n}\}$ is a linearly independent set, the Gram-Schmidt orthogonalization algorithm can be summarized as follows:

STEP 1 Let $\vec{v_1} = \vec{a_1}$.

STEP 2 Let $\vec{v_2} = \vec{a_2} - \dfrac{\vec{a_2} \cdot \vec{v_1}}{(\vec{v_1} \cdot \vec{v_1})}\vec{v_1}$.

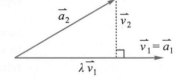

STEP 3 Let $\vec{v_k} = \vec{a_k} - \sum\limits_{i=1}^{k-1} \dfrac{<\vec{a_k}, \vec{v_i}>}{<\vec{v_i}, \vec{v_i}>}\vec{v_i}$ $(3 \leq k \leq n)$.

Figure 5-21 The graph of orthogonalization

STEP 4 Take $\{\dfrac{\vec{v_1}}{|\vec{v_1}|}, \dfrac{\vec{v_2}}{|\vec{v_2}|}, \cdots, \dfrac{\vec{v_n}}{|\vec{v_n}|}\}$.

◀

Based on STEP 3 and the definition of linear combination, we know

$$\text{span}\{\frac{\vec{v_1}}{|\vec{v_1}|}, \frac{\vec{v_2}}{|\vec{v_2}|}, \cdots, \frac{\vec{v_n}}{|\vec{v_n}|}\} = \text{span}\{\vec{a_1}, \cdots, \vec{a_n}\}.$$

Example 13

Let $\vec{a}_1 = -\vec{i} + \vec{j} + \vec{k}$, $\vec{a}_2 = \vec{i} - \vec{j} + \vec{k}$, $\vec{a}_3 = \vec{i} + \vec{j} - \vec{k}$.

(1) Find the corresponding orthonormal set $\{\vec{u}_1, \vec{u}_2, \vec{u}_3\}$.

(2) Make $\vec{A} = \vec{i} + 2\vec{j} + 3\vec{k}$ be written as a linear combination of $\{\vec{u}_1, \vec{u}_2, \vec{u}_3\}$.

Solution

(1) STEP 1 Let $\vec{v}_1 = \vec{a}_1 = (-1, 1, 1)$.

STEP 2 Let $\vec{v}_2 = \vec{a}_2 - \dfrac{<\vec{a}_2, \vec{v}_1>}{<\vec{v}_1, \vec{v}_1>}\vec{v}_1 = \vec{a}_2 - \dfrac{1}{3}\vec{v}_1 = (1, -1, 1) + \dfrac{1}{3}(-1, 1, 1) = (\dfrac{2}{3}, -\dfrac{2}{3}, \dfrac{4}{3})$.

STEP 3 Let $\vec{v}_3 = \vec{a}_3 - \dfrac{<\vec{a}_3, \vec{v}_1>}{<\vec{v}_1, \vec{v}_1>}\vec{v}_1 - \dfrac{<\vec{a}_3, \vec{v}_2>}{<\vec{v}_2, \vec{v}_2>}\vec{v}_2 = \vec{a}_3 - (-\dfrac{1}{3})\vec{v}_1 - (-\dfrac{1}{2})\vec{v}_2$

$= \vec{a}_3 + \dfrac{1}{3}\vec{v}_1 + \dfrac{1}{2}\vec{v}_2 = (1, 1, -1) + \dfrac{1}{3}(-1, 1, 1) + \dfrac{1}{2}(\dfrac{2}{3}, -\dfrac{2}{3}, \dfrac{4}{3}) = (1, 1, 0)$.

STEP 4 Normalization

$$\vec{u}_1 = \frac{\vec{v}_1}{|\vec{v}_1|} = (-\frac{1}{\sqrt{3}}, \frac{1}{\sqrt{3}}, \frac{1}{\sqrt{3}}); \quad \vec{u}_2 = \frac{\vec{v}_2}{|\vec{v}_2|} = (\frac{1}{\sqrt{6}}, -\frac{1}{\sqrt{6}}, \frac{2}{\sqrt{6}});$$

$$\vec{u}_3 = \frac{\vec{v}_3}{|\vec{v}_3|} = (\frac{1}{\sqrt{2}}, \frac{1}{\sqrt{2}}, 0).$$

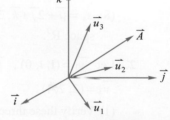

(2) $<\vec{A}, \vec{u}_1> = \dfrac{4}{\sqrt{3}}$, $<\vec{A}, \vec{u}_2> = \dfrac{5}{\sqrt{6}}$, $<\vec{A}, \vec{u}_3> = \dfrac{3}{\sqrt{2}}$,

so $\vec{A} = <\vec{A}, \vec{u}_1>\vec{u}_1 + <\vec{A}, \vec{u}_2>\vec{u}_2 + <\vec{A}, \vec{u}_3>\vec{u}_3$

$= \dfrac{4}{\sqrt{3}}\vec{u}_1 + \dfrac{5}{\sqrt{6}}\vec{u}_2 + \dfrac{3}{\sqrt{2}}\vec{u}_3$, as shown in the figure.   Q.E.D.

5-3 Exercises

Basic questions

1. Determine whether the following set is linearly independent or dependent.
   (1) $s_1 = \{(1, -2, 3)\}$ in the space R^3.
   (2) $s_2 = \{\vec{i}, 2\vec{j}, 3\vec{i} - 4\vec{k}, \vec{i} + \vec{j} + \vec{k}\}$
   in the space R^3.
   (3) $s_3 = \{\vec{i} + 2\vec{j}, 3\vec{i} - 4\vec{k}, 5\vec{i} + 4\vec{j} - 4\vec{k}\}$
   in the space R^3.
   (4) $s_4 = \{(9, -2, 0, 0, 0, 0), (0, 0, 0, 0, 8, 7)\}$ in the space R^6.
   (5) $s_5 = \{(4, 0, 0, 0), (0, 5, 1, 0),$
   $\quad (8, -10, -2, 0)\}$ in the space R^4.
   (6) $s_6 = \{(1, -2), (3, 4), (-5, 8)\}$
   in the space R^2.
   (7) $s_7 = \{(-1, 1, 0, 0, 0), (0, -1, 1, 0, 0),$
   $\quad (0, 1, 1, 1, 0)\}$ in the space R^5.
   (8) $s_8 = \{\vec{i} + 2\vec{j} + \vec{k}, 3\vec{i} - 4\vec{k}, 5\vec{i} + 4\vec{j} - 4\vec{k}\}$ in the space R^3.

2. Let $\vec{v_1} = (1, 1, 0)$, $\vec{v_2} = (2, 0, 1)$,
   $\vec{v_3} = (2, 2, 1)$.
   (1) Verify these three vectors are linearly independent.
   (2) Find a set of these three vectors that is orthogonal.

Advanced questions

1. If the following (1)~(8) are subspaces of R^n, find a basis for each and provide the dimension. If not, please provide the reason.
   (1) s_1 is the set of any vectors in the R^2 plane that include the line $4x + y = 0$.
   (2) s_2 is the set of any vectors in the R^2 plane that include the line $4x + y = 1$.
   (3) s_3 is the set of any vectors in the R^3 space that include the plane $4x + y - z = 0$.
   (4) s_4 is the set of any vectors in the R^3 space that include the plane $4x + y - z = 1$.
   (5) s_5 is the set of any vectors in the R^3 space that include z-axis.
   (6) s_6 is the set of every vectors $(-x, x, y, -3y)$ in the space R^4.
   (7) s_7 is the set of every vectors $(-x, x, y, 2)$ in the space R^4.
   (8) s_8 is the set of every vectors $(x, x - y, x + y - z, z, 0)$ in the space R^5.

2. Reconstructed $\vec{v_1} = (2, -2, -1)$, $\vec{v_2} = (1, 1, 0)$, $\vec{v_3} = (1, -1, 4)$ as a set of orthonormal basis.

3. Reconstructed $\vec{v_1} = (1, 1, 1)$, $\vec{v_2} = (1, 2, 3)$, $\vec{v_3} = (2, 3, 1)$ as a set of normalized orthogonal basis.

4. Reconstructed $\vec{a_1} = (2, 1, 1)$, $\vec{a_2} = (1, 0, 2)$, $\vec{a_3} = (2, 0, 0)$ as a set of orthonormal basis and expressing $\vec{v} = (-1, 2, 3)$ in terms of this orthonormal basis coordinates.

6

Matrix Operations and Linear Algebra

Cayley
(1821~1895, Britain)

From the 17th century until Gauss, the study of matrices in mathematics was primarily focused on solving systems of linear equations. It was not until the British mathematician Arthur Cayley (1821-1895) introduced matrix multiplication and inverse that the theory of modern linear algebra began to take shape. Over time, linear algebra has become an essential tool for solving problems in various fields such as statistical analysis, mechanics, circuit analysis, optics, and quantum physics.

Learning Objectives

6-1
Matrix Definition and Basic Operations

6-1-1 Familiarity with common matrix

6-1-2 Proficiency in basic matrix operations

6-1-3 Recognition of symmetric (anti-symmetric) matrix and Hermitian (anti-Hermitian) matrix

6-2
Matrix Row (Column) Operations and Determinant

6-2-1 Proficiency in basic row operations and their applications

6-2-2 Proficiency in calculating determinants and its properties

6-2-3 Mastering the methods for finding the inverse of square matrices: augmented matrix method, cofactor matrix method

6-3
Solution to Systems of Linear Equations

6-3-1 Proficiency in Gaussian elimination and solving systems of linear equations

6-3-2 Understand the rank of matrix and its relationship with systems of linear equations

6-3-3 Mastering the dimension theorem and the existence of solutions for systems of linear equations

6-3-4 Knowing how to solve systems of linear equations using the "Gaussian elimination"

6-3-5 Knowing how to solve systems of linear equations using the "Cramer's rule"

6-4
Eigenvalues and Eigenvectors

6-4-1 Proficiency in finding eigenvalues and eigenvectors

6-4-2 Understanding the independence of eigenvectors and techniques for observing eigenvalues

6-5
Matrix Diagonalization

6-5-1 Understanding the properties of similar matrix

6-5-2 Mastering the process of matrix diagonalization

6-6
Matrix Functions

6-6-1 Mastering the use of the Cayley-Hamilton theorem to find matrix space bases

6-6-2 Mastering the calculation of matrix functions: long division, Cayley-Hamilton theorem, eigenvalues

6-6-3 Mastering the calculation of matrix functions: diagonalization

ExampleVideo

This chapter will begin with simple matrix operations and introduce the solution of systems of linear equations. We will start with the Gaussian elimination method as a starting point to gain a deeper understanding of the conditions under which a linear system of equations has a solution and whether the solution is unique. We will also introduce Cramer's Rule, which not only solves linear systems of equations but also has many theoretical applications, such as algorithms for computing inverse matrices.

After concluding the discussion on equations, we will proceed to a geometric observation, further abstracting matrices as mappings between two vector spaces. We will introduce the concept of eigenvalues, which prepares a solid foundation for advanced applications in linear algebra. This step is crucial in building the necessary knowledge for higher-level applications in statistics, mechanics, circuit analysis, optics, and quantum physics.

6-1   Matrix Definition and Basic Operations

In our secondary education, we often use the elimination method to solve systems of linear equations. Common types of linear systems include homogeneous systems of constant coefficient linear equations $\begin{cases} x_1 + 2x_2 - x_3 = 0 \\ 2x_1 + 3x_2 + x_3 = 0 \\ x_1 + x_2 + 2x_3 = 0 \end{cases}$ and non-homogeneous systems

$\begin{cases} x_1 + 2x_2 - x_3 = 7 \\ 2x_1 + 3x_2 + x_3 = 14 \\ x_1 + x_2 + 2x_3 = 7 \end{cases}$. We observe that the solutions of these systems are related to the

operations involving the coefficient matrix of the homogeneous system

$A = \begin{bmatrix} 1 & 2 & -1 \\ 2 & 3 & 1 \\ 1 & 1 & 2 \end{bmatrix}$ and the augmented matrix of the non-homogeneous

system $\widetilde{A} = \begin{bmatrix} 1 & 2 & -1 & 7 \\ 2 & 3 & 1 & 14 \\ 1 & 1 & 2 & 7 \end{bmatrix}$. This observation sparked the initial idea among

mathematicians to utilize matrices for solving systems of linear equations. Furthermore, in our daily lives, we frequently encounter the need to handle large amounts of data. For example, analyzing monthly expenses in a household for each year. Let's assume that the important expenses for the first three months of a particular year are as follows:

	January	February	March	...
Rent	5000	5000	5000	...
Food expenses	8000	10000	7000	...
Transport expenses	2000	1500	2500	...
Education expenses	10000	8000	12000	...

If we want to compare expenses for each month or each year, or compare the difference in expenses between two months, a systematic analysis method is required. This directly led to the birth of arrays (now known as matrices). For example, the expenses for the first three months can be represented as

$$A = \begin{bmatrix} 5000 & 5000 & 5000 \\ 8000 & 10000 & 7000 \\ 2000 & 1500 & 2500 \\ 10000 & 8000 & 12000 \end{bmatrix},$$

which first column $\begin{bmatrix} 5000 \\ 8000 \\ 2000 \\ 10000 \end{bmatrix}$, second column $\begin{bmatrix} 5000 \\ 10000 \\ 1500 \\ 8000 \end{bmatrix}$, third column $\begin{bmatrix} 5000 \\ 7000 \\ 2500 \\ 12000 \end{bmatrix}$

respectively represent the particular category of expenditure of January, February, March. For example, the transportation expense for January would be in the third row and first column of matrix A, while the education expense for March would be in the fourth row and third column of matrix A. By utilizing matrix operations, we can perform various data analysis tasks.

6-1-1 Basic Concept

▶ **Definition 6-1-1**

The Definition of Matrix

A **Matrix** (Matrices) is formed by arranging $m \times n$ real numbers (or complex numbers) into a rectangular array with m rows and n columns. It is denoted as $A_{m \times n}$ or $A \equiv [a_{ij}]_{m \times n}$, generally, it will be indicated by the following symbol

$$A = \begin{bmatrix} a_{11} & a_{12} & \cdots & a_{1n} \\ a_{21} & a_{22} & \cdots & a_{2n} \\ \vdots & & & \vdots \\ a_{m1} & a_{m2} & \cdots & a_{mn} \end{bmatrix}_{m \times n}$$

and m represents the number of **rows** of A, n represents the number of **columns** of A, a_{ij} represents the **element** in the i-th row and j-th column of A, which is also called **Matrix coefficient.** It is common to omit $m \times n$ at the bottom right corner if they do not create confusion in the context. ◀

Nouns

We often use R^n represent n-dimensional real number space, C^n represent the n-dimensional complex number space, and take R^m or C^m represent vectors in real m-dimension and complex m-dimension.

(1) Column Matrix (Column vector)

$$X_{m\times1} = \begin{bmatrix} x_1 \\ x_2 \\ \vdots \\ x_m \end{bmatrix} \in \mathrm{R}^n \text{ or } \mathrm{C}^n.$$ $\vec{a} = 2\vec{i} + 3\vec{j} - 4\vec{k}$ can be written as the form of column

matrix $\begin{bmatrix} 2 \\ 3 \\ -4 \end{bmatrix}$.

(2) Row Matrix (Row vector)

$Y_{1\times n} = [y_1 \ y_2 \ \cdots \ y_n] \in \mathrm{R}^n$ or C^n. Similarly, $\vec{a} = 2\vec{i} + 3\vec{j} - 4\vec{k}$ can be expressed as
$[2 \ 3 \ -4]$.

For instance, matrix $A = [a_{ij}]_{m\times n} \equiv \begin{bmatrix} a_{11} & a_{12} & \cdots & a_{1n} \\ a_{21} & \ddots & & \vdots \\ \vdots & & \ddots & \vdots \\ a_{m1} & a_{m2} & \cdots & a_{mn} \end{bmatrix}$ with m row vectors and n

column vectors. In fact, if let $v_i = \begin{bmatrix} a_{1i} \\ a_{2i} \\ \vdots \\ a_{mi} \end{bmatrix}$, then $A_{m\times n} = [v_1 \ v_2 \ \cdots \ v_n]$; similarly,

if let $u_i = [a_{i1} \ a_{i2} \ \cdots \ a_{in}]$, then $A_{m\times n} = \begin{bmatrix} u_1 \\ u_2 \\ \vdots \\ u_m \end{bmatrix}$. Take $A = \begin{bmatrix} 1 & 2 & 3 \\ 4 & 5 & 6 \\ 7 & 8 & 9 \end{bmatrix}$ for

example, then $A = [v_1 \ v_2 \ v_3]$ or $A = \begin{bmatrix} u_1 \\ u_2 \\ u_3 \end{bmatrix}$, where $v_1 = \begin{bmatrix} 1 \\ 4 \\ 7 \end{bmatrix}$, $v_2 = \begin{bmatrix} 3 \\ 5 \\ 8 \end{bmatrix}$,

$v_3 = \begin{bmatrix} 3 \\ 6 \\ 9 \end{bmatrix}$; $u_1 = [1 \ 2 \ 3]$, $u_2 = [4 \ 5 \ 6]$, $u_3 = [7 \ 8 \ 9]$.

(3) Square Matrix

A matrix which its column m = row n, that is

$$A_{n\times n} = \begin{bmatrix} a_{11} & a_{12} & \cdots & a_{1n} \\ a_{21} & a_{22} & & \vdots \\ \vdots & & & \vdots \\ a_{n1} & \cdots & \cdots & a_{nn} \end{bmatrix}_{\text{main diagonal}}$$

For example: $\begin{bmatrix} 1 & 2 \\ 3 & 4 \end{bmatrix}$ is 2 × 2 square matrix; $\begin{bmatrix} 1 & 2 & 3 \\ -1 & 2 & 3 \\ 4 & 5 & 1 \end{bmatrix}$ is 3 × 3 square matrix.

(4) Upper Triangular Matrix

Assuming that square matrix $U = [a_{ij}]_{n \times n}$, if $a_{ij} = 0$, $\forall i > j$, then U is called upper

triangular matrix, that is $U_{n \times n} = \begin{bmatrix} & & \\ i > j & & i \leq j \\ a_{ij} = 0 & & \end{bmatrix}$. For example: $\begin{bmatrix} 1 & 2 & 3 \\ 0 & 2 & 3 \\ 0 & 0 & 1 \end{bmatrix}$.

(5) Lower Triangular Matrix

Assuming that square matrix $L = [a_{ij}]_{n \times n}$, if $a_{ij} = 0$, $\forall i < j$, then L is called lower

triangular matrix, that is $L_{n \times n} = \begin{bmatrix} & & i < j \\ i \geq j & & a_{ij} = 0 \\ & & \end{bmatrix}$.

For example: $\begin{bmatrix} 1 & 0 & 0 \\ 2 & 2 & 0 \\ 3 & 2 & 1 \end{bmatrix}$ is a lower triangular matrix.

(6) Diagonal Matrix

$a_{ij} = 0$, $\forall i \neq j$, then diagonal matrix $D_{n \times n} = \begin{bmatrix} \ddots & & O \\ & & \\ O & & \ddots \end{bmatrix}$. For example: $\begin{bmatrix} 1 & 0 & 0 \\ 0 & 2 & 0 \\ 0 & 0 & 1 \end{bmatrix}$ is a

3×3 diagonal matrix.

In fact, upper triangular, lower triangular, or diagonal matrices do not necessarily need to be square matrices. The definition can still be established if it is not a square matrix. Since most practical physical system applications are based on square matrices, we only consider square matrix in this book as the definition.

(7) Unit Matrix

$A = [a_{ij}]_{n \times n}$, if $a_{ij} = \delta_{ij} \begin{cases} 1, i = j \\ 0, i \neq j \end{cases}$, then $I_n = \begin{bmatrix} 1 & & O \\ & \ddots & \\ O & & 1 \end{bmatrix}$, which is called an

n-order unit matrix.

For example: $I_2 = \begin{bmatrix} 1 & 0 \\ 0 & 1 \end{bmatrix}$, $I_3 = \begin{bmatrix} 1 & 0 & 0 \\ 0 & 1 & 0 \\ 0 & 0 & 1 \end{bmatrix}$ are unit matrices.

(8) Zero Matrix

$a_{ij} = 0$, $\forall i, j$, that is zero matrix $O_{n \times n} = \begin{bmatrix} \ddots & & O \\ & & \\ O & & \ddots \end{bmatrix}$.

For example: $O_{2 \times 2} = \begin{bmatrix} 0 & 0 \\ 0 & 0 \end{bmatrix}$, $O_{3 \times 3} = \begin{bmatrix} 0 & 0 & 0 \\ 0 & 0 & 0 \\ 0 & 0 & 0 \end{bmatrix}$ are zero matrices.

(9) Submatrix

Assuming $A = [a_{ij}]_{n \times n}$, the matrix obtained by deleting certain rows and columns of matrix A is called a submatrix of A. According to this definition, we can deduce the following: ①A is a submatrix of itself; ②among the submatrices of A, those with the same number of rows as columns are called square submatrices of A; ③ assuming $A \equiv [a_{ij}]_{n \times n}$ is square matrix. If we simultaneously remove the same set of rows and columns from A, the resulting square submatrix is called a principal submatrix of A. For instance: $A = \begin{bmatrix} 1 & 4 & 7 \\ 2 & 5 & 8 \\ 3 & 6 & 9 \end{bmatrix}$, then the submatrix of matrix A is:

$[1 \quad 4 \quad 7]$, $[9]$, $\begin{bmatrix} 1 & 4 \\ 2 & 5 \end{bmatrix}$, $\begin{bmatrix} 4 & 7 \\ 5 & 8 \\ 6 & 9 \end{bmatrix}$, then the principal submatrix of matrix A

is: $[1]$, $[5]$, $[9]$, $\begin{bmatrix} 1 & 4 \\ 2 & 5 \end{bmatrix}$, $\begin{bmatrix} 1 & 7 \\ 3 & 9 \end{bmatrix}$, $\begin{bmatrix} 5 & 8 \\ 6 & 9 \end{bmatrix}$ and $\begin{bmatrix} 1 & 4 & 7 \\ 2 & 5 & 8 \\ 3 & 6 & 9 \end{bmatrix}$.

Example 1

Assuming that matrix $A = \begin{bmatrix} -1 & 3 & \pi & 5 \\ 0 & 1 & 0.1 & \sqrt{2} \\ 0 & \frac{1}{2} & -4 & 0 \end{bmatrix}$, please answer the following questions.

(1) What's the element in the second row and third column of matrix A?
(2) Write the order (dimension) of matrix A.
(3) Write the sets formed by all column vectors and row vectors of matrix A.

Solution

(1) $a_{23} = 0.1$.

(2) The order of matrix A is 3×4.

(3) The set formed by all row vectors of A is :

$$\{[-1 \quad 3 \quad \pi \quad 5], [0 \quad 1 \quad 0.1 \quad \sqrt{2}], [0 \quad \frac{1}{2} \quad -4 \quad 0]\};$$

the set formed by all column vectors of matrix A: $\left\{ \begin{bmatrix} -1 \\ 0 \\ 0 \end{bmatrix}, \begin{bmatrix} 3 \\ 1 \\ \frac{1}{2} \end{bmatrix}, \begin{bmatrix} \pi \\ 0.1 \\ -4 \end{bmatrix}, \begin{bmatrix} 5 \\ \sqrt{2} \\ 0 \end{bmatrix} \right\}$. Q.E.D.

6-1-2 The Basic Algebraic Operations of Matrices

1. The Definitions of Equality, Addition, and Scalar Multiplication

(1) The Equality of Matrix

If $A = [a_{ij}]_{m \times n}$, $B = [b_{ij}]_{m \times n}$, define $A = B \Leftrightarrow a_{ij} = b_{ij}$, $\forall i, j$.

(2) The Addition and Subtraction of Matrix

If $A = [a_{ij}]_{m \times n}$, $B = [b_{ij}]_{m \times n}$, define:

$A + B = [a_{ij} + b_{ij}]_{m \times n}$;

$A - B = [a_{ij} - b_{ij}]_{m \times n}$.

(3) The Scalar Multiplication of Matrix

Let $A = [a_{ij}]_{m \times n}$, define $\alpha A = [\alpha a_{ij}]_{m \times n}$.

► Theorem 6-1-1

The Common Theorem of Matrix Operation

Matrix addition and subtraction are only meaningful when matrices A and B have the same size. The matrix addition and scalar multiplication satisfies:

① $A + B = B + A$ (commutative law of addition)

② $(A + B) + C = A + (B + C)$ (associate law of addition)

③ $A + O = O + A = A$ (additive identity element)

④ $A + (-A) = -A + A = O$ (Opposite element of addition)

⑤ $A + B = A + C \Rightarrow B = C$ (Cancellation law)

⑥ $(\alpha + \beta)A = \alpha A + \beta A$

⑦ $\alpha(A + B) = \alpha A + \alpha B$. ◄

For instance: $A = \begin{bmatrix} 1 & -1 & 3 \\ 2 & 4 & 5 \end{bmatrix}$, $B = \begin{bmatrix} 0 & 1 & 1 \\ -1 & -2 & -3 \end{bmatrix}$ then according to the above definition:

$$A + 2B = \begin{bmatrix} 1 & -1 & 3 \\ 2 & 4 & 5 \end{bmatrix} + \begin{bmatrix} 0 & 2 & 2 \\ -2 & -4 & -6 \end{bmatrix} = \begin{bmatrix} 1 & 1 & 5 \\ 0 & 0 & -1 \end{bmatrix}.$$

2. The Transpose of Matrix

Let $A = [a_{ij}]_{m \times n}$, then $A^T = [a_{ji}]_{n \times m}$ is called the transpose of A.

For example: $A = \begin{bmatrix} 1 & 2 & -3 \\ 5 & 0 & 7 \end{bmatrix}$, then the transpose of A is $A^T = \begin{bmatrix} 1 & 5 \\ 2 & 0 \\ -3 & 7 \end{bmatrix}$. By

definition, one can deduce the follows:

(1) $(A^T)^T = A$; (2) $(A+B)^T = A^T + B^T$; (3) $(\alpha A)^T = \alpha A^T$;

(4) $A_{n\times n} = \dfrac{A+A^T}{2} + \dfrac{A-A^T}{2}$.

3. Matrix Conjugation

$A = [a_{ij}]_{m\times n}$, define $\overline{A} = [\overline{a_{ij}}]_{m\times n}$.

Similarly, calculated by definition:

(1) If α, β are real numbers $i = \sqrt{-1}$, $a_{ij} = \alpha + \beta i \Rightarrow$ then $\overline{a_{ij}} = \alpha - \beta i$.

(2) A^H is defined as $\overline{A}^T = A^*$.

(3) $\overline{(\overline{A})} = A$ and $(\overline{A})^T = \overline{(A^T)}$.

For example: $A = \begin{bmatrix} 1-3i & 2 & -3 \\ 5 & i & 7+4i \end{bmatrix}$, then the matrix conjugation of A is

$$\overline{A} = \begin{bmatrix} 1+3i & 2 & -3 \\ 5 & -i & 7-4i \end{bmatrix}.$$

4. Matrix Multiplication

Let $A = [a_{ij}]_{m\times n}$, $B = [b_{ij}]_{n\times \ell}$ (Note that the number of columns of A = the number of rows of B), by the correspondence between elements of rows and columns, matrix multiplication is defined as: $AB = [\sum_{k=1}^{n} a_{ik} b_{kj}]_{m\times \ell}$, this definition can be illustrated as shown in Figure 6-1:

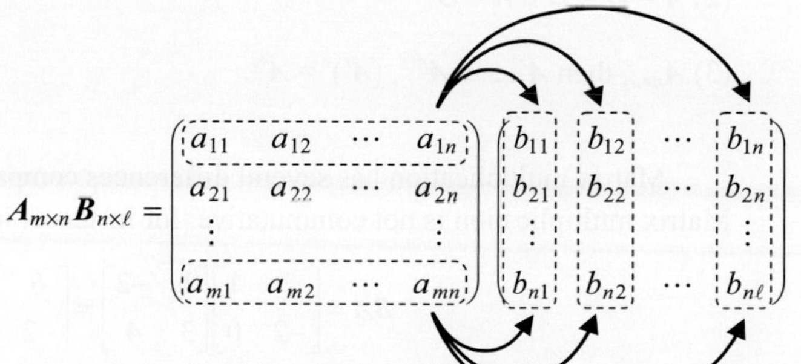

$$= \begin{pmatrix} a_{11}b_{11} + a_{12}b_{21} + \cdots a_{1n}b_{n1} & \cdots & a_{11}b_{1n} + a_{12}b_{2n} + \cdots a_{1n}b_{n\ell} \\ a_{21}b_{11} + a_{22}b_{21} + \cdots a_{2n}b_{n1} & \cdots & a_{21}b_{1n} + a_{22}b_{2n} + \cdots a_{2n}b_{n\ell} \\ \vdots & & \vdots \\ a_{m1}b_{11} + a_{m2}b_{21} + \cdots a_{mn}b_{n1} & \cdots & a_{m1}b_{1n} + a_{m2}b_{2n} + \cdots a_{mn}b_{n\ell} \end{pmatrix}$$

$$= \left(\sum_{k=1}^{n} a_{ik} b_{kj} \right)_{m\times \ell}$$

Figure 6-1 Schematic diagram of matrix multiplication

Example 2

Find the multiplication of two following matrix AB.

(1) $A = \begin{bmatrix} 1 & -2 \\ 3 & 4 \end{bmatrix}$, $B = \begin{bmatrix} 3 & 1 \\ -2 & 0 \end{bmatrix}$.   (2) $A = \begin{bmatrix} 5 & 3 \\ -2 & 1 \\ 0 & 7 \end{bmatrix}$, $B = \begin{bmatrix} 3 & 1 \\ -2 & 0 \end{bmatrix}$.

Solution   Considering Figure 6-1 to get:

(1) $AB = \begin{bmatrix} 1\cdot 3 + (-2)\cdot(-2) & 1\cdot 1 + (-2)\cdot 0 \\ 3\cdot 3 + 4\cdot(-2) & 3\cdot 1 + 4\cdot 0 \end{bmatrix} = \begin{bmatrix} 7 & 1 \\ 1 & 3 \end{bmatrix}$.

(2) $AB = \begin{bmatrix} 5\cdot 3 + 3\cdot(-2) & 5\cdot 1 + 3\cdot 0 \\ (-2)\cdot 3 + 1\cdot(-2) & (-2)\cdot 1 + 1\cdot 0 \\ 0\cdot 3 + 7\cdot(-2) & 0\cdot 1 + 7\cdot 0 \end{bmatrix} = \begin{bmatrix} 9 & 5 \\ -8 & -2 \\ -14 & 0 \end{bmatrix}$.   Q.E.D.

▶ Theorem 6-1-2

The Properties of Matrix Multiplication

Assuming that the matrix multiplication (1) through (6) can be defined, then

(1) $A(B + C) = AB + AC$; $(B + C)A = BA + CA$.   (4) $(AB)^T = B^T A^T$.

(2) $A \times O = O \times A = O$.   (5) $(A^n)^T = (A^T)^n$.

(3) $A_{n\times n}$, then $A^r A^s = A^{r+s}$, $(A^r)^s = A^{rs}$.   (6) $(\overline{AB}) = \overline{A} \times \overline{B}$.   ◀

Matrix multiplication has several differences compared to scalar multiplication: 1. Matrix multiplication is not commutative, for instance: in Example 2 (1),

$$BA = \begin{bmatrix} 3 & 1 \\ -2 & 0 \end{bmatrix}\begin{bmatrix} 1 & -2 \\ 3 & 4 \end{bmatrix} = \begin{bmatrix} 6 & -2 \\ -2 & 4 \end{bmatrix} \neq AB,$$

therefore, the binomial theorem does not hold for matrices, that is $(A + B)^2 = A^2 + AB + BA + B^2$ not necessarily equal to $A^2 + 2AB + B^2$. 2. Matrix multiplication does not possess the cancellation law, for example: $A = \begin{bmatrix} \dfrac{1}{2} & \dfrac{-1}{2} \\ \dfrac{-1}{2} & \dfrac{1}{2} \end{bmatrix} \neq I$ or O but $A^2 = A$; another

example: $A = \begin{bmatrix} 0 & 0 \\ 1 & 0 \end{bmatrix}$, then $A^2 = O$, but

$A \neq O$. Non-zero matrices, when multiplied, may not necessarily result in a non-zero matrix, for instance: $AB = \begin{bmatrix} 2 & 3 \\ 2 & 3 \end{bmatrix}\begin{bmatrix} -3 & -3 \\ 2 & 2 \end{bmatrix} = O$, but $A \neq O, B \neq O$.

5. **The Trace of Square Matrix**

If $A_{n \times n} = [a_{ij}]_{n \times n}$ is a square matrix, defining trace $(A) = tr(A) = a_{11} + a_{22} + \cdots + a_{nn}$ (the sum of its main diagonal elements) is called the trace of the square matrix A.

For example: If $A = \begin{bmatrix} 1 & 5 & 7 \\ 3 & -2 & 9 \\ 2 & 1 & 10 \end{bmatrix}$, then trace$(A) = 1 + (-2) + 10 = 9$.

▶ Theorem 6-1-3

The Properties of the Trace of Square Matrix

The following equations regarding the trace are always true.
(1) tr$(A \pm B) =$ tr$(A) \pm$ tr(B).   (2) tr$(\alpha A) = \alpha \cdot$ tr(A).
(3) tr$(A^T) =$ tr(A).   (4) tr$(A^H) = \overline{\text{tr}(A)}$.
(5) tr$(AB) =$ tr(BA).   ◀

In fact, the trace can be seen as a function from the vector space of matrices to R (or C). From this perspective, the trace function does not preserve multiplication operations, which means tr(AB) is not necessarily equal to tr(A)tr(B), For example, in Example 2, (1), tr$(A) = 5$, tr$(B) = 3$ but tr$(AB) \neq$ tr(A)tr(B). Similarly, tr(A^k) is not necessarily equal to $[\text{tr}(A)]^k$.

6-1-3   Other Common Matrix

1. **Symmetric Matrix and Anti-Symmetric Matrix**

For a square matrix $A_{n \times n} = [a_{ij}]$, if $A^T = A$, that is $a_{ij} = a_{ji}$, then A is called a symmetric matrix. If $A^T = -A$ then is called an anti-symmetric matrix, that is

$a_{ij} = -a_{ji}$, for example: $A = \begin{bmatrix} 1 & 2 & 3 \\ 2 & 5 & 6 \\ 3 & 6 & 9 \end{bmatrix}$ is a symmetric matrix, $B = \begin{bmatrix} 0 & -2 & -3 \\ 2 & 0 & -6 \\ 3 & 6 & 0 \end{bmatrix}$

is an anti-symmetric matrix. It is worth noting that in an anti-symmetric matrix, $a_{ii} = -a_{ii}$, that is $a_{ii} = 0$. Thus, the diagonal elements of an anti-symmetric matrix are always zero.

► **Theorem 6-1-4**

The Decomposition of Matrix

Any square matrix $A_{n \times n} = [a_{ij}]_{n \times n}$ equals to the sum of a symmetric matrix and an anti-symmetric matrix.

Proof

$A_{n \times n} = (\dfrac{A + A^T}{2}) + (\dfrac{A - A^T}{2})$, let $B = \dfrac{A + A^T}{2}$, $C = \dfrac{A - A^T}{2}$. Then

$B^T = (\dfrac{A + A^T}{2})^T = \dfrac{A^T + A}{2} = B$ is a symmetric matrix,

$C^T = (\dfrac{A - A^T}{2})^T = \dfrac{A^T - A}{2} = -C$ is an anti-symmetric matrix. ◄

Example 3

Let $A = \begin{bmatrix} 2 & 2 & -1 \\ 1 & -1 & 0 \\ 0 & 1 & 0 \end{bmatrix}$, find a symmetric matrix B and an anti-symmetric matrix C,

which causes $A = B + C$.

Solution

Refer to the practice of Theorem 6-1-4:

Let $B = (\dfrac{A + A^T}{2}) = \dfrac{1}{2}(\begin{bmatrix} 2 & 2 & -1 \\ 1 & -1 & 0 \\ 0 & 1 & 0 \end{bmatrix} + \begin{bmatrix} 2 & 1 & 0 \\ 2 & -1 & 1 \\ -1 & 0 & 0 \end{bmatrix}) = \begin{bmatrix} 2 & \dfrac{3}{2} & -\dfrac{1}{2} \\ \dfrac{3}{2} & -1 & \dfrac{1}{2} \\ -\dfrac{1}{2} & \dfrac{1}{2} & 0 \end{bmatrix}$,

$C = (\dfrac{A - A^T}{2}) = \dfrac{1}{2}(\begin{bmatrix} 2 & 2 & -1 \\ 1 & -1 & 0 \\ 0 & 1 & 0 \end{bmatrix} - \begin{bmatrix} 2 & 1 & 0 \\ 2 & -1 & 1 \\ -1 & 0 & 0 \end{bmatrix}) = \begin{bmatrix} 0 & \dfrac{1}{2} & -\dfrac{1}{2} \\ -\dfrac{1}{2} & 0 & -\dfrac{1}{2} \\ \dfrac{1}{2} & \dfrac{1}{2} & 0 \end{bmatrix}$. Q.E.D.

2. Hermitian Matrix

If a sqare matrix $A_{n \times n} \in M_{n \times n}(\mathbb{C})$ satisfies $A^H = \overline{A}^T = A$, that is $a_{ij} = \overline{a_{ji}}$, then A is called Hermitian matrix.

Several important properties are as follows: 1. $(AB)^H = B^H A^H$; 2. Real symmetric matrices are Hermitian matrices. By $a_{ii} = \overline{a_{ii}}$, we know that the main diagonal elements of a Hermitian matrix are real numbers, like: $A = \begin{bmatrix} 1 & 2i & 3 \\ -2i & 5 & 6-2i \\ 3 & 6+2i & 9 \end{bmatrix}$.

A Hermitian matrix is also known as a **self-adjoint matrix**.

3. **Skew-Hermitian Matrix**

Compared to a Hermitian matrix, if a square matrix $A_{n \times n}$ acquires an additional negative sign when taking its conjugate transpose, that is $A^H = \overline{A}^T = -A$, then A is called Skew-hermitian matrix. For instance:

$\begin{bmatrix} 0 & 1+i & 3-i \\ -1+i & 2i & 2i \\ -3-i & 2i & 0 \end{bmatrix}$ is a Skew-hermitian matrix.

Similar to the properties of Hermitian matrices, in the case of skew-Hermitian matrices:

1. A skew-real-symmetric matrix is a skew-Hermitian matrix; 2. From $a_{ii} = -\overline{a_{ii}}$ can know the main diagonal elements of a skew-Hermitian matrix are either zero or purely imaginary. In fact, if we compare (skew-) Hermitian matrices with (skew-) real-symmetric matrices, we can observe similar results based on Theorem 6-1-4:

▶ **Theorem 6-1-5**

The Decomposition of Complex Matrix

Any square matrix $A_{n \times n} \in M_{n \times n}(\mathbb{C})$ equals to a Hermitian matrix plus a skew-Hermitian matrix.

Proof   Any square matrix in $M_{m \times n}(\mathbb{C})$ can be written as

$$A = \frac{A + A^H}{2} + \frac{A - A^{II}}{2} = B + C,$$

where $B = \dfrac{A + A^H}{2}$ is a Hermitian matrix, $C = \dfrac{A - A^H}{2}$ is skew-Hermitian matrix. ◀

Example 4

Which of the following are Hermitian matrices.

(1) $\begin{bmatrix} 2 & i \\ -i & 5 \end{bmatrix}$.

(2) $\begin{bmatrix} 1+i & 2 \\ 2 & 5+i \end{bmatrix}$.

(3) $\begin{bmatrix} 1 & 1+i & 5 \\ 1-i & 2 & i \\ 5 & -i & 7 \end{bmatrix}$.

Solution   We can check whether satisfy $A^H = \overline{A}^T = A$.

(1) $\begin{bmatrix} 2 & i \\ -i & 5 \end{bmatrix}^H = \begin{bmatrix} 2 & i \\ -i & 5 \end{bmatrix}$, so it is a Hermitian matrix.

(2) $\begin{bmatrix} 1+i & 2 \\ 2 & 5+i \end{bmatrix}^H = \begin{bmatrix} 1-i & 2 \\ 2 & 5-i \end{bmatrix} \neq \begin{bmatrix} 1+i & 2 \\ 2 & 5+i \end{bmatrix}$, it is not a Hermitian matrix

(3) $\begin{bmatrix} 1 & 1+i & 5 \\ 1-i & 2 & i \\ 5 & -i & 7 \end{bmatrix}^H = \begin{bmatrix} 1 & 1+i & 5 \\ 1-i & 2 & i \\ 5 & -i & 7 \end{bmatrix}$, so it is a Hermitian matrix.   Q.E.D.

6-1   Exercises

Basic questions

1. Find the values of parameters α and β $(\alpha, \beta \in \mathrm{R})$, causes the following two matrices are equal.

(1) $\begin{bmatrix} 2 & \alpha-4 \\ \beta+3 & 1 \end{bmatrix}$, $\begin{bmatrix} 2 & 3\alpha+8 \\ 7 & 1 \end{bmatrix}$.

(2) $\begin{bmatrix} 9 & -2 \\ \beta^3 & 5 \end{bmatrix}$, $\begin{bmatrix} \alpha^2 & -2 \\ 8 & 5 \end{bmatrix}$.

2. Write down the order of matrix A and B which causes the multiplications below to be defined.

(1) $\begin{bmatrix} -4 & -6 \\ 2 & 8 \\ 14 & 4 \end{bmatrix} A \begin{bmatrix} 1 & 2 & 4 \\ -1 & 2 & 1 \\ 5 & 0 & 7 \\ 2 & -1 & 3 \end{bmatrix} = B$.

(2) $\begin{bmatrix} 1 & 2 & -3 & 5 \\ 2 & 0 & 3 & 4 \end{bmatrix} A \begin{bmatrix} 1 \\ 2 \\ -1 \\ 7 \\ 8 \end{bmatrix} = B$.

3. There are two matrix A, B respectively as $A = \begin{bmatrix} 1 & -1 \\ 2 & 1 \\ 3 & 2 \end{bmatrix}$, $B = \begin{bmatrix} 4 & 1 & -1 \\ 1 & -2 & 2 \end{bmatrix}$, find

(1) $4A - 2B^T$.
(2) AB.
(3) BA.
(4) $\mathrm{tr}(AB)$.
(5) $\mathrm{tr}(BA)$.

4. $A = \begin{bmatrix} -2 & -4 \\ 1 & 1 \end{bmatrix}$, $B = \begin{bmatrix} 0 & 2 \\ 1 & -3 \end{bmatrix}$, find $A^3 - B^2$.

5. There are two matrix A, B respectively as $A = \begin{bmatrix} -2 & -4 \\ -3 & 1 \end{bmatrix}$, $B = \begin{bmatrix} 6 & 8 \\ 1 & -3 \end{bmatrix}$, find the following calculation (1) $2A + 3B$. (2) AB. (3) BA. (4) $\mathrm{tr}(BA)$.

Advanced questions

1. $A = \begin{bmatrix} 1 & 2 & 3 \\ 4 & 5 & 6 \\ 7 & 8 & 9 \end{bmatrix}$, $B = \begin{bmatrix} 1 & 2 & 1 \\ 2 & 3 & 2 \\ 3 & 4 & 3 \end{bmatrix}$,

$C = \begin{bmatrix} 1 & 2 \\ 3 & 4 \\ 5 & 6 \end{bmatrix}$, $D = \begin{bmatrix} 1 & 2 & 3 \\ 4 & 5 & 6 \end{bmatrix}$, find

(1) $2A - 3B$.

(2) $A + 2B$.

(3) $C \cdot D$.

(4) $D \cdot C$.

(5) $\text{tr}(AB)$.

(6) $\text{tr}(BA)$.

(7) $\text{tr}(CD)$.

(8) $\text{tr}(DC)$.

2. Let $A = \begin{bmatrix} 2 & 1 & 4 \\ 3 & 2 & 1 \\ 1 & 3 & 2 \end{bmatrix}$, $B = \begin{bmatrix} 5 & 1 & 6 \\ 9 & 2 & -3 \\ -1 & 3 & 7 \end{bmatrix}$,

$C = \begin{bmatrix} 0 & 0 & 0 \\ 2 & 3 & 4 \\ 0 & 0 & 0 \end{bmatrix}$,

please verify $C \neq 0$ and $A \neq B$, but $AC = BC$.

3. Let $A = \begin{bmatrix} 2 & 3 & -1 \\ 1 & -1 & 0 \\ 0 & 1 & 2 \end{bmatrix}$, find a symmetric matrix B

and an anti-symmetric matrix C, causes $A = B + C$.

4. Find a symmetric matrix B and an anti-symmetric matrix C, causes $A = B + C$, where

$A = \begin{bmatrix} 3 & -4 & -1 \\ 6 & 0 & -1 \\ -3 & 13 & -4 \end{bmatrix}$.

5. We know $A = \begin{bmatrix} 1+i & -1 \\ 2i & 3-4i \end{bmatrix}$,

$B = \begin{bmatrix} 2 & 1+i \\ 1-i & 4 \end{bmatrix}$, please verify the following two questions.

(1) $(A+B)^H = A^H + B^H$.

(2) $(AB)^H = B^H A^H$.

6. Which of the following is a Hermitian matrix and which one is a skew-hermitian matrix?

$A = \begin{bmatrix} 1 & 2+i \\ 2-i & -1 \end{bmatrix}$, $B = \begin{bmatrix} i & \dfrac{2}{\sqrt{5}} \\ \dfrac{2}{\sqrt{5}} & -i \end{bmatrix}$,

$C = \begin{bmatrix} 0 & i & 1 \\ i & 0 & -2+i \\ -1 & 2+i & 0 \end{bmatrix}$,

$D = \begin{bmatrix} 3 & 2+i & 3i \\ 2-i & -5 & 7 \\ -3i & 7 & 0 \end{bmatrix}$.

6-2 Matrix Row (Column) Operations and Determinant

In Section 6-1, we introduced column (row) vectors, which allows a matrix A to be seen as composed of column (row) vectors. In fact, the multiplication of two matrices AB can be interpreted as the linear combination of the row elements of A using the row vectors of B. In other words, the i-th row of AB can be computed independently as

follows: $[a_{i1}, a_{i2}, ..., a_{in}] \begin{bmatrix} u_1 \\ \vdots \\ u_n \end{bmatrix} = \sum_{k=1}^{n} a_{ik} u_k$, for example:

$\begin{bmatrix} 1 & 2 \\ 1 & 1 \end{bmatrix} \begin{bmatrix} 1 & 2 \\ 3 & 4 \end{bmatrix} = \begin{bmatrix} 1 \cdot [1 \quad 2] + 2 \cdot [3 \quad 4] \\ 1 \cdot [1 \quad 2] + 1 \cdot [3 \quad 4] \end{bmatrix}$, it represents the left side of matrix $\begin{bmatrix} 1 & 2 \\ 3 & 4 \end{bmatrix}$

multiplied another matrix $\begin{bmatrix} 1 & 2 \\ 1 & 1 \end{bmatrix}$, it performs row vector operations on matrix $\begin{bmatrix} 1 & 2 \\ 3 & 4 \end{bmatrix}$,

that is, matrix $\begin{bmatrix} 1 & 2 \\ 3 & 4 \end{bmatrix}$ conduct row vector addition and subtraction. This directly suggests that solving a system of simultaneous equations of n order variables is equivalent to left-multiplying a specific matrix until the solution to the original system of equations is obtained. This observation directly leads to a major theme in this section: elementary column operations.

The second key point is determinants, which was proposed by the German mathematician Leibniz (1646-1716) in the 17th century, developed and perfected in the 19th century, and extensively used in solving systems of linear equations. The emergence of determinants is closely related to multiple integrals. For example, when performing variable transformations in multivariable calculus, the change in the area of the local integral region is expressed through determinants. In fact, even in the process of solving simultaneous equations in secondary school, determinants have appeared, although the definition was not explicitly formulated at that time. We will discuss this further in Section 6-2-2.

The following will introduce these two key points, preparing for solving systems of simultaneous equations in the next section

6-2-1 Elementary Row Operation and Matrix Reduction

▶ **Definition 6-2-1**

Row Interchange

Interchanging two rows in a matrix is denoted as $r_{ij}(A)$. This operation is also known as the first type of row operation.   ◀

Take $A = \begin{bmatrix} 1 & 2 & 3 \\ 4 & 5 & 6 \\ 7 & 8 & 9 \end{bmatrix}$ for example, $r_{12}(A)$ expresses interchanging the first and second rows of A, it is typically represented as: $\begin{bmatrix} 1 & 2 & 3 \\ 4 & 5 & 6 \\ 7 & 8 & 9 \end{bmatrix} \xrightarrow{r_{12}} \begin{bmatrix} 4 & 5 & 6 \\ 1 & 2 & 3 \\ 7 & 8 & 9 \end{bmatrix}$.

▶ Definition 6-2-2

Row Multiplication by Coefficient

To multiply a row of a matrix by a non-zero number k, it is denoted as $r_i^{(k)}(A)$, $k \neq 0$. This operation is also known as the second type of row operation.   ◀

For instance: $A = \begin{bmatrix} 1 & 2 & 3 \\ 4 & 5 & 6 \\ 7 & 8 & 9 \end{bmatrix}$, then $r_2^{(-3)}(A)$ represents multiply the second row of A by (-3), it is signified as: $\begin{bmatrix} 1 & 2 & 3 \\ 4 & 5 & 6 \\ 7 & 8 & 9 \end{bmatrix} \xrightarrow{r_2^{(-3)}} \begin{bmatrix} 1 & 2 & 3 \\ -12 & -15 & -18 \\ 7 & 8 & 9 \end{bmatrix}$.

▶ Definition 6-2-3

Row Addition

To multiply a row of a matrix by a non-zero number k and add it to another row, it is denoted as $r_{ij}^{(k)}(A)$ or $R_{ij}^{(k)}(A)$, $k \neq 0$. This operation is also known as the third type of row operation.   ◀

For example: $A = \begin{bmatrix} 1 & 2 & 3 \\ 4 & 5 & 6 \\ 7 & 8 & 9 \end{bmatrix}$, then $r_{12}^{(-4)}(A)$ represents multiply the first row of A by (-4) and add it to the second row, it is signified as

$$\begin{bmatrix} 1 & 2 & 3 \\ 4 & 5 & 6 \\ 7 & 8 & 9 \end{bmatrix} \xrightarrow{r_{12}^{(-4)}} \begin{bmatrix} 1 & 2 & 3 \\ 4+(-4) & 5+(-8) & 6+(-12) \\ 7 & 8 & 9 \end{bmatrix}.$$

When we repeatedly perform these three types of row operations, we can systematically eliminate the elements in the same row, except for the diagonal elements, from left to right. As a result, we obtain the following matrix form.

▶ **Definition 6-2-4**

Row Echelon Matrix and Row Reduced Echelon Matrix

(1) A matrix that satisfies the following properties is called a **row echelon matrix**.

  ① Zero columns are below non-zero columns.

  ② The rows containing the leftmost non-zero element (pivot) of each non-zero column are distinct.

  ③ The leftmost pivot of each column moves to the left as the rows go upward.

(2) A column echelon matrix that satisfies the following properties is called a **row reduced echelon matrix**.

  ① In each row containing a pivot element (the leftmost non-zero element), all other elements in that row are zero.

  ② Each pivot element is equal to 1.   ◀

For instance, $A = \begin{bmatrix} \boxed{3} & 1 & 0 & 5 \\ 0 & \boxed{2} & 1 & -4 \\ 0 & 0 & 0 & 0 \end{bmatrix}$ is row echelon matrix, where the pivot

element in the first row is 3, the pivot element in the second row is 2.

$A = \begin{bmatrix} 1 & 0 & 3 & 5 \\ 0 & 1 & 1 & -4 \\ 0 & 0 & 0 & 0 \end{bmatrix}$ is row reduced echelon matrix.

Example 1

Through using elementary row operations to transform the matrix $A = \begin{bmatrix} 1 & 2 & 3 \\ 4 & 5 & 6 \\ 7 & 8 & 9 \end{bmatrix}$:

(1) into a row echelon matrix, (2) into a row reduced echelon matrix.

Solution

(1) $A = \begin{bmatrix} 1 & 2 & 3 \\ 4 & 5 & 6 \\ 7 & 8 & 9 \end{bmatrix} \xrightarrow{r_{12}^{(-4)} r_{13}^{(-7)}} \begin{bmatrix} 1 & 2 & 3 \\ 0 & -3 & -6 \\ 0 & -6 & -12 \end{bmatrix} \xrightarrow{r_{23}^{(-2)}} \begin{bmatrix} 1 & 2 & 3 \\ 0 & -3 & -6 \\ 0 & 0 & 0 \end{bmatrix}$,

we obtain the row echelon matrix of matrix A.

$$(2) \ A = \begin{bmatrix} 1 & 2 & 3 \\ 4 & 5 & 6 \\ 7 & 8 & 9 \end{bmatrix} \xrightarrow{r_{12}^{(-4)} r_{13}^{(-7)}} \begin{bmatrix} 1 & 2 & 3 \\ 0 & -3 & -6 \\ 0 & -6 & -12 \end{bmatrix} \xrightarrow{r_{23}^{(-2)}} \begin{bmatrix} 1 & 2 & 3 \\ 0 & -3 & -6 \\ 0 & 0 & 0 \end{bmatrix}$$

$$\xrightarrow{r_2^{(-1/3)}} \begin{bmatrix} 1 & 2 & 3 \\ 0 & 1 & 2 \\ 0 & 0 & 0 \end{bmatrix} \xrightarrow{r_{21}^{(-2)}} \begin{bmatrix} 1 & 0 & -1 \\ 0 & 1 & 2 \\ 0 & 0 & 0 \end{bmatrix}, \text{ we obtain the row reduced echelon matrix}$$

of matrix A.

Example 2

Through using elementary row operations to transform the matrix

$$A = \begin{bmatrix} -2 & 1 & 4 & 2 \\ 0 & 1 & 16 & 3 \\ 1 & -2 & 4 & 8 \end{bmatrix}$$

into: (1) a row echelon matrix   (2) a row reduced echelon matrix.

Solution

$$(1) \ A = \begin{bmatrix} -2 & 1 & 4 & 2 \\ 0 & 1 & 16 & 3 \\ 1 & -2 & 4 & 8 \end{bmatrix} \xrightarrow{r_{13}} \begin{bmatrix} 1 & -2 & 4 & 8 \\ 0 & 1 & 16 & 3 \\ -2 & 1 & 4 & 2 \end{bmatrix} \xrightarrow{r_{13}^{(2)}} \begin{bmatrix} 1 & -2 & 4 & 8 \\ 0 & 1 & 16 & 3 \\ 0 & -3 & 12 & 18 \end{bmatrix}$$

$$\xrightarrow{r_{23}^{(3)}} \begin{bmatrix} 1 & -2 & 4 & 8 \\ 0 & 1 & 16 & 3 \\ 0 & 0 & 60 & 27 \end{bmatrix}, \text{ we obtain the row echelon matrix of matrix } A.$$

$$(2) \ A = \begin{bmatrix} -2 & 1 & 4 & 2 \\ 0 & 1 & 16 & 3 \\ 1 & -2 & 4 & 8 \end{bmatrix} \xrightarrow{r_{13}} \begin{bmatrix} 1 & -2 & 4 & 8 \\ 0 & 1 & 16 & 3 \\ -2 & 1 & 4 & 2 \end{bmatrix} \xrightarrow{r_{13}^{(2)}} \begin{bmatrix} 1 & -2 & 4 & 8 \\ 0 & 1 & 16 & 3 \\ 0 & -3 & 12 & 18 \end{bmatrix}$$

$$\xrightarrow{r_{23}^{(3)}} \begin{bmatrix} 1 & -2 & 4 & 8 \\ 0 & 1 & 16 & 3 \\ 0 & 0 & 60 & 27 \end{bmatrix} \xrightarrow{r_3^{(1/60)}} \begin{bmatrix} 1 & -2 & 4 & 8 \\ 0 & 1 & 16 & 3 \\ 0 & 0 & 1 & \dfrac{9}{20} \end{bmatrix}$$

$$\xrightarrow{r_{21}^{(2)}} \begin{bmatrix} 1 & 0 & 36 & 14 \\ 0 & 1 & 16 & 3 \\ 0 & 0 & 1 & \dfrac{9}{20} \end{bmatrix} \xrightarrow{r_{32}^{(-16)} r_{31}^{(-36)}} \begin{bmatrix} 1 & 0 & 0 & -\dfrac{11}{5} \\ 0 & 1 & 0 & -\dfrac{21}{5} \\ 0 & 0 & 1 & \dfrac{9}{20} \end{bmatrix},$$

we obtain the row reduced echelon matrix of matrix A.

6-2-2 Determinant

To solve $\begin{cases} a_{11}x + a_{12}y = b_1 \\ a_{21}x + a_{22}y = b_2 \end{cases}$, it is natural to first ask whether a solution exists?

Through the method of substitution and elimination, we can understand that when $a_{11}a_{22} - a_{21}a_{12} \neq 0$ has a unique solution, and if $a_{11}a_{22} - a_{21}a_{12} = 0$, then there are infinitely many solutions. In fact, when the first and second rows are proportional, the graph on R^2 represents two parallel lines. Therefore, the value of the determinant directly affects the state of the solution for the system of equations.

1.　The Definition of Determinant

For square matrix $A_{n \times n} = [a_{ij}]_{n \times n}$, the value of a determinant is defined as

$$|A_{n \times n}| = \begin{vmatrix} a_{11} & a_{12} & \cdots & a_{1n} \\ a_{21} & a_{22} & \cdots & a_{2n} \\ \vdots & & & \\ a_{n1} & \cdots & \cdots & a_{nn} \end{vmatrix} = \sum_{j=1}^{n} a_{ij} C_{ij} = \sum_{i=1}^{n} a_{ij} C_{ij} \quad \text{where} \quad C_{ij} = (-1)^{i+j} M_{ij} \quad \text{is}$$

called cofactor, which is the determinant obtained by removing the ith row and jth column from the original matrix, while $(-1)^{i+j}$ appears as a sequence of positive and negative signs arranged alternately starting from a_{11}: $\begin{vmatrix} + & - & + & \cdots \\ - & + & - & \cdots \\ + & - & & \vdots \\ \vdots & \vdots & \cdots & \vdots \end{vmatrix}$. Hence, a

determinant is a scalar. In fact, a determinant is a function mapping $n \times n$ elements to a single value. For instance, if $A = \begin{bmatrix} a_{11} & a_{12} \\ a_{21} & a_{22} \end{bmatrix}$, then

$$|A| = \begin{vmatrix} a_{11} & a_{12} \\ a_{21} & a_{22} \end{vmatrix} = a_{11} \cdot a_{22} - a_{21} \cdot a_{12} .$$

If $A = \begin{bmatrix} a_{11} & a_{12} & a_{13} \\ a_{21} & a_{22} & a_{23} \\ a_{31} & a_{32} & a_{33} \end{bmatrix}$, according to the definition of a determinant, we expand it

using the elements of the first row, we can obtain

$$|A| = \begin{vmatrix} a_{11} & a_{12} & a_{13} \\ a_{21} & a_{22} & a_{23} \\ a_{31} & a_{32} & a_{33} \end{vmatrix} = a_{11} \cdot M_{11} - a_{12} \cdot M_{12} + a_{13} \cdot M_{13}$$

$$= a_{11} \cdot \begin{vmatrix} a_{22} & a_{23} \\ a_{32} & a_{33} \end{vmatrix} - a_{12} \cdot \begin{vmatrix} a_{21} & a_{23} \\ a_{31} & a_{33} \end{vmatrix} + a_{13} \cdot \begin{vmatrix} a_{21} & a_{22} \\ a_{31} & a_{32} \end{vmatrix}$$

$$= a_{11}a_{22}a_{33} + a_{21}a_{32}a_{13} + a_{31}a_{12}a_{23} - a_{13}a_{22}a_{31} - a_{11}a_{32}a_{23} - a_{21}a_{12}a_{33} .$$

Similarly, readers can also expand the determinant using the elements of the second or third row, and the principle remains the same. By the way, there is a graphical mnemonic for the determinant of a 3×3 matrix:

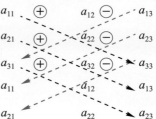

Another example:

$$|A_{4\times4}| = \begin{vmatrix} a_{11} & a_{12} & a_{13} & a_{14} \\ a_{21} & a_{22} & a_{23} & a_{24} \\ a_{31} & a_{32} & a_{33} & a_{34} \\ a_{41} & a_{42} & a_{43} & a_{44} \end{vmatrix}$$

$$= (-1)^{1+1} a_{11}M_{11} + (-1)^{1+2} a_{12}M_{12} + (-1)^{1+3} a_{13}M_{13} + (-1)^{1+4} a_{14}M_{14}$$

(Expanding along the first row)

$$= a_{11}M_{11} - a_{12}M_{12} + a_{13}M_{13} - a_{14}M_{14} \quad \text{(Expanding along the first row)}$$

$$= a_{11}M_{11} - a_{21}M_{21} + a_{31}M_{31} - a_{41}M_{41} \quad \text{(Expanding along the first column)}$$

$$= -a_{21}M_{21} + a_{22}M_{22} - a_{23}M_{23} + a_{24}M_{24}$$

(Expanding along the second column)

$$= \cdots \quad \text{(Expanding along a specific column or row)},$$

where the cofactor is

$$M_{11} = \begin{vmatrix} a_{22} & a_{23} & a_{24} \\ a_{32} & a_{33} & a_{34} \\ a_{42} & a_{43} & a_{44} \end{vmatrix}, \quad M_{12} = \begin{vmatrix} a_{21} & a_{23} & a_{24} \\ a_{31} & a_{33} & a_{34} \\ a_{41} & a_{43} & a_{44} \end{vmatrix}, \quad M_{13} = \begin{vmatrix} a_{21} & a_{22} & a_{24} \\ a_{31} & a_{32} & a_{34} \\ a_{41} & a_{42} & a_{44} \end{vmatrix},$$

$$M_{14} = \begin{vmatrix} a_{21} & a_{22} & a_{23} \\ a_{31} & a_{32} & a_{33} \\ a_{41} & a_{42} & a_{43} \end{vmatrix}.$$

Example 3

Find the determinant value of the following square matrix.

$$A = \begin{bmatrix} 1 & 3 \\ 2 & 4 \end{bmatrix}, \quad B = \begin{bmatrix} 2 & 1 & -3 \\ 3 & 1 & 0 \\ -6 & -4 & 2 \end{bmatrix}.$$

Solution

$$|A| = \begin{vmatrix} 1 & 3 \\ 2 & 4 \end{vmatrix} = 4 - 6 = -2;$$

$$|B| = \begin{vmatrix} 2 & 1 & -3 \\ 3 & 1 & 0 \\ -6 & -4 & 2 \end{vmatrix}$$

$$= 2 \cdot 1 \cdot 2 + 3 \cdot (-4) \cdot (-3) + (-6) \cdot 1 \cdot 0$$

$$- (-3) \cdot 1 \cdot (-6) - 2 \cdot (-4) \cdot 0 - 3 \cdot 1 \cdot 2 = 16.$$

Alternatively, solve using the method of matrix reduction:

$$|B| = \begin{vmatrix} 2 & 1 & -3 \\ 3 & 1 & 0 \\ -6 & -4 & 2 \end{vmatrix} = 2 \cdot C_{11} + 1 \cdot C_{12} + (-3)C_{13} = 2 \cdot (+M_{11}) + 1 \cdot (-M_{12}) + (-3) \cdot (M_{13})$$

$$= 2 \cdot \begin{vmatrix} 1 & 0 \\ -4 & 2 \end{vmatrix} - 1 \cdot \begin{vmatrix} 3 & 0 \\ -6 & 2 \end{vmatrix} + (-3) \cdot \begin{vmatrix} 3 & 1 \\ -6 & -4 \end{vmatrix} = 16 \text{ (using expand along the first row),}$$

where, $M_{11} = \begin{vmatrix} 1 & 0 \\ -4 & 2 \end{vmatrix}$, $M_{12} = \begin{vmatrix} 3 & 0 \\ -6 & 2 \end{vmatrix}$, $M_{13} = \begin{vmatrix} 3 & 1 \\ -6 & -4 \end{vmatrix}$.   Q.E.D.

▶ **Theorem 6-2-1**

Properties of Column (Row) Operations for Determinants

Assuming A is an $n \times n$ square matrix, the following five properties can be directly derived from the definition. Interested readers can refer to the reference books in the appendix or derive them on their own.

(1) Swapping any two columns (rows) changes the sign of the determinant: $|r_{ij}(A)| = -|A|$.

(2) Multiplying any column (row) of a matrix by a scalar $k \neq 0$, results in the determinant being k times the original determinant: $|r_i^{(k)}(A)| = k|A|$.

(3) Multiplying any row (column) of a matrix by a scalar $k \neq 0$ and adding it to another column (row) does not change the determinant: $|r_{ij}^{(k)}(A)| = |A|$.

(4) If a column (row) consists entirely of zeros, the determinant is 0.

(5) If any two columns (rows) are proportional, the determinant is 0.   ◀

Example 4

If $A = \begin{bmatrix} 3 & 1 & 0 \\ -2 & -4 & 3 \\ 5 & 4 & -2 \end{bmatrix}$, $B = \begin{bmatrix} 3 & 1 & 0 \\ -10 & -20 & 15 \\ 5 & 4 & -2 \end{bmatrix}$, $C = \begin{bmatrix} -2 & -4 & 3 \\ 3 & 1 & 0 \\ 5 & 4 & -2 \end{bmatrix}$,

$D = \begin{bmatrix} 3 & 1 & 0 \\ -2 & -4 & 3 \\ 11 & 6 & -2 \end{bmatrix}$, find the determinant value of matrix A, B, C, D.

Solution

$$(1) \ |A| = \begin{vmatrix} 3 & 1 & 0 \\ -2 & -4 & 3 \\ 5 & 4 & -2 \end{vmatrix} = -(1) \cdot \begin{vmatrix} -2 & 3 \\ 5 & -2 \end{vmatrix} + (-4) \cdot \begin{vmatrix} 3 & 0 \\ 5 & -2 \end{vmatrix} - (4) \cdot \begin{vmatrix} 3 & 0 \\ -2 & 3 \end{vmatrix} = -1,$$

the above expression is obtained by expanding along the second column and solving through matrix reduction. (Readers can try using other rows or columns for matrix reduction.)

(2) $B = r_2^{(5)}(A)$, so we know by Theorem 6-2-1 (2), $|B| = 5 \cdot |A| = -5$.

(3) $C = r_{12}(A)$, so we know by Theorem 6-2-1 (1), $|C| = -|A| = -(-1) = 1$.

(4) $D = r_{13}^{(2)}(A)$, so we know by Theorem 6-2-1 (3), $|D| = |A| = -1$.　Q.E.D.

▶ Theorem 6-2-2

Important Properties of Determinant Values

A, B are all $n \times n$ square matrix, then

(1) $|A| = |A^T|$.　　(2) $|AB| = |BA| = |A||B|$.　　(3) $|\bar{A}| = \overline{|A|}$.

(4) $|AA^T| = |A|^2$, $|AA^H| = |A||\bar{A}| = |A||\overline{|A|}| = \|A\|^2 = |A^H A|$.

(5) $|\alpha A| = \alpha^n |A|$; α is a scalar.

(6) Let A is an upper (lower) triangular matrix or a diagonal matrix, then $|A|$ is equal to the product of the diagonal elements of A.　　◀

Example 5

$$A = \begin{bmatrix} a & b & c \\ d & e & f \\ g & h & i \end{bmatrix}, \quad B = \begin{bmatrix} 2 & 1 & -3 \\ 3 & 1 & 0 \\ -6 & -4 & 2 \end{bmatrix} \text{ and } \det(A) = |A| = 5, \text{ find the following}$$

determinant value. (1) $\det(-4A)$　(2) $\det(A^2)$　(3) $\det(A^T)$　(4) $\det(AB)$

Solution

(1) $\det(-4A) = (-4)^3 \det(A) = -64 \times 5 = -320$.

(2) $\det(A^2) = |A|^2 = 5^2 = 25$.

(3) $\det(A^T) = \det(A) = 5$.

$$(4) \ \det(AB) = \det(A) \cdot \det(B) = 5 \cdot \begin{vmatrix} 2 & 1 & -3 \\ 3 & 1 & 0 \\ -6 & -4 & 2 \end{vmatrix} = 5 \cdot 16 = 80.$$　Q.E.D.

2. **Vandermonde Determinant**

$$|A_{n \times n}| = \begin{vmatrix} 1 & x_1 & \cdots & x_1^{n-1} \\ 1 & x_2 & \cdots & x_2^{n-1} \\ \vdots & \vdots & & \vdots \\ 1 & x_n & \cdots & x_n^{n-1} \end{vmatrix} = \begin{vmatrix} 1 & 1 & \cdots & 1 \\ x_1 & x_2 & \cdots & x_n \\ \vdots & & & \\ x_1^{n-1} & x_2^{n-1} & \cdots & x_n^{n-1} \end{vmatrix} = \prod_{i=1}^{n} \prod_{i<j}^{n} (x_j - x_i)$$

Example 6

What's the following determinant value?

$$(1) \begin{vmatrix} 1 & 5 & 25 \\ 1 & 7 & 49 \\ 1 & 9 & 81 \end{vmatrix}. \qquad (2) \begin{vmatrix} 1 & 1 & 1 & 1 \\ 2 & 3 & 4 & 5 \\ 4 & 9 & 16 & 25 \\ 8 & 27 & 64 & 125 \end{vmatrix}.$$

Solution

$$(1) \quad \begin{vmatrix} 1 & 5 & 25 \\ 1 & 7 & 49 \\ 1 & 9 & 81 \end{vmatrix} = \begin{vmatrix} 1 & 5 & 5^2 \\ 1 & 7 & 7^2 \\ 1 & 9 & 9^2 \end{vmatrix} = (7-5) \cdot (9-5) \cdot (9-7) = 16 .$$

$$(2) \quad \begin{vmatrix} 1 & 1 & 1 & 1 \\ 2 & 3 & 4 & 5 \\ 4 & 9 & 16 & 25 \\ 8 & 27 & 64 & 125 \end{vmatrix} = \begin{vmatrix} 1 & 1 & 1 & 1 \\ 2 & 3 & 4 & 5 \\ 2^2 & 3^2 & 4^2 & 5^2 \\ 2^3 & 3^3 & 4^3 & 5^3 \end{vmatrix}$$

$$= (3-2) \cdot (4-2) \cdot (4-3) \cdot (5-2) \cdot (5-3) \cdot (5-4) = 12 . \qquad \boxed{\text{Q.E.D.}}$$

3. **The Determinant of a Square Matrix**

If A, B, C are all square matrix, then

(1) $\det \begin{bmatrix} A & C \\ O & B \end{bmatrix} = \det(A) \cdot \det(B)$.

(2) $\det \begin{bmatrix} A & O \\ C & B \end{bmatrix} = \det(A) \cdot \det(B)$.

(3) $\det \begin{bmatrix} A & B \\ B & A \end{bmatrix} = \det(A + B) \cdot \det(A - B)$.

Example 7

$$A = \begin{bmatrix} 2 & 0 & 0 & 0 \\ 1 & 2 & 0 & 0 \\ 0 & 0 & 0 & 1 \\ 0 & 0 & -6 & 5 \end{bmatrix}, \text{ find } |A|.$$

Solution

$$\det \begin{bmatrix} 2 & 0 \\ 1 & 2 \end{bmatrix} \cdot \det \begin{bmatrix} 0 & 1 \\ -6 & 5 \end{bmatrix} = 4 \cdot 6 = 24.$$

`Q.E.D.`

Example 8

$$A = \begin{bmatrix} \dfrac{1}{2} & -\dfrac{1}{2} & -\dfrac{1}{2} & -\dfrac{1}{2} \\ -\dfrac{1}{2} & \dfrac{1}{2} & \dfrac{1}{2} & \dfrac{1}{2} \\ -\dfrac{1}{2} & -\dfrac{1}{2} & \dfrac{1}{2} & -\dfrac{1}{2} \\ -\dfrac{1}{2} & -\dfrac{1}{2} & -\dfrac{1}{2} & \dfrac{1}{2} \end{bmatrix}, \text{ find } |A|.$$

Solution

$$|A| = \det\left(\begin{bmatrix} \dfrac{1}{2} & -\dfrac{1}{2} \\ -\dfrac{1}{2} & \dfrac{1}{2} \end{bmatrix} + \begin{bmatrix} -\dfrac{1}{2} & -\dfrac{1}{2} \\ -\dfrac{1}{2} & -\dfrac{1}{2} \end{bmatrix} \right) \det\left(\begin{bmatrix} \dfrac{1}{2} & -\dfrac{1}{2} \\ -\dfrac{1}{2} & \dfrac{1}{2} \end{bmatrix} - \begin{bmatrix} -\dfrac{1}{2} & -\dfrac{1}{2} \\ -\dfrac{1}{2} & \dfrac{1}{2} \end{bmatrix} \right)$$

$$= \det \begin{bmatrix} 0 & -1 \\ -1 & 0 \end{bmatrix} \cdot \det \begin{bmatrix} 1 & 0 \\ 0 & 1 \end{bmatrix} = -1 \cdot 1 = -1.$$

`Q.E.D.`

6-2-3 Method for Finding the Inverse Matrix

Assuming we know two square matrices $A_{n \times n}$ and $C_{n \times n}$, and we want to find a matrix $X_{n \times n}$ that satisfies $XA = C$, if A, C and X are scalars, we can directly divide both sides by A, then we can get $X = C/A$, but A is a matrix, we cannot divide by A. Here, the concept of the inverse matrix become crucial. Let's define the inverse matrix and explore methods for solving it, along with some properties of inverse matrices.

▶ **Definition 6-2-5**

Inverse Matrix

For a square matrix $A_{n \times n}$, if there exists another square matrix $B_{n \times n}$, causes $AB = BA = I_n$, where I_n represents the $n \times n$ identity matrix, then the matrix B is called the **inverse matrix** of A, denoted as $B = A^{-1}$. ◀

For instance: if $A = \begin{bmatrix} 1 & 2 \\ 3 & 4 \end{bmatrix}$, $B = \begin{bmatrix} -2 & 1 \\ \frac{3}{2} & \frac{-1}{2} \end{bmatrix}$,

then $AB = \begin{bmatrix} 1 & 2 \\ 3 & 4 \end{bmatrix} \cdot \begin{bmatrix} -2 & 1 \\ \frac{3}{2} & \frac{-1}{2} \end{bmatrix} = \begin{bmatrix} 1 & 0 \\ 0 & 1 \end{bmatrix} = I_2$ and $BA = \begin{bmatrix} -2 & 1 \\ \frac{3}{2} & \frac{-1}{2} \end{bmatrix} \cdot \begin{bmatrix} 1 & 2 \\ 3 & 4 \end{bmatrix} = \begin{bmatrix} 1 & 0 \\ 0 & 1 \end{bmatrix} = I_2$,

so $B = \begin{bmatrix} -2 & 1 \\ \frac{3}{2} & \frac{-1}{2} \end{bmatrix} = A^{-1}$ is the inverse matrix of A. When discussing elementary column operations, we observe that by repeatedly applying various types of column operations, we can ultimately transform a matrix A into a column-echelon form. If we denote the product of these elementary column operations as R, then $RA = B$. If at this point $B = I_n$, we have found the inverse matrix. However, the challenge lies in the need to record this sequence of column operations. Hence, the following augmented matrix method is introduced.

1. **Using the Augmented Matrix Method to Find A^{-1}**

    For a square matrix $A_{n \times n}$, let its augmented matrix be $\begin{bmatrix} A_{n \times n} \mid I_{n \times n} \end{bmatrix}_{n \times (2n)}$. Through elementary column operations, we transform A into the identity matrix, that is $\begin{bmatrix} A_{n \times n} \mid I_{n \times n} \end{bmatrix}_{n \times (2n)} \xrightarrow{r} \begin{bmatrix} I_{n \times n} \mid RI_{n \times n} \end{bmatrix}_{n \times (2n)}$. As a result, the right half of the augmented matrix records the process of column operations, which represents the inverse matrix A^{-1}.

Example 9

$A = \begin{bmatrix} 1 & 0 & 2 \\ 2 & -1 & 3 \\ 4 & 1 & 8 \end{bmatrix}$, please use elementary column operations to find A^{-1}.

Solution

$$[A \mid I] = \begin{bmatrix} 1 & 0 & 2 & 1 & 0 & 0 \\ 2 & -1 & 3 & 0 & 1 & 0 \\ 4 & 1 & 8 & 0 & 0 & 1 \end{bmatrix} \xrightarrow[r_{13}^{(-4)}]{r_{12}^{(-2)}} \begin{bmatrix} 1 & 0 & 2 & 1 & 0 & 0 \\ 0 & -1 & -1 & -2 & 1 & 0 \\ 0 & 1 & 0 & -4 & 0 & 1 \end{bmatrix}$$

$$\xrightarrow{r_2^{(-1)}} \begin{bmatrix} 1 & 0 & 2 & 1 & 0 & 0 \\ 0 & 1 & 1 & 2 & -1 & 0 \\ 0 & 1 & 0 & -4 & 0 & 1 \end{bmatrix} \xrightarrow{r_{23}^{(-1)}} \begin{bmatrix} 1 & 0 & 2 & 1 & 0 & 0 \\ 0 & 1 & 1 & 2 & -1 & 0 \\ 0 & 0 & -1 & -6 & 1 & 1 \end{bmatrix}$$

$$\xrightarrow[r_{31}^{(2)}]{r_{32}^{(1)}} \begin{bmatrix} 1 & 0 & 0 & -11 & 2 & 2 \\ 0 & 1 & 0 & -4 & 0 & 1 \\ 0 & 0 & -1 & -6 & 1 & 1 \end{bmatrix} \xrightarrow{r_3^{(-1)}} \begin{bmatrix} 1 & 0 & 0 & -11 & 2 & 2 \\ 0 & 1 & 0 & -4 & 0 & 1 \\ 0 & 0 & 1 & 6 & -1 & -1 \end{bmatrix}.$$ Q.E.D.

2. Adjoint Matrix

To find the inverse of a matrix, apart from using the augmented matrix method, there is another useful approach using algebraic identities to directly calculate the inverse matrix. For any square matrix $A_{n \times n}$, we define its adjoint matrix as $adj(A) = C_{n \times n}^T = \left[c_{ij} \right]_{n \times n}^T$, where $c_{ij} = (-1)^{i+j} M_{ij}$ is the cofactor of A obtained by taking the determinant of the submatrix formed by removing the ith row and jth column, in the case of a 2×2 matrix:

$$A \cdot adj(A) = \begin{bmatrix} a_{11} & a_{12} \\ a_{21} & a_{22} \end{bmatrix} \begin{bmatrix} C_{11} & C_{12} \\ C_{21} & C_{22} \end{bmatrix}^T = \begin{bmatrix} a_{11} & a_{12} \\ a_{21} & a_{22} \end{bmatrix} \begin{bmatrix} C_{11} & C_{21} \\ C_{12} & C_{22} \end{bmatrix}$$

$$= \begin{bmatrix} a_{11} & a_{12} \\ a_{21} & a_{22} \end{bmatrix} \begin{bmatrix} +a_{22} & -a_{12} \\ -a_{21} & a_{11} \end{bmatrix} = \begin{bmatrix} a_{11}a_{22} - a_{12}a_{21} & 0 \\ 0 & a_{11}a_{22} - a_{21}a_{12} \end{bmatrix}$$

$$= \begin{bmatrix} |A| & 0 \\ 0 & |A| \end{bmatrix} = |A| \begin{bmatrix} 1 & 0 \\ 0 & 1 \end{bmatrix} = |A| I_2$$

We obtain $A^{-1} = \dfrac{adj(A)}{|A|}$. In fact, this holds true for general $n \times n$ matrices. Readers can derive it by comparing the definition of determinants. We will now summarize the steps for finding the inverse matrix using the adjugate matrix for the convenience of quick reference.

▶ Theorem 6-2-3

Solutions of Inverse Matrix

By following the steps (1) to (5) below, the inverse matrix can be obtained.

(1) Calculate the determinant value $|A|$ of A. (if $|A| = 0 \Rightarrow$ then A^{-1} does not exist.)

(2) Calculate the sub-determinant (minor) M_{ij} of A.

(3) Let $C = \begin{bmatrix} C_{11} & C_{12} & \cdots & C_{1n} \\ C_{21} & \cdots & \cdots & C_{2n} \\ C_{n1} & \cdots & \cdots & C_{nn} \end{bmatrix}$, where $C_{ij} = (-1)^{i+j} M_{ij}$.

(4) Define the adjoint matrix of **A** as $adj(A) = C^T$.

(5) then $A^{-1} = \dfrac{adj(A)}{|A|}$. ◀

For a 3×3 matrix $A_{3\times3} = \begin{bmatrix} a_{11} & a_{12} & a_{13} \\ a_{21} & a_{22} & a_{23} \\ a_{31} & a_{32} & a_{33} \end{bmatrix}$, if $|A| \neq 0$, then $A^{-1} = \dfrac{adj(A)}{|A|}$, where

$C_{ij} = (-1)^{i+j} M_{i+j}$, therefore

$$adj(A_{3\times3}) = \begin{bmatrix} +M_{11} & -M_{12} & +M_{13} \\ -M_{21} & +M_{22} & -M_{23} \\ +M_{31} & -M_{32} & +M_{33} \end{bmatrix}^T = \begin{bmatrix} +\begin{vmatrix} a_{22} & a_{23} \\ a_{32} & a_{33} \end{vmatrix} & -\begin{vmatrix} a_{21} & a_{23} \\ a_{31} & a_{33} \end{vmatrix} & +\begin{vmatrix} a_{21} & a_{22} \\ a_{31} & a_{32} \end{vmatrix} \\ -\begin{vmatrix} a_{12} & a_{13} \\ a_{32} & a_{33} \end{vmatrix} & +\begin{vmatrix} a_{11} & a_{13} \\ a_{31} & a_{33} \end{vmatrix} & -\begin{vmatrix} a_{11} & a_{12} \\ a_{31} & a_{32} \end{vmatrix} \\ +\begin{vmatrix} a_{12} & a_{13} \\ a_{22} & a_{23} \end{vmatrix} & -\begin{vmatrix} a_{11} & a_{13} \\ a_{21} & a_{23} \end{vmatrix} & +\begin{vmatrix} a_{11} & a_{12} \\ a_{21} & a_{22} \end{vmatrix} \end{bmatrix}^T .$$

By applying these formulas or utilizing the 3×3 memory aid:

$$\begin{array}{cccccc} a_{11} & a_{12} & a_{13} & a_{11} & a_{12} \\ a_{21} & a_{22} & a_{23} & a_{21} & a_{22} \\ a_{31} & a_{32} & a_{33} & a_{31} & a_{32} \\ a_{11} & a_{12} & a_{13} & a_{11} & a_{12} \\ a_{21} & a_{22} & a_{23} & a_{21} & a_{22} \end{array}^T .$$

Example 10

(1) $A = \begin{bmatrix} 3 & 5 \\ 1 & 3 \end{bmatrix}$, find $adj(A)$ and A^{-1}. (2) $A = \begin{bmatrix} 1 & 0 & 2 \\ 2 & -1 & 3 \\ 4 & 1 & 8 \end{bmatrix}$, find $adj(A)$ and A^{-1}.

Solution

(1) $|A| = 4$, $adj(A) = \begin{bmatrix} 3 & -5 \\ -1 & 3 \end{bmatrix}$, then $A^{-1} = \dfrac{adj(A)}{|A|} = \dfrac{1}{4}\begin{bmatrix} 3 & -5 \\ -1 & 3 \end{bmatrix} = \begin{bmatrix} \dfrac{3}{4} & -\dfrac{5}{4} \\ -\dfrac{1}{4} & \dfrac{3}{4} \end{bmatrix}.$

(2) $|A| = -8 + 4 + 8 - 3 = 1,$

$$adj(A) = \begin{bmatrix} +\begin{vmatrix} -1 & 3 \\ 1 & 8 \end{vmatrix} & -\begin{vmatrix} 2 & 3 \\ 4 & 8 \end{vmatrix} & +\begin{vmatrix} 2 & -1 \\ 4 & 1 \end{vmatrix} \\ -\begin{vmatrix} 0 & 2 \\ 1 & 8 \end{vmatrix} & +\begin{vmatrix} 1 & 2 \\ 4 & 8 \end{vmatrix} & -\begin{vmatrix} 1 & 0 \\ 4 & 1 \end{vmatrix} \\ +\begin{vmatrix} 0 & 2 \\ -1 & 3 \end{vmatrix} & -\begin{vmatrix} 1 & 2 \\ 2 & 3 \end{vmatrix} & +\begin{vmatrix} 1 & 0 \\ 2 & -1 \end{vmatrix} \end{bmatrix}^T = \begin{bmatrix} -11 & 2 & 2 \\ -4 & 0 & 1 \\ 6 & -1 & -1 \end{bmatrix},$$

therefore $A^{-1} = \dfrac{adj(A)}{|A|} = \dfrac{1}{1}\begin{bmatrix} -11 & -4 & 6 \\ 2 & 0 & -1 \\ 2 & 1 & -1 \end{bmatrix}^T = \begin{bmatrix} -11 & 2 & 2 \\ -4 & 0 & 1 \\ 6 & -1 & -1 \end{bmatrix}.$ Q.E.D.

▶ Theorem 6-2-4

The Important Properties of Inverse Matrix

(1) If A^{-1} exists, then $AA^{-1} = I \Rightarrow \det(A^{-1}) = \dfrac{1}{|A|} = \dfrac{1}{\det(A)}.$

(2) $|B^{-1}AB| = |A|$, where square matrix B is invertible.

(3) A is Invertible $\Leftrightarrow |A| \neq 0$; A is not invertible $\Leftrightarrow |A| = 0.$

(4) $|adj(A)| = |A|^{n-1}.$

(5) $(AB)^{-1} = B^{-1}A^{-1}.$ ◀

It is not difficult to prove theorem 6-2-4 (4): By using the identity $A\,adj(A) = |A|I_n$ and the determinant of matrix multipli30

cation

$\det(AB) = \det(A)\det(B)$, we obtain by taking the determinant on both sides yields $|A\|adj(A)| = |A|^n$, therefore $|adj(A)| = |A|^{n-1}$. The other parts (1) to (3) can also be easily derived using the determinant of matrix multiplication. Readers are encouraged to practice and verify these results on their own.

Example 11

Given that $A = \begin{bmatrix} s & t & u \\ v & w & x \\ y & z & r \end{bmatrix}$, if $|A| = -30$, and $|B| \neq 0$ then solve the following questions.

(1) $\det(A^{-1})$.   (2) $\det(B^{-1}AB)$.   (3) $\det(adj(A))$.

Solution

(1) $\det(A^{-1}) = \dfrac{1}{\det(A)} = -\dfrac{1}{30}$. (Theorem 6-2-4(1))

(2) $\det(B^{-1}AB) = \det(ABB^{-1}) = \det(A) = -30$. (Theorem 6-2-4(2))

(3) $\det(adj(A)) = |A|^{(3-1)} = (-30)^2 = 900$. (Theorem 6-2-4(4))   Q.E.D.

Example 12

Given that $adj(A) = \begin{bmatrix} 2 & -2 & 0 \\ 0 & 2 & -1 \\ 0 & 0 & 1 \end{bmatrix}$, solve the following questions.

(1) $|A|$.   (2) A^{-1}.   (3) A.

Solution

(1) $A \cdot adj(A) = |A|I \Rightarrow |A| \cdot |adj(A)| = |A|^3 \Rightarrow |A|^2 = |adj(A)|$
also $|adj(A)| = 4$, $\therefore |A| = \pm 2$.

(2) $A^{-1} = \dfrac{adj(A)}{|A|} = \pm \dfrac{1}{2} \begin{bmatrix} 2 & -2 & 0 \\ 0 & 2 & -1 \\ 0 & 0 & 1 \end{bmatrix} = \pm \begin{bmatrix} 1 & -1 & 0 \\ 0 & 1 & -\dfrac{1}{2} \\ 0 & 0 & \dfrac{1}{2} \end{bmatrix}$.

(3) $A = |A| adj(A)^{-1} = \pm 2 \cdot \dfrac{1}{4} \begin{bmatrix} 2 & 2 & 2 \\ 0 & 2 & 2 \\ 0 & 0 & 4 \end{bmatrix} = \pm \begin{bmatrix} 1 & 1 & 1 \\ 0 & 1 & 1 \\ 0 & 0 & 2 \end{bmatrix}$.   Q.E.D.

6-2   Exercises

Basic questions

1. Using matrix row operations, transform the following matrix into row-echelon form (the answer is not unique).

(1) $\begin{bmatrix} 1 & 2 \\ 3 & 4 \end{bmatrix}$.   (2) $\begin{bmatrix} 0 & 2 \\ 1 & 1 \end{bmatrix}$.

(3) $\begin{bmatrix} 2 & 6 & 1 \\ 1 & 2 & -1 \\ 5 & 7 & -4 \end{bmatrix}$.   (4) $\begin{bmatrix} 2 & -1 & 1 \\ 1 & 1 & 2 \\ 0 & 3 & 3 \end{bmatrix}$.

(5) $\begin{bmatrix} 1 & 2 & 3 \\ 2 & 5 & 8 \\ 3 & 5 & 7 \end{bmatrix}$.

2. Using matrix row operations, transform the following matrix into reduced row-echelon form.

(1) $\begin{bmatrix} 1 & 2 \\ 3 & 4 \end{bmatrix}$.   (2) $\begin{bmatrix} 1 & 1 & -1 \\ 4 & 0 & 1 \\ 0 & 4 & 1 \end{bmatrix}$.

(3) $\begin{bmatrix} 2 & 6 & 1 & 7 \\ 1 & 2 & -1 & -1 \\ 5 & 7 & -4 & 9 \end{bmatrix}$.

3. Find the determinant value and inverse matrix of the following matrix.

(1) $\begin{bmatrix} 1 & 3 \\ 2 & 4 \end{bmatrix}$.   (2) $\begin{bmatrix} 5 & -8 \\ 1 & -3 \end{bmatrix}$.

(3) $\begin{bmatrix} 9 & 1 \\ 1 & 9 \end{bmatrix}$.

4. Find the determinant value and inverse matrix of the following matrix.

(1) $\begin{bmatrix} 1 & 0 & 2 \\ 2 & 1 & 1 \\ 1 & 1 & 1 \end{bmatrix}$.   (2) $\begin{bmatrix} 9 & 2 & 0 \\ 2 & 6 & 0 \\ 0 & 0 & 5 \end{bmatrix}$.

(3) $\begin{bmatrix} 8 & 0 & 1 \\ 3 & -2 & 1 \\ 1 & 4 & 0 \end{bmatrix}$.

5. Find the determinant value of following matrix.

(1) $\begin{vmatrix} 1 & x & x^2 \\ 1 & y & y^2 \\ 1 & z & z^2 \end{vmatrix}$.

(2) $\begin{vmatrix} 1 & 1 & 1 & 1 \\ 3 & 5 & 7 & 11 \\ 9 & 25 & 49 & 121 \\ 27 & 125 & 343 & 1331 \end{vmatrix}$.

Advanced questions

1. $A = \begin{bmatrix} 1 & 0 & -1 \\ 0 & 2 & -1 \\ -1 & 1 & 0 \end{bmatrix}$, please use elementary column operations to find A^{-1}.

2. $A = \begin{bmatrix} 3 & -1 & 1 \\ -15 & 6 & -5 \\ 5 & -2 & 2 \end{bmatrix}$, please use elementary column operations to find A^{-1}.

3. Given that $A = \begin{bmatrix} 2 & 1 & -3 \\ 3 & 1 & 0 \\ -6 & -4 & 2 \end{bmatrix}$, and $|B| \neq 0$, solve following questions

(1) $\det(A)$.   (2) $\det(A^{-1})$.
(3) $\det(B^{-1}AB)$.   (4) $\det(A^T)$.
(5) $\det(adj(A))$.

4. $A = \begin{bmatrix} \cos\theta & 0 & -\sin\theta \\ 0 & 1 & 0 \\ \sin\theta & 0 & \cos\theta \end{bmatrix}$, find $\det(A)$ and A^{-1}.

5. Find the determinant value of following matrix.

$\begin{vmatrix} 6 & 1 & -1 & 5 \\ 2 & -1 & 3 & -2 \\ 1 & 0 & -1 & 0 \\ -4 & 3 & 2 & 1 \end{vmatrix}$

6-3 Solution to Systems of Linear Equations

Solving systems of linear equations is a common problem in daily life, especially in engineering. In secondary school, we learn how to solve linear systems using the method of elimination and substitution. However, these methods are only applicable to systems with a small number of variables. In engineering applications, models often involve numerous unknowns. In such cases, matrix operations come in handy. Consider the following system of linear equations,

$$\begin{cases} -x_1 + x_2 + 2x_3 = 2 \\ 3x_1 - x_2 + x_3 = 6 \\ -x_1 + 3x_2 + 4x_3 = 4 \end{cases}$$

variables are x_1, x_2, x_3, let $A = \begin{bmatrix} -1 & 1 & 2 \\ 3 & -1 & 1 \\ -1 & 3 & 4 \end{bmatrix}$ be the coefficient matrix. By multiplying

the first row of A by 3 and adding it to the second row ($r_{12}^{(3)}(A)$), and simultaneously multiplying the first row by -1 and adding it to the third row ($r_{13}^{(-1)}(A)$), the original system becomes

$$\begin{cases} -x_1 + x_2 + 2x_3 = 2 \\ 2x_2 + 7x_3 = 12 \\ 2x_2 + 2x_3 = 2 \end{cases}. \text{ Again, let } \overline{A} = \begin{bmatrix} -1 & 1 & 2 \\ 0 & 2 & 7 \\ 0 & 2 & 2 \end{bmatrix}, \text{ by multiplying the second row of } A$$

by -1 and adding it to the third row ($r_{23}^{(-1)}(A)$), the system of equations becomes

$$\begin{cases} -x_1 + x_2 + 2x_3 = 2 \\ 2x_2 + 7x_3 = 12 \\ -5x_3 = -10 \end{cases}, \text{ from the third row, we can solve for } x_3 = 2, \text{ substituting this}$$

value into the second row, we can solve for $x_2 = -1$. Finally, substituting these values into the first row, we can obtain $x_1 = 1$.

In abstract terms, we rewrite the original system of linear equations as

$\begin{bmatrix} -1 & 1 & 2 \\ 3 & -1 & 1 \\ -1 & 3 & 4 \end{bmatrix} \begin{bmatrix} x_1 \\ x_2 \\ x_3 \end{bmatrix} = \begin{bmatrix} 2 \\ 6 \\ 4 \end{bmatrix}$, where $A = \begin{bmatrix} -1 & 1 & 2 \\ 3 & -1 & 1 \\ -1 & 3 & 4 \end{bmatrix}$, $X = \begin{bmatrix} x_1 \\ x_2 \\ x_3 \end{bmatrix}$, $B = \begin{bmatrix} 2 \\ 6 \\ 4 \end{bmatrix}$. We then

repeatedly apply row operations to simplify the coefficient matrix A and solve the system.

6-3-1 Gauss Elimination Method

Gaussian elimination is a frequently employed method for solving systems of linear equations. Essentially, it involves reducing a matrix, row by row, from left to right using column operations in a predetermined order.

▶ **Definition 6-3-1**

System of Simultaneous Equations

Given that $A_{m \times n}$, then $AX = B$ represents a system of simultaneous equations, where A is the coefficient matrix and B is the constant matrix., the augmented matrix $[A \mid B]$ is used

to represent this system, that is in $\begin{bmatrix} a_{11} & a_{12} & \cdots & a_{1n} \\ a_{21} & a_{22} & \cdots & a_{2n} \\ \vdots & \vdots & \ddots & \vdots \\ a_{m1} & a_{m2} & \cdots & a_{mn} \end{bmatrix} \begin{bmatrix} x_1 \\ x_2 \\ \vdots \\ x_n \end{bmatrix} = \begin{bmatrix} b_1 \\ b_2 \\ \vdots \\ b_n \end{bmatrix}$, the augmented

matrix is defined as $\begin{bmatrix} a_{11} & a_{12} & \cdots & a_{1n} & b_1 \\ a_{21} & a_{22} & \cdots & a_{2n} & b_2 \\ \vdots & \vdots & \ddots & \vdots & \vdots \\ a_{m1} & a_{m2} & \cdots & a_{mn} & b_n \end{bmatrix}$.   ◀

According to Definition 6-3-1, if $B = O$, the system of equations is called a homogeneous system of equations. Otherwise, it is referred to as a non-homogeneous system of equations. Readers can refer to the chapters so far to point out what is homogeneous and what is non-homogeneous.

▶ **Definition 6-3-2**

Gauss Elimination Method and Gauss-Jordan Elimination Method

(1) Using row operations to transform the augmented matrix of a system of equations into row-echelon form and then solving the system is known as the Gaussian elimination method.

(2) Transforming the augmented matrix into reduced row-echelon form and then solving the system of equations is called the Gauss-Jordan elimination method.   ◀

▶ **Theorem 6-3-1**

Column-Equivalent Homogeneous System of Equations

If the matrix A undergoes elementary column operations and becomes matrix C, which means that A is column-equivalent to C, then $AX = O$ and $CX = O$ have the same solution. In other words, elementary column operations do not change the solution set.   ◀

▶ **Theorem 6-3-2**

Column-Equivalent Non-Homogeneous System of Equations

If the augmented matrix $[A \mid B]$ is column-equivalent to $[A_1 \mid B_1]$, then $AX = B$ and $A_1X = B_1$ have the same solution. ◀

Example 1

Use the Gaussian elimination method to solve the following system of simultaneous equations.

$$\begin{cases} -x_1 + x_2 + 2x_3 = 2 \\ 3x_1 - x_2 + x_3 = 6 \\ -x_1 + 3x_2 + 4x_3 = 4 \end{cases}$$

Solution

$$[A \mid B] = \begin{bmatrix} -1 & 1 & 2 & 2 \\ 3 & -1 & 1 & 6 \\ -1 & 3 & 4 & 4 \end{bmatrix} \xrightarrow{r_{12}^{(3)} r_{13}^{(-1)}} \begin{bmatrix} -1 & 1 & 2 & 2 \\ 0 & 2 & 7 & 12 \\ 0 & 2 & 2 & 2 \end{bmatrix}$$

$$\xrightarrow{r_{23}^{(-1)}} \begin{bmatrix} -1 & 1 & 2 & 2 \\ 0 & 2 & 7 & 12 \\ 0 & 0 & -5 & -10 \end{bmatrix} \Rightarrow \begin{cases} -x_1 + x_2 + 2x_3 = 2 \\ 2x_2 + 7x_3 = 12 \\ -5x_3 = -10 \end{cases}$$

$\Rightarrow x_3 = 2, x_2 = -1, x_1 = 1.$

Q.E.D.

As we have seen in the previous question, the solution of a system of equations is closely related to the augmented matrix. Now, let's introduce the concept of matrix rank and discuss its relationship with systems of equations.

6-3-2 The Rank of Matrix

To determine if the system of equations $AX = B$ has a solution, which means

$$\begin{bmatrix} A_1 & A_2 & A_3 \end{bmatrix} \begin{bmatrix} x_1 \\ x_2 \\ x_3 \end{bmatrix} = x_1A_1 + x_2A_2 + x_3A_3 = B \text{, where } A_1, A_2, A_3 \text{ are the column vectors}$$

of A. The existence of a solution implies that matrix B can be expressed as a linear combination of the row vectors of matrix A. In other words, we are checking if B lies in the range (also known as the column space) of A. If it does, then by the definition of linear combinations, the augmented matrix $[A \mid B]$ will have the same row space as A. In other words, the dimension of the row space of A is equal to the dimension of the row space of $[A \mid B]$. This crucial dimension helps us determine if a system of equations has a solution.

We denote the dimension of the column space of A as $cr(A)$; and the dimension of the row space as $rr(A)$. Here, matrix operations inform us of an important fact: for any matrix A, $cr(A) = rr(A)$. In fact, by performing column operations to transform A into its reduced row-echelon form, we find that the number of pivot (elements where a row has only one "1") is equal to the number of rows. From this observation, we deduce the following definition:

▶ **Definition 6-3-3**

Rank

Given that $A_{m \times n}$, $cr(A)$ ($= rr(A)$) is called the rank of A, referred to as rank(A) or $r(A)$.

◀

For example, $A = \begin{bmatrix} 1 & 2 & 3 \\ 4 & 5 & 6 \\ 7 & 8 & 9 \end{bmatrix} \xrightarrow{r_{12}^{(-4)} r_{13}^{(-7)}} \begin{bmatrix} 1 & 2 & 3 \\ 0 & -3 & -6 \\ 0 & -6 & -12 \end{bmatrix} \xrightarrow{r_{23}^{(-2)}} \begin{bmatrix} 1 & 2 & 3 \\ 0 & -3 & -6 \\ 0 & 0 & 0 \end{bmatrix}$, the

number of linearly independent row vectors in A is 2, in fact, the third row vector in A is linearly dependent on the first and second row vectors. Therefore, the column rank of A is 2, that is $rr(A) = 2$. Now, the rows of A^T are the columns of A, also

$$A^T = \begin{bmatrix} 1 & 4 & 7 \\ 2 & 5 & 8 \\ 3 & 6 & 9 \end{bmatrix} \xrightarrow{r_{12}^{(-2)} r_{13}^{(-3)}} \begin{bmatrix} 1 & 4 & 7 \\ 0 & -3 & -6 \\ 0 & -6 & -12 \end{bmatrix} \xrightarrow{r_{23}^{(-2)}} \begin{bmatrix} 1 & 4 & 7 \\ 0 & -3 & -6 \\ 0 & 0 & 0 \end{bmatrix}$$

The first and second rows of A^T are linearly independent, meaning that the first and second rows of A are linearly independent. Thus, the number of linearly independent row vectors in A is 2. Therefore, the row rank of A is 2, that is $cr(A) = 2$, which means $rr(A) = cr(A) = \text{rank}(A) = 2$. In general, when determining the rank of a matrix, we often refer to the row rank, which is the number of non-zero rows obtained after performing column operations to transform A into row-echelon form. This number represents the rank of A.

Example 2

Find the rank of $A = \begin{bmatrix} 1 & -2 & 1 & 0 \\ 2 & 1 & 1 & 2 \\ 1 & -7 & 2 & -2 \end{bmatrix}$.

Solution

$$A = \begin{bmatrix} 1 & -2 & 1 & 0 \\ 2 & 1 & 1 & 2 \\ 1 & -7 & 2 & -2 \end{bmatrix} \xrightarrow{r_{12}^{(-2)} r_{13}^{(-1)}} \begin{bmatrix} 1 & -2 & 1 & 0 \\ 0 & 5 & -1 & 2 \\ 0 & -5 & 1 & -2 \end{bmatrix} \xrightarrow{r_{23}^{(1)}} \begin{bmatrix} 1 & -2 & 1 & 0 \\ 0 & 5 & -1 & 2 \\ 0 & 0 & 0 & 0 \end{bmatrix},$$

therefore rank(A) = 2.

Q.E.D.

1. The Important Properties of Rank

Let A be $m \times n$ matrix.

(1) $\text{rank}(A_{m \times n}) \leq \min\{m, n\}$.

(2) $\text{rank}(A) = \text{rank}(A^T)$.

(3) A is transformed into B through elementary row operations, then $\text{rank}(A) = \text{rank}(B)$.

(4) If A is an upper (lower) triangular matrix, then the number of non-zero rows in A is equal to the rank of A.

(5) For the system of simultaneous equations $AX = B$, $\text{rank}([A \mid B])$ represents the dimension of the solution space of the system, which is the number of linearly independent equations.

(6) $A_{m \times n}$, $B_{n \times s}$, then $\text{rank}(AB) \leq \min\{\text{rank}(A), \text{rank}(B)\}$, in other words, $\text{rank}(AB) \leq \text{rank}(B)$, $\text{rank}(AB) \leq \text{rank}(A)$, that is the rank decreases as matrix are multiplied.

(7) If the maximum dimension of a non-zero determinant in any submatrix of A is $r \times r$, then $\text{rank}(A) = r$.

Here are some examples illustrating the applications of these properties. Suppose

$$A_{5 \times 1} = \begin{bmatrix} 1 \\ 2 \\ 3 \\ 4 \\ 5 \end{bmatrix}, \text{ then rank}(A) \leq \min\{5, 1\} = 1, \text{ and } A \text{ has only one non-zero column, so}$$

$\text{rank}(A) = 1$; now, $A^T = [1 \quad 2 \quad 3 \quad 4 \quad 5]$, then $B = AA^T$ is a 5×5

square matrix $\begin{bmatrix} 1 & 2 & 3 & 4 & 5 \\ 2 & 4 & 6 & 8 & 10 \\ 3 & 6 & 9 & 12 & 15 \\ 4 & 8 & 12 & 16 & 20 \\ 5 & 10 & 15 & 20 & 25 \end{bmatrix}$, determining the rank of B by directly

applying row operations can be complicated. However, using property (6) and knowing that $\text{rank}(A) = \text{rank}(A^T) = 1$, we know $\text{rank}(B) \leq \min\{\text{rank}(A), \text{rank}(A^T)\} = 1$. Furthermore, $B \neq O$, so $\text{rank}(B) = 1$. Another example involves determining the

dimension of $A_{3 \times 3} = \begin{bmatrix} 1 & -2 & 7 \\ -4 & 8 & 5 \\ 2 & -4 & 3 \end{bmatrix}$, upon observation, we notice that the first

column of A is proportional to the second column, therefore

$$\det(A) = \begin{vmatrix} 1 & -2 & 7 \\ -4 & 8 & 5 \\ 2 & -4 & 3 \end{vmatrix} = 0 \text{ and } 2 \leq \text{rank}(A) \leq 3. \text{ Additionally, by inspecting the}$$

submatrices of A, we find that there exists a non-zero determinant $\begin{vmatrix} 8 & 5 \\ -4 & 3 \end{vmatrix} \neq 0$, so

rank$(A) = 2$.

2. **Properties of Invertible Matrix**

   If matrix A is an $n \times n$ invertible matrix, then

   (1) $\det(A) \neq 0$.

   (2) rank$(A) = n$.

   (3) A has n linearly independent row (column) vectors.

   (4) The dimension of the row (column) space of A is n.

   (5) The homogeneous system $AX = O$ has a unique solution $X_{n \times 1} = A^{-1}O = O$.

   (6) $AX = B$ has a unique non-zero solution $X = A^{-1}B$.

Example 3

Given a circuit with currents $I_1, I_2, I_3,$ simplified using Kirchhoff's law, we obtain

$$\begin{cases} I_1 + I_2 - I_3 = E_1 \\ 4I_1 + I_3 = E_2 \\ 4I_2 + I_3 = E_3 \end{cases},$$

where E_1, E_2, E_3 are external voltage sources.

(1) If there are no external voltage sources, that is E_1, E_2, E_3 are all 0, solve I_1, I_2, I_3.

(2) If the external voltage sources are $E_1 = 0, E_2 = 16, E_3 = 32$, solve I_1, I_2, I_3.

Solution

The original system of equations can be rewritten as $\begin{bmatrix} 1 & 1 & -1 \\ 4 & 0 & 1 \\ 0 & 4 & 1 \end{bmatrix} \begin{bmatrix} I_1 \\ I_2 \\ I_3 \end{bmatrix} = \begin{bmatrix} E_1 \\ E_2 \\ E_3 \end{bmatrix}$, where

$A = \begin{bmatrix} 1 & 1 & -1 \\ 4 & 0 & 1 \\ 0 & 4 & 1 \end{bmatrix}$, $X = \begin{bmatrix} I_1 \\ I_2 \\ I_3 \end{bmatrix}$, $B = \begin{bmatrix} E_1 \\ E_2 \\ E_3 \end{bmatrix}$, also $\det(A) = -24 \neq 0$,

$A^{-1} = \dfrac{1}{-24} \begin{bmatrix} -4 & -5 & 1 \\ -4 & 1 & -5 \\ 16 & -4 & -4 \end{bmatrix}$,

(1) $B = \begin{bmatrix} 0 \\ 0 \\ 0 \end{bmatrix}$, then the homogeneous system of equations has the solution

$$X = \begin{bmatrix} I_1 \\ I_2 \\ I_3 \end{bmatrix} = A^{-1}O = O = \begin{bmatrix} 0 \\ 0 \\ 0 \end{bmatrix}.$$

(2) $B = \begin{bmatrix} 0 \\ 16 \\ 32 \end{bmatrix}$, then the non-homogeneous system of equations has solution

$$X = \begin{bmatrix} I_1 \\ I_2 \\ I_3 \end{bmatrix} = A^{-1}B = \begin{bmatrix} 2 \\ 6 \\ 8 \end{bmatrix}.$$

Q.E.D.

6-3-3 Solution Space of a System of Equations

1. The Solution Space of a Homogeneous System of Equations

Let's take an example of a homogeneous system of equations with two unknowns. By carefully examining the solution process, we can derive the method for solving an n-dimensional homogeneous system of equations. Consider the system of equations $\begin{cases} a_1 x + b_1 y = 0 \\ a_2 x + b_2 y = 0 \end{cases}$. In the x-y plane, this system represents the

intersection of two lines. Based on the coefficients of the equations, there are two possible scenarios:

(1) If $\dfrac{a_1}{a_2} \neq \dfrac{b_1}{b_2}$, the two lines are not parallel, and there is a

unique solution $\begin{cases} x = 0 \\ y = 0 \end{cases}$,

as shown in Figure 6-2, the lines intersect at the origin, and

the zero vector $\begin{bmatrix} x \\ y \end{bmatrix} = \begin{bmatrix} 0 \\ 0 \end{bmatrix}$ represents the trivial solution.

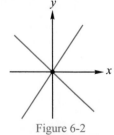

Figure 6-2

(2) If $\dfrac{a_1}{a_2} = \dfrac{b_1}{b_2}$, then the two lines coincide, there are infinitely

many solutions

$\begin{cases} x = t \\ y = -\dfrac{a_1}{b_1} t \end{cases}$, $t \in \mathbb{R}$, as shown in Figure 6-3, the lines overlap.

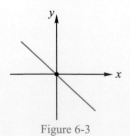

Figure 6-3

Its solution is $\begin{bmatrix} x \\ y \end{bmatrix} = \begin{bmatrix} t \\ -\dfrac{a_1}{b_1}t \end{bmatrix} = t \begin{bmatrix} 1 \\ -\dfrac{a_1}{b_1} \end{bmatrix}$, $t \in$ R, this is a parametric solution.

If we represent the system of equations in matrix form $\begin{bmatrix} a_1 & b_1 \\ a_2 & b_2 \end{bmatrix}\begin{bmatrix} x \\ y \end{bmatrix} = \begin{bmatrix} 0 \\ 0 \end{bmatrix}$, then in case(1), the condition $\dfrac{a_1}{a_2} \neq \dfrac{b_1}{b_2}$ indicates that the two rows of the coefficient matrix $A = \begin{bmatrix} a_1 & b_1 \\ a_2 & b_2 \end{bmatrix}$ are not proportional, which means rank(A) = 2. Therefore, we observe that when rank(A) equals the order n = 2 of the matrix, the system of equations has a unique solution. Similarly, in case (2), the condition $\dfrac{a_1}{a_2} = \dfrac{b_1}{b_2}$ represents rank(A) = 1, which is less than the order of the coefficient matrix. As a result, the system of equations has infinitely many solutions. Can we generalize these two cases when the coefficient matrix is the order n? Let's further discuss.

In R^n, the set of all vectors X that satisfy the homogeneous system of equations $A_{m \times n} X_{n \times 1} = O$ is called the solution space or the null space of the homogeneous system $AX = O$, which is denoted as $N(A)$ or Ker(A), that is

$$N(A) = \text{Ker}(A) = \{X_{n \times 1} \mid A_{m \times n} X_{n \times 1} = O, X \in R^n\}.$$

The dimension of the solution space, nullity(A) = dim($N(A)$) = dim(Ker(A)) , which is called zero null dimension, represents the number of parameters required to describe the solutions of the homogeneous system $A_{m \times n} X_{n \times 1} = O$. In the system of equations, the equations corresponding to the row-reduced echelon form of the matrix are referred to as linearly independent equations.

▶ **Theorem 6-3-3**

Dimension Theorem

In the solution space of the homogeneous system of equations $A_{m \times n} X_{n \times 1} = O$, the number of parameters required to describe the solutions nullity(A), is equal to the number of unknowns, n, minus the number of linearly independent equations, rank(A), which means nullity(A) = n − rank(A).   ◀

Example 4

Consider a homogeneous system of equations $A_{3\times3}X_{3\times1} = O$, where $A_{3\times3} = \begin{bmatrix} 1 & 2 & 3 \\ 2 & 5 & 8 \\ 3 & 5 & 7 \end{bmatrix}$,

$X_{3\times1} = \begin{bmatrix} x_1 \\ x_2 \\ x_3 \end{bmatrix}$, find the rank of A and the solution of $AX = O$.

Solution

(1) $A_{3\times3} = \begin{bmatrix} 1 & 2 & 3 \\ 2 & 5 & 8 \\ 3 & 5 & 7 \end{bmatrix} \xrightarrow{r_{12}^{(-2)} r_{13}^{(-3)}} \begin{bmatrix} 1 & 2 & 3 \\ 0 & 1 & 2 \\ 0 & -1 & -2 \end{bmatrix} \xrightarrow{r_{23}^{(1)}} \begin{bmatrix} 1 & 2 & 3 \\ 0 & 1 & 2 \\ 0 & 0 & 0 \end{bmatrix}$,

then rank(A) = 2, represents there are only two linearly independent equations in this system of equations.

(2) $A_{3\times3}X_{3\times1} = O$ can be reduced through row operations to

$$\begin{bmatrix} 1 & 2 & 3 \\ 0 & 1 & 2 \\ 0 & 0 & 0 \end{bmatrix} \begin{bmatrix} x_1 \\ x_2 \\ x_3 \end{bmatrix} = \begin{bmatrix} 0 \\ 0 \\ 0 \end{bmatrix} \Rightarrow \begin{cases} x_1 + 2x_2 + 3x_3 = 0 \\ x_2 + 2x_3 = 0 \end{cases}.$$

This shows that although there are three unknowns, only two equations are linearly independent.

Through nullity(A) = 3 − rank(A) = 3 − 2 = 1, we can conclude that there is one independent parameter when solving the system.

Let $x_3 = c$, then $x_2 = -2c$, $x_1 = c$, the solution space is $X_{3\times1} = \begin{bmatrix} x_1 \\ x_2 \\ x_3 \end{bmatrix} = \begin{bmatrix} c \\ -2c \\ c \end{bmatrix} = c \begin{bmatrix} 1 \\ -2 \\ 1 \end{bmatrix}$,

which means the solution of $AX = O$ is $X = c \begin{bmatrix} 1 \\ -2 \\ 1 \end{bmatrix}$, where c is an arbitrary constant.

Q.E.D.

2.　The Solutions to a Non-Homogeneous System of Equations

For a non-homogeneous system of equations, we start with the case of two unknowns. Consider the system of equations $\begin{cases} a_1 x + b_1 y = c_1 \\ a_2 x + b_2 y = c_2 \end{cases}$, in the x-y plane, this system also represents the intersection of two lines. However, unlike the homogeneous case, the two lines may not necessarily pass through the origin. Based on the coefficients of the equations, there are three possible situations.

(1) If $\dfrac{a_1}{a_2} \neq \dfrac{b_1}{b_2}$, then the two lines are not parallel, and there

is a unique solution, as shown in Figure 6-4, where

$$\begin{bmatrix} x \\ y \end{bmatrix} = \begin{bmatrix} x_0 \\ y_0 \end{bmatrix}.$$

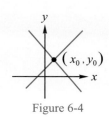

Figure 6-4

(2) If $\dfrac{a_1}{a_2} = \dfrac{b_1}{b_2} \neq \dfrac{c_1}{c_2}$, then the two lines are parallel, and the

system of equations has no solution, as shown in Figure 6-5.

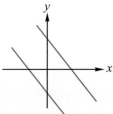

Figure 6-5

(3) If $\dfrac{a_1}{a_2} = \dfrac{b_1}{b_2} = \dfrac{c_1}{c_2}$, the two lines coincide, and the system

has   infinitely many solutions $\begin{cases} x = t \\ y = -\dfrac{a_1}{b_1}t \end{cases}, t \in \mathrm{R}$, as

shown in Figure 6-6.

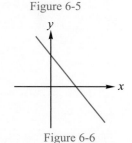

Figure 6-6

Based on the discussion similar to the homogeneous system of equations, we can derive the conclusions for the three cases as follows:

(1) If rank(A) = rank([A | B]) = 2, the system of equations has a unique solution;

(2) If rank(A) = 1 ≠ rank([A | B]) = 2, the system of equations has no solution;

(3) If rank(A) = rank([A | B]) = 1, the system of equations has infinitely many solutions.

▶ Theorem 6-3-4

The Existence of Non-Homogeneous System of Equations

The existence of solutions for the non-homogeneous system of equations $A_{m \times n}X_{n \times 1} = B_{m \times 1}$ is a necessary and sufficient condition given by
rank($A_{m \times n}$) = rank([$A_{m \times n}$ | $B_{n \times 1}$])
In other words, the condition is met when the rank of the matrix A and the augmented matrix [A | B] are the same.   ◀

For example, the augmented matrix corresponding to Example 1's system of

equations is $\begin{bmatrix} -1 & 1 & 2 & | & 2 \\ 3 & -1 & 1 & | & 6 \\ -1 & 3 & 4 & | & 4 \end{bmatrix}$. By performing column operations, we can simplify the

matrix to
$\begin{bmatrix} -1 & 1 & 2 & | & 2 \\ 3 & -1 & 1 & | & 6 \\ -1 & 3 & 4 & | & 4 \end{bmatrix} \xrightarrow{r} \begin{bmatrix} -1 & 1 & 2 & | & 2 \\ 0 & 2 & 7 & | & 12 \\ 0 & 0 & -5 & | & -10 \end{bmatrix}$. Because

rank(A) = 3 = rank([$A \mid B$]), the non-homogeneous system of equations has a solution.

Example 5

Solve $\begin{cases} x_1 - x_2 + x_3 = 2 \\ x_1 + 3x_2 - x_3 = 4 \\ 2x_1 + 2x_2 = c \end{cases}$, where (1) $c = -3$, (2) $c = 6$.

Solution

(1) The original system of equations can be rewritten as $\begin{bmatrix} 1 & -1 & 1 \\ 1 & 3 & -1 \\ 2 & 2 & 0 \end{bmatrix} \begin{bmatrix} x_1 \\ x_2 \\ x_3 \end{bmatrix} = \begin{bmatrix} 2 \\ 4 \\ -3 \end{bmatrix}$,

where $A = \begin{bmatrix} 1 & -1 & 1 \\ 1 & 3 & -1 \\ 2 & 2 & 0 \end{bmatrix}$, $X = \begin{bmatrix} x_1 \\ x_2 \\ x_3 \end{bmatrix}$, $B = \begin{bmatrix} 2 \\ 4 \\ -3 \end{bmatrix}$,

then

$[A \mid B] = \begin{bmatrix} 1 & -1 & 1 & | & 2 \\ 1 & 3 & -1 & | & 4 \\ 2 & 2 & 0 & | & -3 \end{bmatrix} \xrightarrow{r_{12}^{(-1)} r_{13}^{(-2)}} \begin{bmatrix} 1 & -1 & 1 & | & 2 \\ 0 & 4 & -2 & | & 2 \\ 0 & 4 & -2 & | & -7 \end{bmatrix} \xrightarrow{r_{23}^{(-1)}} \begin{bmatrix} 1 & -1 & 1 & | & 2 \\ 0 & 4 & -2 & | & 2 \\ 0 & 0 & 0 & | & -9 \end{bmatrix}$,

since rank(A) = 2 ≠ rank([$A \mid B$]) = 3, the system of equations has no solution.

(2) In the above question, if $c = 6$, we rewrite the third equation as $2x_1 + 2x_2 = 6$, then

$[A \mid B] = \begin{bmatrix} 1 & -1 & 1 & | & 2 \\ 1 & 3 & -1 & | & 4 \\ 2 & 2 & 0 & | & 6 \end{bmatrix} \xrightarrow{r_{12}^{(-1)} r_{13}^{(-2)}} \begin{bmatrix} 1 & -1 & 1 & | & 2 \\ 0 & 4 & -2 & | & 2 \\ 0 & 4 & -2 & | & 2 \end{bmatrix} \xrightarrow{r_{23}^{(-1)}} \begin{bmatrix} 1 & -1 & 1 & | & 2 \\ 0 & 4 & -2 & | & 2 \\ 0 & 0 & 0 & | & 0 \end{bmatrix}$.

This gives rise to two linearly independent equations $\begin{cases} x_1 - x_2 + x_3 = 2 \\ 4x_2 - 2x_3 = 2 \end{cases}$.

Let $x_2 = c$, then $\begin{cases} x_3 = 2c - 1 \\ x_1 = -c + 3 \end{cases}$, we obtain $X = \begin{bmatrix} x_1 \\ x_2 \\ x_3 \end{bmatrix} = \begin{bmatrix} -c+3 \\ c \\ 2c-1 \end{bmatrix} = c \begin{bmatrix} -1 \\ 1 \\ 2 \end{bmatrix} + \begin{bmatrix} 3 \\ 0 \\ -1 \end{bmatrix}$.

Its homogeneous solution is $X_h = c \begin{bmatrix} -1 \\ 1 \\ 2 \end{bmatrix}$, particular solution is $X_p = \begin{bmatrix} 3 \\ 0 \\ -1 \end{bmatrix}$.

It can be observed that the homogeneous solution X_h is actually independent of

the non-homogeneous term $\begin{bmatrix} 2 \\ 4 \\ 6 \end{bmatrix}$, that is, it is the solution of $\begin{cases} x_1 - x_2 + x_3 = 0 \\ x_1 + 3x_2 - x_3 = 0 \\ 2x_1 + 2x_2 = 0 \end{cases}$.

Q.E.D.

6-3-4   Classification of Solutions for Systems of Equations

To solve a system of equations, one can consider the cases where $B = O$ and $B \neq O$ in the equation $AX = B$, respectively.

1. $B = O$: Homogeneous

Because $\text{rank}(A) = \text{rank}([A \mid O])$, the system of equations has solutions, then the classification of solutions can be based on comparing the situation of $\text{rank}(A)$ with the dimension of the original space.

Original space	$r = \text{rank}(A)$	Row operations to convert a matrix to upper triangular form	The situation of solutions
$m = n$	$r = n$		$X = O$
	$r < n$	r-tuple	$X = X_h + X_p$ $= \sum_{i=1}^{n-r} c_i X_i , \ c_i \in \mathbb{R}$
$m > n$	$r = n$		$X = O$
	$r < n$	r-tuple	$X = X_h + X_p$ $= \sum_{i=1}^{n-r} c_i X_i , \ c_i \in \mathbb{R}$
$m < n$	$r \leq m < n$	r-tuple	$X = X_h + X_p$ $= \sum_{i=1}^{n-r} c_i X_i , \ c_i \in \mathbb{R}$

▶ **Theorem 6-3-5**

The Solutions of Homogeneous Linear System of Equations

For $A_{m \times n} X_{n \times 1} = O$, the classification of solutions is as follows:

(1) rank$(A) = n \Leftrightarrow$ There is a unique solution $X = O$.

(2) rank$(A) = r < n \Leftrightarrow$ There are infinitely many solutions; in this case, the solutions take the form of non-zero solutions with $(n - r)$ parameters:

$$X = c_1 X_1 + \cdots + c_{n-r} X_{n-r}.$$ ◀

In theorem, $X_1, X_2, \ldots, X_{n-r}$ are the $(n - r)$ linearly independent vector solutions that satisfying $A_{m \times n} X_{n \times 1} = 0$, as seen in Example 4 in this section, $n = 3$, $r = 2$, there is an one parametric solution, X_1 is $\begin{bmatrix} 1 \\ -2 \\ 1 \end{bmatrix}$ (where c is taken as 1).

2. $B \neq O$: Non-Homogeneous

In contrast to the homogeneous case, one needs to consider both the rank(A), rank $[A \mid B]$ in order to determine the solutions. If they are equal (indicating the existence of solutions), the further classification of solutions can be based on the principles of the homogeneous case. However, if rank$(A) \neq$ rank$[A \mid B]$, then the system has no solution.

Original space	rank$(A) = r$ $= $ rank$([A \mid B])$	Row operations to convert a matrix to upper triangular form	The situation of solutions
$m = n$	$r = n$		$X = A^{-1}B$
	$r < n$		$X = X_h + X_p$ $= \sum_{i=1}^{n-r} c_i X_i + X_p , c_i \in \mathrm{R}$
$m > n$	$r = n$		$X = A^{-1}B$

Original space	rank(A) = r = rank([A | B])	Row operations to convert a matrix to upper triangular form	The situation of solutions
	$r < n$		$X = X_h + X_p$ $= \displaystyle\sum_{i=1}^{n-r} c_i X_i + X_p \, , c_i \in \mathrm{R}$
$m < n$	$r \le m < n$		$X = X_h + X_p$ $= \displaystyle\sum_{i=1}^{n-r} c_i X_i + X_p \, , c_i \in \mathrm{R}$

▶ **Theorem 6-3-6**

The Solutions of Non-Homogeneous Linear System of Equations

For $A_{m \times n} X_{n \times 1} = B_{m \times 1}$, the classification of solutions is as follows:

(1) rank(A) = rank([A \| B]) = r, we can further subdivide into the following two situations:

    ① $r = n \Rightarrow$ There is a unique solution.

    ② $r < n \Rightarrow$ There are infinitely many solutions; the solutions have $(n - r)$ parameters.

(2) rank(A) $\ne$ rank([A \| B]) $\Rightarrow$ There is no solution. ◀

Example 6

Solve by using Gaussian elimination $\begin{cases} x_1 + 2x_2 - x_3 = 7 \\ 2x_1 + 3x_2 + x_3 = 14 \\ x_1 + x_2 + 2x_3 = 7 \end{cases}$.

Solution

The original system of equations can be rewritten as $\begin{bmatrix} 1 & 2 & -1 \\ 2 & 3 & 1 \\ 1 & 1 & 2 \end{bmatrix} \begin{bmatrix} x_1 \\ x_2 \\ x_3 \end{bmatrix} = \begin{bmatrix} 7 \\ 14 \\ 7 \end{bmatrix}$,

where $A = \begin{bmatrix} 1 & 2 & -1 \\ 2 & 3 & 1 \\ 1 & 1 & 2 \end{bmatrix}$, $X = \begin{bmatrix} x_1 \\ x_2 \\ x_3 \end{bmatrix}$, $B = \begin{bmatrix} 7 \\ 14 \\ 7 \end{bmatrix}$,

then

$[A \mid B] = \begin{bmatrix} 1 & 2 & -1 & 7 \\ 2 & 3 & 1 & 14 \\ 1 & 1 & 2 & 7 \end{bmatrix} \xrightarrow{r_{12}^{(-2)} r_{13}^{(-1)}} \begin{bmatrix} 1 & 2 & -1 & 7 \\ 0 & -1 & 3 & 0 \\ 0 & -1 & 3 & 0 \end{bmatrix} \xrightarrow{r_{23}^{(-1)}} \begin{bmatrix} 1 & 2 & -1 & 7 \\ 0 & -1 & 3 & 0 \\ 0 & 0 & 0 & 0 \end{bmatrix}$.

Because $\text{rank}(A) = \text{rank}([A \mid B]) = 2 < 3$, the system of equations has infinitely many

solutions with $3 - \text{rank}(A) = 1$ parameter, through $\begin{cases} x_1 + 2x_2 - x_3 = 7 \\ -x_2 + 3x_3 = 0 \end{cases}$, let $x_3 = c$, then

$x_2 = 3c$, $x_1 = 7 - 5c$.

So the solutions are $X = \begin{bmatrix} x_1 \\ x_2 \\ x_3 \end{bmatrix} = \begin{bmatrix} 7-5c \\ 3c \\ c \end{bmatrix} = c\begin{bmatrix} -5 \\ 3 \\ 1 \end{bmatrix} + \begin{bmatrix} 7 \\ 0 \\ 0 \end{bmatrix}$,

where the homogeneous solution is $X_h = c\begin{bmatrix} -5 \\ 3 \\ 1 \end{bmatrix}$, the particular solution is $X_p = \begin{bmatrix} 7 \\ 0 \\ 0 \end{bmatrix}$.

Q.E.D.

Example 7

Consider the system of equations $A_{3\times3} X_{3\times1} = B_{3\times1}$, where $A_{3\times3} = \begin{bmatrix} 1 & -2 & 3 \\ 2 & k+1 & 6 \\ -1 & 3 & k-2 \end{bmatrix}$,

$B_{3\times1} = \begin{bmatrix} 2 \\ 8 \\ -1 \end{bmatrix}$. Find the value of k, let this system of equations be :

(1) infinitely many solutions, (2) unique solution, (3) no solution.

Solution

$[A \mid B] = \begin{bmatrix} 1 & -2 & 3 & 2 \\ 2 & k+1 & 6 & 8 \\ -1 & 3 & k-2 & -1 \end{bmatrix} \xrightarrow{r_{12}^{(-2)} r_{13}^{(1)}} \begin{bmatrix} 1 & -2 & 3 & 2 \\ 0 & k+5 & 0 & 4 \\ 0 & 1 & k+1 & 1 \end{bmatrix} \xrightarrow{r_{23}} \begin{bmatrix} 1 & -2 & 3 & 2 \\ 0 & 1 & k+1 & 1 \\ 0 & k+5 & 0 & 4 \end{bmatrix}$

$\xrightarrow{r_{23}^{(-k-5)}} \begin{bmatrix} 1 & -2 & 3 & 2 \\ 0 & 1 & k+1 & 1 \\ 0 & 0 & -(k+5)(k+1) & -k-1 \end{bmatrix}$.

(1) For the system to have infinitely many solutions, so
$\text{rank}(A) = \text{rank}([A \mid B]) = r < 3 \Leftrightarrow k = -1$,
now $\text{rank}(A) = \text{rank}([A \mid B]) = 2 \Leftrightarrow$ has a one-parameter solution $\Rightarrow$ the system has infinitely many solutions.

(2) For the system to have a unique solution, then
$\text{rank}(A) = \text{rank}([A \mid B]) = 3 \Rightarrow k \neq -5, -1$.

(3) For the system to have no solution, then $\text{rank}(A) \neq \text{rank}([A \mid B]) \Rightarrow k = -5^{1}$. Q.E.D.

Example 8

If $A = \begin{bmatrix} 1 & 2 & 0 & 1 & 3 \\ 0 & 0 & 1 & 1 & 1 \\ 1 & 2 & 1 & 2 & 4 \end{bmatrix}$, (1) find the solution N of $AX = O$. (2) Find $\dim(N)$.

Solution

(1) $\begin{bmatrix} 1 & 2 & 0 & 1 & 3 \\ 0 & 0 & 1 & 1 & 1 \\ 1 & 2 & 1 & 2 & 4 \end{bmatrix} \xrightarrow{r_{13}^{(-1)}} \begin{bmatrix} 1 & 2 & 0 & 1 & 3 \\ 0 & 0 & 1 & 1 & 1 \\ 0 & 0 & 1 & 1 & 1 \end{bmatrix} \xrightarrow{r_{23}^{(-1)}} \begin{bmatrix} 1 & 2 & 0 & 1 & 3 \\ 0 & 0 & 1 & 1 & 1 \\ 0 & 0 & 0 & 0 & 0 \end{bmatrix}$

so the original homogeneous system of equations can be simplified to

$$\Rightarrow \begin{cases} x_1 + 2x_2 + x_4 + 3x_5 = 0, \text{ take } x_4 = c_1, \ x_5 = c_2, \ x_2 = c_3 \\ x_3 + x_4 + x_5 = 0 \Rightarrow x_3 = -c_1 - c_2, \ x_1 = -2c_3 - c_1 - 3c_2, \end{cases}$$

$$\therefore N = \begin{bmatrix} x_1 \\ x_2 \\ x_3 \\ x_4 \\ x_5 \end{bmatrix} = \begin{bmatrix} -2c_3 - c_1 - 3c_2 \\ c_3 \\ -c_1 - c_2 \\ c_1 \\ c_2 \end{bmatrix} = c_1 \begin{bmatrix} -1 \\ 0 \\ -1 \\ 1 \\ 0 \end{bmatrix} + c_2 \begin{bmatrix} -3 \\ 0 \\ -1 \\ 0 \\ 1 \end{bmatrix} + c_3 \begin{bmatrix} -2 \\ 1 \\ 0 \\ 0 \\ 0 \end{bmatrix}.$$

(2) $\dim(N) = 3^{2}$.   Q.E.D.

Example 9

(1) If $AX = B$ has a unique solution, where $A \in R^{n \times n}$, then $\text{rank}(A) = ?$

(2) Consider $\begin{bmatrix} 1 & 1 & 1 \\ 0 & 0 & 1 \\ 1 & 1 & 0 \end{bmatrix} \begin{bmatrix} x_1 \\ x_2 \\ x_3 \end{bmatrix} = \begin{bmatrix} 2 \\ 1 \\ \alpha \end{bmatrix}$, if $\alpha = 1$, how many solutions are there?

(3) Same as the above question, for what value of α does the system of equations have no solution?

---

[1] When $k = -5$, $[A \mid B] = \begin{bmatrix} 1 & -2 & 3 & 2 \\ 2 & k+1 & 6 & 8 \\ -1 & 3 & k-2 & -1 \end{bmatrix} \xrightarrow{r} \begin{bmatrix} 1 & -2 & 3 & 2 \\ 0 & 1 & -4 & 1 \\ 0 & 0 & 0 & 4 \end{bmatrix}$ the equations become $\begin{cases} x_1 - 2x_2 + 3x_3 = 2 \\ x_2 - 4x_3 = 1 \\ 0 = 4 \end{cases}$, where $0 = 4$ leads to a

contradiction. Therefore, there is no solution.

[2] $\text{rank}(A) = 2$, so $\dim(N) = \text{nullity}(A) = 5 - \text{rank}(A) = 5 - 2 = 3$.

Solution

(1) $AX = B$ has a unique solution $\Rightarrow \text{rank}(A) = n \Rightarrow \det(A) \neq 0$.

(2) $[A \mid B] = \begin{bmatrix} 1 & 1 & 1 & 2 \\ 0 & 0 & 1 & 1 \\ 1 & 1 & 0 & \alpha \end{bmatrix} \longrightarrow \begin{bmatrix} 1 & 1 & 1 & 2 \\ 0 & 0 & 1 & 1 \\ 0 & 0 & -1 & \alpha-2 \end{bmatrix} \longrightarrow \begin{bmatrix} 1 & 1 & 1 & 2 \\ 0 & 0 & 1 & 1 \\ 0 & 0 & 0 & \alpha-1 \end{bmatrix}$.

When $\alpha = 1 \Rightarrow \text{rank}(A) = \text{rank}([A \mid B]) = 2 < 3$, $\therefore$ the system has infinitely many solutions.

$\Rightarrow \begin{cases} x_1 + x_2 + x_3 = 2 \\ x_3 = 1 \end{cases}$, let $x_2 = c_1 \Rightarrow x_1 = 1 - c_1$, $\therefore \begin{bmatrix} x_1 \\ x_2 \\ x_3 \end{bmatrix} = \begin{bmatrix} 1-c_1 \\ c_1 \\ 1 \end{bmatrix} = c_1 \begin{bmatrix} -1 \\ 1 \\ 0 \end{bmatrix} + \begin{bmatrix} 1 \\ 0 \\ 1 \end{bmatrix}$.

(3) If $\alpha \neq 1 \Rightarrow \text{rank}(A) \neq \text{rank}([A \mid B]) \Rightarrow$ the system has no solution.

Example 10

Given that $\begin{bmatrix} 1 & -1 & 2 \\ 2 & 1 & -3 \\ 4 & -1 & 1 \end{bmatrix} \begin{bmatrix} x_1 \\ x_2 \\ x_3 \end{bmatrix} = \begin{bmatrix} 4 \\ -2 \\ 6 \end{bmatrix}$, that is $AX = B$.

(1) Find the rank(A) and nullity(A).

(2) Solve X.

Solution

$[A \mid B] = \begin{bmatrix} 1 & -1 & 2 & 4 \\ 2 & 1 & -3 & -2 \\ 4 & -1 & 1 & 6 \end{bmatrix} \longrightarrow \begin{bmatrix} 1 & -1 & 2 & 4 \\ 0 & 3 & -7 & -10 \\ 0 & 3 & -7 & -10 \end{bmatrix} \longrightarrow \begin{bmatrix} 1 & -1 & 2 & 4 \\ 0 & 3 & -7 & -10 \\ 0 & 0 & 0 & 0 \end{bmatrix}$

(1) rank$(A) = 2$, nullity$(A) = 3 - 2 = 1$.

(2) By(1), we know that

$\Rightarrow \begin{cases} x_1 - x_2 + 2x_3 = 4 \\ 3x_2 - 7x_3 = -10 \end{cases}$. Let $\begin{matrix} x_3 = 3c+1 \ , \ x_2 = 7c-1 \\ x_1 = 4+7c-1-6c-2 = c+1 \end{matrix}$.

$\begin{bmatrix} x_1 \\ x_2 \\ x_3 \end{bmatrix} = \begin{bmatrix} c+1 \\ 7c-1 \\ 3c+1 \end{bmatrix} = c \begin{bmatrix} 1 \\ 7 \\ 3 \end{bmatrix} + \begin{bmatrix} 1 \\ -1 \\ 1 \end{bmatrix}$. Q.E.D.

6-3-5 Cramer's Rule

The concept of using determinants to calculate all solutions of a system of linear equations was first introduced by Gabriel Cramer (1704-1752). Although it may not be the most efficient method in terms of computation, it proves to be relatively useful in many theoretical derivations. Taking a 2×2 system of linear equations as an example:

$$\begin{bmatrix} a_{11} & a_{12} \\ a_{21} & a_{22} \end{bmatrix} \begin{bmatrix} x_1 \\ x_2 \end{bmatrix} = \begin{bmatrix} b_1 \\ b_2 \end{bmatrix}, \text{ if } |A| \neq 0 \text{, then by the method of substitution, we obtain:}$$

$$x_1 = \frac{b_1 a_{11} - b_2 a_{21}}{a_{11} a_{22} - a_{21} a_{12}} = \frac{\begin{vmatrix} b_1 & a_{12} \\ b_2 & a_{22} \end{vmatrix}}{\begin{vmatrix} a_{11} & a_{12} \\ a_{21} & a_{22} \end{vmatrix}} = \frac{\Delta_1}{|A|}, \text{ where } \Delta_1 \text{ represents the determinant value of}$$

matrix A with the first column replaced by B, and similarly for Δ_2,

$$x_2 = \frac{\begin{vmatrix} a_{11} & b_1 \\ a_{21} & b_2 \end{vmatrix}}{\begin{vmatrix} a_{11} & a_{12} \\ a_{21} & a_{22} \end{vmatrix}} = \frac{\Delta_2}{|A|}.$$

▶ **Theorem 6-3-7**

Cramer's Rule

Set a system of equations $\begin{bmatrix} a_{11} & a_{12} & \cdots & a_{1n} \\ a_{21} & a_{22} & \cdots & a_{2n} \\ \vdots & \vdots & \ddots & \vdots \\ a_{n1} & a_{n2} & \cdots & a_{nn} \end{bmatrix} \begin{bmatrix} x_1 \\ x_2 \\ \vdots \\ x_n \end{bmatrix} = \begin{bmatrix} b_1 \\ b_2 \\ \vdots \\ b_n \end{bmatrix}$, where the determinant of

the coefficient matrix $|A_{n \times n}| \neq 0$,

then $x_i = \dfrac{\Delta_i}{|A|}$, where $\Delta_i = b_1 C_{1i} + b_2 C_{2i} + \cdots + b_n C_{ni}$, $i = 1, 2, 3, \cdots, n$.

Proof

Because A is invertible, we have $X = A^{-1}B$; using the formula for computing the adjugate matrix, we obtain:

$$X = \frac{adj(A) \times B}{|A|} = \frac{\Sigma}{|A|}, \text{ where } \Sigma = adj(A) \cdot B = \begin{bmatrix} \Delta_1 \\ \Delta_2 \\ \vdots \\ \Delta_n \end{bmatrix} = \begin{bmatrix} C_{11} & C_{21} & \cdots & C_{n1} \\ C_{12} & C_{22} & \cdots & C_{n2} \\ \vdots & \vdots & \ddots & \vdots \\ C_{1n} & C_{2n} & \cdots & C_{nn} \end{bmatrix} \begin{bmatrix} b_1 \\ b_2 \\ \vdots \\ b_n \end{bmatrix},$$

where $\Delta_1 = b_1 C_{11} + b_2 C_{21} + \cdots + b_n C_{n1}$, $\Delta_2 = b_1 C_{12} + b_2 C_{22} + \cdots + b_n C_{n2}$, $\cdots\cdots$, that is Δ_i represents the determinant value of the matrix obtained by replacing the i-th column of A with the column vector B. ◀

Example 11

Solve following questions by using Cramer's rule.

$$\begin{cases} 3x + 2y + 4z = 1 \\ 2x - y + z = 0 \\ x + 2y + 3z = 1 \end{cases}$$

Solution

$$\Delta = |A| = \begin{vmatrix} 3 & 2 & 4 \\ 2 & -1 & 1 \\ 1 & 2 & 3 \end{vmatrix} = -5 \neq 0 \text{, therefore has a unique solution.}$$

$$\Delta_x = \begin{vmatrix} 1 & 2 & 4 \\ 0 & -1 & 1 \\ 1 & 2 & 3 \end{vmatrix} = 1, \quad \Delta_y = \begin{vmatrix} 3 & 1 & 4 \\ 2 & 0 & 1 \\ 1 & 1 & 3 \end{vmatrix} = 0, \quad \Delta_z = \begin{vmatrix} 3 & 2 & 1 \\ 2 & -1 & 0 \\ 1 & 2 & 1 \end{vmatrix} = -2,$$

from Cramer's rule, we obtain $x = \dfrac{\Delta_x}{\Delta} = -\dfrac{1}{5}$, $y = \dfrac{\Delta_y}{\Delta} = 0$, $z = \dfrac{\Delta_z}{\Delta} = \dfrac{2}{5}$.

Q.E.D.

6-3 Exercises

Basic questions

1. Find the rank of the following matrix and determine the solutions of the homogeneous system of equations $AX = O$.

    (1) $\begin{bmatrix} 5 & -3 \\ 0 & 0 \end{bmatrix}$.

    (2) $\begin{bmatrix} 3 & -3 \\ 1 & -1 \end{bmatrix}$.

    (3) $\begin{bmatrix} 3 & -3 \\ 1 & -2 \end{bmatrix}$.

    (4) $\begin{bmatrix} 1 & -2 \\ 4 & -8 \\ 6 & -1 \\ 4 & 5 \end{bmatrix}$.

    (5) $\begin{bmatrix} 1 & 2 \\ 3 & 6 \\ -1 & 3 \\ 3 & -9 \\ 1 & 7 \end{bmatrix}$.

    (6) $\begin{bmatrix} 4 & 4 & -2 \\ -4 & -4 & 2 \\ -2 & -2 & 1 \end{bmatrix}$.

    (7) $\begin{bmatrix} -9 & 8 & -4 \\ 8 & -9 & -4 \\ -4 & -4 & -32 \end{bmatrix}$.

    (8) $\begin{bmatrix} 3 & 4 & -2 \\ 4 & 3 & -2 \\ -2 & -2 & -1 \end{bmatrix}$.

    (9) $\begin{bmatrix} 4 & -1 & 2 & 1 \\ 2 & -11 & 7 & 8 \\ 0 & 7 & -4 & -5 \\ 2 & 3 & -1 & -2 \end{bmatrix}$.

    (10) $\begin{bmatrix} 1 & 2 & 1 & -1 & 2 \\ 1 & 4 & 5 & -3 & 8 \\ -2 & -1 & 4 & -1 & 5 \\ 3 & 7 & 5 & -4 & 9 \end{bmatrix}$.

2. Solve the following system of equations using Cramer's rule.

    (1) $\begin{cases} x_1 + x_2 = 3 \\ 2x_1 - x_2 = 0 \end{cases}$.

    (2) $\begin{cases} 2x_1 + x_2 - x_3 = 5 \\ x_1 - 3x_2 + x_3 = 2 \\ x_1 + 3x_2 - 3x_3 = 0 \end{cases}$.

Advanced questions

1. Use Gaussian elimination, Gauss-Jordan elimination, and Cramer's rule, respectively, to solve the following system of equations.

    (1) $\begin{cases} x_1 + 2x_2 + 3x_3 = 4 \\ 2x_1 + 5x_2 + 3x_3 = 5 \\ x_1 + 8x_3 = 9 \end{cases}$.

    (2) $\begin{cases} 2x_1 - 4x_2 + 3x_3 = 3 \\ x_1 - x_2 + x_3 = 2 \\ 3x_1 + 2x_2 - x_3 = 4 \end{cases}$.

    (3) $\begin{cases} 2x_1 + 3x_2 - 4x_3 = 1 \\ 3x_1 - x_2 - 2x_3 = 4 \\ 4x_1 - 7x_2 - 6x_3 = -7 \end{cases}$.

2. For the homogeneous system of equations $AX = O$, where A matrix is as follows, find the rank and nullity of the A matrix, and then determine its general solution.

    (1) $\begin{bmatrix} 1 & 1 & 2 \\ 0 & 1 & 1 \\ 1 & 3 & 4 \end{bmatrix}$.

    (2) $\begin{bmatrix} 1 & 2 & 3 \\ 2 & 5 & 3 \\ 1 & 0 & 8 \end{bmatrix}$.

    (3) $\begin{bmatrix} 1 & 2 & -1 & 1 \\ 0 & 1 & -1 & 1 \end{bmatrix}$.

3. For the non-homogeneous system of equations $AX = B$, where A, B matrices are as follows, first check if rank(A) is equal to rank($A \mid B$), if they are equal, then find the general solution of this system of equations.

(1) $A = \begin{bmatrix} 1 & 1 & 1 \\ 1 & -1 & 1 \\ 3 & 1 & 3 \end{bmatrix}$, $B = \begin{bmatrix} 1 \\ 2 \\ 4 \end{bmatrix}$.

(2) $A = \begin{bmatrix} 1 & 0 & 1 & 0 \\ 2 & 2 & 0 & 3 \\ 0 & 4 & -4 & 5 \end{bmatrix}$, $B = \begin{bmatrix} 2 \\ 1 \\ -7 \end{bmatrix}$.

4. If $X_p = \begin{bmatrix} -7 & 8 & 9 & 11 \end{bmatrix}^T$ is one particular solution of the system of equations

$$\begin{cases} x_1 - x_2 + x_3 - x_4 = a \\ -2x_1 + 3x_2 - x_3 + 2x_4 = b \\ 4x_1 - 2x_2 + 2x_3 - 3x_4 = d \end{cases}$$, then what is the

general solution for this system?

5. If $X_p = \begin{bmatrix} 7 & 8 & 9 & 13 \end{bmatrix}^T$ is one particular solution of the system of equations

$$\begin{cases} x_1 + x_3 - x_4 = a \\ -x_1 + x_2 + x_3 + 2x_4 = b \\ x_1 + 2x_2 + 5x_3 + x_4 = d \end{cases}$$,

then what is the general solution for this system?

6. Consider a system of equations

$A_{3\times3} X_{3\times1} = B_{3\times1}$, where

$$A_{3\times3} = \begin{bmatrix} 1 & a & 3 \\ 1 & 2 & 2 \\ 1 & 3 & a \end{bmatrix}, \quad B_{3\times1} = \begin{bmatrix} 2 \\ 3 \\ a+3 \end{bmatrix}$$

find the value of a, causes the system of equations has

(1) a unique solution,

(2) infinitely many solutions,

(3) no solution.

7. Consider a system of equations $A_{3\times3} X_{3\times1} = B_{3\times1}$, where

$$A_{3\times3} = \begin{bmatrix} 0 & a & 1 \\ a & 0 & b \\ a & a & 2 \end{bmatrix}, \quad B_{3\times1} = \begin{bmatrix} b \\ 1 \\ 2 \end{bmatrix},$$

find the value of a, b, causes the system of equations has

(1) a unique solution,

(2) one-parameter solution,

(3) two-parameter solution,

(4) no solution.

6-4 Eigenvalues and Eigenvectors

In engineering applications, many linear systems preserve the original physical quantity's form while only changing its magnitude. One typical example of such a system is the microphone system, which amplifies the physical quantity (sound) of a speaker's voice. In engineering mathematics, we refer to this type of system as an eigenvalue system. Now, let's introduce this type of system.

6-4-1 Basic Definition and Theorem

▶ **Definition 6-4-1**

Eigensystem

Let A be an $n \times n$ square matrix. If X is a non-zero vector in R^n, and there exists a scalar λ such that $A_{n \times n} X_{n \times 1} = \lambda X_{n \times 1}$, then the equation $A_{n \times n} X_{n \times 1} = \lambda X_{n \times 1}$ is called the eigensystem, and X is called an eigenvector of A. The scalar λ is referred to as the eigenvalue corresponding to the eigenvector X. ◀

According to the definition 6-4-1 of eigenvectors, we can understand the following: If L is a line that contains vector X, then the direction of X remains the same before and after the transformation by matrix A. Moreover, $|\lambda|$ represents the scaling factor of vector X along the line L, which means it describes the magnitude of the scale change of the eigenvector when it undergoes the transformation by matrix A. This geometric interpretation is illustrated in Figure 6-7.

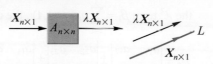

Figure 6-7   The schematic diagram of eigensystem

▶ **Theorem 6-4-1**

Calculation Formula of Eigenvalue

Let A be an $n \times n$ square matrix, λ be an eigenvalue of A if and only if $\det(A - \lambda I) = |A - \lambda I| = 0$.

Proof

Assuming $X \neq O$ is the corresponding eigenvector to λ, then $\{0\} \subsetneq N(A - \lambda I)$. Using the dimension theorem:
nullity$(A - \lambda I) > 1 \Leftrightarrow$ rank$(A - \lambda I) < n \Leftrightarrow \det(A - \lambda I) = |A - \lambda I| = 0$, the original statement is thus proven. ◀

1.　**The Definition of Characteristic Polynomial**

Let A be an $n \times n$ square matrix, then

$$f(\lambda) = \det(A - \lambda I) = |A - \lambda I| = \begin{vmatrix} a_{11} - \lambda & \cdots & a_{1n} \\ \vdots & \ddots & \vdots \\ a_{1n} & \cdots & a_{nn} - \lambda \end{vmatrix}$$ is called the characteristic

polynomial of A; $f(\lambda) = 0$, is referred to as the characteristic equation of A.

▶ **Theorem 6-4-2**

Characteristic Polynomial

Let A be an $n \times n$ square matrix, then β_k, the sum of determinants of all $k \times k$ principal submatrices of A determines the coefficient of λ^{n-k} in the characteristic polynomial. In fact, we have:

$$f(\lambda) = \det(A - \lambda I) = (-1)^n[\lambda^n - \beta_1\lambda^{n-1} + \beta_2\lambda^{n-2} + \cdots + (-1)^n\beta_n].$$ ◀

For example: when $n = 2$, $f(\lambda) = \begin{vmatrix} a_{11} - \lambda & a_{12} \\ a_{21} & a_{22} - \lambda \end{vmatrix} = (-1)^2[\lambda^2 - \beta_1\lambda + \beta_2]$, then

coefficient of the λ term $\beta_1 = a_{11} + a_{22} = tr(A)$, the coefficient of the constant term

$\beta_2 = \begin{vmatrix} a_{11} & a_{12} \\ a_{21} & a_{22} \end{vmatrix} = |A|$; when $n = 3$, the characteristic polynomial

$$|A - \lambda I| = \begin{vmatrix} a_{11} - \lambda & a_{12} & a_{13} \\ a_{21} & a_{22} - \lambda & a_{23} \\ a_{31} & a_{32} & a_{33} - \lambda \end{vmatrix} = (-1)^3[\lambda^3 - \beta_1\lambda^2 + \beta_2\lambda - \beta_3],$$ then coefficient of

the λ^2 term $\beta_1 = \sum_{i=1}^{3} a_{ii} = tr(A)$, the coefficient of the λ term

$\beta_2 = \begin{vmatrix} a_{11} & a_{12} \\ a_{21} & a_{22} \end{vmatrix} + \begin{vmatrix} a_{22} & a_{23} \\ a_{32} & a_{33} \end{vmatrix} + \begin{vmatrix} a_{11} & a_{13} \\ a_{31} & a_{33} \end{vmatrix} = A_{11} + A_{22} + A_{33}$, and the constant term

$$\beta_3 = |A| = \begin{vmatrix} a_{11} & a_{12} & a_{13} \\ a_{21} & a_{22} & a_{23} \\ a_{31} & a_{32} & a_{33} \end{vmatrix}.$$

Example 1

Find the characteristic polynomial and eigenvalues of following square matrix.

(1) $A = \begin{bmatrix} 3 & 1 \\ 1 & 3 \end{bmatrix}$. (2) $A = \begin{bmatrix} 0 & 1 & -2 \\ 2 & 1 & 0 \\ 4 & -2 & 5 \end{bmatrix}$.

Solution

(1) The characteristic polynomial of A is

$$|A-\lambda I|=(-1)^2[\lambda^2-tr(A)\lambda+|A|]=\lambda^2-(3+3)\lambda+(9-1)=\lambda^2-6\lambda+8\,.$$

Through $|A-\lambda I|=\lambda^2-6\lambda+8=0 \Rightarrow \lambda=2,4$,

so the characteristic polynomial of A is $f(\lambda)=|A-\lambda I|=\lambda^2-6\lambda+8$ and eigenvalues are 2, 4.

(2) By using β_k, the sum of determinants of all principal submatrices of A, we can determine the characteristic polynomial of A is:

$$|A-\lambda I|=(-1)^3[\lambda^3-tr(A)\lambda^2+(A_{11}+A_{22}+A_{33})\lambda-|A|]\,,\text{ where tr}(A)=6,$$

$$A_{11}+A_{22}+A_{33}=\begin{vmatrix}1&0\\-2&5\end{vmatrix}+\begin{vmatrix}0&-2\\4&5\end{vmatrix}+\begin{vmatrix}0&1\\2&1\end{vmatrix}=5+8-2=11\,|A|=\begin{vmatrix}0&1&-2\\2&1&0\\4&-2&5\end{vmatrix}=6\,.$$

So the characteristic polynomial of A is $f(\lambda)=|A-\lambda I|=-(\lambda^3-6\lambda^2+11\lambda-6)$,

by $|A-\lambda I|=-(\lambda^3-6\lambda^2+11\lambda-6)=0 \Rightarrow \lambda=1,2,3$, the eigenvalues are 1, 2, 3.

Q.E.D.

2. The Relationship Between Eigenvalues and Coefficients of the Characteristic Polynomial

By using Theorem 6-4-2, we can establish the relationship between eigenvalues and coefficients of the characteristic polynomial. For example, as demonstrated in

Example 1: $A=\begin{bmatrix}0&1&-2\\2&1&0\\4&-2&5\end{bmatrix}$, and the characteristic polynomial of A is

$f(\lambda)=|A-\lambda I|=-(\lambda^3-6\lambda^2+11\lambda-6)$. Solving it yields the eigenvalues $\lambda_1=1$,

$\lambda_2=2$, $\lambda_3=3$, then $\lambda_1+\lambda_2+\lambda_3=1+2+3=\beta_1=\text{tr}(A)=0+1+5$,

$\lambda_1\lambda_2+\lambda_2\lambda_3+\lambda_3\lambda_1=1\times2+2\times3+3\times1=\beta_2=A_{11}+A_{22}+A_{33}=5+8-2$,

$\lambda_1\lambda_2\lambda_3=1\times2\times3=\beta_3=|A|=6$. In fact, we have the following general result:

▶ **Theorem 6-4-3**

The Root and Coefficient of Eigenvalues

Let $\lambda_1,\cdots,\lambda_n$ be the n eigenvalues of A, if the characteristic polynomial is given by

$f(x)=(-1)^n[x^n-\beta_1x^{n-1}+\cdots+(-1)^n\beta_n]$, then $\beta_k=\displaystyle\sum_{i_1<\cdots<i_k}\lambda_{i_1}\cdots\lambda_{i_k}\,.$

Proof

$$f(x)=|A-\lambda I|=(-1)^n[\lambda^n-\beta_1\lambda^{n-1}+\cdots+(-1)^n\beta_n]=(-1)^n\left[(\lambda-\lambda_1)(\lambda-\lambda_2)\cdots(\lambda-\lambda_n)\right]$$

$$=(-1)^n[\lambda^n-(\lambda_1+\lambda_2+\cdots+\lambda_n)\lambda^{n-1}+\cdots+(-1)^n\lambda_1\lambda_2\cdots\lambda_n]$$

Comparing the coefficients, we obtain:

$\beta_1=\lambda_1+\lambda_2+\cdots+\lambda_n$, $\beta_2=\lambda_1\lambda_2+\lambda_1\lambda_3+\cdots+\lambda_{n-1}\lambda_n$, $\cdots$, $\beta_n=\lambda_1\lambda_2\cdots\lambda_n=|A|$.

◀

3. Solution of Eigenvector

Let $\lambda_1, \lambda_2, \cdots, \lambda_n$ be the eigenvalues of A, we substitute $\lambda = \lambda_i$ into $(A - \lambda I)X = O$.

The non-zero solutions X_i obtained in this process are the eigenvectors corresponding to the eigenvalue $\lambda = \lambda_i$.

Example 2

Find the eigenvalues and eigenvectors of square matrix $A = \begin{bmatrix} -5 & 2 \\ 2 & -2 \end{bmatrix}$.

Solution

(1) $|A - \lambda I| = \lambda^2 - (-7)\lambda + 6 = 0$ we obtain the eigenvalues $\lambda = -1, -6$.

(2) The eigenvectors corresponding to the eigenvalues λ is the element in $N(A - \lambda I)$, calculated as follows:

$\boldsymbol{\lambda = -1}$

substituting into $(A - \lambda I)X = O$ then obtained $\begin{bmatrix} -4 & 2 \\ 2 & -1 \end{bmatrix} \begin{bmatrix} x_1 \\ x_2 \end{bmatrix} = O$. Since the rank

of the coefficient matrix is 1, we have only one linearly independent equation, $2x_1 - x_2 = 0$, let $x_1 = c_1$, then $x_2 = 2c_1$,

the eigenvector $X_1 = \begin{bmatrix} x_1 \\ x_2 \end{bmatrix} = \begin{bmatrix} c_1 \\ 2c_1 \end{bmatrix} = c_1 \begin{bmatrix} 1 \\ 2 \end{bmatrix}$, $c_1 \neq 0 \Rightarrow$ also can be expressed as

$X_1 = \text{span}\left\{ \begin{bmatrix} 1 \\ 2 \end{bmatrix} \right\}$.

$\boldsymbol{\lambda = -6}$

substituting into $(A - \lambda I)X = O$ then obtained $\begin{bmatrix} 1 & 2 \\ 2 & 4 \end{bmatrix} \begin{bmatrix} x_1 \\ x_2 \end{bmatrix} = O$, since the rank

of the coefficient matrix is 1.

So only has one linearly independent equation $x_1 + 2x_2 = 0$, let $x_2 = c_2$, then $x_1 = -2c_2$,

the eigenvector $X_2 = \begin{bmatrix} x_1 \\ x_2 \end{bmatrix} = \begin{bmatrix} -2c_2 \\ c_2 \end{bmatrix} = c_2 \begin{bmatrix} -2 \\ 1 \end{bmatrix}$, $c_2 \neq 0 \Rightarrow$ also can be expressed as

$X_2 = \text{span}\left\{ \begin{bmatrix} -2 \\ 1 \end{bmatrix} \right\}$. [3]

Q.E.D.

---

[3]   For the equation $ax_1 + bx_2 = 0$, represents that the two vectors $\begin{bmatrix} x_1 \\ x_2 \end{bmatrix}$ and $\begin{bmatrix} a \\ b \end{bmatrix}$ are orthogonal, therefore $\begin{bmatrix} x_1 \\ x_2 \end{bmatrix} = c \begin{bmatrix} b \\ -a \end{bmatrix}$ or $c \begin{bmatrix} -b \\ a \end{bmatrix}$.

For instance: $x_1 + 2x_2 = 0 \rightarrow \begin{bmatrix} x_1 \\ x_2 \end{bmatrix} = c \cdot \begin{bmatrix} -2 \\ 1 \end{bmatrix}$ or $c \cdot \begin{bmatrix} 2 \\ -1 \end{bmatrix}$.

Example 3

$$A = \begin{bmatrix} 1 & 0 & 0 \\ 3 & 7 & 0 \\ -2 & 4 & -5 \end{bmatrix}$$ (1) Find the eigenvalues of A. (2) Find all linearly independent

eigenvectors.

Solution

(1) By $\det(A - \lambda I) = (1 - \lambda)(7 - \lambda)(-5 - \lambda) = 0$, $\lambda = 1, 7, -5$. It can be deduced that if matrix A is an upper (lower) triangular matrix or a diagonal matrix, the diagonal elements of A are the eigenvalues. The eigenvalues are $1, 7, -5$.

(2) ① When $\lambda_1 = 1$, substituting into $(A - \lambda_1 I)X_1 = O$ then we get

$$\begin{bmatrix} 0 & 0 & 0 \\ 3 & 6 & 0 \\ -2 & 4 & -6 \end{bmatrix} \begin{bmatrix} x_1 \\ x_2 \\ x_3 \end{bmatrix} = O,$$

since the rank of the coefficient matrix is 2, we have two linearly independent

equations $\begin{cases} 3x_1 + 6x_2 = 0 \\ -2x_1 + 4x_2 - 6x_3 = 0 \end{cases}$, representing a line in the three-dimensional

space $\mathbf{R}^3$. By calculating the direction vector

$$x_1 : x_2 : x_3 = \begin{vmatrix} 6 & 0 \\ 4 & -6 \end{vmatrix} : - \begin{vmatrix} 3 & 0 \\ -2 & -6 \end{vmatrix} : \begin{vmatrix} 3 & 6 \\ -2 & 4 \end{vmatrix} = -36 : 18 : 24 = -6 : 3 : 4,$$

so the eigenvector corresponding to $\lambda_1 = 1$ is

$$X_1 = c_1 \begin{bmatrix} -6 \\ 3 \\ 4 \end{bmatrix}, \quad c_1 \neq 0 \quad \text{or} \quad X_1 = \text{span}\left\{ \begin{bmatrix} -6 \\ 3 \\ 4 \end{bmatrix} \right\}.$$

② $\lambda_2 = 7$, substituting into $(A - \lambda_2 I)X_2 = O$ then we get

$$\Rightarrow \begin{bmatrix} -6 & 0 & 0 \\ 3 & 0 & 0 \\ -2 & 4 & -12 \end{bmatrix} \begin{bmatrix} x_1 \\ x_2 \\ x_3 \end{bmatrix} = O.$$

Since the rank of the coefficient matrix is 2, we have two linearly independent

equations $\begin{cases} 3x_1 + 0x_2 + 0x_3 = 0 \\ -2x_1 + 4x_2 - 12x_3 = 0 \end{cases}$,

$$x_1 : x_2 : x_3 = \begin{vmatrix} 0 & 0 \\ 4 & -12 \end{vmatrix} : - \begin{vmatrix} 3 & 0 \\ -2 & -12 \end{vmatrix} : \begin{vmatrix} 3 & 0 \\ -2 & 4 \end{vmatrix} = 0 : 36 : 12 = 0 : 3 : 1,$$

so the eigenvector corresponding to $\lambda_2 = 7$

is $X_2 = c_2 \begin{bmatrix} 0 \\ 3 \\ 1 \end{bmatrix}$, $c_2 \neq 0$ or $X_2 = \text{span}\left\{ \begin{bmatrix} 0 \\ 3 \\ 1 \end{bmatrix} \right\}$.

③ $\lambda_3 = -5$, substituting into $(A - \lambda_3 I)X_3 = O$ then we get

$$\Rightarrow \begin{bmatrix} 6 & 0 & 0 \\ 3 & 12 & 0 \\ -2 & 4 & 0 \end{bmatrix} \begin{bmatrix} x_1 \\ x_2 \\ x_3 \end{bmatrix} = O .$$

Since the rank of the coefficient matrix is 2, we have two linearly independent

equations $\begin{cases} 6x_1 + 0x_2 + 0x_3 = 0 \\ 3x_1 + 12x_2 + 0x_3 = 0 \end{cases}$,

$$x_1 : x_2 : x_3 = \begin{vmatrix} 0 & 0 \\ 12 & 0 \end{vmatrix} : - \begin{vmatrix} 6 & 0 \\ 3 & 0 \end{vmatrix} : \begin{vmatrix} 6 & 0 \\ 3 & 12 \end{vmatrix} = 0 : 0 : 72 = 0 : 0 : 1^4 ,$$

so the eigenvector corresponding to $\lambda_3 = -5$ is $X_3 = c_3 \begin{bmatrix} 0 \\ 0 \\ 1 \end{bmatrix}$, $c_3 \neq 0$ or

$$X_3 = \text{span} \left\{ \begin{bmatrix} 0 \\ 0 \\ 1 \end{bmatrix} \right\} .$$

Q.E.D.

6-4-2 The Important Properties of Eigenvalues and Eigenvectors

Eigenvalues possess many representative properties that play a crucial role in practical calculations and inferences. For example: due to $\det(A) = \det(A^T)$, A and A^T have the same characteristic polynomial, and consequently, they share the same eigenvalues. If $\det(A) = \det(A - 0I) = 0$, it implies that 0 is an eigenvalue of A; If A is an upper (lower) triangular matrix or a diagonal matrix, the n eigenvalues of A are the elements $a_{11}, a_{22}, \cdots, a_{nn}$ along its main diagonal, as defined by the characteristic polynomial. According to the definition of eigenvalues: If λ is an eigenvalue of A, then $\alpha\lambda^m$ is an eigenvalue of αA^m; while the n eigenvalues of A^{-1} are $\lambda_1^{-1}, \lambda_2^{-1}, \cdots, \lambda_n^{-1}$.

---

[4] For the system of simultaneous equations $\begin{cases} a_1 x_1 + b_1 x_2 + c_1 x_3 = 0 \\ a_2 x_1 + b_2 x_2 + c_2 x_3 = 0 \end{cases}$, it can be regarded that $\vec{X} = \begin{bmatrix} x_1 \\ x_2 \\ x_3 \end{bmatrix}$ is orthogonal to $\vec{u} = \begin{bmatrix} a_1 \\ b_1 \\ c_1 \end{bmatrix}$ and

$\vec{v} = \begin{bmatrix} a_2 \\ b_2 \\ c_2 \end{bmatrix}$, in other words $\begin{cases} \vec{X} \cdot \vec{u} = 0 \\ \vec{X} \cdot \vec{v} = 0 \end{cases}$, therefore $\vec{X} = \begin{bmatrix} x_1 \\ x_2 \\ x_3 \end{bmatrix}$ can be obtained by taking the cross product of $\vec{u} = \begin{bmatrix} a_1 \\ b_1 \\ c_1 \end{bmatrix}$ and $\vec{v} = \begin{bmatrix} a_2 \\ b_2 \\ c_2 \end{bmatrix}$ and

dividing each component, that is

$$x_1 : x_2 : x_3 = \begin{vmatrix} b_1 & c_1 \\ b_2 & c_2 \end{vmatrix} : - \begin{vmatrix} a_1 & c_1 \\ a_2 & c_2 \end{vmatrix} : \begin{vmatrix} a_1 & b_1 \\ a_2 & b_2 \end{vmatrix} .$$

▶ Theorem 6-4-4

The Linear Independent Eigenvectors

Eigenvectors corresponding to distinct eigenvalues must be linearly independent.

Proof

Assuming distinct eigenvalues $\lambda_1, \cdots, \lambda_n$ and their corresponding eigenvectors $v_1, \cdots, v_n$. Considering the linear combination $\sum_{i=1}^{n} a_i v_i = \mathbf{O}$, to prove that v_1, v_2, $\cdots$, v_n are linearly independent, we need to demonstrate that $a_1 = a_2 = \cdots = a_n = 0$, using the definition of linear independence, we have

$$(A - \lambda_1 I) \cdots \widehat{(A - \lambda_i I)} \cdots (A - \lambda_n I) \sum_{i=1}^{n} a_i v_i = a_i \prod_{j \neq i} (\lambda_i - \lambda_j) = 0 \,,$$

therefore $a_i = 0$, $i = 1, 2, \cdots, n$, thus v_1, v_2, $\cdots$, v_n are linearly independent.   ◀

The Technique of Observing Eigenvalues

The eigenvalues of matrix A satisfy $|A - \lambda I| = 0$, which means that when we subtract a scalar from each main diagonal element of matrix A, the determinant of the resulting matrix becomes zero. This scalar is precisely the eigenvalue of matrix A. Therefore, when solving for eigenvalues, we can observe the matrix A after subtracting one from its main diagonal elements to check if any row or column becomes all zeros or if any two rows or columns become proportional. Together with the fact that the sum of all eigenvalues is equal to the trace of matrix A ($\mathrm{tr}(A)$), we can identify some eigenvalues without solving high-degree equations. For example: Assuming $A = \begin{bmatrix} 2 & 1 & 0 \\ 2 & 1 & 0 \\ 0 & 0 & 5 \end{bmatrix}$, after subtracting 5 from the main diagonal, the third column becomes a zero column, so one eigenvalue is 5. After subtracting 0 (i.e., not subtracting at all), the first and second columns become proportional, so another eigenvalue is 0. Finally, using the fact that the sum of eigenvalues is equal to $\mathrm{tr}(A) = 2 + 1 + 5 = 8$, we find the third eigenvalue as $8 - 5 - 0 = 3$. This way, we can easily determine all three eigenvalues without solving a cubic equation. Another interesting property worth noting is that if all columns (or rows) of matrix A have the same sum, this common sum is an eigenvalue of A. For instance, $A = \begin{bmatrix} 9 & 1 & 1 \\ 1 & 9 & 1 \\ 1 & 1 & 9 \end{bmatrix}$, then all column sums are 11, so A must have an eigenvalue of 11.

Example 4

Assuming $A \in F^{3 \times 3}$ and its eigenvalues are 1, 2, 3, then

(1) Find the eigenvalues of $2A^{-1} + I$. (2) If $A = \begin{bmatrix} 2 & -1 & 1 \\ 1 & 2 & -1 \\ 1 & -1 & a \end{bmatrix}$, find a.

(3) rank(A^5) =?

Solution

(1) $|A| = 1 \cdot 2 \cdot 3 = 6 \neq 0$, $\therefore A^{-1}$exists, so the eigenvalues of $B = 2A^{-1} + I$

are $2 \cdot \dfrac{1}{1} + 1 = 3$, $2 \cdot \dfrac{1}{2} + 1 = 2$, $2 \cdot \dfrac{1}{3} + 1 = \dfrac{5}{3}$.

(2) $\text{tr}(A) = 4 + a = \lambda_1 + \lambda_2 + \lambda_3 = 6 \Rightarrow a = 2$.

(3) $\det(A^5) = |A|^5 = 6^5 \neq 0$, $\therefore$rank$(A^5) = 3$.

Example 5

Find the eigenvalues and eigenvectors of $A = \begin{bmatrix} 9 & 1 & 1 \\ 1 & 9 & 1 \\ 1 & 1 & 9 \end{bmatrix}$.

Solution

(1) Find the eigenvalues, through $|A - \lambda I| = 0 \Rightarrow \lambda = 8, 8, 11^5$.

(2) Find the eigenvectors

①　$\lambda = 8$ substituting into $(A - \lambda I)X = O$ then we get

$$\begin{bmatrix} 1 & 1 & 1 \\ 1 & 1 & 1 \\ 1 & 1 & 1 \end{bmatrix} \begin{bmatrix} x_1 \\ x_2 \\ x_3 \end{bmatrix} = O \Rightarrow x_1 + x_2 + x_3 = 0,$$

---

[5]

(1) Since the sum of all column (or row) elements is 11, there is an eigenvalue of 11.

(2) As the main diagonal elements of matrix A become proportional after subtracting 8, there is an eigenvalue of 8.

(3) Furthermore, the sum of all eigenvalues is equal to tr$(A) = 27$, so the other eigenvalue is $27 - 11 - 8 = 8$.

let $x_2 = c_1$, $x_3 = c_2$, then $x_1 = -c_1 - c_2$, we obtain $X = c_1 \begin{bmatrix} -1 \\ 1 \\ 0 \end{bmatrix} + c_2 \begin{bmatrix} -1 \\ 0 \\ 1 \end{bmatrix}$,

the eigenvectors are $X_1 = c_1 \begin{bmatrix} -1 \\ 1 \\ 0 \end{bmatrix}$, $X_2 = c_2 \begin{bmatrix} -1 \\ 0 \\ 1 \end{bmatrix}$.

② $\lambda = 11$ substituting into $(A - \lambda I)X = O$ then we obtain

$$\begin{bmatrix} -2 & 1 & 1 \\ 1 & -2 & 1 \\ 1 & 1 & -2 \end{bmatrix} \begin{bmatrix} x_1 \\ x_2 \\ x_3 \end{bmatrix} = O \Rightarrow \begin{cases} -2x_1 + x_2 + x_3 = 0 \\ x_1 - 2x_2 + x_3 = 0 \end{cases},$$

so the eigenvector is $X_3 = c_3 \begin{bmatrix} 1 \\ 1 \\ 1 \end{bmatrix}$, $c_3 \neq 0$.   Q.E.D.

Algebraic Multiplicity and Geometric Multiplicity

In Example 5 above, the eigenvalues of matrix A include repeated roots $\lambda = 8, 8$, which we call algebraic multiplicities. The number of linearly independent eigenvectors corresponding to these repeated roots is called geometric multiplicities.

So, in this example, the algebraic multiplicity of $\lambda = 8$ is 2, and its geometric multiplicity is also 2. The algebraic multiplicity of $\lambda = 11$ is 1, and its geometric multiplicity is also 1. For a square matrix, the geometric multiplicity of distinct eigenvalues is less than or equal to algebraic multiplicities, meaning the number of linearly independent eigenvectors corresponding to distinct eigenvalues is less than or equal to the number of repeated roots of eigenvalues.

6-4 Exercises

Basic questions

Find the eigenvalues and eigenvectors of following square matrix.

1. (1) $\begin{bmatrix} 5 & 4 \\ 1 & 2 \end{bmatrix}$.

(2) $\begin{bmatrix} 2 & 4 \\ 6 & 4 \end{bmatrix}$.

(3) $\begin{bmatrix} -3 & 2 \\ 6 & 1 \end{bmatrix}$.

(4) $\begin{bmatrix} 0 & 0 \\ 0 & 0 \end{bmatrix}$.

2. (1) $\begin{bmatrix} 4 & 0 & 0 \\ 0 & 8 & 0 \\ 0 & 0 & 6 \end{bmatrix}$.

(2) $\begin{bmatrix} 1 & -1 & 0 \\ -1 & 2 & -1 \\ 0 & -1 & 1 \end{bmatrix}$.

(3) $\begin{bmatrix} 3 & 0 & 0 \\ 1 & -2 & -8 \\ 0 & -5 & 1 \end{bmatrix}$.

2. (1) $\begin{bmatrix} 2 & 1 & 1 \\ 1 & 2 & 1 \\ 1 & 1 & 2 \end{bmatrix}$.

(2) $\begin{bmatrix} 0 & 1 & 1 \\ 1 & 0 & 1 \\ 1 & 1 & 0 \end{bmatrix}$.

Advanced questions

Find the eigenvalues and eigenvectors of following square matrix.

1. (1) $\begin{bmatrix} 8 & 0 & 3 \\ 2 & 2 & 1 \\ 2 & 0 & 3 \end{bmatrix}$.

(2) $\begin{bmatrix} -2 & 2 & -3 \\ 2 & 1 & -6 \\ -1 & -2 & 0 \end{bmatrix}$.

(3) $\begin{bmatrix} 13 & 0 & -15 \\ -3 & 4 & 9 \\ 5 & 0 & -7 \end{bmatrix}$.

6-5 Matrix Diagonalization

Matrix diagonalization holds significant value in matrix operations and linear algebra because diagonal matrices are easier to handle. In this section, we will introduce how to diagonalize a matrix using the eigenvalues and eigenvectors obtained from an eigensystem. This process facilitates subsequent computations of high-degree matrix functions.

6-5-1 Similar Matrix

▶ **Definition 6-5-1**

The Definition of Similar Matrix

Let A, B be both $n \times n$ square matrix, if there exists a non-singular matrix Q causes $Q^{-1}AQ = B$, then this transformation is called a similarity transformation, and in this case, we say that A is similar to B, denoted as $A \sim B$. ◀

▶ **Theorem 6-5-1**

The Important Properties of Similar Transformation

If $A \sim B$, then the following facts hold.

(1) $\det(A) = \det(B)$.

(2) $\text{rank}(A) = \text{rank}(B)$.

(3) A and B have the same eigenvalues.

(4) $\text{trace}(A) = \text{trace}(B)$.

Proof

$| B - \lambda I | = | Q^{-1}AQ - \lambda Q^{-1}Q | = | Q^{-1}(A - \lambda I)Q | = | Q^{-1} \| A - \lambda I \| Q | = | A - \lambda I |.$

So similar matrix have the same eigenvalues. ◀

6-5-2 Matrix Diagonalization

▶ **Definition 6-5-2**

The Definition of Diagonalization

If A is an $n \times n$ square matrix, and there exists an invertible matrix P satisfies $P^{-1}AP$ to be a diagonal matrix D, then A is said to be diagonalizable. ◀

▶ **Definition 6-5-3**

Transition Matrix

If A is an $n \times n$ square matrix, and there exists an invertible matrix P satisfies $P^{-1}AP$ to be a diagonal matrix D, then P is called the transition matrix[6] for A. ◀

▶ **Theorem 6-5-2**

Diagonalizable

For an $n \times n$ square matrix A. Then A has n linearly independent eigenvectors if and only if A is similar to a diagonal matrix D, which means A is diagonalizable.

Proof

【⇒】 Let $\lambda_1, \lambda_2, \cdots, \lambda_n$ be the n eigenvalues of matrix A, and let $V_1, V_2, \cdots, V_n$ be the n linearly independent eigenvectors corresponding to these eigenvalues, satisfies $AV_1 = \lambda_1 V_1, AV_2 = \lambda_2 V_2, \cdots, AV_n = \lambda_n V_n$, let $P \equiv [V_1, V_2, \cdots, V_n]$ then

$$AP = A[V_1, V_2, \cdots, V_n] = [AV_1 \quad AV_2 \quad \cdots \quad AV_n] = [\lambda_1 V_1 \quad \lambda_2 V_2 \quad \cdots \quad \lambda_n V_n]$$

$$= [V_1 \quad V_2 \quad \cdots \quad V_n] \begin{bmatrix} \lambda_1 & & & O \\ & \lambda_2 & & \\ & & \ddots & \\ O & & & \lambda_n \end{bmatrix} = PD \Rightarrow AP = PD \Rightarrow P^{-1}AP = D$$

【⇐】 Because $A \sim D$, there exists an invertible matrix P satisfies

$$P^{-1}AP = D \Rightarrow AP = PD.$$

Let $P = [\xi_1 \quad \xi_2 \quad \cdots \quad \xi_n]$, $D = \begin{bmatrix} d_1 & & O \\ & \ddots & \\ O & & d_n \end{bmatrix}$ substituting into $AP = PD$

$$\Rightarrow A[\xi_1 \quad \xi_2 \quad \cdots \quad \xi_n] = [\xi_1 \quad \xi_2 \quad \cdots \quad \xi_n] \begin{bmatrix} d_1 & & O \\ & \ddots & \\ O & & d_n \end{bmatrix} = [d_1 \xi_1 \quad \cdots \quad d_n \xi_n]$$

$\Rightarrow A\xi_k = d_k \xi_k$, $k: 1 \sim n$,

$d_1, d_2, \cdots, d_n$ are the n eigenvalues of A, $\xi_1, \xi_2, \cdots, \xi_n$ are the corresponding eigenvectors, and P is invertible $\Rightarrow \xi_1, \xi_2, \cdots, \xi_n$ are linearly independent. ◀

---

[6] During diagonalization, the arrangement order of eigenvectors in the transition matrix P must be consistent with the diagonal matrix D.

▶ **Theorem 6-5-3**

The Conditions of Diagonalization

If an $n \times n$ square matrix A has n distinct eigenvalues, then A is guaranteed to be diagonalizable.

Proof

Let $\lambda_1, \lambda_2, \cdots, \lambda_n$ be the n distinct eigenvalues, $V_1, V_2, \cdots, V_n$ be their corresponding eigenvectors. By Theorem 6-4-4 , we know that the eigenvectors $\{V_1, V_2, \cdots, V_n\}$ are linearly independent. Moreover, by Theorem 6-5-3, A is similar to a diagonal matrix.

◀

Example 1

$$A = \begin{bmatrix} 5 & 10 \\ 4 & -1 \end{bmatrix}$$

(1) Find a matrix P causes $P^{-1}AP = D$ be a diagonal matrix.

(2) Find this diagonal matrix D.

Solution

(1) From $|A - \lambda I| = 0 \Rightarrow (-1)^2 [\lambda^2 - 4\lambda - 45] = 0 \Rightarrow (\lambda - 9)(\lambda + 5) = 0$,

$\lambda = 9, -5$,   (the sum of the elements in a row of a matrix is 9, then there is an eigenvalue of 9.)

$\lambda = 9 \Rightarrow (A - \lambda I)X = O \Rightarrow \begin{bmatrix} -4 & 10 \\ 4 & -10 \end{bmatrix} \begin{bmatrix} x_1 \\ x_2 \end{bmatrix} = O \Rightarrow X_1 = c_1 \begin{bmatrix} 5 \\ 2 \end{bmatrix}, c_1 \neq 0,$

$\lambda = -5 \Rightarrow (A - \lambda I)X = O \Rightarrow \begin{bmatrix} 10 & 10 \\ 4 & 4 \end{bmatrix} \begin{bmatrix} x_1 \\ x_2 \end{bmatrix} = O \Rightarrow X_2 = c_2 \begin{bmatrix} 1 \\ -1 \end{bmatrix}, c_2 \neq 0,$

$\therefore P = \begin{bmatrix} 5 & 1 \\ 2 & -1 \end{bmatrix}.$

(2) $P^{-1}AP = \begin{bmatrix} 9 & 0 \\ 0 & -5 \end{bmatrix} = D^{\,7}.$

Q.E.D.

---

[7] If $P = \begin{bmatrix} 1 & 5 \\ -1 & 2 \end{bmatrix}$, then $P^{-1}AP = \begin{bmatrix} -5 & 0 \\ 0 & 9 \end{bmatrix} = D$.

Example 2

$$A = \begin{bmatrix} 0 & 1 & 0 \\ 1 & 0 & 0 \\ 0 & 0 & 1 \end{bmatrix}$$

(1) Find the eigenvalues of matrix A.   (2) Find the eigenvactors of matrix A.

(3) Find the matrix P, causes $P^{-1}AP$ become diagonal matrix.   (4) Find the inverse matrix P^{-1} of matrix P.

Solution

(1) By $|A - \lambda I| = 0 \Rightarrow \lambda = -1, 1, 1$ (the sum of the elements in a row of matrix A is 1, then there is an eigenvalue of 1.)

(2) $\lambda = -1 \Rightarrow X_1 = c_1 \begin{bmatrix} 1 \\ -1 \\ 0 \end{bmatrix}, c_1 \neq 0,$

$$\begin{array}{cccccc} & 1 & 1 & 0 & 1 & 1 \\ -1 & \left(1 \right. & 0 & -1 & -1 \\ 0 & 0 & 1 & 0 & 0 \\ 1 & 1 & 0 & 1 & 1 \\ -1 & \left. 1 \right. & 0 & -1 & 1 \end{array}^{T}$$

$\lambda = 1 \Rightarrow X_2 = c_2 \begin{bmatrix} 1 \\ 1 \\ 0 \end{bmatrix}, c_2 \neq 0; \quad X_3 = c_3 \begin{bmatrix} 0 \\ 0 \\ 1 \end{bmatrix}, c_3 \neq 0,$

the eigenvactors can take $\Rightarrow X_1 = \begin{bmatrix} 1 \\ -1 \\ 0 \end{bmatrix}, \quad X_2 = \begin{bmatrix} 1 \\ 1 \\ 0 \end{bmatrix}, \quad X_3 = \begin{bmatrix} 0 \\ 0 \\ 1 \end{bmatrix}.$

(3) $P = \begin{bmatrix} X_1 & X_2 & X_3 \end{bmatrix} = \begin{bmatrix} 1 & 1 & 0 \\ -1 & 1 & 0 \\ 0 & 0 & 1 \end{bmatrix} \Rightarrow P^{-1}AP = D = \begin{bmatrix} -1 & 0 & 0 \\ 0 & 1 & 0 \\ 0 & 0 & 1 \end{bmatrix}.$

(4) $|P| = 2, \therefore P^{-1} = \dfrac{1}{2}\begin{bmatrix} 1 & -1 & 0 \\ 1 & 1 & 0 \\ 0 & 0 & 2 \end{bmatrix}.$ `Q.E.D.`

Both of the examples above are cases where diagonalization is possible. Upon closer observation, it can be noted that the algebraic multiplicities of the distinct eigenvalues are equal to their geometric multiplicities. But if the matrix A rewritten as $A = \begin{bmatrix} 1 & 0 \\ 1 & 1 \end{bmatrix}$, its eigenvalues $\lambda = 1, 1$, then eigenvectors X will satisfy

$$(A - I)X = 0 \Rightarrow \begin{bmatrix} 0 & 0 \\ 1 & 0 \end{bmatrix}\begin{bmatrix} x_1 \\ x_2 \end{bmatrix} = 0 \Rightarrow X = c\begin{bmatrix} 0 \\ 1 \end{bmatrix}.$$

That is, the geometric multiplicity of $\lambda = 1$ is 1, but the algebraic multiplicity is 2. Since the number of eigenvectors is insufficient, the matrix $A = \begin{bmatrix} 1 & 0 \\ 1 & 1 \end{bmatrix}$ can not be diagonalized in this case.

6-5 Exercises

Basic questions

1. For the following(1)~(3) matrix A, write down their transition matrix P and diagonal matrix D which causes $P^{-1}AP = D$ be a diagonal matrix.

(1) $A = \begin{bmatrix} -5 & 2 \\ 2 & -2 \end{bmatrix}$.

(2) $A = \begin{bmatrix} 1 & 0 & 0 \\ 3 & 7 & 0 \\ -2 & 4 & -5 \end{bmatrix}$.

(3) $A = \begin{bmatrix} 9 & 1 & 1 \\ 1 & 9 & 1 \\ 1 & 1 & 9 \end{bmatrix}$.

3. (1) $\begin{bmatrix} 1 & 2 & 2 \\ 1 & 2 & -1 \\ -1 & 1 & 4 \end{bmatrix}$.

(2) $\begin{bmatrix} 5 & 2 & 2 \\ 3 & 6 & 3 \\ 6 & 6 & 9 \end{bmatrix}$.

(3) $\begin{bmatrix} 5 & 1 & 1 \\ 1 & 5 & 1 \\ 1 & 1 & 5 \end{bmatrix}$.

Advanced questions

For the following square matrix, find a matrix P causes $P^{-1}AP = D$ be a diagonal matrix, and find the diagonal matrix D.

1. (1) $\begin{bmatrix} 3 & 4 \\ 2 & -4 \end{bmatrix}$.

(2) $\begin{bmatrix} 1 & 0 \\ 2 & -1 \end{bmatrix}$.

(3) $\begin{bmatrix} 25 & 40 \\ -12 & -19 \end{bmatrix}$.

2. (1) $\begin{bmatrix} 1 & 2 & 1 \\ 6 & -1 & 0 \\ -1 & -2 & -1 \end{bmatrix}$,

(2) $\begin{bmatrix} 2 & 1 & -1 \\ 1 & 4 & 3 \\ -1 & 3 & 4 \end{bmatrix}$.

(3) $\begin{bmatrix} 1 & 1 & -4 \\ 2 & 0 & -4 \\ -1 & 1 & -2 \end{bmatrix}$.

6-6 Matrix Functions

When solving engineering problems using matrices, besides finding eigensystems, we often need to compute functions of square matrices, such as calculating A^{100}. Without the assistance of computers, performing 100 matrix multiplications would be practically impossible.

In 1858, the British mathematician Arthur Cayley (1821-1895) mentioned in his paper "A memoir on the theory of matrices" that he discovered matrices satisfy an equation with a leading coefficient of 1 and a constant term equal to the determinant. Today, we know that he was referring to the characteristic equation. Using this result, we can now compute functions of square matrices, and in the following, we will introduce the related theory and methods.

6-6-1　Cayley-Hamilton Theorem

▶ **Theorem 6-6-1**

Cayley-Hamilton Theorem

Let A be an $n \times n$ square matrix, if the characteristic equation of A is
$f(x) = (-1)^n [x^n - \beta_1 x^{n-1} + \cdots + (-1)^n \beta_n]$, then

$$A^n - \beta_1 A^{n-1} + \cdots + (-1)^n \beta_n I = O .$$ ◀

1. **Example**

(1) If $A = \begin{bmatrix} 2 & 1 \\ 1 & 2 \end{bmatrix}$, then according to definition, the characteristic equation of A is

$\lambda^2 - 4\lambda + 3 = 0$, also

$$A^2 = \begin{bmatrix} 2 & 1 \\ 1 & 2 \end{bmatrix} \begin{bmatrix} 2 & 1 \\ 1 & 2 \end{bmatrix} = \begin{bmatrix} 5 & 4 \\ 4 & 5 \end{bmatrix},$$

therefore $A^2 - 4A + 3I = O$.

so matrix A satisfies its characteristic equation $\lambda^2 - 4\lambda + 3 = 0$, that is

$A^2 - 4A + 3I = O$.

(2) If $A = \begin{bmatrix} 0 & 4 & -1 \\ 1 & 2 & 1 \\ 1 & -1 & 3 \end{bmatrix}$, then the characteristic equation of A is

$\lambda^3 - 5\lambda^2 + 4\lambda + 5 = 0$,

$$\text{also} \quad A^3 = \begin{bmatrix} 0 & 4 & -1 \\ 1 & 2 & 1 \\ 1 & -1 & 3 \end{bmatrix} \begin{bmatrix} 0 & 4 & -1 \\ 1 & 2 & 1 \\ 1 & -1 & 3 \end{bmatrix} \begin{bmatrix} 0 & 4 & -1 \\ 1 & 2 & 1 \\ 1 & -1 & 3 \end{bmatrix} = \begin{bmatrix} 10 & 29 & 9 \\ 11 & 22 & 16 \\ 6 & -1 & 18 \end{bmatrix};$$

$$A^2 = \begin{bmatrix} 0 & 4 & -1 \\ 1 & 2 & 1 \\ 1 & -1 & 3 \end{bmatrix} \begin{bmatrix} 0 & 4 & -1 \\ 1 & 2 & 1 \\ 1 & -1 & 3 \end{bmatrix} = \begin{bmatrix} 3 & 9 & 1 \\ 3 & 7 & 4 \\ 2 & -1 & 7 \end{bmatrix},$$

$$A^3 - 5A^2 + 4A + 5I = O,$$

so matrix A satisfies its characteristic equation $\lambda^3 - 5\lambda^2 + 4\lambda + 5 = 0$.

2. Applications of Finding Bases

Let $A = [a_{ij}]_{n \times n}$ be an $n \times n$ square matrix, then $\{A^{n-1}, A^{n-2}, \cdots, A, I\}$ forms a basis for the space of matrix A's power function. That is, for any positive integer m greater than n or $m = 0$, we have

$$A^m = c_{n-1} A^{n-1} + c_{n-2} A^{n-2} + \cdots + c_1 A + c_0 I$$

For example, for $A = \begin{bmatrix} 2 & 1 \\ 1 & 2 \end{bmatrix}$, the Cayley-Hamilton theorem states that

$A^2 - 4A + 3I = O$, we have $A^2 = 4A - 3I$ affer rearranging, then $A^3 = 4A^2 - 3A = 13A - 12I$. Similarly, expressing A^4 as AA^3, we get $A^4 = 4A^3 - 3A^2 = 4(13A - 12I) - 3(4A - 3I) = 40A - 39I$.

By following this pattern, for general $m > n$, $A^m = c_1 A + c_0 I$, $m = 0, 1, 2, 3, 4, \cdots$, that is $\{A, I\}$ serves as a basis for the space of A's powers function and can be used to represent A^m.

3. The Relationship Between Analytic Functions

For an analytic function $f(x)$ (whose Maclaurin series exists) and an $n \times n$ square matrix A, there exist n constants $c_1, c_2, \cdots, c_n$ within the convergence interval of $f(x)$ such that:

$$f(A) = c_1 A^{n-1} + c_2 A^{n-2} + \cdots + c_n I$$

For instance, if we consider $A = \begin{bmatrix} 2 & 1 \\ 1 & 2 \end{bmatrix}$, according to the Cayley-Hamilton theorem, $A^2 - 4A + 3I = O$, which means that $\{A, I\}$ serves as a basis for the space of A's powers function. Therefore, for the analytic function $f(x) = e^x$, we have

$$f(A) = e^A = c_1 A + c_0 I.$$

6-6-2   The Solution of Matrix Function $f(A)$

1. Using Long Division

Assuming $A = [a_{ij}]_{n \times n}$ is an $n \times n$ square matrix, and A satisfies

$A^n - b_1 A^{n-1} + b_2 A^{n-2} + \cdots + (-1)^n b_n I = O$,

let $\Phi(x) = x^n - b_1 x^{n-1} + b_2 x^{n-2} + \cdots + (-1)^n b_n \Rightarrow \Phi(A) = O$.
If the division of $f(x)$ by $\Phi(x)$ yields a quotient $Q(x)$, and the remainder is
$R(x)$, that is $f(x) = \Phi(x)Q(x) + R(x)$, then $f(A) = \Phi(A)Q(A) + R(A)$,
also $\Phi(A) = O \Rightarrow f(A) = R(A)$.

Example 1

$$A = \begin{bmatrix} 2 & 1 & 1 \\ 1 & 4 & 3 \\ -1 & -1 & 0 \end{bmatrix}, \text{ find } f(A) = A^4 - 3A^3 - 3A^2 + 4A + 2I .$$

Solution

$\det(A - \lambda I) = (-1)^3 (\lambda^3 - 6\lambda^2 + 11\lambda - 6) = 0$,

according to the Cayley-Hamilton theorem, we know $A^3 - 6A^2 + 11A - 6I = O$,
let $f(x) = x^4 - 3x^3 - 3x^2 + 4x + 2$, $\Phi(x) = x^3 - 6x^2 + 11x - 6$,
then $\Phi(A) = O$, using long division, when $f(x)$ is divided by $\Phi(x)$, the quotient is
$Q(x) = x + 3$,
and the remainder is $R(x) = 4x^2 - 23x + 20$, therefore $f(x) = \Phi(x)Q(x) + R(x)$,
then $f(A) = \Phi(A)Q(A) + R(A)$, also $\Phi(A) = O$,

so $f(A) = R(A) = 4A^2 - 23A + 20I = \begin{bmatrix} -10 & -3 & -3 \\ -11 & -16 & -17 \\ 11 & 3 & 4 \end{bmatrix}^{8}$.   Q.E.D.

2. Using Eigenvalues

If $f(x) = c_{n-1} x^{n-1} + c_{n-2} x^{n-2} + \cdots + c_0$ is the polynomial of A, then

$f(A) = c_{n-1} A^{n-1} + c_{n-2} A^{n-2} + \cdots + c_0 I$. Therefore, the problem is now to find the coefficients c_i. Taking a three-order square matrix A as an example, if A has eigenvalues λ_1, λ_2, λ_3, according to the definition of eigenvalues, we have
$f(\lambda) = c_2 \lambda^2 + c_1 \lambda + c_0$.

---

[8] Long division is more suitable for polynomial functions with lower powers when solving.

If $f(x)$ has three distinct roots λ_1, λ_2, λ_3, then solving $\begin{cases} f(\lambda_1) = c_2\lambda_1^2 + c_1\lambda_1 + c_0 \\ f(\lambda_2) = c_2\lambda_2^2 + c_1\lambda_2 + c_0 \\ f(\lambda_3) = c_2\lambda_3^2 + c_1\lambda_3 + c_0 \end{cases}$

will yield c_2, c_1, c_0, which can be used to obtain $f(A)$;

If $f(x)$ has a double root λ_1, λ_1, λ_3, then solving $\begin{cases} f(\lambda_1) = c_2\lambda_1^2 + c_1\lambda_1 + c_0 \\ f'(\lambda_1) = 2c_2\lambda_1 + c_1 \cdot 1 \\ f(\lambda_3) = c_2\lambda_3^2 + c_1\lambda_3 + c_0 \end{cases}$ will

yield c_2, c_1, c_0, which can be used to obtain $f(A)$;

If $f(x)$ has a triple root λ_1, λ_1, λ_1, then solving $\begin{cases} f(\lambda_1) = c_2\lambda_1^2 + c_1\lambda_1 + c_0 \\ f'(\lambda_1) = 2c_2x_1 + c_1 \\ f''(\lambda_1) = 2c_2 \end{cases}$ will yield

c_2, c_1, c_0, which can be used to obtain $f(A)$.

Example 2

$A = \begin{bmatrix} 0 & 1 & 0 \\ 0 & 0 & 1 \\ 0 & 0 & 0 \end{bmatrix}$, find e^A.

Solution

$|A - \lambda I| = 0 \Rightarrow \lambda^3 = 0$, $\therefore A^3 = O$, $\lambda_1 = \lambda_2 = \lambda_3 = 0$, $e^A = \alpha A^2 + \beta A + rI$,

$f(\lambda) = e^\lambda = \alpha\lambda^2 + \beta\lambda + r$, therefore

$$f(0) = 1 = r$$
$$f'(0) = 1 = \beta$$
$$f''(0) = 1 = 2\alpha \Rightarrow \alpha = \frac{1}{2},$$

$\therefore e^A = \frac{1}{2}A^2 + A + I = \frac{1}{2}\begin{bmatrix} 0 & 0 & 1 \\ 0 & 0 & 0 \\ 0 & 0 & 0 \end{bmatrix} + \begin{bmatrix} 1 & 1 & 0 \\ 0 & 1 & 1 \\ 0 & 0 & 1 \end{bmatrix} = \begin{bmatrix} 1 & 1 & \frac{1}{2} \\ 0 & 1 & 1 \\ 0 & 0 & 1 \end{bmatrix}$[9]. 　Q.E.D.

---

[9] In this problem, A cannot be diagonalized, so it is not possible to use diagonalization to calculate e^A.

6-6-3 Diagonalization for Computing Matrix Functions

If matrix A can be diagonalized, the computation of the matrix function $f(A)$ becomes simpler. This is because there exists a transition matrix P such that $P^{-1}AP = D$,

thus $A = PDP^{-1} = P \begin{bmatrix} \lambda_1 & & O \\ & \ddots & \\ O & & \lambda_n \end{bmatrix} P^{-1}$. This $A^k = P \begin{bmatrix} \lambda_1{}^k & & O \\ & \ddots & \\ O & & \lambda_n{}^k \end{bmatrix} P^{-1}$, therefore,

for any polynomial $f(x)$, we have

$$f(A) = P \begin{bmatrix} f(\lambda_1) & & & O \\ & f(\lambda_2) & & \\ & & \ddots & \\ O & & & f(\lambda_n) \end{bmatrix} P^{-1}.$$

Example 3

$A = \begin{bmatrix} 5 & 4 \\ 1 & 2 \end{bmatrix}$, find $A^{100}(7A - 6I) = ?$

Solution

Using Eigenvalues

$|A - \lambda I| = (-1)^2(\lambda^2 - 7\lambda + 6) = 0 \Rightarrow \lambda = 1, 6$, $\therefore A$ satisfies $A^2 - 7A + 6I = O$,

let $f(A) = A^{100}(7A - 6I) = 7A^{101} - 6A^{100} = \alpha A + \beta I$

$\Rightarrow \begin{cases} f(1) = 7 - 6 = \alpha + \beta & \Rightarrow \alpha + \beta = 1 = f(1) \\ f(6) = 7 \cdot 6^{101} - 6 \cdot 6^{100} = 6\alpha + \beta \Rightarrow 6\alpha + \beta = 6^{102} = f(6) \end{cases} \Rightarrow \alpha = \frac{1}{5}(6^{102} - 1)$,

$\beta = \frac{1}{5}(6 - 6^{102})$, $\therefore f(A) = \frac{1}{5}(6^{102} - 1)\begin{bmatrix} 5 & 4 \\ 1 & 2 \end{bmatrix} + \frac{1}{5}(6 - 6^{102})\begin{bmatrix} 1 & 0 \\ 0 & 1 \end{bmatrix}$.

Using Diagonalization

The eigenvector corresponding to $\lambda_1 = 1$ is $V_1 = \begin{bmatrix} 1 \\ -1 \end{bmatrix}$.

The eigenvector corresponding to $\lambda_2 = 6$ is $V_2 = \begin{bmatrix} 4 \\ 1 \end{bmatrix}$.

Then $P^{-1}AP = D = \begin{bmatrix} 1 & 0 \\ 0 & 6 \end{bmatrix} \Rightarrow A = PDP^{-1}$, $P = \begin{bmatrix} 1 & 4 \\ -1 & 1 \end{bmatrix}$, taking $f(x) = 7x^{101} - 6x^{100}$,

then

$$f(A) = P \begin{bmatrix} f(1) & 0 \\ 0 & f(6) \end{bmatrix} P^{-1} = \begin{bmatrix} 1 & 4 \\ -1 & 1 \end{bmatrix} \begin{bmatrix} 1 & 0 \\ 0 & 6^{102} \end{bmatrix} \frac{1}{5} \begin{bmatrix} 1 & -4 \\ 1 & 1 \end{bmatrix}$$

$$= \frac{1}{5} \begin{bmatrix} 1+4\times 6^{102} & -4+4\times 6^{102} \\ -1+6^{102} & 4+6^{102} \end{bmatrix}.$$

Q.E.D.

Example 4

$A = \begin{bmatrix} 5 & -4 & 2 \\ 3 & -2 & 2 \\ 2 & -2 & 3 \end{bmatrix}$, find e^{At}.

Solution

From $\det(A - \lambda I) = 0 \Rightarrow \lambda = 1, 2, 3$,

$\lambda = 1 \Rightarrow$ the eigenvector $V_1 = \begin{bmatrix} 1 \\ 1 \\ 0 \end{bmatrix}$,

$\lambda = 2 \Rightarrow$ the eigenvector $V_2 = \begin{bmatrix} 0 \\ 1 \\ 2 \end{bmatrix}$, $\lambda = 3 \Rightarrow$ the eigenvector $V_3 = \begin{bmatrix} 1 \\ 1 \\ 1 \end{bmatrix}$,

taking $P = \begin{bmatrix} V_1 & V_2 & V_3 \end{bmatrix} = \begin{bmatrix} 1 & 0 & 1 \\ 1 & 1 & 1 \\ 0 & 2 & 1 \end{bmatrix}$, then $P^{-1}AP = D = \begin{bmatrix} 1 & 0 & 0 \\ 0 & 2 & 0 \\ 0 & 0 & 3 \end{bmatrix}$, $A = PDP^{-1}$,

then $e^{At} = P \begin{bmatrix} e^t & 0 & 0 \\ 0 & e^{2t} & 0 \\ 0 & 0 & e^{3t} \end{bmatrix} P^{-1} = \begin{bmatrix} -e^t + 2e^{3t} & 2e^t - 2e^{3t} & -e^t + e^{3t} \\ -e^t - e^{2t} + 2e^{3t} & 2e^t + e^{2t} - 2e^{3t} & -e^t + e^{3t} \\ -2e^{2t} + 2e^{3t} & 2e^{2t} - 2e^{3t} & e^{3t} \end{bmatrix}$.   Q.E.D.

Example 5

$A = \begin{bmatrix} 0 & 1 & 0 \\ 2 & -1 & 3 \\ 0 & 2 & 1 \end{bmatrix}$, use Cayley-Hamilton theorem to find A^{-1}.

Solution

From $|A - \lambda I| = 0 \Rightarrow \lambda^3 - 9\lambda + 2 = 0$, $\therefore$ Through Cayley-Hamilton theorem, we know $A^3 - 9A + 2I = O$,

also $|A| \neq 0 \Rightarrow A^2 - 9I + 2A^{-1} = O \Rightarrow A^{-1} = \frac{1}{2}(9I - A^2) = \frac{1}{2} \begin{bmatrix} 7 & 1 & -3 \\ 2 & 0 & 0 \\ -4 & 0 & 2 \end{bmatrix}$.   Q.E.D.

6-6 Exercises

Basic questions

1. Let $A = \begin{bmatrix} -3 & 2 \\ -10 & 6 \end{bmatrix}$, use Cayley-Hamilton theorem to find A^{-1}.

2. Given that $A = \begin{bmatrix} 2 & 0 \\ 0 & 1 \end{bmatrix}$, find

   (1) A^{20}, (2) e^A, (3) $\cos A$.

3. $A = \begin{bmatrix} 1 & -3 \\ 1 & 1 \end{bmatrix}$, find A^{20}.

4. $A = \begin{bmatrix} 1 & -1 \\ 1 & 3 \end{bmatrix}$, find e^A.

Advanced questions

1. $A = \begin{bmatrix} -3 & 6 & -11 \\ 3 & -4 & 6 \\ 4 & -8 & 13 \end{bmatrix}$, use Cayley-Hamilton theorem to find A^{-1}.

2. $A = \begin{bmatrix} -3 & 0 & 1 \\ -8 & 1 & 2 \\ -16 & 0 & 5 \end{bmatrix}$, find A^{100}.

3. $A = \begin{bmatrix} 1 & 0 & 2 \\ 0 & -1 & 1 \\ 0 & 1 & 0 \end{bmatrix}$, find

   $f(A) = A^6 - 5A^5 - 4A^4 + 3A^2 - 2A + I$.

4. $A = \begin{bmatrix} 1 & 1 & 1 \\ 1 & 1 & 1 \\ 1 & 1 & 1 \end{bmatrix}$, find A^8.

5. $A = \begin{bmatrix} 1 & 1 & 1 \\ -1 & -1 & -1 \\ 1 & 1 & 1 \end{bmatrix}$, find e^A.

6. $A = \begin{bmatrix} 0 & 0 & 1 \\ 0 & 0 & 1 \\ 1 & 1 & 1 \end{bmatrix}$, find A^{100}.

7
Linear Differential Equation System

7-1 The Solution of a System of First-Order Simultaneous Linear Differential Equations

7-2 The Solution of a Homogeneous System of Simultaneous Differential Equations

7-3 Diagonalization of Matrix for Solving Non-Homogeneous System of Simultaneous Differential Equations

Cleve Barry Moler
(1939~, America)

In the late 1970s to early 1980s, Professor Cleve Moler from the University of New Mexico in the United States independently developed the first version of MATLAB to facilitate matrix computations for students. In 1984, he co-founded MathWorks with his friends and extensively utilized the C programming language to create various matrix computation programs. Over time, their toolbox flourished in various fields, becoming an essential program for control system design and analysis, image processing, signal processing, numerical analysis, financial modeling and analysis, and more recently, various artificial intelligence algorithms (AI). Its significant contributions in engineering have been remarkable.

Learning Objectives

7-1
The Solution of a System of First-Order Simultaneous Linear Differential Equations

7-1-1 Mastering the solution methods for systems of differential equations: substitution and elimination method

7-1-2 Mastering the solution methods for systems of differential equations: determinant method

7-1-3 Mastering the solution methods for systems of differential equations: Laplace transform

7-2
The Solution of a Homogeneous System of Simultaneous Differential Equations

7-2-1 Proficiency in the general solution method for systems of first-order linear equations: basic matrix and Laplace transform.

7-2-2 Proficiency in the general solution method for systems of first-order linear equations: diagonalizable and non-diagonalizable

7-2-3 Mastering the solution methods for systems of homogeneous second-order equations

7-3
Diagonalization of Matrix for Solving Non-Homogeneous System of Simultaneous Differential Equations

7-3-1 Mastering the general solution methods for systems of non-homogeneous first-order equations: diagonalization

7-3-2 Mastering the general solution methods for systems of non-homogeneous second-order equations: diagonalization

ExampleVideo

For higher-order differential equations or complex multi-dimensional systems, the most commonly used approach for solving them is to convert the system into a system of first-order ordinary differential equations (ODEs) and then solve it. This concept is extensively employed in numerical ODE solvers, with Matlab being one of the most well-known examples.

For instance, in the case of a second-order vibrational system

$mx'' + cx' + kx = \alpha \cos wt$, it can be organized as follows: $x'' = -\dfrac{c}{m}x' - \dfrac{k}{m}x + \dfrac{\alpha}{m}\cos wt$

Since this system is second-order, we need to introduce two state variables. Let $x_1 = x$ represent displacement, and $x_2 = x'$ represent velocity. This leads to the following system of simultaneous equations $\begin{cases} x_1' = x_2 \\ x_2' = -\dfrac{k}{m}x_1 - \dfrac{c}{m}x_2 + \dfrac{\alpha}{m}\cos wt \end{cases}$. Let

$X = \begin{bmatrix} x_1 \\ x_2 \end{bmatrix}$, then the matrix form of this second-order differential equation is

$$\begin{bmatrix} x_1' \\ x_2' \end{bmatrix} = \begin{bmatrix} 0 & 1 \\ -\dfrac{k}{m} & -\dfrac{c}{m} \end{bmatrix} \begin{bmatrix} x_1 \\ x_2 \end{bmatrix} + \begin{bmatrix} 0 \\ \dfrac{\alpha}{m}\cos wt \end{bmatrix} .$$

That is $X' = AX + B(t)$, where $A = \begin{bmatrix} 0 & 1 \\ -\dfrac{k}{m} & -\dfrac{c}{m} \end{bmatrix}$, $B = \begin{bmatrix} 0 \\ \dfrac{\alpha}{m}\cos wt \end{bmatrix}$. This equation is also referred to as the **dynamic system's state equation** or simply the state equation of the system. In the following chapter, we will introduce how to solve this type of first-order or second-order system of simultaneous differential equations.

7-1   The Solution of a System of First-Order Simultaneous Linear Differential Equations

The most basic direct methods for solving first-order systems of ODEs are elimination methods, which can be further categorized into substitution elimination and determinant elimination. The main idea is to transform the system of ODEs into a single higher-order ODE with a single unknown function and then apply the theory of higher-order ODEs to solve it. To illustrate these two methods, we will provide examples in the following section.

7-1-1 The Elimination Method by Substitution

Example 1

Solve $\begin{cases} 2x' - y' + x + y = -t \\ x' + y' + 4x = 3 \end{cases}$.

Solution

$\begin{cases} 2x' - y' + x + y = -t \cdots ① \\ x' + y' + 4x = 3 \cdots\cdots\cdots ② \end{cases}$

① plus ② obtained $3x' + 5x + y = -t + 3$ that is $y = -3x' - 5x - t + 3$ (called

formula ③),

substituting $y' = -3x'' - 5x' - 1$ into ② results in reducing it to a second-order ODE

involving only the unknown function $x(t)$.

$$x' + (-3x'' - 5x' - 1) + 4x = 3 \Rightarrow x' - 3x'' - 5x' - 1 + 4x = 3 \Rightarrow 3x'' + 4x' - 4x = -4.$$

(1) To find the homogeneous solution $x_h(t)$, let $x = e^{mt}$ substitute then we get a
characteristic equation.

$$3m^2 + 4m - 4 = 0 \Rightarrow (3m - 2)(m + 2) = 0,$$

$$\therefore m = -2, \frac{2}{3}, \quad \therefore x_h(t) = c_1 e^{-2t} + c_2 e^{\frac{2}{3}t} \text{ }^1.$$

(2) To find the particular solution $x_p(t)$: $(3D^2 + 4D - 4)x_p(t) = -4$,

$$x_p(t) = \frac{1}{3D^2 + 4D - 4}(-4) = 1,$$

$$\therefore x(t) = x_h(t) + x_p(t) = c_1 e^{-2t} + c_2 e^{\frac{2}{3}t} + 1,$$

substituting $x(t) = c_1 e^{-2t} + c_2 e^{\frac{2}{3}t} + 1$ into formula ③, we obtain

$$y(t) = -3 \cdot (-2c_1 e^{-2t} + \frac{2}{3}c_2 e^{\frac{2}{3}t}) - 5(c_1 e^{-2t} + c_2 e^{\frac{2}{3}t} + 1) - t + 3 = c_1 e^{-2t} - 7c_2 e^{\frac{2}{3}t} - t - 2.$$

Q.E.D.

---

[1] When solving, it is essential to recognize that this system of simultaneous ODEs is a composition of two first-order ODEs, making it a second-order system with only two independent constants. If any additional independent constants appear in the solutions, it is necessary to substitute these solutions back into the original system of ODEs to derive the relationship between the independent constants. Afterward, the solutions should be simplified using this relationship.

7-1-2 The Determinant Method

Assuming the original system of ODEs can be simplified using the differential operator $L(D)$ to obtain: $\begin{cases} L_1(D)x + L_2(D)y = f(t) \\ L_3(D)x + L_4(D)y = g(t) \end{cases}$.

By applying Cramer's rule, we get $x = \dfrac{\begin{vmatrix} f(t) & L_2(D) \\ g(t) & L_4(D) \end{vmatrix}}{\begin{vmatrix} L_1(D) & L_2(D) \\ L_3(D) & L_4(D) \end{vmatrix}}$, $y = \dfrac{\begin{vmatrix} L_1(D) & f(t) \\ L_3(D) & g(t) \end{vmatrix}}{\begin{vmatrix} L_1(D) & L_2(D) \\ L_3(D) & L_4(D) \end{vmatrix}}$, after

rearrangement, we acquire

(1) $\begin{vmatrix} L_1(D) & L_2(D) \\ L_3(D) & L_4(D) \end{vmatrix} x = \begin{vmatrix} f(t) & L_2(D) \\ g(t) & L_4(D) \end{vmatrix}$ (2) $\begin{vmatrix} L_1(D) & L_2(D) \\ L_3(D) & L_4(D) \end{vmatrix} y = \begin{vmatrix} L_1(D) & f(t) \\ L_3(D) & g(t) \end{vmatrix}$.

Principles for Using the Two Solution Methods

When solving the system, if both unknown functions can be simplified into higher-order single unknown function ODEs using this method, it may result in additional independent constants. It is then necessary to substitute these solutions back into the original system of ODEs to derive the relationship between the independent constants, making the process more complicated. Therefore, it is generally recommended to only find the solution for one unknown function using the determinant method, while the other solution can be obtained using the substitution method.

Example 2

Solve $\begin{cases} x' = 2x + y + 1 \\ y' = 4x + 2y + e^{4t} \end{cases}$.

Solution

Rewriting the system using the differential operator $\begin{cases} Dx = 2x + y + 1 \\ Dy = 4x + 2y + e^{4t} \end{cases}$, after

rearrangement, we have $\begin{vmatrix} D-2 & -1 \\ -4 & D-2 \end{vmatrix} x = \begin{vmatrix} 1 & -1 \\ e^{4t} & D-2 \end{vmatrix}$

$\Rightarrow (D^2 - 4D + 4 - 4)x = -2 + e^{4t} \Rightarrow (D^2 - 4D)x = -2 + e^{4t} \Rightarrow x'' - 4x' = -2 + e^{4t}$.

From the characteristic equation, we obtain $x_h(t) = c_1 + c_2 e^{4t}$. Then, applying the inverse differential operator , we get the particular solution:

$$x_p = \frac{1}{D^2 - 4D}[-2 + e^{4t}] = -2 \cdot \frac{1}{D \cdot (D-4)}e^{0t} + \frac{1}{(D-4)D}e^{4t} = \frac{1}{2}t + \frac{1}{4}te^{4t},$$

$$\therefore x(t) = c_1 + c_2 e^{4t} + \frac{1}{2}t + \frac{1}{4}te^{4t}, \text{ from } x' = 2x + y + 1 \Rightarrow y = x' - 2x - 1,$$

substituting $x(t) = c_1 + c_2 e^{4t} + \frac{1}{2}t + \frac{1}{4}te^{4t}$ into the equation, we have

$$y(t) = (-2c_1 - \frac{1}{2}) - t + (\frac{1}{4} + 2c_2 + \frac{1}{2}t)e^{4t}.$$

Q.E.D.

7-1-3 Solution of Simultaneous ODEs with Constant Coefficients Using Laplace Transform

Laplace transform is a linear transformation, which means that a system of simultaneous linear equations will remain in the form of simultaneous linear equations under its operation. This property allows the theory of solving simultaneous equations in linear algebra to play a significant role. Let's illustrate this with an example of a system of simultaneous first-order ODEs.

Consider a system of two first-order linear ODE:
$$\begin{cases} \dfrac{dx}{dt} = a_{11}x + a_{12}y \\ \dfrac{dy}{dt} = a_{21}x + a_{22}y \end{cases}.$$

First, we apply the Laplace transform separately to each equation
$$\begin{cases} sX(s) - x(0) = a_{11}X(s) + a_{12}Y(s) \\ sY(s) - y(0) = a_{21}X(s) + a_{22}Y(s) \end{cases}, \text{ and after rearranging, we have}$$

$$\begin{bmatrix} a_{11} - s & a_{12} \\ a_{21} & a_{22} - s \end{bmatrix} \begin{bmatrix} X(s) \\ Y(s) \end{bmatrix} = \begin{bmatrix} -x(0) \\ -y(0) \end{bmatrix}, \text{ if } \begin{vmatrix} a_{11} - s & a_{12} \\ a_{21} & a_{22} - s \end{vmatrix} \neq 0, \text{ then by using Cramer's}$$

rule, we obtain

$$X(s) = \frac{\begin{vmatrix} -x(0) & a_{12} \\ -y(0) & a_{22} - s \end{vmatrix}}{\begin{vmatrix} a_{11} - s & a_{12} \\ a_{21} & a_{22} - s \end{vmatrix}}, \quad Y(s) = \frac{\begin{vmatrix} a_{11} - s & -x(0) \\ a_{21} & -y(0) \end{vmatrix}}{\begin{vmatrix} a_{11} - s & a_{12} \\ a_{21} & a_{22} - s \end{vmatrix}}.$$

Finally, by performing the inverse Laplace transform, we obtain the general solution. Let's proceed with the following example to demonstrate its practical application.

Example 3

Use Laplace transform to solve $\begin{cases} \dfrac{dx}{dt} = 2x - 3y \\ \dfrac{dy}{dt} = y - 2x \end{cases}$, $x(0) = 8$, $y(0) = 3$.[2]

Solution

Let $\mathcal{L}\{x(t)\} = \hat{x}(s)$, $\mathcal{L}\{y(t)\} = \hat{y}(s)$, then the original ODE

by using L-T we get $\begin{cases} s\hat{x}(s) - x(0) = 2\hat{x}(s) - 3\hat{y}(s) \\ s\hat{y}(s) - y(0) = \hat{y}(s) - 2\hat{x}(s) \end{cases}$,

the original system change into $\begin{cases} (s-2)\hat{x}(s) + 3\hat{y}(s) = 8 \\ 2\hat{x}(s) + (s-1)\hat{y}(s) = 3 \end{cases}$, then by using Cramer's rule

we obtain $\quad \hat{x}(s) = \dfrac{\begin{vmatrix} 8 & 3 \\ 3 & s-1 \end{vmatrix}}{\begin{vmatrix} s-2 & 3 \\ 2 & s-1 \end{vmatrix}} = \dfrac{5}{s+1} + \dfrac{3}{s-4}$, $\hat{y}(s) = \dfrac{\begin{vmatrix} s-2 & 8 \\ 2 & 3 \end{vmatrix}}{s^2 - 3s - 4} = \dfrac{5}{s+1} + \dfrac{-2}{s-4}$.

$\therefore \qquad x(t) = \mathcal{L}^{-1}\{\hat{x}(s)\} = \mathcal{L}^{-1}\{\dfrac{5}{s+1} + \dfrac{3}{s-4}\} = 5e^{-t} + 3e^{4t}$,

$y(t) = \mathcal{L}^{-1}\{\hat{y}(s)\} = \mathcal{L}^{-1}\{\dfrac{5}{s+1} + \dfrac{-2}{s-4}\} = 5e^{-t} - 2e^{4t}$.

Q.E.D.

---

[2] When using Laplace transform to solve a system of ODEs, if the initial conditions are not provided, you would need to introduce them as constants in the transformed equations.

7-1 Exercises

Basic questions

Using the method of elimination by substitution, solve the following system of simultaneous ODE for questions 1 and 2.

1. $\begin{cases} x_1' + x_2 = 1 \\ 9x_1 + x_2' = 0 \end{cases}$, where $x_1(0) = x_2(0) = 0$.

2. $\begin{cases} x_1' = 4x_1 + x_2 \\ x_2' = 3x_1 + 2x_2 \end{cases}$,

   where $x_1(0) = 0$, $x_2(0) = 1$.

Using Laplace transform to solve following system of simultaneous ODE for questions 3~4.

3. $\begin{cases} x_1' + x_2 = 1 \\ 9x_1 + x_2' = 0 \end{cases}$, where $x_1(0) = x_2(0) = 0$.

4. $\begin{cases} x_1' = 4x_1 + x_2 \\ x_2' = 3x_1 + 2x_2 \end{cases}$,

   where $x_1(0) = 0$, $x_2(0) = 1$.

Advanced questions

Solve following system of simultaneous ODE.

1. $\begin{cases} \dfrac{dx_1}{dt} = -x_1 - 2x_2 + 3 \\ \dfrac{dx_2}{dt} = 3x_1 + 4x_2 + 3 \end{cases}$,

   $x_1(0) = -4, x_2(0) = 5$.

2. $\begin{cases} \dfrac{dy_1}{dt} - 3y_1 = y_2 \\ \dfrac{dy_2}{dt} - y_2 = -y_1 \end{cases}$.

3. $\begin{cases} x_1' = -2x_1 + x_2 \\ x_2' = -x_1 \end{cases}$, $x_1(0) = 1, x_2(0) = 0$.

4. $\begin{cases} x_1' = x_2 + e^{3t} \\ x_2' = x_1 - 3e^{3t} \end{cases}$.

5. $\begin{cases} x_1' + 3x_1 + 4x_2 = 5e^t \\ 5x_1 - x_2' + 6x_2 = 6e^t \end{cases}$,

   $x_1(0) = 1, x_2(0) = 0$.

Using Laplace transform to solve the following system of simultaneous ODE.

6. $\begin{cases} \dfrac{dx}{dt} + 3x + \dfrac{dy}{dt} = \cos t \\ \dfrac{dx}{dt} - x + y = \sin t \end{cases}$,

   $x(0) = 0, y(0) = 4$.

7. $\begin{cases} \dfrac{dx}{dt} - 4x + 2y = 2t \\ \dfrac{dy}{dt} - 8x + 4y = 1 \end{cases}$, $x(0) = 3, y(0) = 5$.

8. $\begin{cases} y_1' + y_2 = 2\cos t \\ y_1 + y_2' = 0 \end{cases}$, $y_1(0) = 0, y_2(0) = -1$.

9. $y_1' = 6y_1 + 9y_2, y_1(0) = -3$,

   $y_2' = y_1 + 6y_2, y_2(0) = -3$.

10. $x' + 2y' - y = 1, 2x' + y = 0$,

    $x(0) = y(0) = 0$.

11. $x' - 2y' = 1, x' + y - x = 0$,

    $x(0) = 0, y(0) = 1$.

7-2 The Solution of a Homogeneous System of Simultaneous Differential Equations

In the previous section, we introduced the methods of elimination by substitution and Laplace transform to solve systems of simultaneous ODEs. However, these methods are more suitable for solving systems with lower order ODEs. If the system has higher-order equations, matrix methods are still required for solving. Now, we will first discuss how to use the concept of elementary matrices to solve homogeneous systems of simultaneous differential equations.

7-2-1 The Solution of a First-Order Homogeneous System of Simultaneous Differential Equations

The generalized first-order constant coefficient ordinary differential equation, as discussed in Section 7-1, can be expressed as
$$\begin{bmatrix} x_1' \\ x_2' \\ \vdots \\ x_n' \end{bmatrix} = \begin{bmatrix} a_{11} & a_{12} & \cdots & a_{1n} \\ a_{21} & a_{22} & & \vdots \\ \vdots & & \ddots & \vdots \\ a_{n1} & \cdots & \cdots & a_{nn} \end{bmatrix} \begin{bmatrix} x_1 \\ x_2 \\ \vdots \\ x_n \end{bmatrix}.$$

This is often written as $X' = AX$. Its solution method is analogous to solving the first-order constant coefficient ordinary differential equation $y' = ay$. When solving $y' = ay$, we assume the solution is $y(x) = ce^{ax}$ and substitute it into the ODE, leading to the characteristic equation. Similarly, when solving $X' = AX$, we assume the solution is $X_{n\times1}(t) = V_{n\times1}e^{\lambda t}$ and substitute it into $X' = AX$, we obtain

$$V_{n\times1}\lambda e^{\lambda t} = AVe^{\lambda t}$$

Rearranging, we get $(A_{n\times n}V_{n\times1} - \lambda V_{n\times1})e^{\lambda t} = O$. Because $e^{\lambda t} \neq 0 \Rightarrow AV = \lambda V$. Hence, the original equation is an eigenvalue system. λ is the eigenvalue, $V_{n\times1}$ is the non-zero eigenvector.

▶ **Theorem 7-2-1**

The Linearly Independent Solutions of an _n_-th Order Homogeneous System of Simultaneous ODE

Assuming A is a diagonalizable $n \times n$ matrix with eigenvalues $\lambda_1, \lambda_2, \cdots, \lambda_n$, and their corresponding linearly independent eigenvectors are $V_1, V_2, \cdots, V_n$, then for the n-th order homogeneous system of simultaneous differential equations $X' = AX$, $V_1e^{\lambda_1 t}, V_2e^{\lambda_2 t}, \cdots, V_ne^{\lambda_n t}$ are n linearly independent solutions. ◀

At $t = 0$, suppose $P = [V_1 \ V_2 \cdots V_n]$ is an invertible square matrix constructed from n linearly independent column vectors $V_1 e^{\lambda_1 t}, V_2 e^{\lambda_2 t}, \cdots, V_n e^{\lambda_n t}$. Since P is invertible, which means $\det(P) \neq 0$. So these n solutions are guaranteed to be linearly independent. Specifically, we define the following:

1. **Fundamental matrix**

   $\Phi = [V_1 e^{\lambda_1 t} \ V_2 e^{\lambda_2 t} \cdots V_n e^{\lambda_n t}]$, is called the fundamental matrix of $X' = AX$.

▶ **Theorem 7-2-2**

Diagonalization for Finding the General Solution

Assuming A is a diagonalizable $n \times n$ matrix, and for the n-th order homogeneous system of simultaneous differential equations $X' = AX$, $\Phi = [V_1 e^{\lambda_1 t} \ V_2 e^{\lambda_2 t} \cdots V_n e^{\lambda_n t}]$ is the fundamental matrix, then $\Phi_{n \times n} C_{n \times 1}$ is the general solution of $X' = AX$.

Proof

Because $V_1 e^{\lambda_1 t}, V_2 e^{\lambda_2 t}, \cdots, V_n e^{\lambda_n t}$ are n linearly independent solutions of $X' = AX$, the solution space spanned by these n linearly independent solutions contains any solution of $X' = AX$. In other words, the general solution of $X' = AX$ can be expressed as $X = c_1 V_1 e^{\lambda_1 t} + c_2 V_2 e^{\lambda_2 t} + \cdots + c_n V_n e^{\lambda_n t}$, the general solution is

$$X = [V_1 e^{\lambda_1 t} \ V_2 e^{\lambda_2 t} \cdots V_n e^{\lambda_n t}] \begin{bmatrix} c_1 \\ c_2 \\ \vdots \\ c_n \end{bmatrix} = \Phi \cdot C. \qquad ◀$$

The fundamental matrix does not have a unique expression, which depends on the choice of the initial state of the system. Additionally, the non-uniqueness of the selection of eigenvectors also leads to the non-uniqueness of the fundamental matrix.

2. **Using Laplace Transform to Solve**

   In fact , the general solution of $X' = AX$ can also be obtained using Laplace transform. By defining the Laplace transform of a matrix function $X(t) = \begin{bmatrix} x_1 \\ x_2 \\ \vdots \\ x_n \end{bmatrix}$ as:

   $\widehat{X}(s) = \begin{bmatrix} \hat{x}_1(s) \\ \hat{x}_2(s) \\ \vdots \\ \hat{x}_n(s) \end{bmatrix}$. Taking $X(0) = C = \begin{bmatrix} c_1 \\ c_2 \\ \vdots \\ c_n \end{bmatrix}$ then taking the Laplace transform of

   $X' = AX$. Emerging and using the formula of differentiation and from Laplace transform, we obtain $s\widehat{X}(s) - X(0) = A\widehat{X}(s)$, then $(sI_{n\times n} - A_{n\times n})\widehat{X}(s) = X(0)$. Rearranging the equation, we get:

$$\widehat{X}(s) = (sI_{n\times n} - A_{n\times n})^{-1} \cdot C = \frac{I}{sI - A} \cdot C .$$

Finally, by performing the inverse Laplace transform, we obtain

$X_{n\times 1}(t) = \mathscr{L}^{-1}\{\widehat{X}(s)\} = \mathscr{L}^{-1}\{\frac{I}{sI - A} \cdot C\} = e^{At} \cdot C$. Therefore, there is a certain

relationship between the fundamental matrix $\Phi = [V_1 e^{\lambda_1 t} \quad V_2 e^{\lambda_2 t} \quad \cdots \quad V_n e^{\lambda_n t}]$ and the matrix function e^{At}.

Example 1

Solve $\begin{bmatrix} x_1' \\ x_2' \end{bmatrix} = \begin{bmatrix} 4 & 2 \\ 2 & 1 \end{bmatrix} \begin{bmatrix} x_1 \\ x_2 \end{bmatrix}$.

Solution

① Using Diagonalization to Solve

Let $A = \begin{bmatrix} 4 & 2 \\ 2 & 1 \end{bmatrix}$, from $|A - \lambda I| = \lambda^2 - 5\lambda = 0$ we get $\lambda = 0, 5$,

$\lambda = 0$ substituting into $(A - \lambda I)V = O \Rightarrow \begin{bmatrix} 4 & 2 \\ 2 & 1 \end{bmatrix} \begin{bmatrix} v_1 \\ v_2 \end{bmatrix} = O \Rightarrow V_1 = c_1 \begin{bmatrix} 1 \\ -2 \end{bmatrix}$,

$\lambda = 5$ substituting into $(A - \lambda I)V = O \Rightarrow \begin{bmatrix} 1 & 2 \\ 2 & -4 \end{bmatrix} \begin{bmatrix} v_1 \\ v_2 \end{bmatrix} = O \Rightarrow V_2 = c_2 \begin{bmatrix} 2 \\ 1 \end{bmatrix}$,

taking fundamental matrix as $\Phi = \begin{bmatrix} 1 \cdot e^{0t} & 2e^{5t} \\ -2 \cdot e^{0t} & 1 \cdot e^{5t} \end{bmatrix}$, so the solution of ODE is

$X = \Phi \cdot C = \begin{bmatrix} 1 \cdot e^{0t} & 2e^{5t} \\ -2 \cdot e^{0t} & 1 \cdot e^{5t} \end{bmatrix} \begin{bmatrix} c_1 \\ c_2 \end{bmatrix} = c_1 \begin{bmatrix} 1 \\ -2 \end{bmatrix} + c_2 e^{5t} \begin{bmatrix} 2 \\ 1 \end{bmatrix}$ therefore

$$\begin{bmatrix} x_1 \\ x_2 \end{bmatrix} = \begin{bmatrix} c_1 + 2c_2 e^{5t} \\ -2c_1 + c_2 e^{5t} \end{bmatrix} .$$

② Using Laplace Transform to Solve

Let $X(0) = \begin{bmatrix} x_1(0) \\ x_2(0) \end{bmatrix} = \begin{bmatrix} c_1 \\ c_2 \end{bmatrix}$ and $\mathscr{L}\{X(t)\} = \widehat{X}(s)$, taking the Laplace transform of

the original ODE $\Rightarrow s\widehat{X}(s) - X(0) = A\widehat{X}(s) \Rightarrow (sI - A)\widehat{X}(s) = X(0)$, we obtain

$$\widehat{X}(s) = (sI - A)^{-1} X(0) = \begin{bmatrix} s-4 & -2 \\ -2 & s-1 \end{bmatrix}^{-1} \begin{bmatrix} c_1 \\ c_2 \end{bmatrix} = \begin{bmatrix} \dfrac{s-1}{s(s-5)} & \dfrac{2}{s(s-5)} \\ \dfrac{2}{s(s-5)} & \dfrac{s-4}{s(s-5)} \end{bmatrix} \begin{bmatrix} c_1 \\ c_2 \end{bmatrix},$$

$$X(t) = \mathscr{L}^{-1}\{\widehat{X}(s)\} = \begin{bmatrix} \mathscr{L}^{-1}\left\{\dfrac{s-1}{s(s-5)}\right\} & \mathscr{L}^{-1}\left\{\dfrac{2}{s(s-5)}\right\} \\ \mathscr{L}^{-1}\left\{\dfrac{2}{s(s-5)}\right\} & \mathscr{L}^{-1}\left\{\dfrac{s-4}{s(s-5)}\right\} \end{bmatrix} \begin{bmatrix} c_1 \\ c_2 \end{bmatrix}$$

$$= \begin{bmatrix} \dfrac{1}{5}+\dfrac{4}{5}e^{5t} & -\dfrac{2}{5}+\dfrac{2}{5}e^{5t} \\ -\dfrac{2}{5}+\dfrac{2}{5}e^{5t} & \dfrac{4}{5}+\dfrac{1}{5}e^{5t} \end{bmatrix} \begin{bmatrix} c_1 \\ c_2 \end{bmatrix} = \begin{bmatrix} (\dfrac{1}{5}c_1-\dfrac{2}{5}c_2)+(\dfrac{4}{5}c_1+\dfrac{2}{5}c_2)e^{5t} \\ -(\dfrac{2}{5}c_1-\dfrac{4}{5}c_2)+(\dfrac{2}{5}c_1+\dfrac{1}{5}c_2)e^{5t} \end{bmatrix},$$

By comparing the results, we can observe that in the Laplace transform solution, $(\dfrac{1}{5}c_1-\dfrac{2}{5}c_2)$ corresponds to " c_1" in the diagonalization solution, and

$(\dfrac{2}{5}c_1+\dfrac{1}{5}c_2)$ corresponds to " c_2" in the diagonalization solution.   Q.E.D.

Example 2

$$X' = AX, \quad X = \begin{bmatrix} x_1 \\ x_2 \\ x_3 \end{bmatrix}, \quad A = \begin{bmatrix} -1 & 1 & 0 \\ 1 & -1 & 0 \\ 0 & 0 & -2 \end{bmatrix}.$$

(1) Find the eigenvalues and eigenvectors of A.   (2)Find the general solution X.

Solution

(1) ① From $|A-\lambda I|=0 \Rightarrow \lambda^3+4\lambda^2+4\lambda=0$, $\lambda=0,-2,-2$,

② $\lambda=0$ substituting into $(A-\lambda I)V=O$

$$\Rightarrow \begin{bmatrix} -1 & 1 & 0 \\ 1 & -1 & 0 \\ 0 & 0 & -2 \end{bmatrix} \begin{bmatrix} v_1 \\ v_2 \\ v_3 \end{bmatrix} = O \Rightarrow V = c_1 \begin{bmatrix} 1 \\ 1 \\ 0 \end{bmatrix},$$

$\lambda=-2$ substituting into $(A-\lambda I)V=O$

$$\Rightarrow \begin{bmatrix} 1 & 1 & 0 \\ 1 & 1 & 0 \\ 0 & 0 & 0 \end{bmatrix} \begin{bmatrix} v_1 \\ v_2 \\ v_3 \end{bmatrix} = O \Rightarrow V = c_2 \begin{bmatrix} 1 \\ -1 \\ 0 \end{bmatrix} + c_3 \begin{bmatrix} 0 \\ 0 \\ 1 \end{bmatrix},$$

take $V_1 = \begin{bmatrix} 1 \\ 1 \\ 0 \end{bmatrix}$, $V_2 = \begin{bmatrix} 1 \\ -1 \\ 0 \end{bmatrix}$, $V_3 = \begin{bmatrix} 0 \\ 0 \\ 1 \end{bmatrix}$.

(2) $X = \begin{bmatrix} x_1 \\ x_2 \\ x_3 \end{bmatrix} = c_1 e^{0t} \begin{bmatrix} 1 \\ 1 \\ 0 \end{bmatrix} + c_2 e^{-2t} \begin{bmatrix} 1 \\ -1 \\ 0 \end{bmatrix} + c_3 e^{-2t} \begin{bmatrix} 0 \\ 0 \\ 1 \end{bmatrix} = \begin{bmatrix} c_1+c_2 e^{-2t} \\ c_1-c_2 e^{-2t} \\ c_3 e^{-2t} \end{bmatrix}$.   Q.E.D.

7-2-2 General Solution and Eigenvalues of the Coefficient Matrix.

▶ **Theorem 7-2-3**

The Eigenvalues of the Coefficient Matrix Are the Solutions of the Homogeneous System of Simultaneous Equations When They Are Complex Numbers.

If in $X' = AX$, the coefficient matrix A has complex conjugate eigenvalues $\lambda = \alpha \pm i\beta$, and their corresponding eigenvectors are $\xi = U \pm iV$, then the general solution can be expressed as

$$e^{\alpha t}\left[c_1^*(U \cos \beta t - V \sin \beta t) + c_2^*(U \sin \beta t + V \cos \beta t) \right] + \cdots.$$

Proof

In general solution $X = c_1 V_1 e^{\lambda_1 t} + c_2 V_2 e^{\lambda_2 t} + \cdots + c_n V_n e^{\lambda_n t}$, substituting $\lambda = \alpha \pm i\beta$ and $\xi = U \pm iV$, then

$$X = c_1(U + iV)e^{(\alpha + i\beta)t} + c_2(U - iV)e^{(\alpha - i\beta)t} + \cdots$$

$$= c_1(U + iV)e^{\alpha t}(\cos \beta t + i \sin \beta t) + c_2(U - iV)e^{\alpha t}(\cos \beta t - i \sin \beta t) + \cdots$$

$$= e^{\alpha t}\left[(c_1 + c_2)(U \cos \beta t - V \sin \beta t) + (c_1 - c_2)i(U \sin \beta t + V \cos \beta t) \right] + \cdots$$

$$= e^{\alpha t}\left[c_1^*(U \cos \beta t - V \sin \beta t) + c_2^*(U \sin \beta t + V \cos \beta t) \right] + \cdots.$$

Because $e^{\alpha t} \cos \beta t$, $e^{\alpha t} \sin \beta t$ are linearly independent from each other,

therefore the two linearly independent solutions corresponding to the complex conjugate eigenvalues $\lambda = \alpha \pm i\beta$ in $X' = AX$ are

$e^{\alpha t}(U \cos \beta t - V \sin \beta t)$, $e^{\alpha t}(U \sin \beta t + V \cos \beta t)$.   ◀

Example 3

Solve $\begin{bmatrix} x_1' \\ x_2' \end{bmatrix} = \begin{bmatrix} 7 & 10 \\ -4 & -5 \end{bmatrix}\begin{bmatrix} x_1 \\ x_2 \end{bmatrix}$.

Solution

Let $A = \begin{bmatrix} 7 & 10 \\ -4 & -5 \end{bmatrix}$, from $|A - \lambda I| = \lambda^2 - 2\lambda + 5 = 0$, $\lambda = 1 \pm 2i \Rightarrow \alpha = 1$, $\beta = 2$,

$\lambda = 1 + 2i$ substituting into $(A - \lambda I)\xi = O \Rightarrow \begin{bmatrix} 6 - 2i & 10 \\ -4 & -6 - 2i \end{bmatrix}\begin{bmatrix} \varsigma_1 \\ \varsigma_2 \end{bmatrix} = O$

$\Rightarrow \xi_1 = \begin{bmatrix} 5 \\ -3 + i \end{bmatrix} = \begin{bmatrix} 5 \\ -3 \end{bmatrix} + i\begin{bmatrix} 0 \\ 1 \end{bmatrix} = U + iV$, where $U = \begin{bmatrix} 5 \\ -3 \end{bmatrix}$, $V = \begin{bmatrix} 0 \\ 1 \end{bmatrix}$,

$\lambda = 1 - 2i$ substituting into $(A - \lambda I)\xi = 0 \Rightarrow \xi_2 = U - iV$,

then the solution of ODE is $X = e^t \{ c_1(U\cos 2t - V\sin 2t) + c_2(U\sin 2t + V\cos 2t) \}$.

Q.E.D.

▶ Theorem 7-2-4

The Solution of the Homogeneous System of Linear Equations with

Real Repeated Eigenvalues That Are Not Diagonalizable

If A has an eigenvalue λ_1 with multiplicity two but is not diagonalizable, then the linearly independent solutions of $X' = AX$ are $\phi_1 = V_1 e^{\lambda_1 t}$, $\phi_2 = V_1 t e^{\lambda_1 t} + V_2 e^{\lambda_1 t}$, where V_1 and V_2 satisfy $(A - \lambda_1 I)V_1 = 0$, $(A - \lambda_1 I)V_2 = V_1$.

Proof

Since there is only one corresponding eigenvector V_1, we can obtain only one linearly independent solution $\phi_1 = V_1 e^{\lambda_1 t}$ for the eigenvalue λ_1 in $X' = AX$. According to the theory of constant coefficient linear ODEs, the other linearly independent solution can be assumed as $\phi_2 = V_1 t e^{\lambda_1 t} + V_2 e^{\lambda_1 t}$. Substituting this into $X' = AX$ satisfies that,

that is $V_1 e^{\lambda_1 t} + V_1 \lambda_1 t e^{\lambda_1 t} + V_2 \lambda_1 e^{\lambda_1 t} = A(V_1 t e^{\lambda_1 t} + V_2 e^{\lambda_1 t})$.

Thus, the problem becomes the need to find a vector V_2 that is linearly independent from V_1. According to the theory of cyclic bases, there exists a vector V_2 that is linearly independent from V_1, satisfying: $(A - \lambda_1 I)V_2 = V_1$

In fact, the above equation contains only V_2 as an unknown, so we can solve for V_2 through a system of simultaneous equations. Therefore, $\phi_1 = V_1 e^{\lambda_1 t}$ and

$\phi_2 = V_1 t e^{\lambda_1 t} + V_2 e^{\lambda_1 t}$ are linearly independent solutions for the repeated eigenvalue λ_1 in $X' = AX$. Thus, the solutions of $X' = AX$ can be written as: $X = c_1\phi_1 + c_2\phi_2 + \cdots$.

◀

▶ Theorem 7-2-5

The Solution of a Homogeneous System of Simultaneous ODEs That Cannot Be Diagonalized.

If A has an eigenvalue λ_1 with multiplicity three and cannot be diagonalized, then the general solution of $X' = AX$ is

$$X = c_1\phi_1 + c_2\phi_2 + c_3\phi_3,$$

where $\phi_1 = V_1 e^{\lambda_1 t}$, $\phi_2 = V_1 t e^{\lambda_1 t} + V_2 e^{\lambda_1 t}$, $\phi_3 = \frac{1}{2}V_1 t^2 e^{\lambda_1 t} + V_2 t e^{\lambda_1 t} + V_3 e^{\lambda_1 t}$, and

V_1, V_2, V_3 satisfy $(A - \lambda_1 I)V_1 = 0$, $(A - \lambda_1 I)V_2 = V_1$, $(A - \lambda_1 I)V_3 = V_2$.

Proof

Assuming V_1 is the corresponding eigenvector for the eigenvalue λ_1 in $X' = AX$, the solution for the eigenvalue λ_1 is $\phi_1 = V_1 e^{\lambda_1 t}$. According to the theory of constant coefficient linear ODEs, another linearly independent solution can be assumed as $\phi_2 = V_1 t e^{\lambda_1 t} + V_2 e^{\lambda_1 t}$, which allows us to find V_2 when substituting it into $X' = AX$.

Similarly, let another linearly independent solution be $\phi_3 = \dfrac{1}{2} V_1 t^2 e^{\lambda_1 t} + V_2 t e^{\lambda_1 t} + V_3 e^{\lambda_1 t}$, and by substituting it into $X' = AX$, we can find V_3. Therefore $\phi_1 = V_1 e^{\lambda_1 t}$, $\phi_2 = V_1 t e^{\lambda_1 t} + V_2 e^{\lambda_1 t}$ and $\phi_3 = \dfrac{1}{2} V_1 t^2 e^{\lambda_1 t} + V_2 t e^{\lambda_1 t} + V_3 e^{\lambda_1 t}$ are linearly independent solutions for the multiplicity three eigenvalue λ_1 in $X' = AX$. Thus, the solutions of $X' = AX$ can be written as $X = c_1 \phi_1 + c_2 \phi_2 + c_3 \phi_3$. ◀

Example 4

Solve the system of simultaneous equations $\begin{cases} \dfrac{dx_1}{dt} = 3x_1 - x_2 \\ \dfrac{dx_2}{dt} = 9x_1 - 3x_2 \end{cases}$.

Solution

Rewriting the original equation as $X' = AX$, where $A = \begin{bmatrix} 3 & -1 \\ 9 & -3 \end{bmatrix}$, $X = \begin{bmatrix} x_1 \\ x_2 \end{bmatrix}$,

from $|A - \lambda I| = 0 \Rightarrow \lambda^2 = 0$, we obtain the eigenvalues $\lambda = 0$,

and substituting $\lambda_1 = 0$ into $(A - \lambda I)V = O \Rightarrow \begin{bmatrix} 3 & -1 \\ 9 & -3 \end{bmatrix} \begin{bmatrix} v_1 \\ v_2 \end{bmatrix} = O$, we get $V_1 = \begin{bmatrix} 1 \\ 3 \end{bmatrix}$,

we obtain one linearly independent solution: $\phi_1 = V_1 e^{\lambda_1 t} = \begin{bmatrix} 1 \\ 3 \end{bmatrix} e^{0t} = \begin{bmatrix} 1 \\ 3 \end{bmatrix}$.

For the other linearly independent solution, we can set $\phi_2 = V_1 t e^{\lambda_1 t} + V_2 e^{\lambda_1 t}$, where $(A - \lambda_1 I)V_2 = V_1$,

that is $\begin{bmatrix} 3 & -1 \\ 9 & -3 \end{bmatrix} \begin{bmatrix} v_1 \\ v_2 \end{bmatrix} = \begin{bmatrix} 1 \\ 3 \end{bmatrix}$, we find that $V_2 = \begin{bmatrix} 0 \\ -1 \end{bmatrix}$. $\phi_2 = \begin{bmatrix} 1 \\ 3 \end{bmatrix} t e^{0t} + \begin{bmatrix} 0 \\ -1 \end{bmatrix} e^{0t} = \begin{bmatrix} t \\ 3t-1 \end{bmatrix}$, the general solution of the ODE is

$$X = \begin{bmatrix} x_1 \\ x_2 \end{bmatrix} = c_1 \phi_1 + c_2 \phi_2 = c_1 \begin{bmatrix} 1 \\ 3 \end{bmatrix} + c_2 \begin{bmatrix} t \\ 3t-1 \end{bmatrix} = \begin{bmatrix} c_1 + c_2 t \\ 3c_1 + c_2(3t-1) \end{bmatrix}.$$

Q.E.D.

7-2-3 The Solution of a Second-Order Homogeneous System of Simultaneous Differential Equations.

We refer to the system of simultaneous equations with the following form as a second-order homogeneous system of simultaneous differential equations:

$$X''_{n\times1} = A_{n\times n}X_{n\times1} .$$

Starting from Section 7-2-1, we defined an n-th order homogeneous system of simultaneous differential equations as $X^{(n)}_{n\times1} = A_{n\times n}X_{n\times1}$. However, once the order exceeds two, applying ODE's solution methods becomes challenging and is beyond the scope of this book. Here, we continue to assume that the matrix $A_{n\times n}$ is diagonalizable.

Similarly, based on the concept of solving the second-order constant coefficient ODE $y'' = ay$, we can set $X = Ve^{wt}$, and compute then obtain $X' = wVe^{wt}$, $X'' = w^2Ve^{wt} \Rightarrow w^2Ve^{wt} = AVe^{wt}$. By rearranging the equation, we get $(A - w^2I)V = O$ which forms an eigenvalue system, where w^2 is the eigenvalue of A, and V is the corresponding eigenvector. If we assume that the matrix A is diagonalizable, and the eigenvalues of A are $\lambda_1, \lambda_2, \cdots, \lambda_n$, and their corresponding linearly independent eigenvectors are $V_1, V_2, \cdots, V_n$, then $w^2 = \lambda_1, \lambda_2 \cdots \lambda_n$. From $w^2 = \lambda_i \Rightarrow w = \pm\sqrt{\lambda_i}$, $\phi_i = V_i(c_ie^{\sqrt{\lambda_i}t} + d_ie^{-\sqrt{\lambda_i}t})$, the general solution of $X' = AX$ is

$$X = \phi_1 + \phi_2 + \cdots + \phi_n = V_1[c_1e^{\sqrt{\lambda_1}t} + d_1e^{-\sqrt{\lambda_1}t}] + \cdots + V_n[c_ne^{\sqrt{\lambda_n}t} + d_ne^{-\sqrt{\lambda_n}t}]^{[3, 4]} .$$

---

[3] If λ_j is a real number but $\lambda_j < 0$, then its corresponding solution can also be written as $\phi_j = V_j[c_j \cos\sqrt{|\lambda_j|}t + d_j \sin\sqrt{|\lambda_j|}t]$.

[4] If $\lambda_j = 0$, then its solution can be written as $\phi_j = V_j[c_j + d_jt]$.

Example 5

Solve $\begin{cases} y_1'' = -5y_1 + 2y_2 \\ y_2'' = 2y_1 - 2y_2 \end{cases}$.

Solution

The original equation write as $\begin{bmatrix} y_1'' \\ y_2'' \end{bmatrix} = \begin{bmatrix} -5 & 2 \\ 2 & -2 \end{bmatrix} \begin{bmatrix} y_1 \\ y_2 \end{bmatrix}$, which means $Y'' = AY$,

where $A = \begin{bmatrix} -5 & 2 \\ 2 & -2 \end{bmatrix}$, $Y = \begin{bmatrix} y_1 \\ y_2 \end{bmatrix}$.

$\therefore$ Find the eigenvalues of A are $w^2 = -1, -6$, and discuss the cases separately below,

$$w^2 = -1 \Rightarrow (A - w^2 I)V = O \Rightarrow w = \pm i \Rightarrow \begin{bmatrix} -4 & 2 \\ 2 & -1 \end{bmatrix} \begin{bmatrix} x_1 \\ x_2 \end{bmatrix} = O \Rightarrow V_1 = \begin{bmatrix} 1 \\ 2 \end{bmatrix},$$

$$w^2 = -6 \Rightarrow (A - w^2 I)V = O \Rightarrow w = \pm\sqrt{6}i \Rightarrow \begin{bmatrix} 1 & 2 \\ 2 & 4 \end{bmatrix} \begin{bmatrix} x_1 \\ x_2 \end{bmatrix} = O \Rightarrow V_2 = \begin{bmatrix} 2 \\ -1 \end{bmatrix},$$

the general solution is $Y = \begin{bmatrix} 1 \\ 2 \end{bmatrix}(c_1 \cos t + d_1 \sin t) + \begin{bmatrix} 2 \\ -1 \end{bmatrix}(c_2 \cos \sqrt{6}\,t + d_2 \sin \sqrt{6}\,t)$.

Q.E.D.

7-2 Exercises

Basic questions

1. In the following (1)~(3) questions, find the fundamental matrix Φ for the system of simultaneous equations $X'' = AX$ and determine its general solution.

    (1) $A = \begin{bmatrix} 1 & 2 \\ 12 & -1 \end{bmatrix}$.

    (2) $A = \begin{bmatrix} 2 & 3 \\ \frac{1}{3} & 2 \end{bmatrix}$; the initial condition $X(0) = \begin{bmatrix} 0 \\ 2 \end{bmatrix}$.

    (3) $A = \begin{bmatrix} 1 & -2 & 2 \\ -2 & 1 & -2 \\ 2 & -2 & 1 \end{bmatrix}$.

2. In the following(1)~(2) questions, solve the system of simultaneous equations $X'' = AX$.

    (1) $A = \begin{bmatrix} -8 & 2 \\ 2 & -5 \end{bmatrix}$.

    (2) $A = \begin{bmatrix} -5 & 2 \\ 2 & -5 \end{bmatrix}$.

Advanced questions

1. Find the fundamental matrix Φ for the system of simultaneous equations $X' = AX$ and determine its general solution, where A as follows.

    (1) $A = \begin{bmatrix} 4 & 1 & 2 \\ 1 & 0 & 0 \\ 2 & 0 & 0 \end{bmatrix}$.

    (2) $A = \begin{bmatrix} 3 & 4 \\ 3 & 2 \end{bmatrix}$, $X(0) = \begin{bmatrix} 6 \\ 1 \end{bmatrix}$.

    (3) $A = \begin{bmatrix} -4 & 1 & 1 \\ 1 & 5 & -1 \\ 0 & 1 & -3 \end{bmatrix}$, $X(0) = \begin{bmatrix} 9 \\ 7 \\ 0 \end{bmatrix}$.

2. Find the system of simultaneous equations $X' = AX$, where A as follows.

    (1) $A = \begin{bmatrix} 1 & 1 \\ -1 & 1 \end{bmatrix}$.

    (2) $A = \begin{bmatrix} 1 & -1 & 0 \\ 0 & 0 & 1 \\ -3 & -1 & 1 \end{bmatrix}$.

3. Solve the system of simultaneous equations $X' = AX$, where $A = \begin{bmatrix} 1 & 3 \\ -3 & 7 \end{bmatrix}$.

4. Solve the system of simultaneous equations $X'' = AX$, where
    $$A = \begin{bmatrix} -37 & 12 \\ 12 & -37 \end{bmatrix};$$
    the initial condition $X(0) = \begin{bmatrix} 2 \\ 1 \end{bmatrix}$, $X'(0) = \begin{bmatrix} 1 \\ 2 \end{bmatrix}$.

7-3 Diagonalization of Matrix for Solving Non-Homogeneous System of Simultaneous Differential Equations

If the source function of the system of simultaneous differential equations is non-zero, we still start by diagonalizing the coefficient matrix, transforming the problem into a non-homogeneous linear ODE problem. Specifically, through diagonalization (considering diagonalizable coefficient matrices throughout this section), we decouple the coupled system of simultaneous differential equations into individual ODEs, each involving a single unknown function. Thus, we can further reduce it to the problem of finding particular solutions for the ODEs, which can be accomplished using methods such as the method of inverse differential operator (as discussed in Chapters 1 and 2). The specific process is as follows.

7-3-1 A System of Simultaneous First-Order Non-Homogeneous ODE

Next, we will introduce how to solve a system of simultaneous differential equations with the following form:

$$
\begin{bmatrix} x_1'(t) \\ x_2'(t) \\ \vdots \\ x_n'(t) \end{bmatrix} = \begin{bmatrix} a_{11} & a_{12} & \cdots & a_{1n} \\ a_{21} & a_{22} & & \vdots \\ \vdots & & \ddots & \vdots \\ a_{n1} & \cdots & \cdots & a_{nn} \end{bmatrix} \begin{bmatrix} x_1(t) \\ x_2(t) \\ \vdots \\ x_n(t) \end{bmatrix} + \begin{bmatrix} b_1(t) \\ b_2(t) \\ \vdots \\ b_n(t) \end{bmatrix}.
$$

The Solution

STEP 1   Let the eigenvalues of A be $\lambda_1, \lambda_2, \cdots, \lambda_n$, and their corresponding eigenvectors be $V_1, V_2, \cdots, V_n$. Let the transition matrix $P = [V_1 \ V_2 \cdots V_n]$,

then $P^{-1}AP = D = \begin{bmatrix} \lambda_1 & & O \\ & \ddots & \\ O & & \lambda_n \end{bmatrix}$.

STEP 2   Next, we perform a change of coordinates using the transition matrix to achieve decoupling. Therefore let $X = PY$, where
$Y = [y_1(t) \quad y_2(t) \quad \cdots \quad y_n(t)]^T$. Differentiating both sides, we get $X' = PY'$
and substituting this into the original equation $X' = AX + B(t)$
$\Rightarrow PY' = APY + B(t) \Rightarrow Y' = P^{-1}APY + P^{-1}B(t)$
$\Rightarrow Y' = DY + P^{-1}B(t)$, we obtain

$$\begin{bmatrix} y_1' \\ y_2' \\ \vdots \\ y_n' \end{bmatrix} = \begin{bmatrix} \lambda_1 & & & O \\ & \lambda_2 & & \\ & & \ddots & \\ O & & & \lambda_n \end{bmatrix} \begin{bmatrix} y_1 \\ y_2 \\ \vdots \\ y_n \end{bmatrix} + \begin{bmatrix} b_1^*(t) \\ b_2^*(t) \\ \vdots \\ b_n^*(t) \end{bmatrix}.$$

To solve the decoupled system, we can use the methods for solving ordinary differential equations

$$\begin{cases} y_1(t) = c_1 e^{\lambda_1 t} + \xi_1(t) \\ y_2(t) = c_2 e^{\lambda_2 t} + \xi_2(t) \\ \vdots \\ y_n(t) = c_n e^{\lambda_n t} + \xi_n(t) \end{cases},$$

where $\xi_k(t) = \dfrac{1}{D - \lambda_k} b_k^*(t) = e^{\lambda_k t} \displaystyle\int e^{-\lambda_k t} \cdot b_k^*(t) dt$; $k = 1, 2, 3, \cdots, n$.

STEP 3 Therefore $X = PY = \begin{bmatrix} V_1 & V_2 & \cdots & V_n \end{bmatrix} \begin{bmatrix} c_1 e^{\lambda_1 t} + \xi_1(t) \\ \vdots \\ c_n e^{\lambda_n t} + \xi_n(t) \end{bmatrix}$,

that is $X = c_1 V_1 e^{\lambda_1 t} + \cdots + c_n V_n e^{\lambda_n t} + V_1 \xi_1(t) + \cdots + V_n \xi_n(t)$,

the homogeneous solution is $X_h = c_1 V_1 e^{\lambda_1 t} + \cdots + c_n V_n e^{\lambda_n t}$, and the particular solution is $X_p = V_1 \xi_1(t) + \cdots + V_n \xi_n(t)$.

By observing the homogeneous solution $X_h = c_1 V_1 e^{\lambda_1 t} + \cdots + c_n V_n e^{\lambda_n t}$, we can see that it is consistent with the results obtained in the previous section.

Example 1

Use matrix diagonalization to solve $\begin{bmatrix} x_1' \\ x_2' \end{bmatrix} = \begin{bmatrix} 4 & 2 \\ 2 & 1 \end{bmatrix} \begin{bmatrix} x_1 \\ x_2 \end{bmatrix} + \begin{bmatrix} 3e^t \\ e^t \end{bmatrix}$.

Solution

Let $X = \begin{bmatrix} x_1 \\ x_2 \end{bmatrix}$, $A = \begin{bmatrix} 4 & 2 \\ 2 & 1 \end{bmatrix}$, $B = \begin{bmatrix} 3e^t \\ e^t \end{bmatrix}$ then we obtain the form $X' = AX + B(t)$.

Therefore, following the discussions in the previous text, we proceed to solve step by step as follows:

(1) From $|A - \lambda I| = 0 \Rightarrow \lambda^2 - 5\lambda = 0 \Rightarrow \lambda = 0, 5$ (thus A is diagonalizable).

(2) $\lambda = 0$ substituting into $(A - \lambda I)V = O \Rightarrow \begin{bmatrix} 4 & 2 \\ 2 & 1 \end{bmatrix} \begin{bmatrix} v_1 \\ v_2 \end{bmatrix} = O \Rightarrow$ we take the

eigenvector $V_1 = \begin{bmatrix} 1 \\ -2 \end{bmatrix}$,

$\lambda = 5$ substituting into $(A - \lambda I)V = O \Rightarrow \begin{bmatrix} -1 & 2 \\ 2 & -4 \end{bmatrix} \begin{bmatrix} v_1 \\ v_2 \end{bmatrix} = O \Rightarrow$ taking the

eigenvector $V_2 = \begin{bmatrix} 2 \\ 1 \end{bmatrix}$.

(3) As a result, we obtain the transition matrix $P = \begin{bmatrix} 1 & 2 \\ -2 & 1 \end{bmatrix}$, and $P^{-1} = \frac{1}{5} \begin{bmatrix} 1 & -2 \\ 2 & 1 \end{bmatrix}$, so

$Y' = P^{-1}APY + P^{-1}B(t) = \begin{bmatrix} 0 & 0 \\ 0 & 5 \end{bmatrix} \begin{bmatrix} y_1 \\ y_2 \end{bmatrix} + \frac{1}{5} \begin{bmatrix} 1 & -2 \\ 2 & 1 \end{bmatrix} \begin{bmatrix} 3e^t \\ e^t \end{bmatrix}$, and using the formula,

we have

$$\begin{bmatrix} x_1 \\ x_2 \end{bmatrix} = X = c_1 \begin{bmatrix} 1 \\ -2 \end{bmatrix} + c_2 \begin{bmatrix} 2 \\ 1 \end{bmatrix} e^{5t} + \begin{bmatrix} -\dfrac{1}{2}e^t \\ -\dfrac{3}{4}e^t \end{bmatrix} = \begin{bmatrix} c_1 + 2c_2 e^{5t} - \dfrac{1}{2}e^t \\ -2c_1 + c_2 e^{5t} - \dfrac{3}{4}e^t \end{bmatrix}.$$

Q.E.D.

7-3-2   A Second-Order System of Simultaneous Differential Equations

Here, we further consider the second-order situation

$$X''_{n\times 1} = A_{n\times n}X_{n\times 1} + B(t).$$

The Solution

Let the eigenvalues of A be $\lambda_1, \lambda_2, \cdots, \lambda_n$, and their corresponding eigenvectors be $V_1, V_2, \cdots, V_n$. Let $P = [V_1 \; V_2 \; \cdots \; V_n]$ be the transition matrix and $X = PY$, then

$$\begin{bmatrix} y_1'' \\ y_2'' \\ \vdots \\ y_n'' \end{bmatrix} = \begin{bmatrix} \lambda_1 & & & O \\ & \lambda_2 & & \\ & & \ddots & \\ O & & & \lambda_n \end{bmatrix} \begin{bmatrix} y_1 \\ y_2 \\ \vdots \\ y_n \end{bmatrix} + \begin{bmatrix} b_1^*(t) \\ b_2^*(t) \\ \vdots \\ b_n^*(t) \end{bmatrix}.$$ According to the theory of second-order

linear ODEs,

$$\begin{cases} y_1(t) = c_1 e^{\sqrt{\lambda_1}t} + d_1 e^{-\sqrt{\lambda_1}t} + \xi_1(t) \\ \qquad \vdots \\ y_n(t) = c_n e^{\sqrt{\lambda_n}t} + d_n e^{-\sqrt{\lambda_n}t} + \xi_n(t) \end{cases}$$, where ξ_i is the particular solution. The eigenvalues λ

appear inside the square root, and we can classify them into three cases $\lambda > 0$, $\lambda = 0$, $\lambda < 0$:

(1) $\lambda > 0$

$$X = (c_1 e^{\sqrt{\lambda_1}t} + d_1 e^{-\sqrt{\lambda_1}t})V_1 + \cdots + (c_n e^{\sqrt{\lambda_n}t} + d_n e^{-\sqrt{\lambda_n}t})V_n$$
$$+ \xi_1(t)V_1 + \xi_2(t)V_2 + \cdots + \xi_n(t)V_n$$

where $X_h = (c_1 e^{\sqrt{\lambda_1}t} + d_1 e^{-\sqrt{\lambda_1}t})V_1 + \cdots + (c_n e^{\sqrt{\lambda_n}t} + d_n e^{-\sqrt{\lambda_n}t})V_n$ is the homogeneous solution and the particular solution is

$$X_p = \xi_1(t)V_1 + \xi_2(t)V_2 + \cdots + \xi_n(t)V_n.$$

(2) $\lambda_j < 0$

The homogeneous solution is the same as in the previous section, but the coefficient functions for V_j are replaced by the following functions

$$(c_j \cos\sqrt{|\lambda_j|}\,t + d_j \sin\sqrt{|\lambda_j|}\,t).$$

(3) $\lambda_j = 0$

The homogeneous solution is the same as in the previous section, but the coefficient functions for V_j are replaced by the following functions

$$(c_j + d_j t).$$

Example 2

Use matrix diagonalization to solve $\begin{cases} y_1'' = -3y_1 - y_2 + \sin^2 t \\ y_2'' = -2y_1 - 2y_2 + \cos^2 t \end{cases}$.

Solution

(1) From $\begin{bmatrix} y_1'' \\ y_2'' \end{bmatrix} = \begin{bmatrix} -3 & -1 \\ -2 & -2 \end{bmatrix}\begin{bmatrix} y_1 \\ y_2 \end{bmatrix} + \begin{bmatrix} \sin^2 t \\ \cos^2 t \end{bmatrix}$, we obtain the form $Y'' = AY + B(t)$.

(2) $A = \begin{bmatrix} -3 & -1 \\ -2 & -2 \end{bmatrix} \Rightarrow |A - \lambda I| = 0 \Rightarrow \lambda^2 + 5\lambda + 4 = 0,\ \lambda = -1, -4$.

(3) $\lambda = -1$ substituting into $(A - \lambda I)V = O \Rightarrow \begin{bmatrix} -2 & -1 \\ -2 & -1 \end{bmatrix}\begin{bmatrix} v_1 \\ v_2 \end{bmatrix} = O \Rightarrow$ we take the

eigenvector $V_1 = \begin{bmatrix} 1 \\ -2 \end{bmatrix}$,

$\lambda = -4$ substituting into $(A - \lambda I)V = O \Rightarrow \begin{bmatrix} 1 & -1 \\ -2 & 2 \end{bmatrix}\begin{bmatrix} v_1 \\ v_2 \end{bmatrix} = O \Rightarrow$ we take the

eigenvector $V_2 = \begin{bmatrix} 1 \\ 1 \end{bmatrix}$.

(4) We obtain the transition matrix $P = \begin{bmatrix} 1 & 1 \\ -2 & 1 \end{bmatrix}$, then $P^{-1} = \dfrac{1}{3}\begin{bmatrix} 1 & -1 \\ 2 & 1 \end{bmatrix}$,

and $Z'' = P^{-1}APZ + P^{-1}B(t) = \begin{bmatrix} -1 & 0 \\ 0 & -4 \end{bmatrix}\begin{bmatrix} z_1 \\ z_2 \end{bmatrix} + \begin{bmatrix} \dfrac{1}{3}(\sin^2 t - \cos^2 t) \\ \dfrac{1}{3}(2\sin^2 t + \cos^2 t) \end{bmatrix}$, therefore, from the formulas given in the discussion for eigenvalues $\lambda_j < 0$,

$$\begin{cases} z_1(t) = c_1 \cos t + d_1 \sin t + \dfrac{1}{9}\cos 2t \\ z_2(t) = c_2 \cos 2t + d_2 \sin 2t + \dfrac{1}{8} - \dfrac{1}{24} t \sin 2t \end{cases}$$

Therefore, we have

$$Y = PZ = \begin{bmatrix} 1 & 1 \\ -2 & 1 \end{bmatrix} \begin{bmatrix} c_1 \cos t + d_1 \sin t + \dfrac{1}{9}\cos 2t \\ c_2 \cos 2t + d_2 \sin 2t + \dfrac{1}{8} - \dfrac{1}{24} t \sin 2t \end{bmatrix}$$

$$= (c_1 \cos t + d_1 \sin t) \begin{bmatrix} 1 \\ -2 \end{bmatrix} + (c_2 \cos 2t + d_2 \sin 2t) \begin{bmatrix} 1 \\ 1 \end{bmatrix} + \begin{bmatrix} \dfrac{1}{8} + \dfrac{1}{9}\cos 2t - \dfrac{1}{24} t \sin 2t \\ \dfrac{1}{8} - \dfrac{2}{9}\cos 2t - \dfrac{1}{24} t \sin 2t \end{bmatrix}.$$

Q.E.D.

7-3 Exercises

Basic questions

Use matrix diagonalization to solve the following system of simultaneous ODE.

1. $\begin{bmatrix} x_1' \\ x_2' \end{bmatrix} = \begin{bmatrix} 0 & -1 \\ -9 & 0 \end{bmatrix} \begin{bmatrix} x_1 \\ x_2 \end{bmatrix} + \begin{bmatrix} 1 \\ 0 \end{bmatrix}$.

2. $\begin{cases} x_1' = x_2 \\ x_2' = -6x_1 + 5x_2 + e^{4t} \end{cases}$;

    the initial condition $x_1(0) = 0$, $x_2(0) = 1$.

3. $\begin{bmatrix} x_1' \\ x_2' \end{bmatrix} = \begin{bmatrix} 3 & 3 \\ 1 & 5 \end{bmatrix} \begin{bmatrix} x_1 \\ x_2 \end{bmatrix} + \begin{bmatrix} 8 \\ 4e^{3t} \end{bmatrix}$.

4. $\begin{cases} y_1' = y_1 + y_2 \\ y_2' = -2y_1 + 4y_2 + 1 \end{cases}$ and $\begin{cases} y_1(0) = 1 \\ y_2(0) = 0 \end{cases}$.

5. $\begin{bmatrix} x_1' \\ x_2' \\ x_3' \end{bmatrix} = \begin{bmatrix} 1 & 1 & 0 \\ 1 & 1 & 0 \\ 0 & 0 & 3 \end{bmatrix} \begin{bmatrix} x_1 \\ x_2 \\ x_3 \end{bmatrix} + \begin{bmatrix} e^t \\ e^{2t} \\ te^{3t} \end{bmatrix}$.

Advanced questions

Use matrix diagonalization to solve the following system of simultaneous ODE.

1. $\begin{bmatrix} x_1' \\ x_2' \end{bmatrix} = \begin{bmatrix} -2 & 1 \\ -4 & 3 \end{bmatrix} \begin{bmatrix} x_1 \\ x_2 \end{bmatrix} + \begin{bmatrix} 0 \\ 10\cos t \end{bmatrix}$.

2. $\begin{bmatrix} x_1' \\ x_2' \end{bmatrix} = \begin{bmatrix} -3 & 1 \\ 1 & -3 \end{bmatrix} \begin{bmatrix} x_1 \\ x_2 \end{bmatrix} + \begin{bmatrix} -6e^{-2t} \\ 2e^{-2t} \end{bmatrix}$.

3. $\begin{bmatrix} x_1' \\ x_2' \end{bmatrix} = \begin{bmatrix} 1 & 1 \\ 1 & 1 \end{bmatrix} \begin{bmatrix} x_1 \\ x_2 \end{bmatrix} + \begin{bmatrix} 6e^{3t} \\ 4 \end{bmatrix}$.

4. $\begin{bmatrix} x_1' \\ x_2' \end{bmatrix} = \begin{bmatrix} 2 & 1 \\ 4 & -1 \end{bmatrix} \begin{bmatrix} x_1 \\ x_2 \end{bmatrix} + \begin{bmatrix} e^t \\ -e^t \end{bmatrix}$,

    $x_1(0) = 1$, $x_2(0) = 0$.

5. $\begin{bmatrix} x_1'' \\ x_2'' \end{bmatrix} = \begin{bmatrix} -5 & 2 \\ 2 & -5 \end{bmatrix} \begin{bmatrix} x_1 \\ x_2 \end{bmatrix} + \begin{bmatrix} 1 \\ 9 \end{bmatrix}$.

8 Vector Function Analysis

Heaviside
(1850~1925, Britain)

Vector function analysis, also known as vector calculus, was developed by Josiah Willard Gibbs and Oliver Heaviside in the late 19th century. It has been widely applied in the fields of physics and engineering, particularly in electromagnetism, gravitational fields, and fluid dynamics.

Learning Objectives

8-1
Vector Functions and Differentiation

8-1-1 Mastering the fundamental properties of vector functions: continuity, differentiability

8-1-2 Mastering coordinate systems in $\mathbb{R}^3$: polar coordinates, cylindrical coordinates, spherical coordinates.

8-1-3 Mastering the fundamental properties of curves in $\mathbb{R}^3$: physical interpretation, arc length of curves.

8-2
Directional Derivative

8-2-1 Mastering the fundamental properties and operations of operators ∇: gradient, divergence, curl

8-2-2 Mastering directional derivatives: calculation formulas, extrema

8-3
Line Integral

8-3-1 Mastering curve classification: smooth curves, piecewise-smooth curves, self-intersections, closed curves

8-3-2 Mastering the calculation of scalar function line integrals in $\mathbb{R}^3$

8-3-3 Mastering the calculation of vector function line integrals in $\mathbb{R}^3$

8-3-4 Understanding the properties of conservative vector fields : existence, integrals in simply (multiply) connected domains

8-4
Multiple Integral

8-4-1 Proficiency in double integrals on $\mathbb{R}^3$

8-4-2 Proficiency in triple integrals on $\mathbb{R}^3$

8-4-3 Proficienct in variable transformations for triple integrals: cylindrical coordinates

8-4-4 Proficient in variable transformations for triple integrals: spherical coordinates

8-5
Surface Integral

8-5-1 Understanding surface integrals: definition, physical properties

8-5-2 Proficiency in algorithms for dA: vector form, level curves (contour lines), and determining positive/negative orientation

8-6
Green's Theorem

8-6-1 Mastering Green's theorem: general computation, physical significance

8-6-2 Mastering the application of Green's theorem: area calculation

| **8-7** Gauss's Divergence Theorem | 8-7-1 | Mastering Gauss's divergence theorem: general computation, flux across a closed surface algorithm, physical significance |
| | 8-7-2 | Mastering the other two forms of Gauss's divergence theorem |

| **8-8** Stokes' Theorem | 8-8-1 | Understanding basic terminology: circulation, vorticity, and curl |
| | 8-8-2 | Mastering the calculation of Stokes' theorem: general computation, physical significance |

ExampleVideo

This chapter begins with vector functions and introduces the concepts of calculus. It covers the calculus of vector functions and introduces the common operator ∇(Del) used in vector functions. Then, it discusses the gradient, divergence, and curl of vector functions. Lastly, it introduces the three fundamental integral theorems in vector calculus (Green's theorem, Gauss's divergence theorem, and Stokes' theorem) in a progressive manner, allowing readers to gain a complete understanding of vector function analysis.

8-1   Vector Functions and Differentiation

In this section, after a brief review of the concept of vector functions, we will introduce different coordinate systems and their relationship to the Cartesian coordinates. Finally, we will discuss how to define the differentiation of vector functions based on the differentiation of single-variable functions, and how to use it to calculate the arc length of curves in space.

8-1-1   Vector Functions

In physical systems, fields are divided into scalar fields and vector fields. A scalar field only has magnitude without direction. For example, the temperature field $T(x, y, z)$ at any position within a classroom is a scalar function that depends on the position (x, y, z), and it is called a scalar field. On the other hand, if the field represents a physical quantity with both magnitude and direction, it is called a vector field. For instance, when air flows inside the classroom, the velocity field of air particles $\vec{V}(x, y, z)$ is a vector field that depends on the position, and it is a function with both magnitude and direction, known as a vector function.

Next, we will introduce this type of vector function, which can be further classified into single-variable vector functions and multi-variable vector functions based on the number of independent variables.

▶ **Definition 8-1-1**

Single Variable Vector Functions

Assuming that $V_1(t)$, $V_2(t)$ and $V_3(t)$ are single-variable scalar functions defined on a certain interval I. Then, $\vec{V}(t) = V_1(t)\,\vec{i} + V_2(t)\,\vec{j} + V_2(t)\vec{k}$ is called a single-variable vector function, denoted as $\vec{V}(t)$.   ◀

In terms of function behavior, the continuity of a vector function is consistent with its components and possesses the following properties.

1. Continuity

Given the aforementioned notation, its components $V_1(t)$, $V_2(t)$, and $V_3(t)$ is continuous in the interval I if and only if $\vec{V}(t)$ is continuous in the interval I.

▶ **Definition 8-1-2**

Derivative

Assuming that the first order derivatives of $V_1(t)$, $V_2(t)$, $V_3(t)$ exist, the differentiation of the vector function is defined as: $\dfrac{d\vec{V}(t)}{dt} = \lim\limits_{\Delta t \to 0} \dfrac{\vec{V}(t+\Delta t) - \vec{V}(t)}{\Delta t} = \dfrac{dV_1}{dt}\vec{i} + \dfrac{dV_2}{dt}\vec{j} + \dfrac{dV_3}{dt}\vec{k}$.

Usually, the notation $\dfrac{d\vec{V}}{dt}$ or $\vec{V}'(t)$ is used to represent the differentiation of a vector function.   ◀

By utilizing the linearity and multiplication rules of differentiation, among other properties, it is not difficult to observe that many of the properties related to differentiation that we have learned in calculus courses also hold true for vector functions. Readers can use the properties of differentiation for scalar functions to prove the following theorem:

▶ **Theorem 8-1-1**

Properties of Single-Variable Vector Functions

1. $\dfrac{d}{dt}(\vec{A}(t) \pm \vec{B}(t)) = \dfrac{d\vec{A}(t)}{dt} \pm \dfrac{d\vec{B}(t)}{dt}$.

2. $\dfrac{d}{dt}(\phi(t)\vec{B}(t)) = \dfrac{d\phi(t)}{dt}\vec{B}(t) + \phi(t)\dfrac{d\vec{B}(t)}{dt}$.

3. $\dfrac{d}{dt}(\vec{A}(t) \cdot \vec{B}(t)) = \dfrac{d}{dt}\vec{A}(t) \cdot \vec{B}(t) + \vec{A}(t) \cdot \dfrac{d\vec{B}(t)}{dt}$.

4. $\dfrac{d}{dt}(\vec{A}(t) \times \vec{B}(t)) = \dfrac{d\vec{A}(t)}{dt} \times \vec{B}(t) + \vec{A}(t) \times \dfrac{d\vec{B}(t)}{dt}$.

5. $\dfrac{d\vec{A}(t)}{ds} = \dfrac{d\vec{A}(t)}{dt} \cdot \dfrac{dt}{ds}$.   ◀

For example, in the familiar helix vector function $\overrightarrow{A}(t) = \sin t\overrightarrow{i} + \cos t\overrightarrow{j} + 3t\overrightarrow{k}$,

$\overrightarrow{A}(0) = \sin 0\overrightarrow{i} + \cos 0\overrightarrow{j} + 3 \cdot 0\overrightarrow{k} = \overrightarrow{j}$,

$\overrightarrow{A}(\dfrac{\pi}{2}) = \sin\dfrac{\pi}{2}\overrightarrow{i} + \cos\dfrac{\pi}{2}\overrightarrow{j} + 3 \cdot \dfrac{\pi}{2}\overrightarrow{k} = \overrightarrow{i} + \dfrac{3\pi}{2}\overrightarrow{k}$,

and at every point on the curve $\overrightarrow{A}(t)$, there exists a natural

tangent vector $\dfrac{d\overrightarrow{A}(t)}{dt} = \dfrac{d(\sin t)}{dt}\overrightarrow{i} + \dfrac{d(\cos t)}{dt}\overrightarrow{j} + \dfrac{d(3t)}{dt}\overrightarrow{k}$,

for instance $\dfrac{d\overrightarrow{A}(0)}{dt} = \cos 0\overrightarrow{i} - \sin 0\overrightarrow{j} + 3\overrightarrow{k} = \overrightarrow{i} + 3\overrightarrow{k}$, as

shown in Figure 8-1.

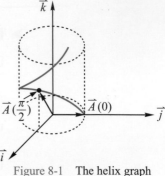

Figure 8-1   The helix graph

Example 1

Let $\overrightarrow{A} = \overrightarrow{A}(t)$, and $|\overrightarrow{A}(t)| = $ constants, $\dfrac{d\overrightarrow{A}}{dt} \neq \overrightarrow{0}$, prove that $\dfrac{d\overrightarrow{A}}{dt} \perp \overrightarrow{A}$. If $\overrightarrow{A}(t)$ represents

the position vector of an object, then $\overrightarrow{A}$ and its change vector $\dfrac{d\overrightarrow{A}}{dt}$ are perpendicular,

which represents uniform circular motion kinematically.

Solution

Let $|\overrightarrow{A}| = c \Rightarrow \overrightarrow{A} \cdot \overrightarrow{A} = c^2 \Rightarrow \dfrac{d}{dt}(\overrightarrow{A} \cdot \overrightarrow{A}) = 0$

$\Rightarrow \dfrac{d\overrightarrow{A}}{dt} \cdot \overrightarrow{A} + \overrightarrow{A} \cdot \dfrac{d\overrightarrow{A}}{dt} = 0 \Rightarrow 2\dfrac{d\overrightarrow{A}}{dt} \cdot \overrightarrow{A} = 0 \Rightarrow \dfrac{d\overrightarrow{A}}{dt} \cdot \overrightarrow{A} = 0$,

and $\left|\dfrac{d\overrightarrow{A}}{dt}\right| \neq 0 \Rightarrow \dfrac{d\overrightarrow{A}}{dt} \perp \overrightarrow{A}$.

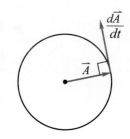

Q.E.D.

▶ Definition 8-1-3

Mutivariable Vector Functions

If $F_1(u, v, w)$, $F_2(u, v, w)$, $F_3(u, v, w)$ are three-variable scalar function defined on a certain interval I, then $\overrightarrow{F} = \overrightarrow{F}(u, v, w) = F_1(u, v, w)\overrightarrow{i} + F_2(u, v, w)\overrightarrow{j} + F_3(u, v, w)\overrightarrow{k}$ is called a three-variable vector function.   ◀

In general, when the variables of coefficient functions exceed two, we collectively refer to $\overrightarrow{F}$ as multi-variable vector functions. Similar to the case of single-variable functions, we have the following results.

2. Continuity and Differentiation

Following the above notation, F_1, F_2, F_3 are continuous in the interval I if and only if $\vec{F}(t)$ is continuous in the interval I.

Unlike single-variable functions, here, when we talk about differentiation, it naturally refers to partial derivatives.

▶ ## Definition 8-1-4

Partial Derivatives

Assuming that the first order partial derivatives of $F_1(t)$, $F_2(t)$, $F_3(t)$ exist, the definition of the partial derivative of the vector function $\vec{F}$ is:

$$\vec{F_u} = \frac{\partial \vec{F}}{\partial u} = \lim_{\Delta u \to 0} \frac{\vec{F}(u + \Delta u, v, w) - \vec{F}(u, v, w)}{\Delta u} \equiv (\frac{\partial \vec{F}}{\partial u})_{v,\ w\ \text{are constants}}\text{,}$$

$$\vec{F_v} = \frac{\partial \vec{F}}{\partial v} = \lim_{\Delta v \to 0} \frac{\vec{F}(u, v + \Delta v, w) - \vec{F}(u, v, w)}{\Delta v} \equiv (\frac{\partial \vec{F}}{\partial v})_{u,\ w\ \text{are constants}}\text{,}$$

$$\vec{F_w} = \frac{\partial \vec{F}}{\partial w} = \lim_{\Delta w \to 0} \frac{\vec{F}(u, v, w + \Delta w) - \vec{F}(u, v, w)}{\Delta w} \equiv (\frac{\partial \vec{F}}{\partial w})_{u,\ v\ \text{are constants}}\text{.}$$

◀

▶ ## Theorem 8-1-2

Properties of Partial Derivative

Partial derivatives possess the following properties. Readers can verify them using the definition of differentiation.

(1) $\dfrac{\partial}{\partial u}(\vec{A} \pm \vec{B}) = \dfrac{\partial \vec{A}}{\partial u} \pm \dfrac{\partial \vec{B}}{\partial u}$.   (2) $\dfrac{\partial}{\partial u}(\phi\vec{B}) = \dfrac{\partial \phi}{\partial u}\vec{B} + \phi\dfrac{\partial \vec{B}}{\partial u}$.

(3) $\dfrac{\partial}{\partial u}(\vec{A} \cdot \vec{B}) = \dfrac{\partial \vec{A}}{\partial u} \cdot \vec{B} + \vec{A} \cdot \dfrac{\partial \vec{B}}{\partial u}$.   (4) $\dfrac{\partial}{\partial u}(\vec{A} \times \vec{B}) = \dfrac{\partial \vec{A}}{\partial u} \times \vec{B} + \vec{A} \times \dfrac{\partial \vec{B}}{\partial u}$.

◀

▶ ## Definition 8-1-5

Total Derivative

Assuming that $\vec{f} = \vec{f}(u,\ v,\ w)$, and $\dfrac{\partial \vec{f}}{\partial u}$, $\dfrac{\partial \vec{f}}{\partial v}$, $\dfrac{\partial \vec{f}}{\partial w} \in C$, then

$$d\vec{f} = \frac{\partial \vec{f}}{\partial u}du + \frac{\partial \vec{f}}{\partial v}dv + \frac{\partial \vec{f}}{\partial w}dw$$ is called total derivative of vector function.

◀

Example 2

Assuming that $\vec{F}(x, y, z) = \sin(xy)\vec{i} + e^{xyz}\vec{j} + x \cdot \cos z\vec{k}$, find $\vec{F}_x, \vec{F}_y, \vec{F}_z$.

Solution

$$\vec{F}_x = (\frac{\partial \vec{F}}{\partial x})_{y, z \text{ constants}} = y \cdot \cos(xy)\vec{i} + yz \cdot e^{xyz}\vec{j} + \cos z\vec{k} \ ,$$

$$\vec{F}_y = (\frac{\partial \vec{F}}{\partial y})_{x, z \text{ constants}} = x \cdot \cos(xy)\vec{i} + xz \cdot e^{xyz}\vec{j} + 0\vec{k} \ ,$$

$$\vec{F}_z = (\frac{\partial \vec{F}}{\partial z})_{x, y \text{ constants}} = 0\vec{i} + xy \cdot e^{xyz}\vec{j} - x \cdot \sin z\vec{k} \ .$$

Q.E.D.

8-1-2 Coordinate System

Naturally, we use Cartesian coordinates to describe any point P: $\vec{r}_p = x\vec{i} + y\vec{j} + z\vec{k}$ in 3-dimensional Euclidean space R^3, which is called a position vector, as shown in Figure 8-2. However, from the experience of calculus, we understand that this may not be the most convenient coordinate system for calculus operations. Depending on the specific physical characteristics, we often choose coordinates that are more suitable for describing the particular physical system. For example, if the physical behavior of a system involves describing cylinders, we would use cylindrical coordinates. Below, we list several commonly used coordinate systems.

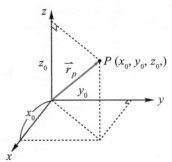

Figure 8-2   Position Vector in Space

1. **Polar Coordinate**

Suppose a point P in the x-y plane has Cartesian coordinates (x, y) , let $r = \sqrt{x^2 + y^2}$, where the angle between the direction vector (x, y) and the x-axis is θ. Then, we define the polar coordinates of point P as (r, θ). From a vector perspective, the relationship between the components of the position vector $\vec{r}_p = x\vec{i} + y\vec{j}$ and the Cartesian coordinates is: $\begin{cases} x = r\cos\theta \\ y = r\sin\theta \end{cases}$.

Furthermore, $\{\vec{e_r}, \vec{e_\theta}\} = \{\cos\theta\,\vec{i} + \sin\theta\,\vec{j}, -\sin\theta\,\vec{i} + \cos\theta\,\vec{j}\}$ forms a set of normalized orthogonal (orthonormal) basis vectors along polar coordinate on R^2, as shown in Figure 8-3.

2. Cylindrical Coordinate

Suppose a point P in 3-dimensional space R^3 has Cartesian coordinates (x, y, z). If we change the x and y components to polar coordinates, we obtain the cylindrical coordinates (r, θ, z). From a vector perspective, the relationship between the components of the position vector $\vec{r}_p = x\,\vec{i} + y\,\vec{j} + z\,\vec{k}$ and the Cartesian coordinates

is:
$$\begin{cases} x = r\cos\theta \\ y = r\sin\theta \\ z = z \end{cases}$$
, where we need to constrain $0 \le \theta \le 2\pi$ [1].

Additionally, $\{\vec{e}_r, \vec{e}_\theta, \vec{e}_z\} = \{\cos\theta\,\vec{i} + \sin\theta\,\vec{j}, -\sin\theta\,\vec{i} + \cos\theta\,\vec{j}, \vec{k}\}$ forms a set of normalized orthogonal (orthonormal) basis vectors in R^3, as shown in Figure 8-4.

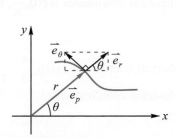

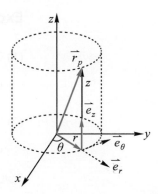

Figure 8-3   Polar Coordinates in the Plane   Figure 8-4   Cylindrical Coordinate

3. Spherical Coordinate

Suppose a point P in 3-dimensional space R^3 has Cartesian coordinates (x, y, z), let $r = \sqrt{x^2 + y^2 + z^2}$, where the direction vector (x, y, z) makes an angle θ with the x-axis and an angle ϕ with the z-axis. Then, we define the spherical coordinates as (r, ϕ, θ). From a vector perspective, the relationship between the components of the position vector $\vec{r}_p = x\,\vec{i} + y\,\vec{j} + z\,\vec{k}$ and the Cartesian coordinates is:

$$\begin{cases} x = r\sin\phi\cos\theta \\ y = r\sin\phi\sin\theta \\ z = r\cos\phi \end{cases}$$
, where $0 \le \theta \le 2\pi$, $0 \le \phi \le \pi$; Additionally, $\{\vec{e}_r, \vec{e}_\phi, \vec{e}_\theta\}$ forms a

set of normalized orthogonal (orthonormal) basis vectors moving along the spherical coordinates in R^3, where $\vec{e}_r = \sin\phi\cos\theta\,\vec{i} + \sin\phi\sin\theta\,\vec{j} + \cos\phi\,\vec{k}$,

---

[1] When projected onto the x-y plane, the cylindrical coordinates become polar coordinates. The representation of a point in cylindrical coordinates is $P(r, \theta, z)$.

The first component represents the distance from the symmetry axis, the second component represents the angle with the positive x-axis, and the third component represents the projection onto the z-axis.

$$\vec{e}_\phi = \cos\phi\cos\theta\,\vec{i} + \cos\phi\sin\theta\,\vec{j} - \sin\phi\,\vec{k}\ ,$$

$$\vec{e}_\theta = -\sin\theta\,\vec{i} + \cos\theta\,\vec{j}\ ,\text{ as shown in Figure 8-5}^2.$$

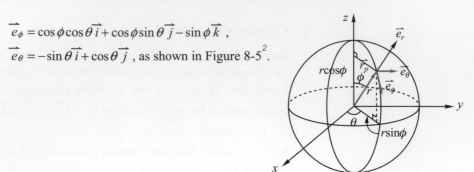

Figure 8-5 Spherical coordinate diagram

Example 3

(1) The point P in cylindrical coordinates is $(10, \dfrac{\pi}{6}, 8)$, please convert it to Cartesian coordinates.

(2) The point P in spherical coordinates is $(10, \dfrac{\pi}{3}, \dfrac{\pi}{4})$, please convert it to Cartesian coordinates and cylindrical coordinates.

Solution

$$(1)\ \begin{cases} x = r\cos\theta = 10 \cdot \cos\dfrac{\pi}{6} = 5\sqrt{3} \\[2mm] y = 10\sin\dfrac{\pi}{6} = 5 \\[2mm] z = 8 \end{cases}$$, so the point P in Cartesian coordinates is $(5\sqrt{3}, 5, 8)$.

$$(2)\ \begin{cases} x = 10\sin\dfrac{\pi}{3}\cos\dfrac{\pi}{4} = \dfrac{5\sqrt{6}}{2} \\[2mm] y = 10\sin\dfrac{\pi}{3}\sin\dfrac{\pi}{4} = \dfrac{5\sqrt{6}}{2} \\[2mm] z = 10\cos\dfrac{\pi}{3} = 5 \end{cases}$$, so the point P in Cartesian coordinates is $(\dfrac{5\sqrt{6}}{2}, \dfrac{5\sqrt{6}}{2}, 5)$,

then transform into cylindrical coordinates, where

$$r = 10\sin\dfrac{\pi}{3} = 5\sqrt{3}\ ,\quad \theta = \dfrac{\pi}{4}\ ,\quad z = 10\cos\dfrac{\pi}{3} = 5\ .$$

Therefore, the cylindrical coordinates of point P are $(5\sqrt{3}, \dfrac{\pi}{4}, 5)$. Q.E.D.

---

[2] The expression for a point P in spherical coordinates is $P(r, \phi, \theta)$.

The first component represents the distance from the origin (center of the sphere), the second component represents the angle with the z-axis, and the third component represents the angle formed with the x-axis after projecting onto the xy-plane.

8-1-3 Curves in Space

With the representation of various coordinate systems, describing curves and surfaces in $\mathbb{R}^3$ becomes much easier. Let's start with curves in space. Suppose the position vector $\vec{r}_P(t)$ of a point P in space is a function of parameter t, as the parameter t varies, it traces out a curve C in space, as shown in Figure 8-6. We denote it as

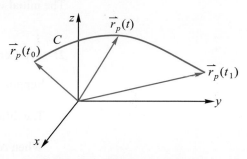

Figure 8-6   The curves in space

$$\vec{r}_P(t) = x(t)\vec{i} + y(t)\vec{j} + z(t)\vec{k}$$

where $t \in [t_0, t_1]$.

1. Physical Meaning

In correspondence with kinematics in physics, the first derivative of the position vector $\vec{r}_P(t) = x(t)\vec{i} + y(t)\vec{j} + z(t)\vec{k}$ with respect to time gives the velocity vector

$\vec{v}_P(t) = \dfrac{d\,\vec{r}_P(t)}{dt} = \dfrac{dx(t)}{dt}\vec{i} + \dfrac{dy(t)}{dt}\vec{j} + \dfrac{dz(t)}{dt}\vec{k}$, and the second derivative gives the

acceleration vector $\vec{a}_P(t) = \dfrac{d^2\,\vec{r}_P(t)}{dt^2} = \dfrac{d^2x(t)}{dt^2}\vec{i} + \dfrac{d^2y(t)}{dt^2}\vec{j} + \dfrac{d^2z(t)}{dt^2}\vec{k}$.

For example, if the position vector function is $\vec{r}_P(t) = \cos t\,\vec{i} + \sin t\,\vec{j} + t\,\vec{k}$, $t \in [0, 2\pi]$, then the velocity vector function is

$$\vec{v}_P(t) = \frac{d\,\vec{r}_P(t)}{dt} = \frac{d(\cos t)}{dt}\vec{i} + \frac{d(\sin t)}{dt}\vec{j} + \frac{d(t)}{dt}\vec{k} = -\sin t\,\vec{i} + \cos t\,\vec{j} + \vec{k} ,$$

and the acceleration vector function is

$$\vec{a}_P(t) = \frac{d^2\,\vec{r}_P(t)}{dt^2} = \frac{d^2(\cos t)}{dt^2}\vec{i} + \frac{d^2(\sin t)}{dt^2}\vec{j} + \frac{d^2(t)}{dt^2}\vec{k} = -\cos t\,\vec{i} - \sin t\,\vec{j} .$$

Example 4

There is a position vector function $\vec{r}(t) = \sin 2t\,\vec{i} + e^{-2t}\,\vec{j} + t^2\,\vec{k}$, $t \in [0, 2\pi]$. Find its corresponding velocity and acceleration functions, and determine its initial velocity and acceleration.

Solution

The velocity vector function is:

$$\vec{v}(t) = \frac{d\,\vec{r}(t)}{dt} = \frac{d(\sin 2t)}{dt}\vec{i} + \frac{d(e^{-2t})}{dt}\vec{j} + \frac{d(t^2)}{dt}\vec{k} = 2\cos 2t\,\vec{i} - 2e^{-2t}\,\vec{j} + 2t\,\vec{k} .$$

The acceleration vector function is:

$$\vec{a}(t) = \frac{d^2\vec{r}(t)}{dt^2} = \frac{d^2(\sin 2t)}{dt^2}\vec{i} + \frac{d^2(e^{-2t})}{dt^2}\vec{j} + \frac{d^2(t^2)}{dt^2}\vec{k} = -4\sin 2t\,\vec{i} + 4e^{-2t}\,\vec{j} + 2\vec{k}\ .$$

The initial velocity $\vec{v}(0) = 2\vec{i} - 2\vec{j}$; the initial acceleration $\vec{a}(0) = 4\vec{j} + 2\vec{k}$.

Q.E.D.

2. Arc Length of Space Curves

Supposing $\vec{r}(t) = x(t)\vec{i} + y(t)\vec{j} + z(t)\vec{k}$ is a curve C defined on the interval $[a, b]$.

The differential displacement vector on C is defined as $\Delta\vec{r} = [\vec{r}(t + \Delta t) - \vec{r}(t)] = \dfrac{d\vec{r}}{dt}\Delta t$.

When $\Delta t \to 0$, $|\Delta\vec{r}|$ approximates the small arc length ds, therefore, $ds = |d\vec{r}|$,

which is known as the **differential arc length**, and in reality, $\dfrac{d\vec{r}(t)}{dt}$ represents the

tangent vector to the curve C, as shown in Figure 8-7. Now, since

$|d\vec{r}| = |\dfrac{d\vec{r}}{dt}dt| = |\dfrac{d\vec{r}}{dt}\|dt|$, we generally take dt to be positive

$\Rightarrow ds = |\dfrac{d\vec{r}}{dt}|dt$, and we obtain the arc length of the curve as follows:

$$s = \int ds = \int_{t=a}^{b} |\frac{d\vec{r}}{dt}|\,dt$$

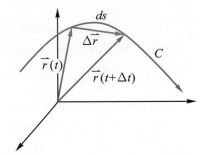

Figure 8-7   Differential arc length in space

Furthermore, in terms of parametric equations, since $|\dfrac{d\vec{r}}{dt}| = \sqrt{(\dfrac{dx}{dt})^2 + (\dfrac{dy}{dt})^2 + (\dfrac{dz}{dt})^2}$, we

obtain the operational formula for calculating the arc length of the curve

$s = \int_{t=a}^{b}\sqrt{(\dfrac{dx}{dt})^2 + (\dfrac{dy}{dt})^2 + (\dfrac{dz}{dt})^2}\,dt$. If we consider b as a variable, the arc length of the

curve can also be seen as a parameter describing the curve. However, this technique is more abstract, and we won't delve into it further here.

3. Arc Length of a Plane Curve

In the case of a space curve with $z(t) = 0$, as shown in Figure 8-8, we obtain the

differential arc length of the curve in the plane as $ds = \sqrt{(\dfrac{dx}{dt})^2 + (\dfrac{dy}{dt})^2}\, dt$, and the

arc length of the curve as

$$s = \int_{t=a}^{b} \sqrt{(\frac{dx}{dt})^2 + (\frac{dy}{dt})^2}\, dt$$

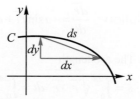

Figure 8-8   Differential arc length in Cartesian coordinates on the plane

If the curve itself has a functional form $y = f(x)$, then its parametric form is
$\vec{r}(x) = x\,\vec{i} + y\,\vec{j} = x\,\vec{i} + f(x)\,\vec{j}$. In this case, the tangent vector to a point on the curve is

$\dfrac{d\vec{r}}{dx} = \vec{i} + f'(x)\,\vec{j}$. Substituting this into the arc length formula, we get:

$$s = \int_{x=a}^{b} \sqrt{1 + (\frac{df}{dx})^2}\, dx$$

Now, consider the polar coordinate parametric form of the curve

$\vec{r}(\theta) = \rho(\theta)\cos\theta\,\vec{i} + \rho(\theta)\sin\theta\,\vec{j}$, then the tangent

vector to a point on the curve is

$\dfrac{d\vec{r}}{d\theta} = (\dfrac{d\rho}{d\theta}\cos\theta - \rho\sin\theta)\,\vec{i} + (\dfrac{d\rho}{d\theta}\sin\theta + \rho\cos\theta)\,\vec{j}$.

Therefore, we obtain the differential arc length as

$ds = \sqrt{\rho^2 + (\dfrac{d\rho}{d\theta})^2}\, d\theta$ [3], as shown in Figure 8-9.

Consequently, we derive the arc length formula
in polar coordinates as:

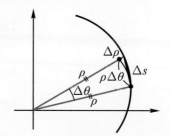

Figure 8-9   Differential Arc Length
in Polar Coordinates on the Plane

$$s = \int_{\theta=\theta_1}^{\theta_2} \sqrt{(\frac{d\rho}{d\theta})^2 + \rho^2}\, d\theta$$

---

[3]   The curve C is given as: $\rho = \rho(\theta)$. The differential arc length can also be obtained using geometric methods.

$$(\Delta s)^2 \approx (\Delta\rho)^2 + (\rho\Delta\theta)^2 = [(\frac{\Delta\rho}{\Delta\theta})^2 + \rho^2](\Delta\theta)^2 \Rightarrow (ds)^2 = [(\frac{d\rho}{d\theta})^2 + \rho^2](d\theta)^2 \Rightarrow ds = \sqrt{(\frac{d\rho}{d\theta})^2 + \rho^2}\, d\theta .$$

Example 5

Let C: $\vec{r}(t) = a\cos t\,\vec{i} + a\sin t\,\vec{j} + ct\,\vec{k}$.

(1) Find the arc length of curve C when $t \in [0, 2\pi]$.

(2) If $a = 1$, $c = 1$, find the arc length of the curve from $t = 0$ to $t = \pi$.

(3) Find $\vec{r}(s)$ (the position vector of a parameter as the arc length s).

Solution $\quad \dfrac{d\vec{r}}{dt} = -a\sin t\,\vec{i} + a\cos t\,\vec{j} + c\vec{k}$, $\therefore \dfrac{d\vec{r}}{dt} \cdot \dfrac{d\vec{r}}{dt} = a^2 + c^2$.

(1) The arc length

$$s = \int_0^{2\pi} \sqrt{\vec{r}\,' \cdot \vec{r}\,'}\; dt = \int_0^{2\pi} \sqrt{a^2 + c^2}\; dt = 2\pi \cdot \sqrt{a^2 + c^2}\ .$$

(2) $s = \int_0^{\pi} \sqrt{1^2 + 1^2}\; dt = \sqrt{2}\pi$.

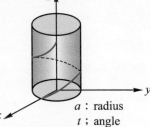

a : radius
t ; angle

(3) $s(t) = \int_0^t \sqrt{\vec{r}\,' \cdot \vec{r}\,'}\; dt = \int_0^t \sqrt{a^2 + c^2}\; dt = \sqrt{a^2 + c^2}\; t \Rightarrow t = \dfrac{s}{\sqrt{a^2 + c^2}}$,

therefore we obtain $\quad \vec{r}(s) = a\cos(\dfrac{s}{w})\vec{i} + a\sin(\dfrac{s}{w})\vec{j} + \dfrac{cs}{w}\vec{k}$

where $w = \sqrt{a^2 + c^2}$.

Q.E.D.

Example 6

Let the cardiac curve be defined as $\rho(\theta) = a(1 - \cos\theta)$, $\theta \in [0, 2\pi]$, $a > 0$. Find its arc length.

Solution

First, $\dfrac{d\rho}{d\theta} = a\sin\theta$,

therefore we get $\quad ds = \sqrt{\rho^2 + (\dfrac{d\rho}{d\theta})^2}\; d\theta$

$$= \sqrt{a^2(1 - \cos\theta)^2 + a^2 \sin^2\theta}\; d\theta$$

$$= \sqrt{2}a\sqrt{1 - \cos\theta}\; d\theta$$

$$= \sqrt{2}a \cdot \sin\dfrac{\theta}{2} \cdot \sqrt{2}\; d\theta$$

$$= 2a\sin\dfrac{\theta}{2}\, d\theta,$$

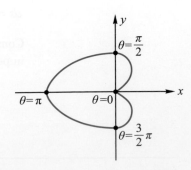

$\therefore s = \int_0^{2\pi} ds = \int_0^{2\pi} 2a\sin(\dfrac{\theta}{2})d\theta = -4a\cos\dfrac{\theta}{2}\Big|_0^{2\pi} = 8a$.

Q.E.D.

8-1   Exercises

Basic questions

Find the velocity and acceleration at $t = 0$ for the position functions $\vec{r}(t)$ in the following three questions.

1. $\vec{r}(t) = t^2\,\vec{i} + 2\,\vec{j}$.

2. $\vec{r}(t) = t\,\vec{i} + e^t\,\vec{j}$.

3. $\vec{r}(t) = \vec{i} + t\,\vec{j} + \sin(t)\vec{k}$.

Find the arc length of the curve C in the given range for the following questions 4~5.

4. $C: \vec{r}(t) = \vec{i} + t\,\vec{j}$, $t \in [0,1]$.

5. $C: \vec{r}(t) = \sin(t)\,\vec{i} + \cos(t)\,\vec{j}$, $t \in [0, 2\pi]$.

6. Given a vector field $\vec{f} = x^2\,\vec{i} + y\,\vec{j} + e^z\,\vec{k}$, find $\vec{f}_x$, $\vec{f}_y$, $\vec{f}_z$.

Advanced questions

1. The position function $\vec{r}(t)$ is defined as follows. Find its velocity and acceleration vectors at $t = 0$.

(1) $\vec{r}(t) = 3t^2\,\vec{i} + \sin t\,\vec{j} - 2t^2\,\vec{k}$.

(2) $\vec{r}(t) = \vec{i} - 2\cos(t)\vec{j} + t\vec{k}$.

(3) $\vec{r}(t) = \sinh(t)\vec{i} - 2t^2\,\vec{k}$.

2. If vector field $\vec{f}$ as follows, find $\vec{f}_x$, $\vec{f}_y$, $\vec{f}_z$, $\vec{f}_{xx}$, $\vec{f}_{xy}$.

(1) $\vec{f} = 4x\,\vec{i} + 5xy\,\vec{j} - z\,\vec{k}$.

(2) $\vec{f} = e^x\,\vec{i} - 3x^2yz\,\vec{j}$.

(3) $\vec{f} = 2xy\,\vec{i} + y\sin(x)\,\vec{j} + \cos(z)\vec{k}$.

3. Find the arc length of the curve defined by the following parametric equations.

(1) $C: \begin{cases} x = 3\sin(t) \\ y = 3\cos t \\ z = 2t \end{cases}$; $t \in [0, 2\pi]$.

(2) $C: \begin{cases} x = 2t^2 \\ y = t^2 \\ z = 2t^2 \end{cases}$; $t \in [0, 1]$.

(3) $C: \begin{cases} x = t^3 \\ y = t^3 \\ z = t^3 \end{cases}$; $t \in [-2, 2]$.

4. The position vector function $\vec{r}(t)$ is defined as follows. Find its velocity and acceleration vector functions.

(1) $\vec{r}(t) = 3t\,\vec{i} - 2\,\vec{j} + t^2\,\vec{k}$.

(2) $\vec{r}(t) = 2\cos t\,\vec{i} + 2\sin t\,\vec{j} - 3t\,\vec{k}$.

(3) $\vec{r}(t) = t\sin t\,\vec{i} - 2e^{-3t}\,\vec{j} + e^{-t}\cos t\,\vec{k}$.

5. If $\vec{f}(x, y) = e^{xy}\,\vec{i} + (x + y)\,\vec{j} + x\cos y\,\vec{k}$, find $\vec{f}_x$, $\vec{f}_y$, $\vec{f}_{xx}$.

6. The curve C is defined as follows. Find the arc length.

(1) $C : \vec{r}(t) = \cos t\,\vec{i} + \sin t\,\vec{j} + \dfrac{1}{3}t\,\vec{k}$, $-4\pi \leq t \leq 4\pi$.

(2) $C : \vec{r}(t) = 2t\,\vec{i} + t^2\,\vec{j} + \ln t\,\vec{k}$, $1 \leq t \leq 2$.

(3) $C : \vec{r}(t) = t^2\,\vec{i} + t^2\,\vec{j} + \dfrac{1}{2}t^2\,\vec{k}$, $1 \leq t \leq 3$.

(4) $C : \rho(\theta) = 4\sin\theta$, $0 \leq \theta \leq \pi$.

7. The position of point P in cylindrical coordinates is given as follows. Convert it to Cartesian coordinates.

(1) $(10, \dfrac{3\pi}{4}, 5)$.   (2) $(\sqrt{3}, \dfrac{\pi}{3}, -4)$.

8. The position of point P in spherical coordinates is given as follows. Convert it to Cartesian coordinates and cylindrical coordinates.

(1) $(\dfrac{2}{3}, \dfrac{\pi}{2}, \dfrac{\pi}{6})$.   (2) $(8, \dfrac{\pi}{4}, \dfrac{3\pi}{4})$.

8-2 **Directional Derivative**

The derivative or partial derivative of a single-variable scalar function does not have directionality. However, in practical physical systems, most functions are multivariable, and the rate of change naturally depends on the angle at which the change is observed. Therefore, it can vary depending on the direction along which it is measured. It's similar to how the change in height on a mountain can be different depending on the direction you climb. Climbing from the southwest side may have the smallest change in slope and be easier to climb, but it results in the longest distance traveled; while climbing from the northeast side may have the greatest change in slope, making it harder to climb, but results in the shortest distance traveled. For example, in height fields, temperature fields, etc., the rate of change along different directions can differ.

This section will use directional derivatives to characterize such problems. In simple terms, we first find a curve on the surface, and then take a derivative along the curve at a specified point, resulting in a rate of change represented as components of a fixed vector.

8-2-1 ∇ Operator

In practical computations, mathematicians discovered that the rate of change for multivariable functions can be represented as components of a fixed-form vector. For the convenience of describing this vector, we introduce the Delta operator (commonly known as the Del).

$$\nabla \equiv \vec{i}\frac{\partial}{\partial x} + \vec{j}\frac{\partial}{\partial y} + \vec{k}\frac{\partial}{\partial z}$$

Here, we regard $\dfrac{\partial}{\partial x}$, $\dfrac{\partial}{\partial y}$ and $\dfrac{\partial}{\partial z}$ as partial derivative operators for functions with respect to the coordinates in R^3. Consequently, the length of the vector ∇ is given by:

$$\nabla \cdot \nabla = \nabla^2 = \frac{\partial^2}{\partial x^2} + \frac{\partial^2}{\partial y^2} + \frac{\partial^2}{\partial z^2}$$

is called the Laplacian operator. When applied to a smooth function ϕ, it yields:

$$\nabla^2 \phi \equiv \frac{\partial^2 \phi}{\partial x^2} + \frac{\partial^2 \phi}{\partial y^2} + \frac{\partial^2 \phi}{\partial z^2} = 0$$

which is called **Laplace's equation** (Laplacian of ϕ), and a function ϕ that satisfies $\nabla^2 \phi = 0$ is commonly referred to as a harmonic function. For smooth scalar field ϕ and vector field $\vec{u} = u_1 \vec{i} + u_2 \vec{j} + u_3 \vec{k}$, there are three major physical quantities that can be described using the Delta operator ∇ as follows:

1. Gradient

$$\nabla \phi \equiv \frac{\partial \phi}{\partial x} \vec{i} + \frac{\partial \phi}{\partial y} \vec{j} + \frac{\partial \phi}{\partial z} \vec{k} = \mathrm{grad}(\phi)$$

In general, in physical systems, the gradient field $\nabla \phi$ represents the vector with the maximum rate of change of the scalar field ϕ.

2. Divergence

$$\nabla \cdot \vec{u} = \frac{\partial u_1}{\partial x} + \frac{\partial u_2}{\partial y} + \frac{\partial u_3}{\partial z} = \mathrm{div}(\vec{u})$$

In general, in physical systems, the divergence field $\nabla \cdot \vec{u}$ represents the quantity of the vector field $\vec{u}$ that is outwardly diverging. In the context of fluid flow, electric fields, or magnetic fields, the divergence represents the net flow of a certain small volume, indicating the amount entering or leaving that volume.

3. Curl

$$\nabla \times \vec{u} \equiv \begin{vmatrix} \vec{i} & \vec{j} & \vec{k} \\ \dfrac{\partial}{\partial x} & \dfrac{\partial}{\partial y} & \dfrac{\partial}{\partial z} \\ u_1 & u_2 & u_3 \end{vmatrix} = \mathrm{curl}(\vec{u})$$

In general, in physical systems, the curl field $\nabla \times \vec{u}$ represents the rotational vector of the vector field $\vec{u}$. In the context of fluid flow, electric fields, or magnetic fields, the curl represents the circulation of a certain small closed curve, indicating the amount of rotation around that curve.

▶ **Definition 8-2-1**

Common Vector Fields

---

(1) $\mathrm{div}(\vec{u}) > 0$ indicates an outwardly diverging tendency of the vector field $\vec{u}$, it is called a **source**. Conversely, if $\mathrm{div}(\vec{u}) < 0$ indicates an inwardly converging tendency of the vector field $\vec{u}$, it is called a **sink**.

(2) Zero divergence field, that is $\mathrm{div}(\vec{u}) = 0$, irrotational field, that is $\mathrm{curl}(\vec{u}) = 0$.

(3) From the perspective of fluid mechanics, $\nabla \cdot \vec{u} = \mathrm{div}(\vec{u}) = 0$, $\vec{u}$ is called an **incompressible flow field**, otherwise, it is compressible. In electromagnetic theory, a vector field with $\nabla \cdot \vec{u} = 0$ is called a **solenoidal field**. ◀

Example 1

Let the scalar field be $f(x, y, z) = x^4 + y^4 + z$,

and the vector field be $\vec{V}(x, y, z) = (x+y)^2 \vec{i} + z^2 \vec{j} + 2yz \vec{k}$,

$\vec{F}(x, y, z) = 2x\vec{i} - y\vec{j} - z\vec{k}$.

(1) Find grad(f) at $(4, -1, 3)$.

(2) Calculate $\mathrm{div}(\vec{V})$, $\mathrm{curl}(\vec{V})$ and $\nabla^2 f$.

(3) Prove that $\vec{F}$ is incompressible.

Solution

(1) $\nabla f = \dfrac{\partial f}{\partial x}\vec{i} + \dfrac{\partial f}{\partial y}\vec{j} + \dfrac{\partial f}{\partial z}\vec{k} = 4x^3\vec{i} + 4y^3\vec{j} + \vec{k}$,

so $\left.\mathrm{grad}(f)\right|_{(4,-1,3)} = 256\vec{i} - 4\vec{j} + \vec{k}$.

(2) $\mathrm{div}(\vec{V}) = \nabla \cdot \vec{V} = \dfrac{\partial (x+y)^2}{\partial x} + \dfrac{\partial (z^2)}{\partial y} + \dfrac{\partial (2yz)}{\partial z} = 2(x+y) + 2y = 2x + 4y$

$\nabla^2 f = \dfrac{\partial^2 f}{\partial x^2} + \dfrac{\partial^2 f}{\partial y^2} + \dfrac{\partial^2 f}{\partial z^2} = 12x^2 + 12y^2$

$\mathrm{curl}(\vec{V}) = \nabla \times \vec{V} = \begin{vmatrix} \vec{i} & \vec{j} & \vec{k} \\ \dfrac{\partial}{\partial x} & \dfrac{\partial}{\partial y} & \dfrac{\partial}{\partial z} \\ (x+y)^2 & z^2 & 2yz \end{vmatrix}$

$= (2z - 2z)\vec{i} - 0\vec{j} + (0 - 2(x+y))\vec{k} = -2(x+y)\vec{k}$.

(3) $\mathrm{div}(\vec{F}) = \nabla \cdot \vec{F} = \dfrac{\partial(2x)}{\partial x} + \dfrac{\partial(-y)}{\partial y} + \dfrac{\partial(-z)}{\partial z} = 2 - 1 - 1 = 0$, so $\vec{F}$ is incompressible.

Q.E.D.

4. **The Basic Properties of** ∇

Assuming that the second-order partial derivatives of functions ϕ and φ exist and are continuous, then for smooth vector functions $\vec{u}$, $\vec{v}$, we have the following properties. These properties can be proved using the definitions of calculus along with the rules of scalar functions and vector operations. The details are left as an exercise for the reader.

(1) $\nabla(\phi + \varphi) = \nabla\phi \pm \nabla\varphi$ (4) $\nabla(\phi\varphi) = \varphi\nabla\phi + \phi\nabla\varphi$

(2) $\nabla \cdot (\vec{u} \pm \vec{v}) = \nabla \cdot \vec{u} \pm \nabla \cdot \vec{v}$ (5) $\nabla \cdot (\phi\vec{u}) = \nabla\phi \cdot \vec{u} + \phi\nabla \cdot \vec{u}$

(3) $\nabla \times (\vec{u} \pm \vec{v}) = \nabla \times \vec{u} \pm \nabla \times \vec{v}$ (6) $\nabla \times (\phi\vec{u}) = \nabla\phi \times \vec{u} + \phi\nabla \times \vec{u}$

▶ **Definition 8-2-2**

Irrotational Vector Field

If $\vec{v}$ is a smooth vector field defined on a region D, and within D, $\nabla \times \vec{v} = \vec{0}$, then $\vec{v}$ is called a irrotational vector field on D. ◀

▶ **Theorem 8-2-1**

The Curl of the Gradient Is Zero

Assuming ϕ is a smooth scalar field, and $\nabla \times (\nabla\phi) = 0$, that is, the gradient field is an irrotational vector field.

Proof

$$\nabla \times (\nabla\phi) = \nabla \times (\frac{\partial\phi}{\partial x}\vec{i} + \frac{\partial\phi}{\partial y}\vec{j} + \frac{\partial\phi}{\partial z}\vec{k}) = \begin{vmatrix} \vec{i} & \vec{j} & \vec{k} \\ \dfrac{\partial}{\partial x} & \dfrac{\partial}{\partial y} & \dfrac{\partial}{\partial z} \\ \dfrac{\partial\phi}{\partial x} & \dfrac{\partial\phi}{\partial y} & \dfrac{\partial\phi}{\partial z} \end{vmatrix} = \vec{0} .$$ ◀

▶ **Theorem 8-2-2**

The Existence Theorem for Gradient Fields

If $\vec{v}$ is an irrotational vector field defined on a region D, then there must exist a scalar potential function ϕ such that $\vec{v} = \nabla\phi$. In other words, irrotational vector fields are always gradient fields. ◀

▶ **Theorem 8-2-3**

The Curl Field Does Not Have Divergence

$$\nabla \cdot (\nabla \times \vec{u}) = 0.$$

Proof

From the definition of curl and divergence, we know

$$\nabla \cdot \nabla \times \vec{u} = \nabla \cdot \begin{vmatrix} \vec{i} & \vec{j} & \vec{k} \\ \dfrac{\partial}{\partial x} & \dfrac{\partial}{\partial y} & \dfrac{\partial}{\partial z} \\ u_1 & u_2 & u_3 \end{vmatrix} = \nabla \cdot \left[\left(\dfrac{\partial u_3}{\partial y} - \dfrac{\partial u_2}{\partial z} \right) \vec{i} - \left(\dfrac{\partial u_3}{\partial x} - \dfrac{\partial u_1}{\partial z} \right) \vec{j} + \left(\dfrac{\partial u_2}{\partial x} - \dfrac{\partial u_1}{\partial y} \right) \vec{k} \right]$$

$$= \dfrac{\partial}{\partial x} \left(\dfrac{\partial u_3}{\partial y} - \dfrac{\partial u_2}{\partial z} \right) - \dfrac{\partial}{\partial y} \left(\dfrac{\partial u_3}{\partial x} - \dfrac{\partial u_1}{\partial z} \right) + \dfrac{\partial}{\partial z} \left(\dfrac{\partial u_2}{\partial x} - \dfrac{\partial u_1}{\partial y} \right) = 0.$$

◀

▶ **Theorem 8-2-4**

The Existence Theorem for Curl Fields

If the first-order partial derivatives of $\vec{v}$ exist and are continuous in a region D, and $\nabla \cdot \vec{v} = 0$ within D, then $\vec{v}$ is called a solenoidal vector field in D. In this case, there must exist a vector function $\vec{u}$ such that $\vec{v} = \nabla \times \vec{u}$ is a solenoidal vector field.

◀

▶ **Theorem 8-2-5**

Judgment of Harmonic Function

If $\vec{v}$ is both irrotational vector field and a solenoidal vector field in region D, then the scalar potential function ϕ for $\vec{v}$ must be a harmonic function in D, that is $\nabla^2 \phi = 0$.

Proof

$\because \nabla \times \vec{v} = 0$, hence, by the "Existence Theorem for Gradient Fields," there must exist a scalar function ϕ, which causes $\vec{v} = \nabla \phi$.

$\because \nabla \cdot \vec{v} = 0$, therefore $\nabla \cdot \vec{v} = \nabla \cdot \nabla \phi = \nabla^2 \phi = 0$.

◀

▶ **Theorem 8-2-6**

The Divergence and Curl of the Position Vector

Assuming that $\vec{r} = x\vec{i} + y\vec{j} + z\vec{k}$, then $\nabla \cdot \vec{r} = 3$, $\nabla \times \vec{r} = \vec{0}$.

Proof

$$\nabla \cdot \vec{r} = \frac{\partial x}{\partial x} + \frac{\partial y}{\partial y} + \frac{\partial z}{\partial z} = 3 \quad \text{(exactly calculated the dimension).}$$

$$\nabla \times \vec{r} = \begin{vmatrix} \vec{i} & \vec{j} & \vec{k} \\ \dfrac{\partial}{\partial x} & \dfrac{\partial}{\partial y} & \dfrac{\partial}{\partial z} \\ x & y & z \end{vmatrix} = \vec{0} \quad \text{(represents } \vec{r} \text{ is irrotational vector field).} \quad ◀$$

▶ **Theorem 8-2-7**

Differential Quantity = Component of the Gradient in the Direction of the Tangent

Let $\phi = \phi(x, y, z)$, then $d\phi = \nabla \phi \cdot d\vec{r}$, where $d\vec{r} = dx\,\vec{i} + dy\,\vec{j} + dz\,\vec{k}$.

Proof

From the definition of total differential, we know

$$d\phi = \frac{\partial \phi}{\partial x}dx + \frac{\partial \phi}{\partial y}dy + \frac{\partial \phi}{\partial z}dz = (\vec{i}\frac{\partial \phi}{\partial x} + \vec{j}\frac{\partial \phi}{\partial y} + \vec{k}\frac{\partial \phi}{\partial z}) \cdot (dx\,\vec{i} + dy\,\vec{j} + dz\vec{k}) = \nabla \phi \cdot d\vec{r} .$$

◀

Example 2

Assuming $\vec{r} = x\vec{i} + y\vec{j} + z\vec{k}$, and $R = |\vec{r}| = \sqrt{x^2 + y^2 + z^2}$, prove that $\nabla R = \dfrac{\vec{r}}{R}$.

Solution

$$\nabla R = \frac{\partial R}{\partial x}\vec{i} + \frac{\partial R}{\partial y}\vec{j} + \frac{\partial R}{\partial z}\vec{k} ,$$

through $\dfrac{\partial R}{\partial x} = \dfrac{x}{\sqrt{x^2 + y^2 + z^2}} = \dfrac{x}{R}$, similarly: $\dfrac{\partial R}{\partial y} = \dfrac{y}{R}$, $\dfrac{\partial R}{\partial z} = \dfrac{z}{R}$,

$\therefore \nabla R = \dfrac{x}{R}\vec{i} + \dfrac{y}{R}\vec{j} + \dfrac{z}{R}\vec{k} = \dfrac{\vec{r}}{R}$.　　Q.E.D.

Example 3

Assuming $\phi = \phi(u)$, $u = u(x, y, z)$, prove that $\nabla\phi = \phi'(u)\nabla u$.

Solution

$$\nabla\phi = \dfrac{\partial\phi}{\partial x}\vec{i} + \dfrac{\partial\phi}{\partial y}\vec{j} + \dfrac{\partial\phi}{\partial z}\vec{k}$$

$$= \dfrac{\partial\phi}{\partial u}\left(\dfrac{\partial u}{\partial x}\vec{i} + \dfrac{\partial u}{\partial y}\vec{j} + \dfrac{\partial u}{\partial z}\vec{k}\right)$$

$$= \dfrac{\partial\phi}{\partial u}\nabla u = \phi'(u)\nabla u .$$　　Q.E.D.

Example 4

Let $f = f(R)$ be a smooth function, find $\nabla f(R)$, ∇R^n, $\nabla\left(\dfrac{1}{R}\right)$, where

$R = |\vec{r}| = \sqrt{x^2 + y^2 + z^2}$.

Solution

(1) $\nabla f(R) = f'(R) \cdot \nabla R = f'(R) \cdot \dfrac{\vec{r}}{R}$, where $\vec{r} = x\vec{i} + y\vec{j} + z\vec{k}$.

(2) Let $f(R) = R^n$, then $\nabla R^n = (R^n)'\nabla R = n \cdot R^{n-1} \cdot \dfrac{\vec{r}}{R} = n \cdot R^{n-2}\vec{r}$.

(3) Let $f(R) = \left(\dfrac{1}{R}\right)$, then $\nabla\left(\dfrac{1}{R}\right) = \left(\dfrac{1}{R}\right)'\nabla R = -\dfrac{1}{R^2}\dfrac{\vec{r}}{R} = -\dfrac{\vec{r}}{R^3}$.　　Q.E.D.

Example 5

Find (1) $\nabla \cdot (f(R)\vec{r})$; (2) $\nabla \times (f(R)\vec{r})$.

Solution

(1) $\nabla \cdot (f(R)\vec{r}) = \nabla f(R) \cdot \vec{r} + f(R) \nabla \cdot \vec{r} = f'(R)\dfrac{\vec{r}}{R} \cdot \vec{r} + f(R) \cdot 3 = Rf'(R) + 3f(R)$.

(2) $\nabla \times (f(R)\vec{r}) = \nabla f(R) \times \vec{r} + f(R) \nabla \times \vec{r} = f'(R)\dfrac{\vec{r}}{R} \times \vec{r} + f(R) \cdot 0 = 0 + 0 = 0$.   Q.E.D.

In physical systems, electromagnetic fields (or gravitational fields) $\vec{F} = \dfrac{A}{R^3}\vec{r}$, (the common gravitational fields $\vec{F} = \dfrac{GMm}{R^2}\vec{e}_r$, where $\vec{e}_r = \dfrac{\vec{r}}{R}$). From the previous example, we know that $\nabla \times \vec{F} = \nabla \times \dfrac{A}{R^3}\vec{r} = \mathbf{0}$. Therefore, gravitational fields are non-rotational vector fields, simultaneously, $\nabla \cdot \vec{F} = \nabla \cdot \dfrac{A}{R^3}\vec{r} = \mathbf{0}$, so electromagnetic fields are solenoidal.

8-2-2   Direction Derivative

The rate of change of a multivariable function can vary depending on the perspective (as illustrated in the mountain climbing scenario mentioned at the beginning of this chapter). Next, we will use mathematical notation to precisely describe this concept.

► Definition 8-2-3

The Directional Derivative of a Scalar Field

Let $F(\vec{r})$ be a scalar field defined in domain D. If $\vec{r}_p$ is the position vector of a point P in D, and $\vec{e}$ is a unit vector at $\vec{r}_p$, then the directional derivative of the scalar field $F(\vec{r})$ at point $\vec{r}_p$ along with the direction of $\vec{e}$ is given by:

$$D_{\vec{e}}F(\vec{r}_p) = \lim_{\Delta\ell \to 0} \frac{F(\vec{r}_p + \Delta\ell\vec{e}) - F(\vec{r}_p)}{\Delta\ell} .$$

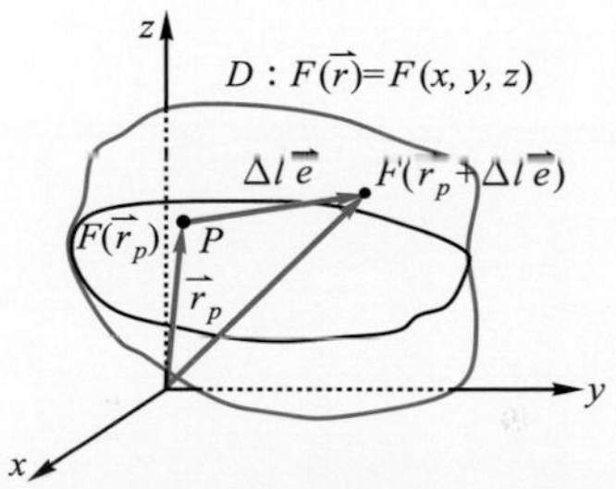

Figure 8-10   The schematic diagram of directional derivative

This is called the directional derivative of the scalar field $F(\vec{r})$ at point $\vec{r}_p$ along with the direction of $\vec{e}$, as shown in Figure 8-10.   ◄

For a function $F(\vec{r}) = F(x, y, z)$ defined on $\mathbf{R}^3$, we can write $\vec{r}_p = P(x_0, y_0, z_0)$, $\vec{u} = u_1\vec{i} + u_2\vec{j} + u_3\vec{k}$ a unit vector, which is $\vec{e}$ in the definition 8-2-3, therefore we get

$$D_{\vec{u}}F(P) = \lim_{\Delta\ell \to 0} \frac{F(x_0 + \Delta\ell u_1, \ y_0 + \Delta\ell u_2, \ z_0 + \Delta\ell u_3) - F(x_0, y_0, z_0)}{\Delta l}$$

From the previously learned partial derivatives in calculus, we can derive the following special cases:

(1) $\vec{u} = \vec{i} \Rightarrow D_{\vec{u}}F(P) = F_x(P)$;

(2) $\vec{u} = \vec{j} \Rightarrow D_{\vec{u}}F(P) = F_y(P)$;

(3) $\vec{u} = \vec{k} \Rightarrow D_{\vec{u}}F(P) = F_z(P)$.

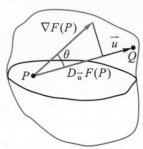

Next is the most important theorem about directional derivatives, which explains that the directional derivative is actually the projection of the gradient onto the direction of differentiation, as shown in Figure 8-11. We will explore the reasoning behind this from the perspective of "level curves."

Figure 8-11 The solution of directional derivative

1. The Calculation of Directional Derivative

Level surface is a type of surface. In space, if every point has a set value (constant value). This value may be temperature, speed, pressure, potential. The level surfaces within an infinitesimally small region of the scalar field $F(x, y, z)$ can be considered as planes of constant value, as shown in Figure 8-12. The directional derivative of F at point P along $\vec{e}$ is

$$\lim_{\Delta\ell \to 0} \frac{\Delta F}{\Delta\ell} = \lim_{\Delta l \to 0} \frac{\varepsilon}{\Delta\ell} = \lim_{\Delta\ell \to 0} \frac{\varepsilon}{\Delta\ell_n} \frac{\Delta\ell_n}{\Delta\ell} = \lim_{\Delta\ell \to 0} \frac{\varepsilon}{\Delta\ell_n} \cos\theta = \lim_{\Delta\ell \to 0} \frac{\varepsilon}{\Delta\ell_n} \vec{n} \cdot \vec{e} \ .$$

So, along the direction of the normal vector, the distance Δl_n is minimized, that is

$$\nabla F(P) = \lim_{\Delta\ell \to 0} \frac{\varepsilon}{\Delta\ell_n} \vec{n} \ ,$$

which is the maximum rate of change. In other words, the gradient $\nabla F(P)$ represents the maximum rate of change of the scalar field $F(x, y, z)$ at point P; it can also be symbolically represented as

$$\nabla F(P) = \text{grad}(F(P)) \triangleq \lim_{\Delta\ell \to 0} \left(\frac{F(\vec{r}_p + \Delta\vec{\ell} e_{\max}) - F(\vec{r}_p)}{\Delta\ell} \right) \vec{e}_{\max}$$

$\vec{e}_{\max}$: The direction of the maximum rate of change
The vector of the maximum rate of change.

Therefore, we can state the following important theorem.

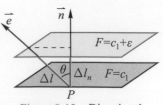

Figure 8-12 Directional Derivative Micrograph

▶ **Theorem 8-2-8**

The Differential Quantity = the Component of the Gradient in the

Tangent Direction

For a first-order continuously differentiable scalar field $F(x, y, z)$

(1) $\nabla F(P) = \text{grad}(F(P))$

$$= \lim_{\Delta \ell \to 0} \left(\frac{F(\vec{r}_p + \Delta \, \vec{\ell e}_{\max}) - F(\vec{r}_p)}{\Delta \ell} \right) \vec{e}_{\max}$$

(2) The directional derivative of the scalar field F at point P along the direction of a unit vector $\vec{e}$ is $\nabla F(P) \cdot \vec{e}$.

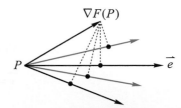

Figure 8-13   Directional Derivatives in Different Directions

◀

▶ **Theorem 8-2-9**

The Calculation of Directional Derivative

Let $F(\vec{r})$ be a scalar field defined in $D \subseteq \mathrm{R}^3$. If P is a point in D, Q is any point in the neighborhood of P, and $F(\vec{r})$ is a first-order continuously differentiable function at point P, then the directional derivative along the direction $\vec{u} = \overrightarrow{PQ}$ is

$$D_{\vec{u}} F(P) = \nabla F(P) \cdot \frac{\vec{u}}{|\vec{u}|} = |\nabla F(P)| \cos \theta .$$

Specifically, the directional derivative of F at point P along the direction of the unit vector $\vec{e}$ is $\nabla F(P) \cdot \vec{e}$.

◀

Example 6

If $f(x, y, z) = 3x^2 - y^3 + 2z^2$, compute the directional derivative of the function $f(x, y, z)$ at the point $f(x, y, z)$ along the direction of vector $\vec{v} = 2\vec{i} + 2\vec{j} - \vec{k}$.

Solution

$\nabla f = 6x\vec{i} - 3y^2\vec{j} + 4z\vec{k}$,

$\nabla f(P) = 6\vec{i} - 3\vec{j} + 8\vec{k}$,

$D_{\vec{v}} f(P) = \nabla f(P) \cdot \frac{\vec{v}}{|\vec{v}|} = (6\vec{i} - 3\vec{j} + 8\vec{k}) \cdot \frac{(2\vec{i} + 2\vec{j} - \vec{k})}{\sqrt{2^2 + 2^2 + 1^2}} = -\frac{2}{3}$.

Q.E.D.

Example 7

Find the directional derivative of the function $f(x, y) = 2x^2y^3 + 6xy$ at point $P(2, 1)$, along the direction $\dfrac{\pi}{6}$ that makes an angle with the positive x-axis.

Solution

$\nabla f(x, y) = (4xy^3 + 6y)\vec{i} + (6x^2y^2 + 6x)\vec{j}$,

$\nabla f(2, 1) = (8 + 6)\vec{i} + (24 + 12)\vec{j} = 14\vec{i} + 36\vec{j}$,

$\vec{v} = \cos\dfrac{\pi}{6}\vec{i} + \sin\dfrac{\pi}{6}\vec{j} = \dfrac{\sqrt{3}}{2}\vec{i} + \dfrac{1}{2}\vec{j}$,

$\therefore D_{\vec{v}}f(2, 1) = (14\vec{i} + 36\vec{j}) \cdot (\dfrac{\sqrt{3}}{2}\vec{i} + \dfrac{1}{2}\vec{j}) = 7\sqrt{3} + 18$.   Q.E.D.

2. The Extremum of the Directional Derivative

In $\mathbf{R}^3$, according to the formula for directional derivatives, we have the following results.

▶ **Theorem 8-2-10**

The Extremum of the Directional Derivative

(1) If $\vec{u} = \dfrac{\nabla F(P)}{|\nabla F(P)|}$, that is $\theta = 0$, the directional derivative attains its maximum value,

and $\max\{D_{\vec{u}}F(P)\} = \nabla F(P) \cdot \dfrac{\nabla F(P)}{|\nabla F(P)|} = |\nabla F(P)|$.

(2) If $\vec{u} = -\dfrac{\nabla F(P)}{|\nabla F(P)|}$, that is $\theta = \pi$, the directional derivative attains its minimum value,

and $\min\{D_{\vec{u}}F(P)\} = \nabla F(P) \cdot (\dfrac{-\nabla F(P)}{|\nabla F(P)|}) = -|\nabla F(P)|$.   ◀

Example 8

Find the direction in which the function $f(x, y, z) = x^2 + y^2 + z^2$ increases most rapidly at $P_0(1, 1, 1)$ and determine this rate of change?

Solution $\nabla f = 2x\vec{i} + 2y\vec{j} + 2z\vec{k}$, $\nabla f(1, 1, 1) = 2\vec{i} + 2\vec{j} + 2\vec{k}$.

(1) The direction of the maximum directional derivative is $\dfrac{\nabla f(P_0)}{|\nabla f(P_0)|} = \dfrac{1}{\sqrt{3}}\vec{i} + \dfrac{1}{\sqrt{3}}\vec{j} + \dfrac{1}{\sqrt{3}}\vec{k}$.

(2) The maximum rate of change is $|\nabla f(P_0)| = 2\sqrt{3}$.

▶ Theorem 8-2-11

Normal Vector of the Surface

Suppose $F(x, y, z)$ is a smooth level surface in $\mathbb{R}^3$, and $P(x_0, y_0, z_0) = \vec{r}(0)$ is a point on the surface. Then, the normal vector at point P is $\nabla F(P)$. In other words, for any tangent vector $\vec{r}'(0)$ passing through $P(x_0, y_0, z_0)$, we have

$$\nabla F(P) \cdot \frac{d\vec{r}}{dt}(0) = 0.$$

Proof

① From the parameter curve

Consider curve $\vec{r}(t) = (x(t), y(t), z(t))$ and $P = \vec{r}(0)$. Then, in the contour surface $F(x, y, z) = c$, according to the chain rule, we know,

$$\left. \frac{dF(x, y, z)}{dt} \right|_{t=0} = \frac{\partial F}{\partial x} x'(0) + \frac{\partial F}{\partial y} y'(0) + \frac{\partial F}{\partial z} z'(0)$$

$$= \left(\left. \frac{\partial F}{\partial x} \right|_P, \left. \frac{\partial F}{\partial y} \right|_P, \left. \frac{\partial F}{\partial z} \right|_P \right) \cdot (x'(0), y'(0), z'(0))$$

$$= \nabla F(P) \cdot \frac{d\vec{r}}{dt}(0).$$

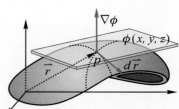

Hence, ∇F is the surface normal vector at point P, as illustrated in Figure 8-14.

Figure 8-14　Surface Normal Vector Diagram in Space

② From total differential

Assuming there is a family of surfaces described by $\phi(x, y, z) = c$, the total differential formula offers another approach for the proof:

$$d\phi(x, y, z) = \frac{\partial \phi}{\partial x} dx + \frac{\partial \phi}{\partial y} dy + \frac{\partial \phi}{\partial z} dz$$

$$= \left(\frac{\partial \phi}{\partial x} \vec{i} + \frac{\partial \phi}{\partial y} \vec{j} + \frac{\partial \phi}{\partial z} \vec{k} \right) \cdot (dx\,\vec{i} + dy\,\vec{j} + dz\,\vec{k})$$

$$= \nabla \phi \cdot d\vec{r} = 0.$$

where $d\vec{r}$ is the tangent vector to the surface, therefore $\nabla \phi$ is the normal vector of the surface. In other words, the unit normal vector of the surface $\phi(x, y, z) = c$ is

$$\vec{e}_n = \pm \frac{\nabla \phi}{|\nabla \phi|}.$$

◀

Example 9

Find the normal vector directional derivative of the function $f(x, y, z) = x + 3y^2 + 4z^3$ at the point $P(\frac{1}{2}, \frac{1}{2}, 2)$ along the surface $z = 4x^2 + 4y^2$.

Solution

$\nabla f = \vec{i} + 6y\,\vec{j} + 12z^2\,\vec{k}$, $\nabla f(P) = \vec{i} + 3\,\vec{j} + 48\,\vec{k}$, let the surface function be

$\phi = 4x^2 + 4y^2 - z = 0$,

$\nabla \phi = 8x\,\vec{i} + 8y\,\vec{j} - \vec{k}$, $\nabla \phi(P) = 4\,\vec{i} + 4\,\vec{j} - \vec{k}$, then the normal vector of the surface ϕ is

$$\vec{n} = \pm \frac{\nabla \phi(P)}{|\nabla \phi(P)|} = \pm(\frac{4\,\vec{i} + 4\,\vec{j} - \vec{k}}{\sqrt{33}}),$$

$$\therefore D_{\vec{u}} f(P) = \nabla f(P) \cdot \vec{n} = \pm \frac{32}{\sqrt{33}}.$$

Q.E.D.

Example 10

(1) Find the unit normal vector and the normal parametric equation of the surface $x^2 + y^2 + z^2 = 3$ at the point $P(1, 1, 1)$.

(2) Find the tangent plane of the surface at point $P(1, 1, 1)$.

Solution

(1) Let the surface function $\phi = x^2 + y^2 + z^2 - 3 = 0$,

then $\nabla \phi = 2x\,\vec{i} + 2y\,\vec{j} + 2z\,\vec{k}$,

$\nabla \phi(1, 1, 1) = 2\,\vec{i} + 2\,\vec{j} + 2\,\vec{k}$,

we can take the normal vector of the surface be $\vec{N} = \vec{i} + \vec{j} + \vec{k}$,

the unit normal vector is

$$\pm \frac{\nabla \phi}{|\nabla \phi|} = \pm \frac{\vec{i} + \vec{j} + \vec{k}}{\sqrt{3}}.$$

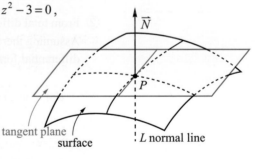

The normal parametric equation is $L : \begin{cases} x = 1 + t \\ y = 1 + t, t \in R. \\ z = 1 + t \end{cases}$

(2) $\vec{n} = (1, 1, 1) \Rightarrow$ tangent plane is

$(x - 1) + (y - 1) + (z - 1) = 0 \Rightarrow x + y + z = 3$.

Q.E.D.

tangent plane / surface / L normal line

8-2   Exercises

Basic questions

1. Supposing $\vec{F} = 2x\vec{i} + y\vec{j} + z\vec{k}$,
    please find $\nabla \cdot \vec{F}$ and $\nabla \times \vec{F}$.

2. Supposing $\phi(x, y, z) = 2x - 2y$,
    please find $\nabla\phi$ and verify $\nabla \times (\nabla\phi) = \vec{0}$.

3. Whether $\vec{F} = x\vec{i} + y\vec{j} - 2z\vec{k}$ is a solenoidal vector
    field?

4. Let $\phi(x, y, z) = xyz$, calculate its gradient and the
    Laplacian.

5. Supposing $f(x, y) = 100 - 2x^2 - y^2$, calculate
    $\nabla f(1, 2)$.

6. Find the directional derivative of
    $f(x, y, z) = x + y + z$ at point $P(1, 0, 0)$ in the
    direction of vector $\vec{a} = \vec{i} + \vec{j}$.

7. Find the directional derivative of $f(x, y, z) = xyz$ at
    point $P(1, 1, 0)$ along the direction of vector
    $\vec{a} = \vec{i} + 2\vec{j} - 2\vec{k}$.

8. Find the directional derivative of
    $f(x, y, z) = 2x^2 + 3y^2 + z^2$ at point $P(2, 1, 3)$ along
    the direction of vector $\vec{a} = \vec{i} - 2\vec{k}$.

9. Find the directional derivative of $f(x, y) = (xy + 1)^2$
    at point $P(3, 2)$ along the direction formed by
    vectors $(3, 2)$ and $(5, 3)$.

10. Find the direction in which the scalar field
    $F(x, y) = x^3 - y^3$ decreases most rapidly at point
    $P(2, -2)$ and determine this rate of change.

Advanced questions

1. In the following questions, please find
    $\nabla \cdot \vec{F}$ and $\nabla \times \vec{F}$, and verify $\nabla \cdot (\nabla \times \vec{F}) = 0$
    (1) $\vec{F} = 2xy\vec{i} + x^2 e^y\vec{j} + 2z\vec{k}$.
    (2) $\vec{F} = \cosh(xyz)\vec{j}$.

2. Find the gradient $\nabla\phi$ of following functions, and
    verify $\nabla \times (\nabla\phi) = 0$.
    (1) $\phi(x, y, z) = \cos(xz)$.
    (2) $\phi(x, y, z) = xyz + e^x$.

3. Please determine the following vector fields, which
    one is a solenoidal vector field (incompressible field)?
    (1) $\vec{F} = 3xy^2\vec{i} - y^3\vec{j} + e^{xyz}\vec{k}$.
    (2) $\vec{F} = \sin(y)\vec{i} + \cos(x)\vec{j} + z\vec{k}$.
    (3) $\vec{F} = (z^2 - 3x)\vec{i} - 3x\vec{j} + (3z)\vec{k}$.

4. Please determine the following vector fields, which
    one is irrotational vector field?
    (1) $\vec{F} = x\vec{i} + y\vec{j} + z\vec{k}$.
    (2) $\vec{F} = yz\vec{i} + xz\vec{j} + xy\vec{k}$.
    (3) $\vec{F} = xy\vec{i} + xy\vec{j} + z^2\vec{k}$.
    (4) $\vec{F} = y^3\vec{i} + (3xy^2 - 4)\vec{j} + z\vec{k}$.

5. Calculate the gradient and the Laplacian of
    following functions.
    (1) $\phi(x, y, z) = \dfrac{xy^2}{z^3}$.
    (2) $\phi(x, y, z) = xy \cos(yz)$.

6. Calculate the divergence and curl of the following
    vector functions.
    (1) $\vec{V}(x, y, z) = xz\vec{i} + yz\vec{j} + xy\vec{k}$.
    (2) $\vec{V}(x, y, z) = 10yz\vec{i} + 2x^2z\vec{j} + 6x^3\vec{k}$.
    (3) $\vec{V}(x, y, z) = xe^{-z}\vec{i} + 4yz^2\vec{j} + 3ye^{-z}\vec{k}$.

7. Calculate the gradient of each function at point P.
    (1) $f(x, y, z) = 2x^2 + 3y^2 + z^2$, $P(2, 1, 3)$.
    (2) $f(x, y, z) = x^2z + yz^2$, $P(1, 0, 2)$.

8. Find the directional derivative of the function
    $F(x, y, z) = xy^2 - 4x^2y + z^2$ at point $P(1, -1, 2)$
    along the direction of $\vec{u} = 6\vec{i} + 2\vec{j} + 3\vec{k}$.

9. Given the function $f(x, y, z) = x + y + z$ and the line
    $S: x = t, y = 2t, z = 3t$, find the directional derivative
    of f at point $P(1, 2, 3)$ along the direction of the line S.

10. Find the directional derivative of the function
$f(x, y) = 5x^3y^6$ at point $P(-1, 1)$ along the direction
that makes an angle $\dfrac{\pi}{6}$ with the positive x-axis.

11. Find the directional derivative of
$F(x, y, z) = x^2y^2(2z + 1)^2$ at point $P(1, -1, 1)$ along
the direction of the line formed by the points
$(1, -1, 1)$ and $(0, 3, 3)$.

12. If there is a scalar field F and a point P on it, find
the direction in which the scalar field F increases
most rapidly at point P, and determine this rate of
change.

(1) $F(x, y) = e^{2x} \sin y$, $P(0, \dfrac{\pi}{4})$.

(2) $F(x, y, z) = x^2 + 4xz + 2yz^2$, $P(1, 2, -1)$.

13. If there is a scalar field F and a point P on it, find
the direction in which the scalar field F decreases
most rapidly at point P, and determine this rate of
change. $F(x, y, z) = \ln(\dfrac{xy}{z})$, $P(\dfrac{1}{2}, \dfrac{1}{6}, \dfrac{1}{3})$.

14. Find the tangent plane and the normal parametric
equation of the surface $x^2 + y^2 + z^2 = 4$ at the point
$P(1, 1, \sqrt{2})$.

15. Find the tangent plane and the normal vector of the
surface $z = xy^2$ at the point $P(1, 1, 1)$.

16. If $\vec{V} = (5x - 7)\vec{i} + (3y + 13)\vec{j} - 4\alpha z\vec{k}$,
what value of α makes $\vec{V}$ an incompressible flow
field or a solenoidal electromagnetic field?

17. If there is a temperature field given by
$f(x, y, z) = x^2 + y^2 + z^2$, and your current position is
$(1, 0, 3)$, in which direction should you walk to feel
cooler as quickly as possible?

18. Calculate the directional derivative of the function
$T(x, y, z) = x^2 + 2y^2 + 3z^2$ at point $P(0, 1, 2)$ along
the direction of the line $S: x = t, y = t + 1, z = t + 2$.

19. Find $(1)\nabla^2(f(R))$ $(2)\nabla^2 R^n$, where
$R = \sqrt{x^2 + y^2 + z^2}$, $\vec{r} = x\vec{i} + y\vec{j} + z\vec{k}$.

20. Find the directional derivative of
$f(x, y, z) = x^2 + y^2 + z^2$ at point $P(2, 2, 2)$
along the direction of the line $\vec{a} = \vec{i} + 2\vec{j} - 3\vec{k}$.

8-3 Line Integrals

This section introduces line integrals in vectors, which includes scalar function line integrals and vector function line integrals. The integrals discussed in this section are frequently used to calculate the work done by force fields. However, before discussing line integrals, it is necessary to understand the various forms of curves C along which the line integral is taken. Several common forms will be introduced first, followed by an explanation of how line integrals are computed.

8-3-1 The Classification of Curves

1. **Smooth Curve**

   If the tangent vector $\vec{e}_t = \dfrac{d\,\vec{c}(t)}{dt}$ of the curve C is continuous, then the curve C is called a smooth curve, as shown in Figure 8-15.

2. **Piecewise Smooth Curve**

   If the curve C is the sum of a finite number of smooth curves, and the tangent vector $\vec{e}_t$ is piecewise continuous, then the curve C is called a piecewise smooth curve, as shown in Figure 8-16.

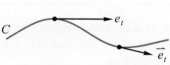

Figure 8-15 Smooth Curve

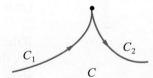

Figure 8-16 Piecewise Smooth Curve

3. **Classify Based on the Number of Self-Intersection Points of the Curve**

   (1) Multiple Point

   The self-intersection points of the curve are called multiple points, as shown in Figure 8-17.

   (2) Simple Curve

   A curve that does not have multiple points is called a simple curve, as shown in Figure 8-18.

   (3) Regular Curve

   A piecewise smooth and simple curve is called a regular curve.

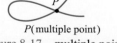

Figure 8-17 multiple point

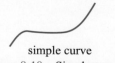

Figure 8-18 Simple curve

4. Closed Curve

If the curve C is $\vec{r}(t) = x(t)\vec{i} + y(t)\vec{j} + z(t)\vec{k}$, $t \in [a, b]$, and it satisfies $\vec{r}(a) = \vec{r}(b)$ then the curve is called a closed curve. Depending on whether the closed curve has self-intersection points, it can be classified into a simple closed curve and a complex closed curve, as shown in Figures 8-19 and 8-20.

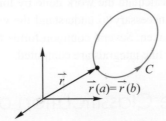
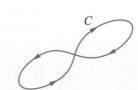

Figure 8-19   simple closed curve   Figure 8-20   A complex closed curve (a closed curve that has multiple points)

8-3-2   Line Integral of a Scalar Function

For the line integral of a scalar function, we can define it using the concept of Riemann integration.

▶ **Definition 8-3-1**

Line Integral of a Scalar Function

If the function $F(x, y, z)$ is a continuous function on $\mathbf{R}^3$, then the line integral of the function $F(x, y, z)$ along the curve C is defined as:

$$\int_C F(x, y, z)ds = \lim_{n \to \infty} \sum_{i=1}^{n} F(x_i, y_i, z_i) \cdot \Delta s_i$$

where max $(\Delta s_i) \to 0$, as shown in Figure 8-21.

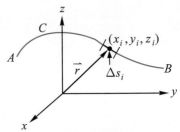

Figure 8-21   Line integral of a curve in space

◀

1. The Properties of Line Integral

The line integral is defined by the Riemann integral, and the following properties should come as no surprise to readers familiar with calculus. Naturally, their proofs can be directly derived from the original definition.

(1) $\int_C [aF_1(x, y, z) + bF_2(x, y, z)]ds = a\int_C F_1(x, y, z)ds + b\int_C F_2(x, y, z)ds$.

(2) $\int_C F(x, y, z)ds = -\int_{-C} F(x, y, z)ds$, where $(-C)$ is the path that is opposite in direction to C.

(3) $\int_{C_1+C_2} F(x,\,y,\,z)ds = \int_{C_1} F(x,\,y,\,z)ds + \int_{C_2} F(x,\,y,\,z)ds$.

(4) If $F(x,y,z)=1$, then $\int_C F(x,\,y,\,z)ds$ express the arc length of curve C.

(5) If $F(x,y,z)$ represents the density function of the curve C,

then $\int_C F(x,\,y,\,z)ds$ represents the mass of the curve C.

2. **Explicit Calculation of Line Integral**

   According to the original definition, to calculate the line integral, we need to express the infinitesimal variable Δs in terms of the coordinate system, as mentioned in the previous section. Therefore, the results are listed as follows. It is important to note that the curve here may not be smooth over the entire interval, so when performing the integral, it may need to be divided into several smooth segments and calculate them separately.

   (1) Space curve C: $\vec{r}(t)=x(t)\vec{i}+y(t)\vec{j}+z(t)\vec{k}$, parameter $t\in[a,b]$,

   the differential arc length of the curve is $ds = \sqrt{(\dfrac{dx}{dt})^2 +(\dfrac{dy}{dt})^2 +(\dfrac{dz}{dt})^2}\, dt$,

   we obtain $\int_C F(x,\,y,\,z)ds = \int_a^b F(x(t),\,y(t),\,z(t))\sqrt{(\dfrac{dx}{dt})^2 +(\dfrac{dy}{dt})^2 +(\dfrac{dz}{dt})^2}\, dt$.

   (2) Space curve C: $\vec{r}(t)=t\,\vec{i}+y(t)\vec{j}+z(t)\vec{k}$, parameter $t\in[a,b]$, the differential arc length of the curve is $ds = \sqrt{1+(\dfrac{dy}{dt})^2 +(\dfrac{dz}{dt})^2}\, dt$,

   we obtain $\int_C F(x,\,y,\,z)ds = \int_a^b F(t,\,y(t),\,z(t))\sqrt{1+(y')^2 +(z')^2}\, dt$.

Example 1

Find $\int_C (x^2 + y^2 + z^2)ds$, where C is: $\vec{r}(t)=\cos t\,\vec{i}+\sin t\,\vec{j}+3t\,\vec{k}$, and the curve from point $(1,0,0)$ to point $(1,0,6\pi)$.

Solution

In $\vec{r}(t)=\cos t\,\vec{i}+\sin t\,\vec{j}+3t\,\vec{k}$, let $\begin{cases} x=\cos t \\ y=\sin t \\ z=3t \end{cases}$, then from $(1,0,0)$ to $(1,0,6\pi)$, we know

$t\in[0,2\pi]$. According to the formula for the differential arc length, we have

$ds = \sqrt{(\dfrac{dx}{dt})^2 +(\dfrac{dy}{dt})^2 +(\dfrac{dz}{dt})^2}\, dt$, where $\dfrac{dx}{dt}=-\sin t$, $\dfrac{dy}{dt}=\cos t$, $\dfrac{dz}{dt}=3$,

then substituting into the arc length formula, we obtain

$\int_C (x^2 + y^2 + z^2)ds = \int_0^{2\pi} [(\cos t)^2 +(\sin t)^2 +(3t)^2]\cdot \sqrt{\sin^2 t + \cos^2 t + 3^2}\, dt$

$= \int_0^{2\pi} \sqrt{10}(1+9t^2)dt = \sqrt{10}(2\pi+24\pi^3)$.

Q.E.D.

Example 2

Find line integral $\int_C xy^3 ds$, where C is the curve that $y = 2x$ from point $(-1, -2)$ to point $(1, 2)$.

Solution

According to the formula for the differential arc length: $ds = \sqrt{1 + (y')^2}\, dx = \sqrt{5}\, dx$,

$$\therefore \int_C xy^3\, ds = \int_{-1}^{1} x \cdot (2x)^3 \cdot \sqrt{5} dx = \int_{-1}^{1} 8\sqrt{5} x^4 dx = 8\sqrt{5} \cdot \frac{1}{5} x^5 \Big|_{-1}^{1} = \frac{16\sqrt{5}}{5}.$$

Q.E.D.

Example 3

Find line integral $\int_C (3x^2 + 3y^2) ds$, where C is the path for the following (1)~(2):

(1) $x + y = 1$, the straight path from $(0, 1)$ to $(1, 0)$.

(2) $x^2 + y^2 = 1$, the counterclockwise path from $(0, 1)$ to $(1, 0)$.

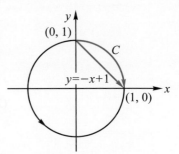

Solution

(1) The path is $y = -x + 1$, so the differential arc length $ds = \sqrt{1 + (\frac{dy}{dx})^2}\, dx = \sqrt{2}\, dx$,

$$\therefore \int_C (3x^2 + 3y^2)\, ds = \int_0^1 [3x^2 + 3(-x+1)^2]\sqrt{2} dx = \int_0^1 (6x^2 - 6x + 3) \cdot \sqrt{2} dx = 2\sqrt{2}.$$

(2) The path is: $x^2 + y^2 = 1$ (counterclockwise), therefore, polar coordinate parametric equations can be utilized.

$x = \cos\theta$, $y = \sin\theta$, we obtain the differential arc length $ds = d\theta$, where $\theta \in [\frac{\pi}{2}, 2\pi]$,

so the original equation $\int_C (3x^2 + 3y^2)\, ds = \int_{\frac{\pi}{2}}^{2\pi} 3 d\theta = \frac{9\pi}{2}$.

Q.E.D.

8-3-3   The Line Integral of Vector Functions

Curve C: $\vec{r} = x\vec{i} + y\vec{j} + z\vec{k}$ which is a smooth curve between two endpoints A and B, and the vector function $\vec{F}(x, y, z) = F_1(x, y, z)\vec{i} + F_2(x, y, z)\vec{j} + F_3(x, y, z)\vec{k}$ is continuous at any point on the corresponding curve C. Then, the line integral of the vector function $\vec{F}(x, y, z) = F_1(x, y, z)\vec{i} + F_2(x, y, z)\vec{j} + F_3(x, y, z)\vec{k}$ along the curve C is defined as

$$\int_C \vec{F}(x, y, z) \cdot d\vec{r} = \int_C F_1(x, y, z)dx + F_2(x, y, z)dy + F_3(x, y, z)dz = \int_C [\vec{F}(x, y, z) \cdot \frac{d\vec{r}}{dt}]dt$$

▶ **Theorem 8-3-1**

The Properties of Line Integral (Same as Scalar Function)

(1) $\int_C \vec{F} \cdot d\vec{r} = -\int_{-C} \vec{F} \cdot d\vec{r}$

(2) $\int_{C_1 + C_2} \vec{F} \cdot d\vec{r} = \int_{C_1} \vec{F} \cdot d\vec{r} + \int_{C_2} \vec{F} \cdot d\vec{r}$

   (C_1, C_2 are two non-overlapping smooth curves)

(3) $\int_C (k_1\vec{F}_1 + k_2\vec{F}_2) \cdot d\vec{r} = k_1 \int_C \vec{F}_1 \cdot d\vec{r} + k_2 \int_C \vec{F}_2 \cdot d\vec{r}$

(4) $\int_C (\vec{F} \cdot d\vec{r}) = \int_C [\vec{F} \cdot \frac{d\vec{r}}{dt}]dt$ ◀

In the line integral of a vector function,

$$\vec{F} \cdot d\vec{r} = |\vec{F}| \cdot |d\vec{r}| \cdot \cos\theta = |\vec{F}| ds \cos\theta = |\vec{F}| \cos\theta ds,$$ so if $\vec{F}(x, y, z)$ represents a force field, $d\vec{r}$ represents displacement, then $\vec{F} \cdot d\vec{r}$ represents the work done by the force field along the curve C for a small displacement. Therefore, $\int_C \vec{F}(x, y, z) \cdot d\vec{r}$ represents the total work done by the force field $\vec{F}(x, y, z)$ while moving along the curve C, as shown in Figure 8-22.

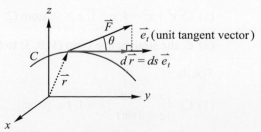

Figure 8-22   The schematic diagram of work

Example 4

$$\vec{F} = xy^2\vec{i} + (x^2y + y^3)\vec{j}\ ,\ C:\quad \vec{r}(t) = \cos t\,\vec{i} + \sin t\,\vec{j} + t\,\vec{k}\ ,$$

$0 \le t \le \dfrac{3}{2}\pi$, calculate $\displaystyle\int_C \vec{F} \cdot d\vec{r}$.

Solution

$$C:\ \begin{cases} x = \cos t \\ y = \sin t\ ,\quad d\vec{r} = (-\sin t\,\vec{i} + \cos t\,\vec{j} + \vec{k})dt\ , \\ z = t \end{cases}$$

$$\vec{F} \cdot d\vec{r} = (-\sin t \cdot \cos t \sin^2 t + \cos t \cdot (\cos^2 t \sin t + \sin^3 t))dt = \sin t \cdot \cos^3 t\ ,$$

$$\therefore \int_C \vec{F} \cdot d\vec{r} = \int_0^{\frac{3\pi}{2}} \sin t \cos^3 t\, dt = -\frac{1}{4}\cos^4 t \Big|_0^{\frac{3\pi}{2}} = \frac{1}{4}\ .$$

Q.E.D.

Example 5

$$\vec{F} = xy\,\vec{i} + x\,\vec{j}\ ,\ C_1: y = x,\ C_2: y = x^2,\ \text{from } (0,0) \text{ to } (1,1),$$

find $\displaystyle\int_{C_1} \vec{F} \cdot d\vec{r}$ and $\displaystyle\int_{C_2} \vec{F} \cdot d\vec{r}$.

Solution

(1) $C_1: x = y,\ d\vec{r} = dx\,\vec{i} + dy\,\vec{j} = (\vec{i} + \vec{j})dx\ ,$

$$\therefore \int_C \vec{F} \cdot d\vec{r} = \int_0^1 (xy + x)dx = \int_0^1 (x^2 + x)dx = \frac{1}{3} + \frac{1}{2} = \frac{5}{6}\ .$$

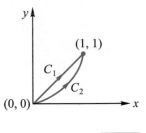

(2) $C_2: y = x^2,\ dy = 2xdx,\ d\vec{r} = dx\,\vec{i} + 2xdx\,\vec{j} = (\vec{i} + 2x\vec{j})dx\ ,$

$$\vec{F} = x^3\,\vec{i} + x\,\vec{j}\ ,\ \therefore \int_{C_2} \vec{F} \cdot d\vec{r} = \int_0^1 (x^3 + 2x^2)dx = \frac{11}{12}\ .$$

Q.E.D.

Example 6

$$\vec{F} = 4\vec{i} - 3x\,\vec{j} + z^2\,\vec{k}\ ,\ \text{find } \int_C \vec{F} \cdot d\vec{r}\ ,\ \text{where}$$

(1) $C: x^2 + z^2 = 4,\ y = 1,\ z \ge 0,\ \text{from } (2, 1, 0) \text{ to } (-2, 1, 0).$

(2) $C:$ the straight line from $(1, 0, 3)$ to $(2, 1, 1)$.

Solution

(1) $C:\ \begin{cases} x = 2\cos t \\ z = 2\sin t \end{cases},\ y = 1,\ t \in [0, \pi];$

$$d\vec{r} = (-2\sin t\,\vec{i} + 0\,\vec{j} + 2\cos t\,\vec{k})dt\ ;$$

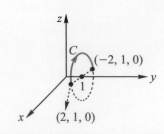

$$\vec{F} = 4\vec{i} - 6\cos t\,\vec{j} + 4\sin^2 t\,\vec{k}\ ,$$

$$\therefore \int_C \vec{F}\cdot d\vec{r} = \int_C 4dx - 3xdy + z^2 dz = \int_0^\pi (-8\sin t - 3\cdot 2\cos t\cdot 0 + 4\sin^2 t\cdot 2\cos t)dt$$

$$= \int_0^\pi (-8\sin t + 8\sin^2 t\cdot\cos t)dt = +8\cos t\Big|_0^\pi + \frac{8}{3}\sin^3 t\Big|_0^\pi = -16\ .$$

(2) C: $\begin{cases} x = 1+t \\ y = t \\ z = 3-2t \end{cases}$, $t \in [0, 1]$; $\vec{F} = 4\vec{i} - 3(1+t)\,\vec{j} + (3-2t)^2\,\vec{k}$;

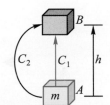

$$\int_C \vec{F}\cdot d\vec{r} = \int_C 4dx - 3xdy + z^2 dz$$

$$= \int_0^1 (4 - 3t - 2(9 - 12t + 4t^2))\,dt$$

$$= \int_0^1 (-14 + 21t - 8t^2)\,dt$$

$$= -14 + \frac{21}{2} - \frac{8}{3} = -\frac{37}{6}\ .$$

Q.E.D.

8-3-4 Conservative Field

From the previous examples, we can observe that general vector and scalar function line integrals depend on the integration path. That is, when integrating along different paths, even with the same starting and ending points, the integral values may differ. However, there exists a type of vector field in physical systems where the integral is independent of the path. The most well-known example is the gravitational field. In a gravitational field, when we

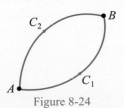

Figure 8-23　Diagram of work in a conservative field

raise an object from point A to point B, the increase in potential energy corresponds to the amount of work done. According to the law of energy conservation in physics, regardless of the path taken, the work done is always equal to mgh, as shown in Figure 8-23. Therefore, the line integral that represents work done is path-independent. Now, we are going to introduce this type of vector field.

1. The Definition of Conservative Field

If $\vec{F}$ is a vector field defined in a region D, and its first-order partial derivatives exist and are continuous in D, and there exists a scalar function ϕ such that for any regular curve C starting at A and ending at B in D, $\int_C \vec{F}\cdot d\vec{r} = \phi(B) - \phi(A)$, then $\int_C \vec{F}\cdot d\vec{r}$ is called path-independent in D, as shown in Figure 8-24. Also, $\vec{F}$, it is referred to as a conservative vector field within D. In general physical systems, this scalar function ϕ is called the potential energy function or potential function.

Figure 8-24

▶ **Theorem 8-3-2**

The Existence of Conservative Field

If $\vec{F}$ is a vector field defined in a simply connected region D, and its first-order partial derivatives exist and are continuous in D, and if $\nabla \times \vec{F} = 0$ within D, then $\vec{F}$ is a conservative vector field within D.

Proof

Because $\nabla \times \vec{F} = 0$, $\vec{F}$ is irrotational vector field. According to the content in section 8-1, there must exist a scalar function ϕ causes $\vec{F} = \nabla \phi$, then

$$\int_C \vec{F} \cdot d\vec{r} = \int_C \nabla \phi \cdot d\vec{r} = \int_C (\frac{\partial \phi}{\partial x} dx + \frac{\partial \phi}{\partial y} dy + \frac{\partial \phi}{\partial z} dz) = \int_C d\phi = \phi \Big|_A^B = \phi(B) - \phi(A).$$

Therefore, this integral is independent of the integration path C and only depends on the starting and ending points. Hence, when solving the line integral of a conservative vector field, it is necessary to first find the scalar function ϕ, and when finding ϕ, it can be obtained through $\vec{F} = \nabla \phi$, and then the partial integration is performed to find ϕ. Further explanations will be provided in the following examples. ◀

▶ **Theorem 8-3-3**

The Line Integral of a Conservative Field over a Closed Loop Is 0

If $\vec{F}$ is a conservative vector field defined in a simply connected region D (as explained on the next page), then for any simple closed curve C inside D (as shown in Figure 8-25), we have $\oint_C \vec{F} \cdot d\vec{r} = 0$, which means the line integral work done over a closed loop inside the conservative field is zero.

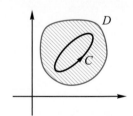

Figure 8-25 Simple closed

Proof

As shown in Figure 8-24, connecting any two curves C_1, C_2 between points A and B, we take $C = C_1 + C_2^*$, where $C_2^* = -C_2$. Then, C becomes a simple closed curve inside D, and by elementary calculus, we know:

$$\int_{C_2^*} \vec{F} \cdot d\vec{r} = \int_{-C_2} \vec{F} \cdot d\vec{r} = -\int_{C_2} \vec{F} \cdot d\vec{r}.$$

Also, since $\vec{F}$ is a conservative field, its line integral of the vector function is independent of the path, that is $\int_{C_1} \vec{F} \cdot d\vec{r} = \int_{C_2} \vec{F} \cdot d\vec{r}$. Therefore,

$$\int_C \vec{F} \cdot d\vec{r} = \int_{C_1} \vec{F} \cdot d\vec{r} + \int_{C_2^*} \vec{F} \cdot d\vec{r} = \int_{C_1} \vec{F} \cdot d\vec{r} - \int_{C_1} \vec{F} \cdot d\vec{r} = 0.$$ ◀

2. Connected Region

In this theorem, we mentioned the concept of a connected region. What does a connected region mean? Its definition is as follows: If any two distinct points within the region can be connected by a piecewise smooth curve entirely contained in that region, then the region is called a connected region. As shown in Figure 8-26, the region with the diagonal lines represents the entire region D, where D_1 is a connected region, while D_2 is not a connected region.

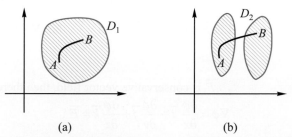

(a)   (b)

Figure 8-26   The schematic diagram of connected (a) and unconnected (b) region.

3. Simply Connected and Multiply Connected Region

If a connected region inwards contracts any simple closed curve C to a point, and the area swept by this process remains entirely within the connected region, then the region is called simply connected. Otherwise, it is called multiply connected. As shown in Figure 8-27, the region with the diagonal lines represents the entire region D, where D_1 is a simply connected region, while D_2 is a multiply connected region. In general, for plane regions, if there are holes within the region, it is a multiply connected region.

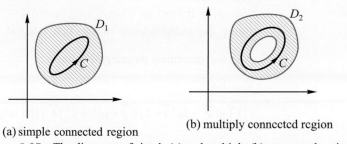

(a) simple connected region   (b) multiply connected region

Figure 8-27   The diagrams of simple (a) and multiply (b) connected region

Example 7

Find $\int_C (e^x \cos y\, dx - e^x \sin y\, dy)$, C: any piecewise smooth curve from $(0, 0)$ to $(2, \dfrac{\pi}{4})$.

Solution

Let $\overrightarrow{F} = e^x \cos y\, \overrightarrow{i} - e^x \sin y\, \overrightarrow{j}$, then $\nabla \times \overrightarrow{F} = \begin{vmatrix} \overrightarrow{i} & \overrightarrow{j} & \overrightarrow{k} \\ \dfrac{\partial}{\partial x} & \dfrac{\partial}{\partial y} & \dfrac{\partial}{\partial z} \\ e^x \cos y & -e^x \sin y & 0 \end{vmatrix} = 0$,

$\therefore \overrightarrow{F}$ is conservative vector field, then there must exist a scalar function ϕ, which causes

$$\nabla \phi = \frac{\partial \phi}{\partial x}\overrightarrow{i} + \frac{\partial \phi}{\partial y}\overrightarrow{j} + \frac{\partial \phi}{\partial z}\overrightarrow{k} = \overrightarrow{F},$$

$$\therefore \begin{cases} \dfrac{\partial \phi}{\partial x} = e^x \cos y \\ \dfrac{\partial \phi}{\partial y} = -e^x \sin y \end{cases} \Rightarrow \begin{matrix} \phi = e^x \cos y + f(y) \\ \phi = e^x \cos y + g(x) \end{matrix}$$

comparing above two formulas then we can choose $\phi(x, y) = e^x \cos y + c$,

therefore, we obtain $\int_C \overrightarrow{F} \cdot d\overrightarrow{r} = e^x \cos y + c \Big|_{(0,0)}^{(2,\frac{\pi}{4})} = \dfrac{e^2}{\sqrt{2}} - 1$.   **Q.E.D.**

Example 8

Prove the independence of the line integral $\int_{(0,2,1)}^{(2,0,1)} ze^x dx + 2yz\, dy + (e^x + y^2)dz$ of the path and determine its integral value.

Solution

(1) Let $\overrightarrow{F} = ze^x \overrightarrow{i} + 2yz\, \overrightarrow{j} + (e^x + y^2)\overrightarrow{k}$, the original integral is $\int_C \overrightarrow{F} \cdot d\overrightarrow{r}$,

then $\nabla \times \overrightarrow{F} = \begin{vmatrix} \overrightarrow{i} & \overrightarrow{j} & \overrightarrow{k} \\ \dfrac{\partial}{\partial x} & \dfrac{\partial}{\partial y} & \dfrac{\partial}{\partial z} \\ ze^x & 2yz & e^x + y^2 \end{vmatrix} = 0$, so $\overrightarrow{F}$ is conservative vector field. Its line

integral is independent of the path.

(2) Let $\overrightarrow{F} = \nabla \phi$, that is $\begin{cases} \dfrac{\partial \phi}{\partial x} = ze^x \\ \dfrac{\partial \phi}{\partial y} = 2yz \\ \dfrac{\partial \phi}{\partial z} = e^x + y^2 \end{cases} \Rightarrow \begin{matrix} \phi = ze^x + f(x, y) \\ \phi = y^2 z + g(x, y) \\ \phi = ze^x + y^2 z + h(x, y) \end{matrix}$,

$\therefore \phi(x, y, z) = ze^x + y^2 z + c$,

$$\therefore \int_{(0,2,1)}^{(2,0,1)} ze^x dx + 2yz dy + (e^x + y^2) dz = \int_{(0,2,1)}^{(2,0,1)} \overrightarrow{F} \cdot d\overrightarrow{r} = ze^x + y^2 z + c \Big|_{(0,2,1)}^{(2,0,1)} = e^2 - 5 .$$

Q.E.D.

Example 9

(1) Prove that $\overrightarrow{F} = (y^2 \cos x + z^3)\overrightarrow{i} + (2y \sin x - 4)\overrightarrow{j} + (3xz^2 + 2)\overrightarrow{k}$ is a conservative vector field.

(2) Find the potential function of $\overrightarrow{F}$.

(3) Calculate the work done by the force $\overrightarrow{F}$ on an object along any piecewise smooth curve from $(0, 1, -1)$ to $(\frac{\pi}{2}, -1, 2)$.

Solution

(1)(2) $\nabla \times \overrightarrow{F} = 0 \Rightarrow \overrightarrow{F}$ is a conservative vector field, then there must exist a scalar function $\phi(x, y, z)$, which causes $\overrightarrow{F} = \nabla \phi(x, y, z)$,

then
$$\begin{cases} \dfrac{\partial \phi}{\partial x} = y^2 \cos x + z^3 \\ \dfrac{\partial \phi}{\partial y} = 2y \sin x - 4 \\ \dfrac{\partial \phi}{\partial z} = 3xz^2 + 2 \end{cases} \Rightarrow \begin{cases} \phi = y^2 \sin x + z^3 x + f(y, z) \\ \phi = y^2 \sin x - 4y + g(x, z) \\ \phi = xz^3 + 2z + h(x, y) \end{cases} ,$$

we obtain after comparing $\phi(x, y, z) = y^2 \sin x + xz^3 - 4y + 2z + c$.

(3) The work done $= \int_C \overrightarrow{F} \cdot d\overrightarrow{r} = \phi(\dfrac{\pi}{2}, -1, 2) - \phi(0, 1, -1) = 15 + 4\pi$.

Q.E.D.

4. In a Simply-Connected Domain, a Vector Field May Have Discontinuous Points Known as Sources or Sinks (Optional Reading)

We already know that if a vector field $\overrightarrow{F}$ in a simply-connected domain has continuous first-order partial derivatives and is conservative within that domain, satisfying $\nabla \times \overrightarrow{F} = 0$, then the line integral of the field's work along a closed path within that region $\oint_C \overrightarrow{F} \cdot d\overrightarrow{r} = 0$. However, if the vector field

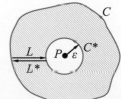

Figure 8-28   The vector field is discontinuous at point P

$\overrightarrow{F}$ has discontinuous points within that connected region, as shown in Figure 8-28, we cannot directly use the conservative approach for calculation. Instead, we must create a small circle around the discontinuous point, forming a multiply-connected domain. Then, we divide this multiply-connected domain into simply-connected regions, allowing us to reduce the calculation to the case of a conservative field.

5. Calculation Method

Let's assume that the vector function is discontinuous at point P. We then take a small circle C^* with a radius ε centered at P: $x^2 + y^2 = \varepsilon^2$. Next, we take a curved path for the line integral $\Gamma = C + L + (-C^*) + L^*$ where two paths L and L^* are two identical straight lines but with opposite directions. The region $\Gamma = C + L + (-C^*) + L^*$ enclosed by these paths is a simply-connected domain with the discontinuous point P removed. In this region $\Gamma = C + L + (-C^*) + L^*$, the vector field $\overrightarrow{F}$ is continuous and conservative, so the line integral is zero: $\oint_C \overrightarrow{F} \cdot d\overrightarrow{r} = 0$, then

$$\overrightarrow{F} \cdot d\overrightarrow{r} = \int_C \overrightarrow{F} \cdot d\overrightarrow{r} + \int_L \overrightarrow{F} \cdot d\overrightarrow{r} + \int_{-C^*} \overrightarrow{F} \cdot d\overrightarrow{r} + \int_{L^*} \overrightarrow{F} \cdot d\overrightarrow{r} = 0.$$

Since L and L^* overlap but have opposite directions, we have:

$$\int_L \overrightarrow{F} \cdot d\overrightarrow{r} = -\int_{L^*} \overrightarrow{F} \cdot d\overrightarrow{r}.$$

$$\oint_C \overrightarrow{F} \cdot d\overrightarrow{r} = \int_C \overrightarrow{F} \cdot d\overrightarrow{r} + \int_{-C^*} \overrightarrow{F} \cdot d\overrightarrow{r} = 0; \quad \int_C \overrightarrow{F} \cdot d\overrightarrow{r} = -\int_{-C^*} \overrightarrow{F} \cdot d\overrightarrow{r} = \int_{C^*} \overrightarrow{F} \cdot d\overrightarrow{r}.$$

Therefore, $\oint_C \overrightarrow{F} \cdot d\overrightarrow{r} = \oint_C \overrightarrow{F} \cdot d\overrightarrow{r}$.

Example 10

Assuming that $\overrightarrow{F} = \dfrac{-y\overrightarrow{i} + x\overrightarrow{j}}{x^2 + y^2}$, find $\oint_C \overrightarrow{F} \cdot d\overrightarrow{r}$,

where C is any arbitrary simple closed curve.

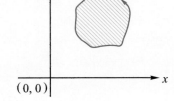

(1) The origin $(0, 0)$ is outside C.

(2) The origin $(0, 0)$ is inside C.

Solution

$$\nabla \times \overrightarrow{F} = \begin{vmatrix} \overrightarrow{i} & \overrightarrow{j} & \overrightarrow{k} \\ \dfrac{\partial}{\partial x} & \dfrac{\partial}{\partial y} & \dfrac{\partial}{\partial z} \\ \dfrac{-y}{x^2 + y^2} & \dfrac{x}{x^2 + y^2} & 0 \end{vmatrix} = 0, \text{ but } \overrightarrow{F} \text{ is discontinuous at } (0, 0), \text{ so we consider}$$

two cases.

(1) When the origin $(0, 0)$ is outside C, because $\nabla \times \overrightarrow{F} = 0$, and the first-order partial derivatives of $\overrightarrow{F}$ exist and are continuous within C, the field $\overrightarrow{F}$ is conservative:

$$\oint_C \overrightarrow{F} \cdot d\overrightarrow{r} = 0.$$

(2) When the origin (0, 0) is inside C, because $\overrightarrow{F}$ is not differentiable at (0, 0), we take
$C^*: x^2 + y^2 = \varepsilon^2$, $x = \varepsilon\cos\theta$, $y = \varepsilon\sin\theta$, $\theta \in [0, 2\pi]$,

∴the first-order partial derivatives of $\overrightarrow{F}$ exist and are
continuous at the region enclosed by C and C^*, hence

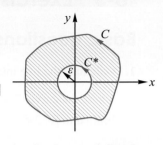

$$\oint_C \overrightarrow{F}\cdot d\overrightarrow{r} = \oint_C \overrightarrow{F}\cdot d\overrightarrow{r} = \oint_C [\frac{-y}{x^2+y^2}dx + \frac{x}{x^2+y^2}dy]$$

$$= \int_0^{2\pi}[\frac{-\varepsilon\sin\theta}{\varepsilon^2}(-\varepsilon\sin\theta d\theta) + \frac{\varepsilon\cos\theta}{\varepsilon^2}(\varepsilon\cos\theta d\theta)]$$

$$= \int_0^{2\pi} d\theta = 2\pi,$$

$$\therefore \oint_C \overrightarrow{F}\cdot d\overrightarrow{r} = 2\pi .$$

8-3 Exercises

Basic questions

1. Calculate the line integral $\int_C (x^2 + y^2 + z^2)ds$, where the curve C is $\vec{r}(t) = \sin t\,\vec{i} + \cos t\,\vec{j} + 1\vec{k}$ limited to $[0, 2\pi]$.

2. Calculate the line integral $\int_C xyds$, where the curve C is the line $y = x$ at the xy-plane and is limited to $0 \le x \le 1$.

3. Find $\int_C (xy + z^2)ds$, where

   $C:\ \vec{r}(t) = \cos t\,\vec{i} + \sin t\,\vec{j} + t\vec{k}$,

   the curve from $(1, 0, 0)$ to $(-1, 0, \pi)$.

4. Find $\int_C (xy\,dx + x^2 dy)$, where $C: y = x^3, -1 \le x \le 2$.

5. Find the work done by the vector function $\vec{F} = x\vec{i} + y\vec{j} + z\vec{k}$ along the curve $\vec{r}(t) = \sin t\,\vec{i} + \cos t\,\vec{j} + 1\vec{k}$ in the region $[0, 2\pi]$.

6. Given the vector function $\vec{F} = x\vec{i} + y\vec{j} + z\vec{k}$.

   (1)Prove that $\vec{F}$ is conservative.   (2) Find the potential function of $\vec{F}$.   (3) Calculate the work done by an object under the force $\vec{F}$ along any curve from point $(1, 0, 0)$ to point $(1, 1, 2)$.

7. $\vec{F} = (yz^2 - 1)\vec{i} + (xz^2 + e^y)\vec{j} + (2xyz + 1)\vec{k}$,

   the curve C is a straight line from $(1, 1, 1)$ to $(-2, 1, 3)$, find $\int_C \vec{F} \cdot d\vec{r}$.

8. (1) Prove that

   $\vec{F} = (2xy + z^3)\vec{i} + (x^2)\vec{j} + (3xz^2)\vec{k}$

   is a conservative vector field.

   (2) Find the potential function of $\vec{F}$.

   (3) Calculate the work done by an object under the force $\vec{F}$ from point $(1, -2, 1)$ to point $(3, 1, 4)$.

Advanced questions

1. The curve parameterization of a rope is

   $C:\ \vec{r}(t) = 2\cos t\,\vec{i} + 2\sin t\,\vec{j} + 3\vec{k}$,

   $t \in [0, \dfrac{\pi}{2}]$, and the linear density of the rope is

   $\rho = xy^2$. Find the mass of the rope.

2. Find $\int_C 4xyzds$, where $C:\ x = \dfrac{1}{3}t^3, y = t^2, z = 2t$,

   the curve from $t \in [0, 1]$.

3. Find $\int_{(0,0,0)}^{(6,8,5)} (ydx + zdy + xdz)$, where the integral curve C as follows.

   (1) The curve C consists of the line segment from $(0, 0, 0)$ to $(2, 3, 4)$ and the line segment from $(2, 3, 4)$ to $(6, 8, 5)$.

   (2) $C: x = 3t, y = t^3,\ z = \dfrac{5}{4}t^2$, from $t \in [0, 2]$.

4. $\vec{F} = x\vec{i} - yz\,\vec{j} + e^z\vec{k}$, the curve $C: x = t^3, y = -t$, $z = t^2$, from $t \in [1, 2]$, find $\int_C \vec{F} \cdot d\vec{r} = ?$

5. $\vec{F} = xy\,\vec{i} - \cos(yz)\vec{j} + xz\vec{k}$, the curve C is a straight line from $(1, 0, 3)$ to $(-2, 1, 3)$, find $\int_C \vec{F} \cdot d\vec{F}$.

6. $\vec{F} = 3xy\,\vec{i} - 5z\,\vec{j} + 10x\vec{k}$, the curve $C: x = t^2 + 1, y = 2t^2, z = t^3$, from $t \in [1, 2]$, find $\int_C \vec{F} \cdot d\vec{r}$.

7. $\vec{F} = xy^2\vec{i} + (x^2 y + y^3)\vec{j}$, the curve

   $C:\ \vec{r}(t) = \cos t\,\vec{i} + \sin t\,\vec{j} + \vec{k}$, $t \in [0, \dfrac{3\pi}{2}]$

   (1) Find the arc length of curve C.

   (2) Find the line intrgral $\int_C \vec{F} \cdot d\vec{r}$.

8. (1) Prove that
$$\vec{F} = (2xz^3 + 6y)\vec{i} + (6x - 2yz)\vec{j}$$
$$+(3x^2z^2 - y^2)\vec{k} \text{ is a conservative vector field.}$$

(2) Find the potential function of $\vec{F}$.

(3) Calculate the work done by an object under the force $\vec{F}$ from $(1, -1, 1)$ to $(2, 1, -1)$.

9. Find $\int_C (2xy - 1)dx + (x^2 + 1)dy$,

C: any piecewise smooth curve from $(0, 1)$ to $(2, 3)$.

10. Find $\int_C 2xdx + 3y^2zdy + y^3dz$,

C: the straight line from$(0, 0, 0)$ to $(2, 2, 3)$.

11. Find $\int_C [2xyz^2 dx + (x^2z^2 + z\cos yz)dy$

$$+(2x^2yz + y\cos yz)dz]\,,$$

C: the straight line from $(0, 0, 1)$ to $(1, \dfrac{\pi}{4}, 2)$.

12. Find the value of the integral along any closed path
$$\oint (yze^{xyz} - 4x)dx + (xze^{xyz} + z)dy + (xye^{xyz} + y)dz \,.$$

13. Find $\int_C [(y + yz)dx + (x + 3z^3 + xz)dy$

$$+(9yz^2 + xy - 1)dz\,,$$
C: the curve $x^2 = z$,
from $(1, 1, 1)$ to $(2, 1, 4)$ at $y = 1$ plane.

14. $\vec{F} = y\cos z\vec{i} + x\cos z\vec{j} - xy\sin z\vec{k}$, the curve

C: $\vec{r} = \dfrac{t^2}{\sqrt{2}}\vec{i} + (t + 1)\vec{j} + \dfrac{t^3}{3}\vec{k}$,

from $t \in [0, 1]$, find $\int_C \vec{F} \cdot d\vec{r}$.

15. Find $\int_C (6xy - 4e^x)dx + (3x^2)dy$,

C: any piecewise smooth curve from $(0, 0)$ to $(-2, 1)$.

16. $\vec{F} = (4y^3 - 8x)\vec{i} + 12xy^2\vec{j} - 8z\vec{k}$, the curve C is a straight line from $(0, 0, 0)$ to $(2, 2, 10)$, find $\int_C \vec{F} \cdot d\vec{r}$.

17. Find the value of the integral along the closed path
$$x^{\frac{2}{3}} + y^{\frac{2}{3}} = a^{\frac{2}{3}}:$$
$$\oint (x^2 y\cos x + 2xy\sin x - y^2e^x)dx$$
$$+(x^2 \sin x - 2ye^x)dy.$$

18. Find the line integral
$$\int_C 2xyz^2 dx + (x^2z^2 + z\cos yz)dy$$

$$+(2x^2yz + y\cos yz)dz \,.$$

where C is any curve from $(0, 0, 1)$ to $(1, \dfrac{\pi}{4}, 2)$.

19. Calculate the work done by the force field
$$\vec{F} = 4xy\vec{i} - 8y\vec{j} + z\vec{k} \text{ along}$$

(1) the curve $y = 2x$, $z = 2$, from $(0, 0, 2)\sim(3, 6, 2)$.

(2) the curve $y = 2x$, $z = 2x$, from $(0, 0, 0)\sim(3, 6, 6)$.

(3) the curve $x^2 + y^2 = 4$, $z = 0$, from$(2, 0, 0)\sim(-2, 0, 0)$.

8-4 Multiple Integral

In this chapter, we first review double integrals in the Cartesian coordinate system, and then discuss double integrals in different coordinate systems. The key here lies in determining the limits of integration for each variable. Moreover, if the limits of integration are not easily given, the focus of this chapter is on how to transform them into suitable coordinate systems.

If readers are already familiar with double integrals from the calculus courses, you can skip this section and move directly to the next topic, surface integrals.

8-4-1 Double Integrals

▶ **Definition 8-4-1**

Riemann Sum Defines the Double Integral

Assuming R is a closed region with a piecewise continuous boundary, if $f(x, y)$ is an integrable function in R, then the double integral of $f(x, y)$ with respect to R is defined

as $\iint_R f(x, y)dA = \iint_R f(x, y)dxdy = \lim_{n \to \infty} \sum_{i=1}^{n} f(x_i, y_i)\Delta A_i$, where max $\Delta A_i \to 0$, as

shown in Figure 8-29 to 8-30.

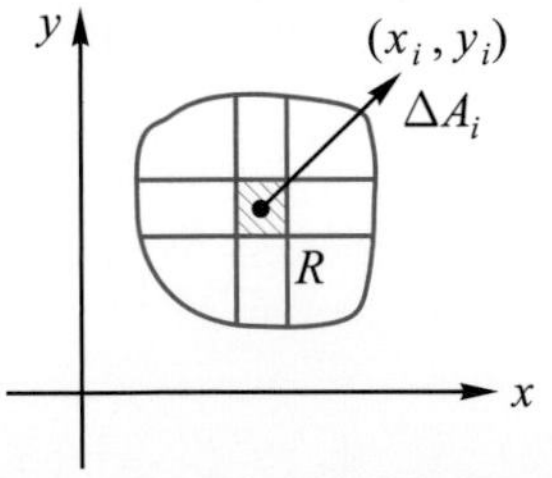

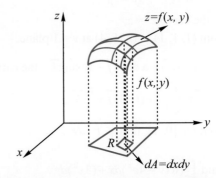

Figure 8-29　The double integral in the plane

Figure 8-30　Diagram illustrating the volume enclosed by a surface and the xy-plane in space

◀

1. **The Properties of Double Integrals**

   (1) $\iint_R [af + bg]\,dxdy = a \iint_R f\,dxdy + b \iint_R g\,dxdy$ (linear).

   (2) If $R = R_1 + R_2 + \cdots$ is the area where the intersection area of each other is 0,
   then $\iint_R f\,dxdy = \iint_{R_1} f\,dxdy + \iint_{R_2} f\,dxdy + \cdots$.

   (3) $f = 1$ is a constant function, then $\iint_R dxdy = A$ represents the area of the region R.

   (4) The centroid coordinates $(\overline{x}, \overline{y})$ of region R are: $\overline{x} = \dfrac{1}{A} \iint_R x\,dxdy$,

   $\overline{y} = \dfrac{1}{A} \iint_R y\,dxdy$.

   (5) If $z = f(x, y)$ represents a surface in space, then the volume enclosed by it and the
   x-y-plane is $V = \iint_R \underbrace{f(x, y)}_{\text{height}}\,\underbrace{dxdy}_{dA}$.

2. **The Method for Determining the Limits of Integration**

   The determination of the upper and lower limits for the double integral is crucial,
   and the principle is to assign points for the outer integral and lines for the inner
   integral, as shown below.

   $$\iint_R f(x, y)\,dxdy = \int_{x=a}^{x=b} \left[\int_{y=y_1(x)}^{y=y_2(x)} f(x, y)\,dy \right] dx = \int_{y=c}^{y=d} \left[\int_{x=x_1(y)}^{x=x_2(y)} f(x, y)\,dx \right] dy$$

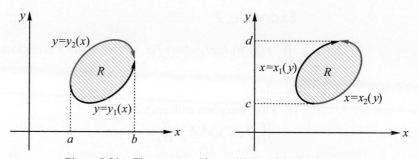

Figure 8-31   The upper and lower limits of double integrals

Example 1

Find $\iint_R f(x, y)dxdy$, where $f(x, y) = xy$ while R is enclosed by $y = 0$, $y = x$, $x + y = 2$.

Solution

① First, integrate with respect to y

$$\iint_R f(x, y)dxdy = \int_{x=0}^{x=1}(\int_{y=0}^{y=x} xydy)dx + \int_{x=1}^{x=2}(\int_{y=0}^{y=2-x} xydy)dx$$

$$= \int_0^1 \frac{1}{2}x^3 dx + \int_1^2 \frac{1}{2}x \cdot (2-x)^2 dx$$

$$= \frac{1}{8}x^4 \Big|_0^1 + \frac{1}{2}(2x^2 - \frac{4}{3}x^3 + \frac{1}{4}x^4)\Big|_1^2$$

$$= \frac{1}{3}.$$

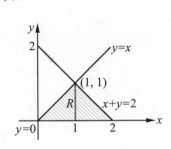

② Integrate with respect to x first

$$\iint_R f(x, y)dxdy = \int_{y=0}^{y=1}(\int_{x=y}^{x=2-y} xydx)dy$$

$$= \frac{1}{3}.$$

Q.E.D.

Example 2

$\iint_R f(x, y)dxdy$, where $f(x, y) = x^2$, while R is enclosed by $y = x$, $y = 0$, $x = 8$, $xy = 16$.

Solution

① First, integrate with respect to y

$$\iint_R f(x, y)dxdy = \iint_R x^2 \cdot dxdy$$

$$= \int_{x=0}^{x=4}\int_{y=0}^{y=x} x^2 dydx + \int_{x=4}^{x=8}\int_{y=0}^{y=\frac{16}{x}} x^2 dydx$$

$$= \int_0^4 x^3 dx + \int_4^8 16x dx$$

$$= 448 .$$

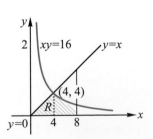

② First, integrate with respect to x

$$\iint_R f(x, y)dxdy = \int_{y=0}^{y=2}\int_{x=y}^{x=8} x^2 dxdy + \int_{y=2}^{y=4}\int_{x=y}^{x=\frac{16}{y}} x^2 dxdy$$

$$= 448 .$$

 Q.E.D.

Example 3

Find $\displaystyle\int_0^4 \int_{\frac{x}{2}}^2 e^{y^2}\, dydx$.

Solution

$$\int_{x=0}^{x=4}\int_{y=\frac{x}{2}}^{y=2} e^{y^2}\, dydx = \int_0^2 \int_{x=0}^{x=2y} e^{y^2}\, dxdy$$

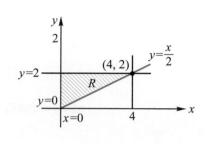

$$= \int_0^2 e^{y^2}\, x\Big|_0^{2y}\, dy$$

$$= \int_0^2 2ye^{y^2}\, dy$$

$$= e^{y^2}\Big|_0^2 = e^4 - e^0$$

$$= e^4 - 1 .$$

Q.E.D.

Example 4

Find $\displaystyle\int_{y=0}^{y=8} \int_{x=\sqrt[3]{y}}^{x=2} \frac{y}{\sqrt{16+x^7}}\, dxdy$.

Solution

$$\int_{y=0}^{y=8} \int_{x=\sqrt[3]{y}}^{x=2} \frac{y}{\sqrt{16+x^7}}\, dxdy = \int_{x=0}^{x=2} \int_{y=0}^{y=x^3} \frac{y}{\sqrt{16+x^7}}\, dydx$$

$$= \int_0^2 \frac{1}{2}\, \frac{y^2}{\sqrt{16+x^7}}\Bigg|_0^{x^3}\, dx$$

$$= \int_0^2 \frac{\frac{1}{2}x^6}{\sqrt{16+x^7}}\, dx$$

$$= \int_0^2 (16+x^7)^{-\frac{1}{2}} \times \frac{1}{14}\, d(16+x^7)$$

$$= 2 \times \frac{1}{14}(16+x^7)^{\frac{1}{2}}\Bigg|_0^2$$

$$= \frac{12-4}{7} = \frac{8}{7} .$$

Q.E.D.

8-4-2 Triple Integrals

Let D be a closed region with a piecewise smooth boundary. If $f(x, y, z)$ is an integrable function over D, then its triple integral (also known as volume integral) is defined as follows:

$$\iiint_D f(x, y, z)dV = \iiint_D f(x, y, z)dxdydz$$

$$= \lim_{n\to\infty} \sum_{i=1}^{n} f(x_i, y_i, z_i)\, \Delta V_i$$

where $\max \Delta V_i \to 0$, as shown in Figure 8-32.

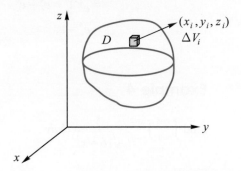

Figure 8-32   The schematic diagram of triple integrals

1. **The Properties and Application of Triple Integrals**

(1) If $f(x, y, z) = 1$, then $\iiint_D dxdydz = V$ represents the **volume** of region D.

(2) The **centroid** of the spatial region D

$$\bar{x} = \frac{1}{V}\iiint_D xdV , \quad \bar{y} = \frac{1}{V}\iiint_D ydV , \quad \bar{z} = \frac{1}{V}\iiint_D zdV .$$

(3) If $f(x, y, z) = \rho(x, y, z)$ represents density, then $\iiint_D \rho dV = M$ represents the **quality** of region D, then the **center of mass**

$$\bar{x} = \frac{1}{M}\iiint_D x\rho dV , \quad \bar{y} = \frac{1}{M}\iiint_D y\rho dV , \quad \bar{z} = \frac{1}{M}\iiint_D z\rho dV .$$

(4) The **moment of inertia** of the spatial region D is $I = \iiint_D r^2\rho dV$.

(5) $\iiint_D (k_1 f_1 + k_2 f_2)dV = k_1\iiint_D f_1 dV + k_2\iiint_D f_2 dV$.

(6) $\iiint_{D_1+D_2} f(x, y, z)dV = \iiint_{D_1} f(x, y, z)dV + \iiint_{D_2} f(x, y, z)dV$.

2. The Upper and Lower Limits of Integrals

The determination of the upper and lower limits for a triple integral can follow the principle of "assigning points for the outermost integral, lines for the second integral, and surfaces for the innermost integral," as demonstrated in the example below (Example 5).

Example 5

The region D is the tetrahedron formed by the plane $20x + 15y + 12z = 60$ and the first quadrant, then calculate the volume using six different integration methods.

Solution

$$V = \iiint_D dxdydz = \int_{x=0}^{x=3} \int_{y=0}^{y=\frac{12-4x}{3}} \int_{z=0}^{z=\frac{1}{12}(60-20x-15y)} dzdydx$$

$$= \int_{y=0}^{y=4} \int_{x=0}^{x=\frac{1}{4}(12-3y)} \int_{z=0}^{z=\frac{1}{12}(60-20x-15y)} dzdxdy$$

$$= \int_{x=0}^{x=3} \int_{z=0}^{z=\frac{1}{3}(15-5x)} \int_{y=0}^{y=\frac{1}{15}(60-20x-12z)} dydzdx$$

$$= \int_{z=0}^{z=5} \int_{x=0}^{x=\frac{1}{5}(15-3z)} \int_{y=0}^{y=\frac{1}{15}(60-20x-12z)} dydxdz$$

$$= \int_{y=0}^{y=4} \int_{z=0}^{z=\frac{1}{4}(20-5y)} \int_{x=0}^{x=\frac{1}{20}(60-15y-12z)} dxdzdy$$

$$= \int_{z=0}^{z=5} \int_{y=0}^{y=\frac{1}{5}(20-4z)} \int_{x=0}^{x=\frac{1}{20}(60-15y-12z)} dxdydz \ .$$

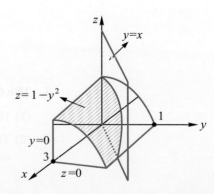

$(0,0,5)$

$5y+4z=20$

$5x+3z=15$

D #

$(0,4,0)$ y

$4x+3y=12$

$(3,0,0)$ $20x+15y+12z=60$

x

Q.E.D.

Example 6

Calculate the volume enclosed by $z = 1 - y^2$, $y = x$, $x = 3$, $y = 0$, $z = 0$ in the first quadrant.

Solution

$$V = \int_{y=0}^{y=1} \int_{x=y}^{x=3} \int_{z=0}^{z=1-y^2} dzdxdy$$

$$= \int_{y=0}^{y=1} \int_{x=y}^{x=3} (1-y^2)dxdy$$

$$= \int_0^1 (x - xy^2) \Big|_y^3 dy$$

$$= \int_0^1 (3 - 3y^2 - y + y^3)dy$$

$$= 3y - y^3 - \frac{1}{2}y^2 + \frac{1}{4}y^4 \Big|_0^1$$

$$= 3 - 1 - \frac{1}{2} + \frac{1}{4} = \frac{7}{4} .$$

z

$y=x$

$z=1-y^2$ 1 y

$y=0$ 3

x $z=0$

8-4-3 Cylindrical Coordinate Double Integral

Many physical systems have geometries in cylindrical form, making integration in Cartesian coordinates challenging. Here, we will introduce how to perform volume integrals using cylindrical coordinates or double integrals in polar coordinates. For the relationship between common coordinates and Cartesian coordinates, please refer to Section 8-1-2 in this chapter.

1. **The Relationship Between Cylindrical Coordinates and Cartesian Coordinates**

    $$\begin{cases} x = \rho\cos\theta \\ y = \rho\sin\theta \\ z = z \end{cases}$$, as shown in Figure 8-33.

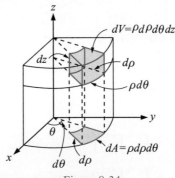

Figure 8-33 cylindrical coordinate system

2. **Coordinate Transformation in Integrals**

    Let $F(x, y, z) = F(\rho\cos\theta, \rho\sin\theta, z) = f(\rho, \theta, z)$

    (1) Double Integral in Polar Coordinates on the xy-Plane

    $$\iint_R f(x, y)dxdy = \int_{\theta_1}^{\theta_2} \int_{\rho_1}^{\rho_2} f(\rho, \theta)\rho d\rho d\theta$$

    (2) Triple Integral in Cylindrical Coordinates in Space

    $$\iiint_D F(x, y, z)dxdydz = \int_{\theta_1}^{\theta_2} \int_{\rho_1}^{\rho_2} \int_{z_1}^{z_2} f(\rho, \theta, z)\rho dz d\rho d\theta$$

    as shown in Figure 8-34.

Figure 8-34

3. **The Occasions to Use Polar or Cylindrical Coordinate Transformations**

    (1) The integrand involves terms such as $x^2 + y^2$, $y^2 + z^2$, $x^2 + z^2$, etc.

    (2) The integration region is symmetric with respect to an axis, or a cylindrical or disk-shaped region.

Example 7

Find $\displaystyle\int_0^2 \int_x^{\sqrt{8-x^2}} \frac{1}{5+x^2+y^2}\,dydx$.

Solution

Notice that the denominator of the integrand contains $x^2 + y^2$, so we adopt polar coordinate transformation: $x = \rho\cos\theta$, $y = \rho\sin\theta$,

$$\int_{x=0}^{x=2} \int_{y=x}^{y=\sqrt{8-x^2}} \frac{1}{5+x^2+y^2}\,dydx$$

$$= \int_{\theta=\frac{\pi}{4}}^{\theta=\frac{\pi}{2}} \int_{\rho=0}^{\rho=2\sqrt{2}} \frac{1}{5+\rho^2}\,\rho\,d\rho\,d\theta$$

$$= \int_{\frac{\pi}{4}}^{\frac{\pi}{2}} \frac{1}{2}\ln(5+\rho^2)\Big|_0^{2\sqrt{2}}\;d\theta$$

$$= \frac{\pi}{8}\ln\left(\frac{13}{5}\right).$$ Q.E.D.

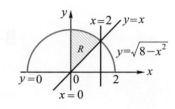

Example 8

Find $\displaystyle\iint_R f(x,\,y)dxdy$, where $f(x,\,y) = \cos(x^2+y^2)$, the region R is $x^2+y^2 \le \dfrac{\pi}{2}$, $x \ge 0$.

Solution

Notice that the integrand contains $x^2 + y^2$, so we adopt polar coordinate transformation: $x = \rho\cos\theta$, $y = \rho\sin\theta$,

$$\iint_R f(x,\,y)dxdy = \int_{\theta=-\frac{\pi}{2}}^{\theta=\frac{\pi}{2}} \int_{\rho=0}^{\rho=\sqrt{\frac{\pi}{2}}} \cos(\rho^2)\times\rho\,d\rho\,d\theta$$

$$= \int_{-\frac{\pi}{2}}^{\frac{\pi}{2}} d\theta \times \int_0^{\sqrt{\frac{\pi}{2}}} \cos(\rho^2)\rho\,d\rho$$

$$= \pi\cdot\frac{1}{2} = \frac{\pi}{2}.$$ Q.E.D.

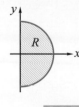

Example 9

Find the moment of inertia of the cone in the attached figure relative to the z-axis $\displaystyle\iiint_D (x^2+y^2)dxdydz$.

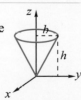

Solution

This is a triple integral with the integrand being $x^2 + y^2$. Therefore, no special treatment is required for the z-direction.

Let $x = \rho \cos \theta, y = \rho \sin \theta, z = z$,

$$\iiint_D (x^2 + y^2)dxdydz = \int_{z=0}^{h} \int_{\theta=0}^{2\pi} \int_{\rho=0}^{\frac{b}{h}z} \rho^2 \times \rho d\rho d\theta dz$$

$$= \int_0^{2\pi} d\theta \int_0^h \int_{\rho=0}^{\frac{b}{h}z} \rho^2 \times \rho d\rho dz$$

$$= \frac{\pi}{10} b^4 h .$$

Q.E.D.

8-4-4 Spherical Coordinate Triple Integral

Apart from cylindrical coordinates, another commonly used coordinate system is spherical coordinates. In physical systems with spherical geometry, using spherical coordinates makes solving problems easier.

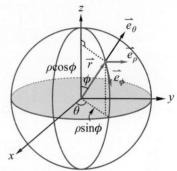

Figure 8-35   the spherical coordinates

1. **The Relationship Between Spherical Coordinates and Cartesian Coordinates**

$$\begin{cases} x = \rho \sin\phi \cos\theta \\ y = \rho \sin\phi \sin\theta \\ z = \rho \cos\phi \end{cases}$$, as shown in Figure 8-35.

2. **Coordinate Transformation in Integrals**

Let $F(x, y, z) = F(\rho \sin\phi \cos\theta, \rho \sin\phi \sin\theta, \rho \cos\phi) = f(\rho, \phi, \theta)$

(1) $\iiint_V F(x, y, z)dxdydz = \int_{\theta_1}^{\theta_2} \int_{\phi_1}^{\phi_2} \int_{\rho_1}^{\rho_2} f(\rho, \phi, \theta)\rho^2 \sin\phi d\rho d\phi d\theta$

(2) The Occasions to Use Spherical Coordinates in Multiple Integrals

① The integrand involves terms such as $x^2 + y^2 + z^2$.

② The integration region is symmetric with respect to a point, as shown in Figure 8-36.

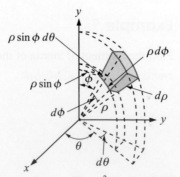

Figure 8-36 $dV = \rho^2 \sin\phi d\rho d\phi d\theta$

Example 10

Find $\iiint_D \dfrac{1}{\sqrt{1-x^2-y^2-z^2}} \, dx\,dy\,dz$, where $D = \{(x,y,z) \mid x^2+y^2+z^2 \leq 1\}$.

Solution

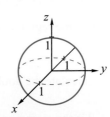

As the integrand contains the term $1-(x^2+y^2+z^2)$,

we use spherical coordinate transformation, let $\begin{cases} x = \rho\sin\phi\cos\theta \\ y = \rho\sin\phi\sin\theta \\ z = \rho\cos\phi \end{cases}$

where the integration region D is : $x^2+y^2+z^2 \leq 1 \Rightarrow 0 \leq \rho \leq 1, \, 0 \leq \phi \leq \pi, \, 0 \leq \theta \leq 2\pi$,

$$\therefore \iiint_D \frac{1}{\sqrt{1-x^2-y^2-z^2}} \, dx\,dy\,dz$$

$$= \iiint \frac{1}{\sqrt{1-\rho^2}} \rho^2 \sin\phi \, d\rho \, d\phi \, d\theta$$

$$= \int_0^{2\pi} \int_0^{\pi} \int_0^1 \frac{\rho^2}{\sqrt{1-\rho^2}} \sin\phi \, d\rho \, d\phi \, d\theta$$

$$= \int_0^{2\pi} 1 \, d\theta \cdot \int_0^{\pi} \sin\phi \, d\phi \cdot \int_0^1 \frac{\rho^2}{\sqrt{1-\rho^2}} \, d\rho$$

$$= 2\pi \cdot (-\cos\phi)\Big|_0^{\pi} \cdot \int_0^{\frac{\pi}{2}} \frac{\sin^2 t}{\cos t} \cdot \cos t \, dt$$

$$= 2\pi \cdot 2 \cdot \int_0^{\frac{\pi}{2}} (\frac{1}{2} - \frac{\cos 2t}{2}) \, dt$$

$$= 4\pi \cdot \frac{1}{2}(t - \frac{1}{2}\sin 2t)\Big|_0^{\frac{\pi}{2}}$$

$$= 2\pi \cdot \frac{\pi}{2} = \pi^2 .$$

Q.E.D.

8-4 Exercises

Basic questions

1. Calculate the double integral of $\iint_R x^2 dxdy$, where

R is the region enclosed by $\begin{cases} x+y=2 \\ x=0 \\ y=x \end{cases}$.

2. Calculate the double integral $\iint_R 2xydxdy$, where

R is the region enclosed by $\begin{cases} y=x^2 \\ y=x \end{cases}$.

3. Calculate the double integral $\int_0^1 \int_x^1 e^{y^2} dydx$.

4. Calculate the double integral $\iint_R x^2 ydxdy$, where R

is the region enclosed by $\begin{cases} x^2+y^2 \le 9 \\ y \ge 0 \end{cases}$.

5. Please use the double integral of

$\iint_R z(x, y)dxdy$ to calculate the volume of the

region enclosed by the planes:
$2x+y+z=4, x=0, y=0, z=0$ in the first octant.

Advanced questions

1. Calculate the double integral $\iint_R f(x, y)dxdy$ as

follows.
(1) $f(x, y) = x^2 + y^2$, while R is enclosed by
$y=x, y=x+a, y=a$ and $y=3a$, where $a>0$.
(2) $f(x, y) = 3x^2y$, while R is enclosed by
$x=\sqrt{y}$ and $y=-x$.

2. Calculate following double integral.

(1) $\int_0^2 \int_{y^2}^4 y\cos x^2 dxdy$.

(2) $\int_0^\pi \int_x^\pi \frac{\sin y}{y} dydx$.

3. Please calculate the following triple integral

$\iiint_D xdxdydz$, where

$D = \{x+2y+z=4,$ forming a tetrahedron in the
first octant$\}$.

4. Calculate following integrals.

(1) $\int_0^3 \int_0^{\sqrt{9-x^2}} \sin(x^2+y^2)dydx$.

(2) $\int_0^1 \int_0^{\sqrt{1-y^2}} e^{-(x^2+y^2)}dxdy$.

5. Calculate following double integral
$\iint_R f(x, y)dxdy$.

(1) $f(x, y) = y$, while R is enclosed by
$1 \le x^2 + y^2 \le 2$.
(2) $f(x, y) = e^{x^2+y^2}$, while R is enclosed by
$x^2+y^2 \le 1, 0 \le y \le x$.

6. Please calculate the following triple integral

$\iiint_D x^2 dxdydz$, where $D=\{x^2+y^2+z^2 \le 1\}$.

7. Please calculate the following triple integral

$\iiint_D (x^2+y^2+z^2)dxdydz$,

where $D = \{x^2+y^2+z^2 \le a^2\}$.

8-5 Surface integral

Surface integral (also known as area integral or surface integration) is a definite integral taken over a certain surface in space. Given a surface, it is possible to perform integration with respect to a scalar field or a vector field defined on that surface. Surface integrals find extensive use in physics, particularly in fluid dynamics and electromagnetics. Vector field surface integrals, in particular, represent the flux of a physical quantity passing through the surface, making it a fundamental concept in physics. The following section will explain how to perform surface integrals.

8-5-1 The Definition and Properties

A surface S is considered a **smooth surface** if at each point on the surface, the unit normal vector $\vec{n}$ is nonzero and continuous. On the other hand, if a surface S is composed of a finite number of smooth surfaces $S_1, S_2, \cdots, S_n$, then it is referred to as a **piecewise smooth surface.**

▶ **Definition 8-5-1**

The Definition of Surface Integral

The surface integral $\iint\limits_D f(x, y, z)dA$ of a scalar function $f(x, y, z)$ over the smooth surface S is defined as the limit of the following Riemann sum: $\sum\limits_{i=1}^{n} f(x_i, y_i, z_i)\Delta A_i$, where max $\Delta A_i \to 0$ follows $n \to \infty$, and the relation between the infinitesimal region ΔA_i and the integration region is illustrated in Figure 8-37.

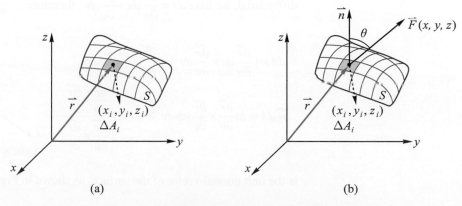

(a) (b)

Figure 8-37 Illustration of surface integral in space

Let $\vec{n}$ be the unit normal vector on the smooth surface S. Then, $\vec{F}(x, y, z)$, the surface integral of the vector function $\vec{F}$ over the surface S is expressed as

$$\iint_S \vec{F}(x, y, z) \cdot d\vec{A} = \iint_S \vec{F}(x, y, z) \cdot \vec{n}\, dA$$

The Physical Properties of Surface Integrals

If $f(x, y, z) = 1$, then from the definition we know $\iint_S dA$ represents the **area** of the spatial surface S; If $f(x, y, z)$ represents the density function of the surface S, then $\iint_S f(x, y, z)\, dA$ represents **the mass of the surface**; if $\vec{F}(x, y, z)$ represents a physical quantity, then the **flux** of that physical quantity across the surface S is $\phi = \iint_S \vec{F}(x, y, z) \cdot d\vec{A}$, where $\vec{F} \cdot d\vec{A} = \vec{F} \cdot \vec{n}\, dA = |\vec{F}| \cdot |\vec{n}|\, dA \cdot \cos\theta = |\vec{F}| \cos\theta \cdot dA$,

represents the flux of $\vec{F}$ flowing outward along the surface normal direction, and the unit normal vector of the surface S is $\vec{n}$.

8-5-2 The Method to Find dA

Therefore, the computation of surface integrals relies on determining the infinitesimal area beforehand. Below, we present several important calculation methods, each corresponding to different representations of surfaces: parametric form, vector form, and level curves.

1. Parametric Form

If the parameterization of the surface S is $S : \begin{cases} x = x(u, v) \\ y = y(u, v) \\ z = z(u, v) \end{cases}$, then from the total

differential, we have $d\vec{r} = \dfrac{\partial \vec{r}}{\partial u} du + \dfrac{\partial \vec{r}}{\partial v} dv$, therefore

$$dA = |\frac{\partial \vec{r}}{\partial u} du \times \frac{\partial \vec{r}}{\partial v} dv| = |\frac{\partial \vec{r}}{\partial u} \times \frac{\partial \vec{r}}{\partial v}|\, dudv,$$

$$\vec{n}\, dA = \pm(\frac{\partial \vec{r}}{\partial u} \times \frac{\partial \vec{r}}{\partial v})dudv \; ; \; \vec{n} = \pm\frac{\dfrac{\partial \vec{r}}{\partial u} \times \dfrac{\partial \vec{r}}{\partial v}}{|\dfrac{\partial \vec{r}}{\partial u} \times \dfrac{\partial \vec{r}}{\partial v}|}$$

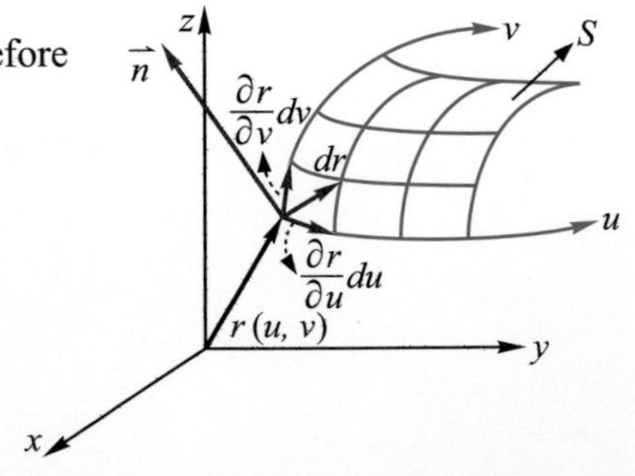

Figure 8-38 Find dA in the surface space

is the unit normal vector of the surface, as shown in Figure 8-38.

2. Vector Form

Assuming that the surface $z = f(x, y)$, then $\vec{r}(x, y) = x\vec{i} + y\vec{j} + f(x, y)\vec{k}$, therefore

$$\frac{\partial \vec{r}}{\partial x} \times \frac{\partial \vec{r}}{\partial y} = \begin{vmatrix} \vec{i} & \vec{j} & \vec{k} \\ 1 & 0 & f_x \\ 0 & 1 & f_y \end{vmatrix} = -f_x\vec{i} - f_y\vec{j} + \vec{k}$$

$$dA = \left| \frac{\partial \vec{r}}{\partial x} \times \frac{\partial \vec{r}}{\partial y} \right| dxdy = \sqrt{(f_x)^2 + (f_y)^2 + 1^2}\, dxdy \; ,$$

$$\vec{n} = \pm \frac{\dfrac{\partial \vec{r}}{\partial x} \times \dfrac{\partial \vec{r}}{\partial y}}{\left| \dfrac{\partial \vec{r}}{\partial x} \times \dfrac{\partial \vec{r}}{\partial y} \right|} = \pm \frac{(-f_x\vec{i} - f_y\vec{j} + \vec{k})}{\sqrt{(f_x)^2 + (f_y)^2 + 1^2}} \; .$$

Therefore, the infinitesimal area is obtained as $d\vec{A} = dA \times \vec{n} = \pm(f_x\vec{i} + f_y\vec{j} - \vec{k})dxdy$.

For example, if the surface S is $z = f(x, y) = x^2 + y^2$,

then $\vec{r}(x, y) = x\vec{i} + y\vec{j} + (x^2 + y^2)\vec{k}$;

$$\vec{n} = + \frac{\dfrac{\partial f}{\partial x}\vec{i} + \dfrac{\partial f}{\partial y}\vec{j} - \vec{k}}{\sqrt{1 + (\dfrac{\partial f}{\partial x})^2 + (\dfrac{\partial f}{\partial y})^2}} = \frac{2x\vec{i} + 2y\vec{j} - \vec{k}}{\sqrt{1 + 4x^2 + 4y^2}} \; .$$

Hence, the infinitesimal area

$$dA = \sqrt{1 + (\frac{\partial f}{\partial x})^2 + (\frac{\partial f}{\partial y})^2}\, dxdy$$

is equal to $\sqrt{1 + 4x^2 + 4y^2}\, dxdy$

(taking the positive value since the z-component is negative) , as shown in Figure 8-39.

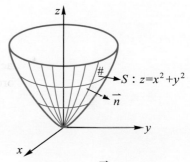

Figure 8-39 $\vec{n}$ is positive, z-component is negative

3. Level Curves

If according to surface equation $z = f(x, y)$, let $\phi(x, y, z) = z - f(x, y)$, then the surface can be represented as $\phi(x, y, z) = 0$, and $\vec{n}$ is the unit normal vector on the infinitesimal area dA of the surface. The angle between and the z-axis is γ. In this case, we can use the projection method. Assuming that the projection of dA onto the x-y plane is dA^*, then $dA^* = dxdy = dA \cdot |\cos \gamma| = dA |\vec{n} \cdot \vec{k}|$, where the projections of onto the three coordinate planes are:

$$dA = \frac{dxdy}{|\overrightarrow{n} \cdot \overrightarrow{k}|} \quad \text{(projection onto the } x-y \text{ plane)},$$

$$dA = \frac{dydz}{|\overrightarrow{n} \cdot \overrightarrow{i}|} \quad \text{(projection onto the } y-z \text{ plane)},$$

$$dA = \frac{dxdz}{|\overrightarrow{n} \cdot \overrightarrow{j}|} \quad \text{(projection onto the } x-z \text{ plane)}.$$

For the projection onto the x-y plane, the unit normal vector of the surface obtained from the gradient is $\overrightarrow{n} = \pm \dfrac{\nabla \phi}{|\nabla \phi|}$; Therefore, the infinitesimal area is

$$dA = \frac{dxdy}{|\overrightarrow{n} \cdot \overrightarrow{k}|} = \frac{dxdy}{|\frac{\nabla \phi}{|\nabla \phi|} \cdot \overrightarrow{k}|} = \frac{|\nabla \phi|}{|\nabla \phi \cdot \overrightarrow{k}|} dxdy = \frac{\sqrt{(\frac{\partial \phi}{\partial x})^2 + (\frac{\partial \phi}{\partial y})^2 + (\frac{\partial \phi}{\partial z})^2}}{|\frac{\partial \phi}{\partial z}|} dxdy ;$$

$$\overrightarrow{n}dA = \frac{\pm \nabla \phi}{|\nabla \phi|} \frac{|\nabla \phi|}{|\nabla \phi \cdot \overrightarrow{k}|} dxdy = \frac{\pm \nabla \phi}{|\nabla \phi \cdot \overrightarrow{k}|} dxdy ,$$

similarly, $\overrightarrow{n}dA = \dfrac{\pm \nabla \phi}{|\nabla \phi \cdot \overrightarrow{i}|} dydz = \dfrac{\pm \nabla \phi}{|\nabla \phi \cdot \overrightarrow{j}|} dxdz$.

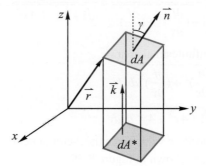

Figure 8-40   Projection of the infinitesimal area of a surface in space

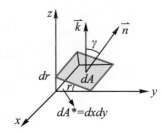

Figure 8-41   The projection of dA onto $x-y$ plane

4. Determined of Positive and Negative Signs

To determine the sign, we typically adjust it to be positive for an outward-facing surface. It is often convenient to use the z-component to determine whether the normal vector points upward (positive) or downward (negative). For example, for the equation $\phi(x, y, z) = x^2 + y^2 + z^2 = a^2$, $z > 0$ of the upper hemisphere surface, the unit normal vector of the surface is $\overrightarrow{n}$, also

$$\overrightarrow{n} = \pm \frac{\nabla \phi}{|\nabla \phi|} = \frac{2x \overrightarrow{i} + 2y \overrightarrow{j} + 2z \overrightarrow{k}}{\sqrt{(2x)^2 + (2y)^2 + (2z)^2}} = \frac{x \overrightarrow{i} + y \overrightarrow{j} + z \overrightarrow{k}}{a} = \frac{\overrightarrow{r}}{a} ,$$

$$dA = \frac{dxdy}{|\vec{n}\cdot\vec{k}|} = \frac{dxdy}{\frac{z}{a}} = \frac{a}{z}dxdy \text{ (projection onto the } x\text{–}y \text{ plane)},$$

$$\vec{n}\,dA = \frac{x\vec{i}+y\vec{j}+z\vec{k}}{a}\cdot\frac{a}{z}dxdy$$

$$= \frac{x\vec{i}+y\vec{j}+z\vec{k}}{z}dxdy \ ,$$

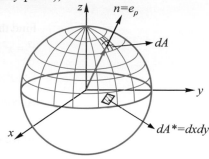

Figure 8-42

Example 1

Calculate $I = \iint_S \frac{xy}{z}\,dA$, where in S: $z = x^2 + y^2$

is the portion of the surface in the first quadrant with
$4 \le x^2 + y^2 \le 9$.

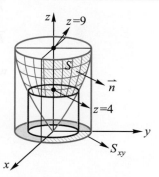

Solution

Let $z = x^2 + y^2 = f(x,y)$, according to the vector
form formula in this section, the infinitesimal area is:

$$dA = \sqrt{(f_x)^2 + (f_y)^2 + 1^2}\ dxdy = \sqrt{(2x)^2 + (2y)^2 + 1^2}\ dxdy = \sqrt{4x^2 + 4y^2 + 1^2}\ dxdy \ ,$$

$$\phi = x^2 + y^2 - z \ , \quad \vec{n} = \frac{\nabla\phi}{|\nabla\phi|} = \frac{2x\vec{i}+2y\vec{j}-\vec{k}}{\sqrt{4x^2+4y^2+1}}\ ,$$

therefore $\quad I = \iint_S \frac{xy}{z}\,dA \ = \iint_{S_{xy}} \frac{xy}{x^2+y^2}\sqrt{4x^2+4y^2+1}\ dxdy$.

Due to the nature of the surface being integrated, we will use the polar coordinate
transformation $x = r\cos\theta, y = r\sin\theta, \ \theta : 0 \sim \dfrac{\pi}{2}$,

$r = 2\sim3$ $(z = 4 \rightarrow r = 2; z = 9 \rightarrow r = 3)$,

we obtain $\quad I = \displaystyle\int_0^{\frac{\pi}{2}} \int_2^3 \frac{r^2\sin\theta\cos\theta}{r^2}\sqrt{1+4r^2}\,rdrd\theta$

$$= \int_0^{\frac{\pi}{2}}\sin\theta\cos\theta d\theta \cdot \int_2^3 \sqrt{1+4r^2}\,rdr$$

$$= \frac{1}{2}\cdot\int_2^3 (1+4r^2)^{\frac{1}{2}}\cdot d(1+4r^2)\cdot\frac{1}{8}$$

$$= \frac{1}{16}\cdot\frac{2}{3}(1+4r^2)^{\frac{3}{2}}\Big|_2^3$$

$$= \frac{1}{24}(37^{\frac{3}{2}} - 17^{\frac{3}{2}})\ .$$

Q.E.D.

Example 2

Find the flux of the vector $\vec{F} = z\,\vec{i} + y\,\vec{j} + x\,\vec{k}$ passing through the surface of the cone $z^2 = x^2 + y^2, 0 < z < 1$.

Solution

According to the definition, the flux is $\iint_S \vec{F} \cdot \vec{n}\, dA$,

so we calculate $\vec{n}\, dA = \dfrac{\nabla \phi}{|\nabla \phi \cdot \vec{k}|} dxdy$;

$\phi = x^2 + y^2 - z^2$, $\nabla \phi = 2x\,\vec{i} + 2y\,\vec{j} - 2z\,\vec{k}$;

$\vec{n}\, dA = \dfrac{x\,\vec{i} + y\,\vec{j} - z\,\vec{k}}{z} dxdy$,

$\vec{F} = z\,\vec{i} + y\,\vec{j} + x\,\vec{k}$; and $z = \sqrt{x^2 + y^2}$,

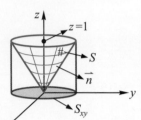

we obtain $\quad \phi = \iint_S \vec{F} \cdot \vec{n}\, dA = \iint_{S_{xy}} \dfrac{xz + y^2 - xz}{z} dxdy$

$$= \iint_{S_{xy}} \dfrac{y^2}{\sqrt{x^2 + y^2}} dxdy = \int_0^{2\pi} \int_0^1 \dfrac{r^2 \sin^2 \theta}{r} r\,dr\,d\theta$$

$$= \dfrac{1}{3} r^3 \Big|_0^1 \cdot \int_0^{2\pi} \sin^2 \theta\, d\theta = \dfrac{1}{3} \int_0^{2\pi} \dfrac{1 - \cos 2\theta}{2} d\theta = \dfrac{\pi}{3}.$$

Q.E.D.

Example 3

Find the flux of $\vec{F} = xz\,\vec{i} - y\,\vec{k}$ passing through S: $x^2 + y^2 + z^2 = 4, z > 1$
(not contain the plane $z = 1$).

Solution

Let $\phi = x^2 + y^2 + z^2 - 4$, $\vec{n}\, dA = \dfrac{\nabla \phi}{|\nabla \phi \cdot \vec{k}|} dxdy$,

$\nabla \phi = 2x\,\vec{i} + 2y\,\vec{j} + 2z\,\vec{k}$,
therefore

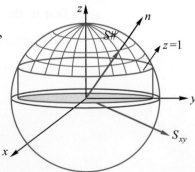

$$\vec{n}\, dA = \dfrac{2x\,\vec{i} + 2y\,\vec{j} + 2z\,\vec{k}}{2z} dxdy$$

$$= \dfrac{x\,\vec{i} + y\,\vec{j} + z\,\vec{k}}{z} dxdy,$$

$$\vec{F} \cdot \vec{n}\, dA = \dfrac{x^2 z - yz}{z} dxdy = (x^2 - y) dxdy.$$

$S_{xy} : x^2 + y^2 \leq 3$

Substituting the given information into the formula, we have

$$\iint_S \vec{F} \cdot \vec{n} \, dA = \iint_{S_{xy}} (x^2 - y) dx dy$$

$$= \int_0^{2\pi} \int_0^{\sqrt{3}} (r^2 \cos^2 \theta - r \sin \theta) r \, dr \, d\theta$$

$$= \int_0^{2\pi} \left(\frac{1}{4} r^4 \cos^2 \theta - \frac{1}{3} r^3 \sin \theta \right) \Big|_0^{\sqrt{3}} d\theta$$

$$= \frac{9}{4} \int_0^{2\pi} \cos^2 \theta - \sqrt{3} \int_0^{2\pi} \sin \theta \, d\theta$$

$$= \frac{9}{4} \int_0^{2\pi} \frac{1 + \cos 2\theta}{2} d\theta = \frac{9}{4} \pi .$$ Q.E.D.

Example 4

Assuming that $\vec{f} = 18z \vec{i} - 12 \vec{j} + 3y \vec{k}$, and S is the region in the first octant bounded by the plane $2x + 3y + 6z = 12$, find $\iint_S \vec{f} \cdot \vec{n} \, dA$.

Solution

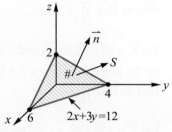

Let $\phi = 2x + 3y + 6z - 12$, $\nabla \phi = 2 \vec{i} + 3 \vec{j} + 6 \vec{k}$,

$$\vec{n} = \frac{\nabla \phi}{|\nabla \phi|} = \frac{2}{7} \vec{i} + \frac{3}{7} \vec{j} + \frac{6}{7} \vec{k}, \quad dA = \frac{dx dy}{|\vec{n} \cdot \vec{k}|} = \frac{7}{6} dx dy ,$$

$$\vec{n} \, dA = \frac{1}{6} (2 \vec{i} + 3 \vec{j} + 6 \vec{k}) dx dy .$$

Substituting the given information into the formula, we have

$$\iint_S \vec{f} \cdot \vec{n} \, dA = \iint_{S_{xy}} (18z \vec{i} - 12 \vec{j} + 3y \vec{k}) \cdot (2 \vec{i} + 3 \vec{j} + 6 \vec{k}) \frac{1}{6} dx dy$$

$$= \iint_{R_{xy}} (6z - 6 + 3y) dx dy$$

$$= \iint_{R_{xy}} (12 - 2x - 3y - 6 + 3y) dx dy$$

$$= \iint_{R_{xy}} (6 - 2x) dx dy$$

$$= \int_{y=0}^{y=4} \int_{x=0}^{x=\frac{12-3y}{2}} (6 - 2x) dx dy = 24 .$$ Q.E.D.

8-5   Exercises

Basic questions

1. Calculate the surface integral
    $\iint_S (x^2 + 2y + z - 1)\, dA$, where the surface S is the
    part of the plane $2x + 2y + z = 2$ in the first
    quadrant.

2. Continuing from the previous question, calculate
    the flux of $\vec{F} = y\vec{i} + z\vec{j}$ passing through the surface
    S: $\iint_S \vec{F} \cdot \vec{n}\, dA$, where $\vec{n}$ is the outward unit normal
    vector to the surface S.

3. Let $\vec{V} = x\vec{i} + y\vec{j} - z\vec{k}$, and S is the region bounded
    by the plane $x + 2y + z = 8$ in the first quadrant,
    find its flux $\iint_S \vec{V} \cdot \vec{n}\, dA$.

4. Let $\vec{V} = x^2\vec{i} + 2y^2\vec{z}$, and S is the region bounded
    by the plane $3x + 2y + z = 6$ in the first quadrant,
    find its flux $\iint_S \vec{V} \cdot \vec{n}\, dA$.

5. Let $\vec{F} = x^2\vec{i} + e^y\vec{j} + \vec{k}$, and S is the region
    bounded by the plane $x + y + z = 1$ in the first
    quadrant, find its flux $\iint_S \vec{F} \cdot \vec{n}\, dA$.

Advanced questions

1. Find the flux of the vector $\vec{F} = 2z\vec{i} + (x - y - z)\vec{k}$
    passing through the surface
    S: $z = x^2 + y^2$, $x^2 + y^2 \leq 6$.

2. Find the area of the surface S using surface
    integration, where
    S: $z = 1 - x - y$, $0 \leq x \leq 1$; $0 \leq y \leq 1$; $0 \leq z \leq 1$.

3. Find the surface area of $x + 2y + z = 4$ inside the
    cylinder $x^2 + y^2 = 1$.

4. Let $\vec{F} = y^3\vec{i} - x^3\vec{j}$, and the surface S is
    $1 \leq x^2 + y^2 \leq 4$, $z = 0$, find $\iint_S (\text{curl}\,\vec{F}) \cdot \vec{n}\, dA$.

5. Let $\vec{F} = z\vec{i} + x\vec{j} - 3y^2z\vec{k}$, and S is the region of the
    surface $x^2 + y^2 = 16$, $0 < z < 5$ in the first octant,
    find its flux $\iint_S \vec{F} \cdot \vec{n}\, dA$.

6. Let $\vec{F} = y\vec{i} - z\vec{j} + yz\vec{k}$, and S is the region of the
    surface $x = \sqrt{y^2 + z^2}$, $y^2 + z^2 \leq 1$, find its flux
    $\iint_S \vec{F} \cdot \vec{n}\, dA$.

7. Let $\vec{F} = y^3\vec{i} + x^3\vec{j} + z^3\vec{k}$, and S is the region of
    $x^2 + 4y^2 = 4$, $x \geq 0$, $y \geq 0$, $0 \leq z \leq 1$, find its flux
    $\iint_S \vec{F} \cdot \vec{n}\, dA$. Hint: projection onto xz-plane
    $$dA = \frac{dx\,dz}{|\vec{n} \cdot \vec{j}|} .$$

8. $\vec{v} = y\vec{i} - z\vec{j} + yz\vec{k}$, find $\phi = \iint_S \vec{v} \cdot \vec{n}\, dA$.
    S: $x = \sqrt{y^2 + z^2}$, $y^2 + z^2 \leq 1$, not contain the plane
    $x = 1$.

9. Let $\vec{F} = z^2\vec{k}$,
    find $\iint_R \vec{F} \cdot \vec{n}\, dA$,
    while R represents the
    lateral surface of a cone,
    as shown in the diagram.

    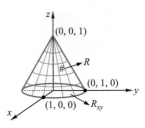

10. Calculate the flux that
    $\vec{V} = x\vec{i} + y\vec{j} + z\vec{k}$ passing through the surface
    S: $x^2 + y^2 + z^2 = 4$, $1 \leq z \leq 2$
    (not contain the plane $z = 1$, as shown in the
    following figure).

    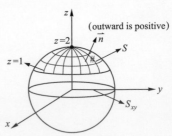

8-6 Green's Theorem

In the following section from section 8-6 to 8-8, we will introduce the three most important integral theorems in vector calculus. These theorems include Green's theorem, Gauss's divergence theorem, and Stokes' theorem. They involve the techniques of vector function differentiation and integration that were previously discussed. These three theorems are extensively used in courses related to fluid mechanics, electromagnetics, and other relevant disciplines.

First, we will introduce Green's theorem. This theorem establishes the relationship between the work done (line integral) around a closed curve in a two-dimensional xy-plane (line integral) and the double integral in the plane .

8-6-1 Green's Theorem

▶ **Theorem 8-6-1**

Green's Theorem (In a Plane Region)

Let $f(x, y)$, $g(x, y)$ be continuous and integrable functions in the region R and its boundary C, then

$$\iint_R (\frac{\partial g}{\partial x} - \frac{\partial f}{\partial y})dxdy = \oint_C fdx + gdy \ ,$$

where R is a simply (multiply) connected region in the xy-plane, and C is the boundary curve of R. The curve C is a regular, closed curve, and the orientation of C is chosen to be counterclockwise relative to R (commonly considered as positive orientation, which is opposite to the direction of time).

Proof

When $a \leq x \leq b$, the corresponding range for y is:
$h_1(x) \leq y \leq h_2(x)$,
therefore, according to the definition of a double integral:

$$\iint_R \frac{\partial f}{\partial y} dxdy = \int_{x=a}^{x=b} \int_{y=h_1(x)}^{y=h_2(x)} \frac{\partial f}{\partial y} dydx = \int_a^b f(x, y)\Big|_{h_1(x)}^{h_2(x)} dx$$

$$= \int_a^b [f(x, h_2(x)) - f(x, h_1(x))]dx$$

$$= -\int_b^a f(x, h_2(x))dx - \int_a^b f(x, h_1(x))dx$$

$$= -[\int_{C_2} f(x, y)dx + \int_{C_1} f(x, y)dx]$$

$$= -[\oint_C f(x, y)dx] \ ;$$

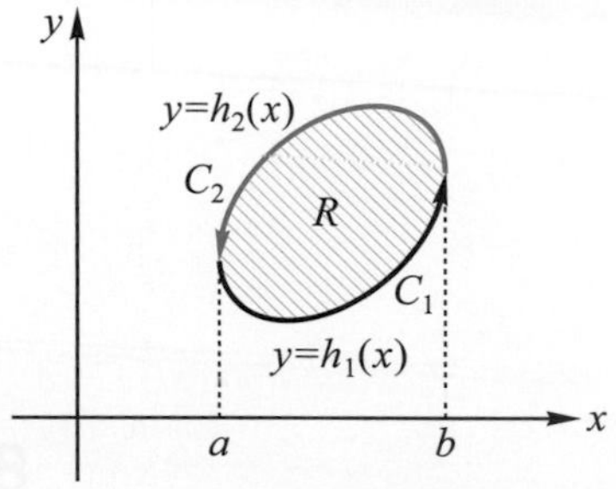

similarly, $\iint_R \frac{\partial g}{\partial x} dxdy = \oint_C gdy$, thus $\iint_R [\frac{\partial g}{\partial x} - \frac{\partial f}{\partial y}]dxdy = \int_C fdx + gdy$.   ◀

$C = C_1 + C_2$ is the closed boundary of the region

Figure 8-43   The integration limits for Green's theorem

Example 1

Find $\oint_C (x^2 + 2y)dx + (4x + y^2)dy$, $C: x^2 + y^2 = 1$ (clockwise).

Solution

$$\oint_C (x^2 + 2y)dx + (4x + y^2)dy$$

$$= -\iint_R [\frac{\partial}{\partial x}(4x + y^2) - \frac{\partial}{\partial y}(x^2 + 2y)]dxdy$$

$$= -\iint_R [4 - 2]dxdy$$

$$= -2\iint_R dxdy$$

$$= -2\pi \cdot 1^2 = -2\pi .$$

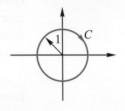

Q.E.D.

Example 2

Calculate the work done by the particle as it moves counterclockwise around the circle $x^2 + y^2 = a^2$ for one complete revolution in the force field
$$\vec{F}(x, y) = (\sin x - y)\vec{i} + (e^y - x^2)\vec{j} .$$

Solution

$$w = \int_C \vec{F} \cdot d\vec{r} = \int_C (\sin x - y)dx + (e^y - x^2)dy$$

$$= \iint_R [\frac{\partial}{\partial x}(e^y - x^2) - \frac{\partial}{\partial y}(\sin x - y)]dxdy$$

$$= \iint_R (-2x + 1)dxdy$$

$$= \int_0^{2\pi} \int_0^a (-2r\cos\theta + 1)rdrd\theta$$

$$= \int_0^{2\pi} (-\frac{2}{3}r^3\cos\theta + \frac{1}{2}r^2)\Big|_0^a d\theta$$

$$= (-\frac{2}{3}a^3\sin\theta + \frac{1}{2}a^2\theta)\Big|_0^{2\pi} = a^2\pi .$$

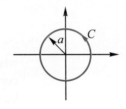

Q.E.D.

8-6-2　The Application of Green's Theorem

As shown in Figure 8-44, we want to calculate the area of region R. This involves finding f and g which causes $A = \iint_R dxdy$. Therefore, there are three ways to achieve this, in Green's theorem

$$\iint_R (\frac{\partial g}{\partial x} - \frac{\partial f}{\partial y})dxdy = \oint_C (fdx + gdy) .$$

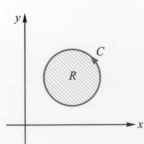

Figure 8-44　Region of plane integration

(1) Let $f = 0$, $g = x$, then the area of R is $A = \oint_C x\,dy$.

(2) Let $f = -y$, $g = 0$, then the area of R is $A = -\oint_C y\,dx$.

(3) Let $g = \dfrac{x}{2}$, $f = -\dfrac{y}{2}$, then the area of R is $A = \dfrac{1}{2}\oint x\,dy - y\,dx$.

Interestingly, by performing a polar coordinate transformation in (3) , $\begin{cases} x = r(\theta)\cos\theta \\ y = r(\theta)\sin\theta \end{cases}$

we obtain $\begin{cases} dx = -r\sin\theta\,d\theta + \cos\theta\,dr \\ dy = r\cos\theta\,d\theta + \sin\theta\,dr \end{cases}$, and substituting it into the area formula from (3),

then we get $A = \dfrac{1}{2}\oint_C r^2(\theta)\,d\theta$, we arrive at the familiar area formula in polar coordinates

that we learned in calculus. Hence, it can be said that Green's theorem reveals the connection between area and boundary curves.

Example 3

Given a smooth, closed curve C and the region R enclosed by it.
(1) Please describe what Green's theorem is.

(2) Please prove the area of region R is $A = \dfrac{1}{2}\oint_C (x\,dy - y\,dx)$.

(3) Please prove the area of region R is $A = -\oint_C y\,dx = \oint_C x\,dy$.

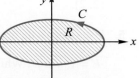

(4) Utilize the above-proven result to calculate the area of $\dfrac{x^2}{a^2} + \dfrac{y^2}{b^2} = 1$.

Solution

(1) $\oint_C f\,dx + g\,dy = \iint_R (\dfrac{\partial g}{\partial x} - \dfrac{\partial f}{\partial y})\,dx\,dy$.

(2) $\dfrac{1}{2}\oint_C (x\,dy - y\,dx) = \dfrac{1}{2}\iint_R (\dfrac{\partial x}{\partial x} - \dfrac{\partial(-y)}{\partial y})\,dx\,dy = \iint_R dx\,dy = A$.

(3) $-\oint_C y\,dx = -\iint_R -\dfrac{\partial(y)}{\partial y}\,dx\,dy = \iint_R dx\,dy = A$, $\oint_C x\,dy = \iint_R 1\,dx\,dy = A$.

(4) C: $\begin{cases} x = a\cos\theta \\ y = b\sin\theta \end{cases}$; $\theta : \theta \in [0, 2\pi] \Rightarrow A = -\oint_C y\,dx$,

$\therefore A = -\displaystyle\int_0^{2\pi} b\sin\theta \cdot (-a\sin\theta)\,d\theta = \int_0^{2\pi} ab \cdot \dfrac{1 - \cos 2\theta}{2}\,d\theta = \pi ab$.   Q.E.D.

8-6 Exercises

Basic questions

In the following questions 1 to 4, please use Green's theorem to calculate $\oint_C \vec{F} \cdot d\vec{r}$, where C is oriented counterclockwise.

1. $\vec{F} = 3y\,\vec{i} - 2x\,\vec{j}$, C is the circular arc with center $(2, 3)$ and a radius of 2.

2. $\vec{F} = 3xy^2\,\vec{i} + 3x^2 y\,\vec{j}$, C is the ellipse with center $(2, 3)$, a major axis of 6, and a minor axis of 4.

3. $\vec{F} = x^2\,\vec{i} - 2xy\,\vec{j}$, C is the triangle with vertices $(0, 0)$, $(1, 0)$, and $(0, 1)$.

4. $\vec{F} = (x^3 - y)\,\vec{i} + (\cos(2y) + e^{y^2} + 3x)\,\vec{j}$,
   C is the square with vertices $(0, 0)$, $(0, 2)$, $(2, 0)$, $(2, 2)$.

5. Use Green's theorem to calculate $\oint_C \vec{F} \cdot d\vec{r}$, where
   $\vec{F} = y^2\,\vec{i} + \vec{j}$. While the closed curve C is the boundary of the triangle formed by the points $O\,(0, 0)$, $A\,(1, 0)$, and $B\,(0, 2)$, along the path $O\text{-}A\text{-}B\text{-}O$.

Advanced questions

1. Use Green's theorem to calculate the line integral
   $\oint_C y\,dx + x^2 y\,dy$, where C is the region enclosed by the curves $y^2 = 2x$ and $y^3 = 4x$ between the point $(0, 0)$ and $(2, 2)$.

2. If the cardioid is represented in polar coordinates as $r = 2(1 + \cos\theta)$, find the area enclosed by this cardioid. (Hint: the area can be calculated as
   $A = \dfrac{1}{2}\oint_C r^2\,d\theta$)

3. Calculate the line integral $\oint_C \vec{F} \cdot d\vec{r}$, where
   $$\vec{F} = (\frac{x^2 + y^2}{2} + 2y)\vec{i} + (xy - ye^y)\vec{j}$$
   and the closed curve C is
   $$C : \begin{cases} y = \pm 1,\ -1 \le x \le 1 \\ x = \pm 1,\ -1 \le y \le 1 \end{cases}.$$

4. Use Green's theorem to calculate $\oint_C \vec{F} \cdot d\vec{r}$, where
   $\vec{F} = (e^{\sin y} - y)\vec{i} + (\sinh y^3 - 4x)\vec{j}$, and the closed curve C is the counterclockwise circular path with center $(-8, 0)$ and a radius of 2.

5. Use Green's theorem to calculate $\oint_C \vec{F} \cdot d\vec{r}$, where
   $\vec{F} = \dfrac{2x}{x^2 + y^2}\vec{i} + \dfrac{2y}{x^2 + y^2}\vec{j}$, and the closed curve C is the counterclockwise circular path
   $(x - 2)^2 + (y - 1)^2 = 1$.

6. Use Green's theorem to calculate $\oint_C \vec{F} \cdot d\vec{r}$, where
   $\vec{F} = x^2 y\,\vec{i} - xy^2\,\vec{j}$, and the closed curve C is the counterclockwise boundary of the region R defined by: $x^2 + y^2 \le 4$, $x \ge 0$, $y \ge 0$.

8-7 Gauss's Divergence Theorem

This theorem will introduce the relationship of transformation between the flux of a vector function across a closed surface in space and the volume integral of the divergence of the vector function over the region enclosed by the surface.

8-7-1 The Definition and Theorem

As shown in Figure 8-45, let $\vec{V}$ be a vector function in the region D in space, which is integrable and continuous within D. Let P be any point within D. Then, the divergence is defined as the outflow rate per unit volume at point P. If we denote ΔV as a small volume element within the region D and ΔS as the surface of this small volume element, the **divergence** can be symbolically expressed as:

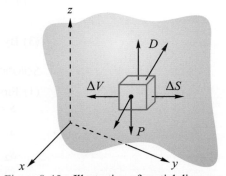

Figure 8-45 Illustration of spatial divergence

$$\nabla \cdot \vec{V}\,|_P = \lim_{\Delta V \to 0} \frac{\oiint_{\Delta S} \vec{V} \cdot \vec{n}\,dA}{\Delta V}$$

the divergence $\nabla \cdot \vec{V}\,|_P$ represents the flux of the vector field $\vec{V}$ through a unit volume at point P.

1. Divergence Theorem of Gauss

Suppose the vector function $\vec{V}$ is a continuous and integrable function within the region D and its boundary surface S. Then, Gauss's divergence theorem states

$$\iiint_D \nabla \cdot \vec{V}\,dV = \oiint_S \vec{V} \cdot \vec{n}\,dA$$

where D is a simply (multiply)connected region, $\vec{n}$ is the outward unit normal vector on S that is orientable and points outward, $\iiint (\nabla \cdot \vec{V})dV$ representing the flux outward from the region D (surface flux). The physical interpretation states that the amount of outward flux from the region D is equal to the amount of flux leaving through the boundary surface of D. In other words, "the flux running out of the surface = the physical quantity produced by the source's strength." The general Gauss's divergence theorem is used to handle the computation of the surface flux through a closed surface S, $\oiint_S \vec{V} \cdot \vec{n}\,dA$, which is not easily done through surface integrals. Therefore, this theorem allows us to convert it into a more manageable volume integral $\iiint_D (\nabla \cdot \vec{V})dV$.

Example 1

$\vec{F} = (xy - 1)\vec{i} + yz\vec{j} + xz\vec{k}$. In space, the region D is a cube defined by: $0 \leq x \leq 1$,
$0 \leq y \leq 1$, $0 \leq z \leq 1$, and let S be the closed surface that encloses D,
with the outward unit normal vector denoted by $\vec{n}$.

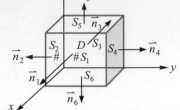

(1) Calculate $\oiint_S \vec{F} \cdot \vec{n}\, dA$ by using surface integral.

(2) Calculate $\iiint_D (\nabla \cdot \vec{F}) dV$.

(3) By comparing (1) and (2), which theorem we can verify?

Solution

(1) First, divide the area to be calculated into six smaller regions.
$S = S_1 + S_2 + S_3 + S_4 + S_5 + S_6$, where

S_1 : $x = 1$, $0 \leq y \leq 1$, $0 \leq z \leq 1$, and the outward unit normal vector is $\vec{n_1} = \vec{i}$

S_2 : $y = 0$, $0 \leq x \leq 1$, $0 \leq z \leq 1$, and the outward unit normal vector is $\vec{n_2} = -\vec{j}$

S_3 : $x = 0$, $0 \leq y \leq 1$, $0 \leq z \leq 1$, and the outward unit normal vector is $\vec{n_3} = -\vec{i}$

S_4 : $y = 1$, $0 \leq x \leq 1$, $0 \leq z \leq 1$, and the outward unit normal vector is $\vec{n_4} = \vec{j}$

S_5 : $z = 1$, $0 \leq x \leq 1$, $0 \leq y \leq 1$, and the outward unit normal vector is $\vec{n_5} = \vec{k}$

S_6 : $z = 0$, $0 \leq x \leq 1$, $0 \leq y \leq 1$, and the outward unit normal vector is $\vec{n_6} = -\vec{k}$

Calculate the surface integrals for each S_i using their respective definitions, we obtain

$$\iint_{S_1} \vec{F} \cdot \vec{n}\, dA = \iint_{S_1} (xy - 1) dy dz = \int_{z=0}^{1} \int_{y=0}^{y=1} (y - 1) dy dz = -\frac{1}{2},$$

$$\iint_{S_2} \vec{F} \cdot \vec{n}\, dA = \iint_{S_2} -yz\, dx dz = \iint_{S_1} 0 \cdot z\, dx dz = 0 ,$$

$$\iint_{S_3} \vec{F} \cdot \vec{n}\, dA = \iint_{S_3} -(xy - 1) dy dz = \int_{z=0}^{1} \int_{y=0}^{y=1} (1) dy dz = 1 ,$$

$$\iint_{S_4} \vec{F} \cdot \vec{n}\, dA = \iint_{S_4} (yz) dy dz = \int_{z=0}^{1} \int_{x=0}^{1} (z) dx dz = \frac{1}{2} ,$$

$$\iint_{S_5} \vec{F} \cdot \vec{n}\, dA = \iint_{S_5} (xz) dx dy = \int_{y=0}^{1} \int_{x=0}^{1} (x) dx dy = \frac{1}{2} ,$$

$$\iint_{S_6} \vec{F} \cdot \vec{n}\, dA = \iint_{S_6} (-xz) dx dy = \int_{y=0}^{1} \int_{x=0}^{1} (0) dx dy = 0 ,$$

so $$\iint_{S = S_1 + S_2 + S_3 + S_4 + S_5 + S_6} \vec{F} \cdot \vec{n}\, dA = \iint_{S_1} \vec{F} \cdot \vec{n}\, dA + \iint_{S_2} \vec{F} \cdot \vec{n}\, dA + \iint_{S_3} \vec{F} \cdot \vec{n}\, dA$$

$$+ \iint_{S_4} \vec{F} \cdot \vec{n}\, dA + \iint_{S_5} \vec{F} \cdot \vec{n}\, dA + \iint_{S_6} \vec{F} \cdot \vec{n}\, dA$$

$$= \frac{3}{2}.$$

(2) Because $\nabla \cdot \vec{F} = y + z + x$,

$$\iiint_D (\nabla \cdot \vec{F})dV = \iiint_D (x+y+z)\,dV = \int_{z=0}^1 \int_{y=0}^1 \int_{x=0}^1 (x+y+z)dxdydz = \frac{3}{2}.$$

(3) From (1) and (2), we know $\iint_S \vec{F} \cdot \vec{n}\, dA = \iiint_D (\nabla \cdot \vec{F})dV = \frac{3}{2}$.

The Gauss's divergence theorem has been verified.   Q.E.D.

Example 2

If $\vec{F} = 5x\vec{i} - 3y\vec{j} + 2z\vec{k}$, S is a closed spherical surface $x^2 + y^2 + z^2 = 9$. Using Gauss's divergence theorem, calculate $\iint_S \vec{F} \cdot \vec{n}dA$, where $\vec{n}$ is the outward unit normal vector of the surface S and dA represents the differential area element.

Solution

$$\iint_S \vec{F} \cdot \vec{n}dA = \iiint_D (\nabla \cdot \vec{F})dV = \iiint_D (4)dV = 4\iiint_D dV$$

$$= 4 \times \frac{4}{3}\pi \times 3^3 = 144\pi.$$ Q.E.D.

Example 3

Suppose that the function $\vec{F} = x\vec{i} + y\vec{j} + z\vec{k}$, and consider the surface is the union of S: $z^2 = (x^2 + y^2)$, $0 \le z \le 1$ and the top of the unit disk $x^2 + y^2 = 1$, composed by $z = 1$, calculate the two following:

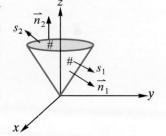

(1) Compute $\oiint_S \vec{F} \cdot \vec{n}\, dA$ using surface integral, where $\vec{n}$ is the outward unit normal vector of the surface S.

(2) Recalculate the result from (1) using Gauss's divergence theorem.

Solution

(1) Seperate $S = S_1 + S_2$; where S_2: $x^2 + y^2 = 1$, $z = 1$; S_1: $z = \sqrt{x^2 + y^2}$, $0 \le z < 1$, then suppose that $\phi = \sqrt{x^2 + y^2} - z$,

we obtain $\nabla\phi = \frac{x}{\sqrt{x^2+y^2}}\vec{i} + \frac{y}{\sqrt{x^2+y^2}}\vec{j} - \vec{k}$, $|\nabla\phi| = \sqrt{2}$.

$\therefore ① \; \vec{n_1} = \frac{1}{\sqrt{2}}(\frac{x}{\sqrt{x^2+y^2}}\vec{i} + \frac{y}{\sqrt{x^2+y^2}}\vec{j} - \vec{k})$.

② $\iint_{S_1} \vec{F} \cdot \vec{n} \, dA = \frac{1}{\sqrt{2}} \iint_{S_1} (\frac{x^2}{\sqrt{x^2+y^2}} + \frac{y^2}{\sqrt{x^2+y^2}} - z) dA$

$= \frac{1}{\sqrt{2}} \iint_{S_1} (\sqrt{x^2+y^2} - z) dA = 0$.

③ $\vec{n_2} = \vec{k} \Rightarrow \iint_{S_2} \vec{F} \cdot \vec{n} \, dA = \iint_{S_2} z \, dA = \iint_{S_2} dA = \pi \cdot 1^2 = \pi$

∴ $\iint_{S} \vec{F} \cdot \vec{n} \, dA = \iint_{S_1} \vec{F} \cdot \vec{n} \, dA + \iint_{S_2} \vec{F} \cdot \vec{n} \, dA = \pi$.

(2) $\iint_{S} \vec{F} \cdot \vec{n} \, dA = \iiint_{D} (\nabla \cdot \vec{F}) dV = 3 \iiint_{D} dV = 3 \cdot \frac{1}{3} \pi \cdot 1^2 \cdot 1 = \pi$.

(The volume of a cone $= \frac{1}{3} \times$ the base area $\times$ the height)   Q.E.D.

2. The Method of Surface Completion

We can complete an open surface by adding the missing boundary surface. In this way, we can use Gauss's theorem to calculate the desired physical quantity. Let S be an open surface, and we can add a surface S^* causes $S + S^*$ forms a closed surface, as shown in Figure 8-46. Then, we have $\iiint_{D} \nabla \cdot \vec{V} dV = \oiint_{S} \vec{V} \cdot \vec{n} dA + \oiint_{S^*} \vec{V} \cdot \vec{n} \, dA$,

and after rearranging

$$\iint_{S} \vec{V} \cdot \vec{n} \, dA = \iiint_{D} \nabla \cdot \vec{V} dV - \iint_{S^*} \vec{V} \cdot \vec{n}^* dA$$

Notably, when the divergence of the vector field is zero, Gauss's divergence theorem becomes $\iint_{S} \vec{V} \cdot \vec{n} \, dA = -\iint_{S^*} \vec{V} \cdot \vec{n}^* dA$. In other words, the surface integral over the completed surface and the original surface differ only by a negative sign.

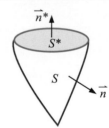

Figure 8-46   Illustration of the integration over the completed surface in the divergence theorem.

Example 4

$\vec{V}(x, y, z) = (x^3 + 7y^2 z + 2z^3)\vec{i} + (4x - 3x^2 y + 2yz)\vec{j} + (x^2 + y^2 - z^2)\vec{k}$

(1) Calculate the flux of $\vec{V}$ pass through the upper hemisphere $x^2 + y^2 + z^2 = a^2, z > 0$.

(2) Calculate the flux of $\vec{V}$ pass through the lower hemisphere $x^2 + y^2 + z^2 = a^2, z < 0$.

Solution

(1) Consider the region of the flux:

S_1: $x^2 + y^2 + z^2 = a^2$, $z > 0$, we use the completed surface (as shown in the attached figure in the blue region)

S_2: $x^2 + y^2 + z^2 = a^2$, $z = 0$, merge S_1 to form a closed surface S, enclosing the region D. Now since $\nabla \cdot \overrightarrow{V} = 3x^2 - 3x^2 + 2z - 2z = 0$,

substituting into Gauss's divergence theorem, which yields:

$$\iint_{S_1+S_2} \overrightarrow{V} \cdot \overrightarrow{n}\, dA = \iint_{S_1} \overrightarrow{V} \cdot \overrightarrow{n}\, dA + \iint_{S_2} \overrightarrow{V} \cdot \overrightarrow{n}\, dA = \iiint_{D} \nabla \cdot \overrightarrow{V} dV = 0,$$

that is $\iint_{S_1} \overrightarrow{V} \cdot \overrightarrow{n}\, dA = -\iint_{S_2} \overrightarrow{V} \cdot (-\overrightarrow{k})dA = \iint_{S_2} (x^2 + y^2 - z^2)\, dxdy$

$$= \iint (x^2 + y^2)\, dxdy = \int_0^{2\pi} \int_0^a r^2 r\, dr d\theta$$

$$= \frac{1}{4} r^4 \bigg|_0^a \cdot 2\pi = \frac{1}{2} \pi a^4 .$$

(2) Following the approach of (1), taking $S^* = S_1 + S_3$ as the closed surface enclosing the region, D^*, which lies inside the sphere, then

$$\iint_{S_1+S_3} \overrightarrow{V} \cdot \overrightarrow{n}\, dA = \iint_{S_1} \overrightarrow{V} \cdot \overrightarrow{n}\, dA + \iint_{S_3} \overrightarrow{V} \cdot \overrightarrow{n}\, dA = \iiint_{D^*} \nabla \cdot \overrightarrow{V} dV = 0$$

$$\iint_{S_3} \overrightarrow{V} \cdot \overrightarrow{n}\, dA = -\iint_{S_1} \overrightarrow{V} \cdot \overrightarrow{n}\, dA = -\frac{1}{2} \pi a^4 .$$

Q.E.D.

8-7-2   The Two Alternative Forms of Gauss's Divergence Theorem

▶ **Theorem 8-7-1**

Divergence Theorem of Gauss in $\mathbb{R}^3$

Suppose $\overrightarrow{F} = F_1 \overrightarrow{i} + F_2 \overrightarrow{j} + F_3 \overrightarrow{k}$ is a vector function within the region D and its boundary surface S (denoted as $\overline{D} = D + S$) is continuous and integrable. Then, Gauss's divergence theorem can be expressed as follows:

$$\iint_S F_1 dydz + F_2 dzdx + F_3 dxdy = \iiint_D (\frac{\partial F_1}{\partial x} + \frac{\partial F_2}{\partial y} + \frac{\partial F_3}{\partial z}) dxdydz .$$ ◀

If we take $\{\overrightarrow{i}, \overrightarrow{j}, \overrightarrow{k}\}$ as a coordinate system in $\mathbb{R}^3$, then the divergence

$\nabla \cdot \overrightarrow{F} = \frac{\partial F_1}{\partial x} + \frac{\partial F_2}{\partial y} + \frac{\partial F_3}{\partial z}$. As shown in Figure 8-47, the outward unit normal vector of

the surface is $\overrightarrow{n} = \cos\alpha \overrightarrow{i} + \cos\beta \overrightarrow{j} + \cos\gamma \overrightarrow{k}$. Therefore, based on the algorithm from Section 8-5 for calculating dA, we have

$$\vec{n}\,dA = (\cos\alpha\,\vec{i} + \cos\beta\,\vec{j} + \cos\gamma\,\vec{k})dA$$
$$= [\cos(\alpha)dA]\,\vec{i} + [\cos(\beta)dA]\,\vec{j} + [\cos(\gamma)dA]\,\vec{k}$$
$$= dydz\,\vec{i} + dxdz\,\vec{j} + dxdy\,\vec{k}$$

Therefore, substituting into the general form of

Gauss's divergence theorem, the proof is obtained.

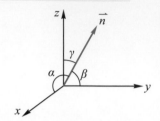

Figure 8-47   Illustration of the directio
angles of the vector $\vec{n}$

▶ Theorem 8-7-2

The Second Form

Suppose ϕ has second-order partial derivatives that exist and are continuous within the

region D and its boundary S. Then $\iiint_D (\nabla^2\phi)\,dV = \iint_S \nabla\phi\cdot\vec{n}\,dA = \iint_S \dfrac{\partial\phi}{\partial n}\,dA$.

Proof

In the statement of Theorem 8-7-1, let $\vec{F} = \nabla\phi$. Then, according to the definition of

divergence, we have $\iiint_D (\nabla^2\phi)\,dV = \iint_S \nabla\phi\cdot\vec{n}\,dA$. Now, using the fact that the

directional derivative is the dot product of the gradient and the direction vector, we have

$\nabla\phi\cdot\vec{n} = \dfrac{\partial\phi}{\partial n}$, therefore $\iint_S \nabla\phi\cdot\vec{n}\,dA = \iint_S \dfrac{\partial\phi}{\partial n}\,dA$.   ◀

Example 5

S is the surface of the cylinder $x^2 + y^2 = 4$, and
$0 \le z \le 3$(including the top and bottom).

(1) Find $\iint_S (x^3 dydz + (x^2 y + 1)dzdx + x^2 z dxdy)$.

(2) Find $\iint_S \dfrac{\partial\phi}{\partial n}\,dA$,where $\phi = e^x \cos y + x^3 + xz$.

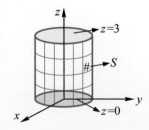

Solution

(1) Let $\vec{F} = x^3\vec{i} + (x^2y+1)\vec{j} + (x^2z)\vec{k}$, then according to the problem statement, we choose the second form of Gauss's divergence theorem, which allows us to avoid the computation of dA. Substituting the relevant functions into the original expression, we obtain

$$\iint_S \vec{F} \cdot \vec{n}\, dA = \iiint_D \nabla \cdot \vec{F} dV = \iiint_D [3x^2 + x^2 + x^2]\, dxdydz$$

$$= \iiint_D 5x^2\, dxdydz = \int_{\theta=0}^{\theta=2\pi} \int_{r=0}^{r=2} \int_{z=0}^{z=3} 5 \cdot (r\cos\theta)^2\, rdrd\theta dz$$

$$= \int_0^{2\pi} \int_0^2 5r^3 \cos^2\theta \cdot 3drd\theta = \int_0^{2\pi} \frac{15}{4}r^4 \Big|_0^2 \cos^2\theta d\theta$$

$$= \int_0^{2\pi} 60 \cdot \frac{1+\cos 2\theta}{2} d\theta = 60\pi.$$

(2) $\nabla^2\phi = \dfrac{\partial^2\phi}{\partial x^2} + \dfrac{\partial^2\phi}{\partial y^2} + \dfrac{\partial^2\phi}{\partial z^2} = 6x$, from the second form of Gauss's divergence theorem

we obtain $\displaystyle\iint_S \frac{\partial\phi}{\partial n}\, dA = \iiint_D 6x\, dx\, dy\, dz = \int_{\theta=0}^{2\pi} \int_{r=0}^2 \int_{z=0}^3 6r\cos\theta \cdot rdrd\theta dz = 0.$

Q.E.D.

8-7 Exercises

Basic questions

In the following 1 to 4, please use Gauss's divergence theorem to calculate the surface flux $\displaystyle\oiint_S \vec{F} \cdot \vec{n} dA$.

1. $\vec{F} = 2x\vec{i} + 3y\vec{j} + 4z\vec{k}$, the surface S, which is the closed surface formed by the tetrahedron with vertices at $(0, 0, 0)$, $(1, 0, 0)$, $(0, 1, 0)$, $(0, 0, 1)$.

2. $\vec{F} = 4x\vec{i} - 6y\vec{j} + z\vec{k}$, the surface S, which is the sphere centered at $(0, 0, 0)$ with a radius of 3.

3. $\vec{F} = (4x + e^{yz})\vec{i} + (2y + e^{xz})\vec{j} + (e^{xy} - 6z)\vec{k}$, the surface S, which is the closed surface formed by the upper hemisphere with a radius of 3 centered at $(0, 0, 0)$ and including the bottom surface $z = 0$.

4. $\vec{F} = y^3 \cos(yz)\vec{i} + 3y\vec{j} + x^3 \sinh(xy)\vec{k}$,

the surface S is the closed cylindrical surface defined by $x^2 + y^2 = 1$, $-1 \le z \le 1$, and includes $z = 1$ and $z = -1$.

5. For the vector function $\vec{F} = x^3\vec{i} + y^3\vec{j} + z^3\vec{k}$, and the surface S: $x^2 + y^2 + z^2 = 4$, please using Gauss's divergence theorem to calculate $\displaystyle\oiint_S \vec{F} \cdot \vec{n} dA$.

6. For the vector function $\vec{F} = x\vec{i} + y\vec{j} + 2z\vec{k}$, the surface S: $x + y + z = 1$ is the surface of tetrahedron formed by the first quadrant of the coordinate plane, calculate $\displaystyle\oiint_S \vec{F} \cdot \vec{n} dA$ using Gauss's divergence theorem.

Advanced questions

1. For the vector function $\vec{F} = x^2\vec{i} + 2y\vec{j} + 4z^2\vec{k}$, the surface S is the surface of the cylinder $x^2 + y^2 \le 4$, $0 \le z \le 2$ (including the top and bottom), calculate $\displaystyle\oiint_S \vec{F} \cdot \vec{n} dA$ using Gauss's divergence theorem.

2. For the vector function
$\overrightarrow{F} = x^2 yz\,\overrightarrow{i} + xy^2 z\,\overrightarrow{j} - 2xyz^2\,\overrightarrow{k}$, the surface S which

is the ellipsoidal surface $x^2 + y^2 + \dfrac{z^2}{9} = 1$, calculate

$\oiint_S \overrightarrow{F} \cdot \overrightarrow{n}\, dA$ using Gauss's divergence theorem.

3. Please using Gauss's divergence theorem to
calculate $\oiint_S (7x\,\overrightarrow{i} - z\,\overrightarrow{k}) \cdot \overrightarrow{n}\, dA$, the surface S is the
spherical surface $x^2 + y^2 + z^2 = 4$.

4. For the vector function $\overrightarrow{F} = e^x\,\overrightarrow{i} - ye^x\,\overrightarrow{j} + 3z^2\,\overrightarrow{k}$,
and the surface S which is the surface of the
cylinder $x^2 + y^2 \le 4$, $0 \le z \le 5$ (including the top and
bottom), please using Gauss's divergence theorem
to calculate $\oiint_S \overrightarrow{F} \cdot \overrightarrow{n}\, dA$.

5. For the vector function $\overrightarrow{F} = xy^2\,\overrightarrow{i} + y^3\,\overrightarrow{j} + 4x^2 z\,\overrightarrow{k}$,
and the surface S which is the surface of the cone
$x^2 + y^2 \le z$, $0 \le z \le 4$ (including $z = 4$), calculate
$\oiint_S \overrightarrow{F} \cdot \overrightarrow{n}\, dA$ using Gauss's divergence theorem.

6. Verify Gauss's divergence theorem, where the
vector function $\overrightarrow{F} = z^2\,\overrightarrow{k}$, and the closed region in
space is
$D: x^2 + y^2 \le (2z)^2$, $0 \le z \le 3$ (including $z = 3$).
(1) Calculate the triple integral $\iiint_D (\nabla \cdot \overrightarrow{F})\, dV$.

(2) Calculate the surface integral $\oiint_S \overrightarrow{F} \cdot \overrightarrow{n}\, dA$,
where
$S = S_1 + S_2$;
$S_1 : x^2 + y^2 = 36$, $z = 3$;
$S_2 : x^2 + y^2 = 4z^2$, $0 \le z \le 3$,
(not include $z = 3$).

7. The region D is the sphere $x^2 + y^2 + (z-1)^2 = 9$
containing the plane $z = 1$ and $1 \le z \le 4$, if $\overrightarrow{F}$ is
$\overrightarrow{F} = x\,\overrightarrow{i} + y\,\overrightarrow{j} + (z-1)\,\overrightarrow{k}$, verify the divergence
theorem.

8. Please derive the divergence theorem for a plane
$\oint_C \overrightarrow{F} \cdot \overrightarrow{n}\, ds = \oiint_R (\nabla \cdot \nabla \cdot \overrightarrow{F})\, dA$, where R is the

plane region with boundary C, and $\overrightarrow{n}$ is the unit
normal vector to the curve C and is perpendicular
to the tangent vector $\overrightarrow{e_t}$.

9. Let ϕ, φ have second-order partial derivatives that
exist and are continuous within the region D and its
boundary S. Prove the Green's first identity:
$$\iiint_D [\nabla \phi \cdot \nabla \varphi + \phi \nabla^2 \varphi]\, dV = \oiint_S \phi \frac{\partial \varphi}{\partial n}\, dA$$
and the Green's second identity:
$$\iiint_D [\phi \nabla^2 \varphi - \varphi \nabla^2 \phi]\, dV = \oiint_S \left[\phi \frac{\partial \varphi}{\partial n} - \varphi \frac{\partial \phi}{\partial n} \right] dA .$$

8-8 Stokes' Theorem

This theorem describes the relationship between the circulation on its boundary and the total amount of its curl, when a vector field passing through a closed surface. Unlike the general approach in vector analysis, here we understand how to use line integrals to describe local rotational effects from the perspective of fluid mechanics. Later, we will see how to calculate the total rotational effect of a region through path integrals of the vector field along the boundary of the region, which is known as Stokes' theorem.

8-8-1 Circulation, Vorticity, and Curl

The **circulation** of a vector field $\vec{V}$ along the curve C is defined as the path integral of $\vec{V}$ along curve C:

$$\Gamma \equiv \oint_C \vec{V} \cdot \vec{dr} = \oint_C \vec{V} \cdot \vec{e_t}\, ds$$

where $\vec{e_t}$ is the unit tangent vector on the curve C as shown in Figure 8-48. The rate of change of circulation per unit area is called **vorticity**, denoted by

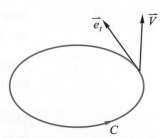

Figure 8-48 The unit tangent vector of the curve in the circulation field

$$\lim_{\Delta A \to 0} \frac{1}{\Delta A} \oint_C \vec{V} \cdot \vec{dr}$$

where $\vec{n}$ points in the direction satisfying the right-hand screw rule with respect to the orientation of the curve C, as shown in Figure 8-49.

Now, we move on to the main focus of this section.

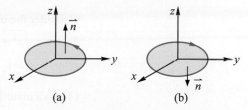

(a)   (b)

Figure 8-49 Direction of circulation

Curl

Let $\vec{V}$ be a function with continuous first-order partial derivatives on the surface S, and P be a point on the surface S. Then, at point P, the curl is defined as the component of vorticity on the unit normal vector of the surface S at point P, as shown in Figure 8-50. Mathematically, it is expressed as:

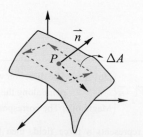

Figure 8-50 Diagram of curl

$$\nabla \times \vec{V}\Big|_P \cdot \vec{n} = (\lim_{\Delta A \to 0} \oint_C \frac{1}{\Delta A} \vec{V} \cdot d\vec{r}).$$

Its physical meaning is $\vec{V}$ at the maximum rotational rate vector on the unit area at point P (i.e., the rotational intensity in the direction of the normal vector) [4].

8-8-2 Stokes' Theorem

▶ **Theorem 8-8-1**

Stokes' Theorem

Suppose $\vec{V}$ be a function with continuous first-order partial derivatives on the surface S and its boundary. Then, $\iint_S (\nabla \times \vec{V}) \cdot \vec{n} dA = \oint_C \vec{V} \cdot d\vec{r}$

where the orientation of the curve C and the direction of $\vec{n}$ are chosen according to the right-hand screw rule.   ◀

This theorem is commonly used when line integrals are difficult to evaluate, as it allows us to convert them into easier-to-handle surface integrals.

Physical Interpretation

Stokes' theorem states that the total rotational effect of a physical quantity $\vec{V}$ on the surface S is equal to the circulation of the quantity along the boundary curve C. [5]

Example 1

Consider the closed surface $\Sigma = \Sigma_1 + \Sigma_2$,

Σ_1: $z = \sqrt{x^2 + y^2}$, $x^2 + y^2 \le 1$; Σ_2: $x^2 + y^2 \le 1$, $z = 1$,

C: $x^2 + y^2 = 1$, $z = 1$, $\vec{F} = -y\vec{i} + x\vec{j} - xyz\vec{k}$

(1) Calculate $\oint_C \vec{F} \cdot d\vec{r}$.

(2) Calculate $\iint_{\Sigma_1} (\nabla \times \vec{F}) \cdot \vec{n}_1 \, dA$, $\vec{n}_1$ is the unit normal

vector to Σ_1.

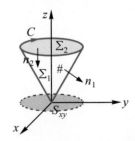

---

[4] $\nabla \times \vec{V}\Big|_p \cdot \vec{n}$ $\begin{cases} \text{Direction: The direction where the circulation value per unit area is maximized at} \\ \qquad \text{point along the normal direction of the surface.} \\ \text{Magnitude: The corresponding magnitude is the circulation value per unit area.} \end{cases}$

[5] If $\vec{V}$ represents a force field, then the circulation is equivalent to the work done along the closed curve, if $\vec{V}$ is a conservative field, then $\nabla \times \vec{V} = 0$, the work done along the closed curve $\oint_C \vec{V} \cdot d\vec{r} = \iint_S (\nabla \times \vec{V}) \cdot \vec{n} dA = 0$.

(3) Calculate $\iint_{\Sigma_2} (\nabla \times \vec{F}) \cdot \vec{n_2} \, dA$, $\vec{n_2}$ is the unit normal vector to Σ_2.

(4) Determine whether the values obtained from (1), (2), and (3) are the same, and if so, explain the mathematical theory behind it.

Solution

(1) $C: x^2 + y^2 = 1$ and $z = 1 \Rightarrow \begin{cases} x = \cos\theta \\ y = \sin\theta \; ; \; \theta = 2\pi\sim 0 \\ z = 1 \end{cases}$

then $\qquad \vec{F} = -\sin\theta \vec{i} + \cos\theta \vec{j} - \sin\theta\cos\theta \vec{k}$, $d\vec{r} = (-\sin\theta \vec{i} + \cos\theta \vec{j})d\theta$,

therefore $\qquad \oint_C \vec{F} \cdot d\vec{r} = \int_C (\sin^2\theta + \cos^2\theta) \, d\theta = \int_{2\pi}^{0} 1 \cdot d\theta = -2\pi$.

(2) $\nabla \times \vec{F} = \begin{vmatrix} \vec{i} & \vec{j} & \vec{k} \\ \dfrac{\partial}{\partial x} & \dfrac{\partial}{\partial y} & \dfrac{\partial}{\partial z} \\ -y & x & -xyz \end{vmatrix} = -xz\vec{i} + yz\vec{j} + 2\vec{k}$, from Σ_1 we know, let $\phi = x^2 + y^2 - z^2$

then $\qquad \nabla\phi = 2x\vec{i} + 2y\vec{j} - 2z\vec{k}$, $\vec{n}dA = \dfrac{\nabla\phi}{|\nabla\phi \cdot \vec{k}|} dxdy = \dfrac{x\vec{i} + y\vec{j} - z\vec{k}}{z} dxdy$,

therefore $\quad \iint_{\Sigma_1} (\nabla \times \vec{F}) \cdot \vec{n}dA = \iint_{S_{xy}} \dfrac{-x^2 z + y^2 z - 2z}{z} dxdy = \iint_{S_{xy}} \dfrac{-x^2 + y^2 - 2}{1} dxdy$

$\qquad\qquad = \int_{\theta=0}^{2\pi} \int_{r=0}^{1} (-r^2 \cos^2\theta + r^2 \sin^2\theta - 2) r \, dr d\theta = -2\pi$.

(3) $\nabla \times \vec{F} = \begin{vmatrix} \vec{i} & \vec{j} & \vec{k} \\ \dfrac{\partial}{\partial x} & \dfrac{\partial}{\partial y} & \dfrac{\partial}{\partial z} \\ -y & x & -xyz \end{vmatrix} = -xz\vec{i} + yz\vec{j} + 2\vec{k}$; $\vec{n_2} = -\vec{k}$,

so $\qquad \iint_{\Sigma_2} (\nabla \times \vec{F}) \cdot \vec{n_2} \, dA = \iint_{\Sigma_2} -2 dA = -2 \cdot \pi = -2\pi$.

(4) The results of (1), (2), and (3) are all the same. This result verifies the Stokes' theorem is independent of the choice of the surface. That is

$$\oint_C \vec{F} \cdot d\vec{r} = \iint_{\Sigma_1} (\nabla \times \vec{F}) \cdot \vec{n_1} \, dA = \iint_{\Sigma_2} (\nabla \times \vec{F}) \cdot \vec{n_2} \, dA ,$$

when using Stokes' theorem to convert a line integral to a surface integral, as long as the boundary curve encloses the same surface, the value of the surface integral will remain the same. Therefore, $\iint_{\Sigma_1} (\nabla \times \vec{F}) \cdot \vec{n_1} \, dA = \iint_{\Sigma_2} (\nabla \times \vec{F}) \cdot \vec{n_2} \, dA$.   Q.E.D.

Example 2

If $\vec{F} = y\,\vec{i} + (x - 2xz)\,\vec{j} - (xy + 3)\,\vec{k}$, calculate $\iint_S (\nabla \times \vec{F}) \cdot \vec{n}\,ds$, where

$S: x^2 + y^2 + z^2 = a^2, z \geq 0$, $\vec{n}$ is the unit normal vector to the surface S pointing outward.

Solution

$S: x^2 + y^2 + z^2 = a^2, z \geq 0$; $S^*: z = 0, x^2 + y^2 \leq a^2$; $\vec{n_2} = +\vec{k}$, from Stokes' theorem we

know $\qquad \iint_S (\nabla \times \vec{F}) \cdot \vec{n_1}\,ds = \oint_C \vec{F} \cdot d\vec{r} = \iint_{S^*} (\nabla \times \vec{F}) \cdot \vec{n_2}\,ds\ (\,\vec{n_2} = \vec{k}\,),$

where $\qquad (\nabla \times \vec{F}) \cdot \vec{n_2} = \begin{vmatrix} 0 & 0 & 1 \\ \dfrac{\partial}{\partial x} & \dfrac{\partial}{\partial y} & \dfrac{\partial}{\partial z} \\ y & x - 2xz & -(xy + 3) \end{vmatrix} \cdot \vec{k} = -2z ,$

$$\iint_{S^*} (\nabla \times \vec{F}) \cdot \vec{n_2}\,ds = \iint_{S^*} (\nabla \times \vec{F}) \cdot \vec{k}\,ds = \iint_{S^*} (-2z)\,ds = 0$$

(invariance under change of surface), so $\qquad \iint_S (\nabla \times \vec{F}) \cdot \vec{n}\,ds = 0$. $\qquad$

8-8   Exercises

Basic questions

In the following questions 1~4, please use Stokes'
theorem $\oint_C \vec{V} \cdot d\vec{r} = \iint_S (\nabla \times \vec{V}) \cdot \vec{n} \, dA$ calculate the
following line integrals, where C is the boundary of the
closed surface S and $\vec{n}$ is the normal vector to the
surface S and $\vec{r} = x\,\vec{i} + y\,\vec{j} + z\,\vec{k}$.

1. $\vec{V} = 3x\,\vec{i} - 2y\,\vec{j} + z\,\vec{k}$, where C is the boundary
   $x^2 + y^2 = 4$ of the upper hemisphere S centered at
   (0, 0, 0) with a radius of 2 on the xy-plane.

2. $\vec{V} = (y+z)\,\vec{i} + (x+z)\,\vec{j} + (x+y)\vec{k}$, where C and S
   are the same as in the previous question.

3. $\vec{V} = -y\,\vec{i} + x\,\vec{j}$, $S: x^2 + y^2 \leq 1$,
   $C: x^2 + y^2 = 1$ (counterclockwise).

4. $\vec{V} = xy\,\vec{i} + yz\,\vec{j} + xz\,\vec{k}$, $S: x + y + z = 1$
   in the first octant, the slanted plane and C is the
   boundary of this plane.

5. If $\vec{V} = y\,\vec{i}$, please calculate $\iint_S (\nabla \times \vec{V}) \cdot \vec{n} \, dA$ where
   $S: x^2 + y^2 + z^2 = 1$ is the surface above the xy-plane.

Advanced questions

1. Please verify Stokes' theorem, where
   $\vec{V} = -y\,\vec{i} + x\,\vec{j} - z\,\vec{k}$, the surface
   $S: z = \sqrt{x^2 + y^2}$, $0 \leq z < 2$ (not include $z = 2$),
   and the boundary C of surface S is
   $x^2 + y^2 = 4$ on $z = 2$.

2. Calculate $\oint_C \vec{F} \cdot d\vec{r}$, where $\vec{F} = [\ 5y,\ 4x,\ z]$
   C is $x^2 + y^2 = 4$, oriented counterclockwise on
   $z = 1$.

3. Please verify Stokes' theorem, where
   $\vec{F} = x\,\vec{i} + x\,\vec{j} + 2xy\vec{k}$, the surface S is
   $x^2 + y^2 + z^2 = 4$, $z < 0$ (not include $z = 0$), the curve
   C is $x^2 + y^2 = 4$, and it is closed by rotating
   counterclockwise around $z = 0$.
   (1) Calculate $\nabla \times \vec{F}$.
   (2) Find line integral $\oint_C \vec{F} \cdot d\vec{r}$.
   (3) Calculate $\iint_S (\nabla \times \vec{F}) \cdot \vec{n} \, dA$.

4. Using Stokes' theorem to calculate $\oint_C \vec{F} \cdot d\vec{r}$,
   where $\vec{F} = y\,\vec{i} + xz^3\,\vec{j} - zy^3\,\vec{k}$, $\vec{r} = x\,\vec{i} + y\,\vec{j} + z\,\vec{k}$,
   C is $x^2 + y^2 = 4$, oriented counterclockwise on
   $z = -3$.

5. Using Stokes' theorem to calculate $\oint_C \vec{F} \cdot d\vec{r}$,
   where $\vec{F} = xy\,\vec{i} + yz\,\vec{j} + xz\,\vec{k}$, $\vec{r} = x\,\vec{i} + y\,\vec{j} + z\,\vec{k}$,
   C is $z = 1 - x^2$, $0 \leq x \leq 1$, $-2 \leq y \leq 2$ the boundary
   of the surface , and it is oriented counterclockwise.

6. Utilize $\vec{F} = 3y\,\vec{i} - xz\,\vec{j} + yz^2\,\vec{k}$,
   $S: x^2 + y^2 = 2z$, $0 \leq z \leq 2$,
   please verify Stokes' theorem.

7. Please derive the planar Stokes' theorem
   $\oint_C \vec{V} \cdot d\vec{r} = \iint_S (\nabla \times \vec{V}) \cdot \vec{k} \, dA$ where S is the region
   in the xy-plane with C as its boundary.

9 Orthogonal Functions and Fourier Analysis

Fourier
(1768~1830, France)

Fourier (also known as Joseph Fourier) was a French mathematician and physicist who introduced the concept of Fourier series and applied it to the theory of heat conduction and vibration. The subsequent application of his work in signal processing led to the development of the Fourier transform, which is named after him. He is also credited with being one of the early discoverers of the greenhouse effect.

Learning Objectives

9-1

Orthogonal Functions

9-1-1 Understanding orthogonal functions and the Sturm-Liouville boundary value problem

9-1-2 Mastering the properties of solutions to the Sturm-Liouville boundary value problem

9-1-3 Understanding common orthogonal bases: expansion using sine (cosine) functions

9-2

Fourier Series

9-2-1 Familiarity with periodic functions

9-2-2 Proficiency in the Fourier series of general periodic functions

9-2-3 Understanding the Gibbs phenomenon and convergence of Fourier series

9-2-4 Mastering even (odd) function Fourier series

9-2-5 Mastering half-range even (odd) expansion

9-2-6 Mastering the applications of Fourier series: calculating infinite series sum

9-3

Complex Fourier Series and Fourier Integral

9-3-1 Understanding complex Fourier series

9-3-2 Mastering Fourier integrals and their convergence

9-4

Fourier Transform

9-4-1 Understanding Fourier transform and its existence

9-4-2 Mastering the fundamental formulas of Fourier transform

9-4-3 Understanding the physical applications of Fourier transform: Parseval's theorem

9-4-4 Understanding Fourier integral for periodic functions

9-4-5 Mastering Fourier sine (cosine) transform and its applications: transformation formulas for derivatives

ExampleVideo

French mathematician Fourier, while investigating heat conduction and mechanical vibration problems, discovered that certain functions can be expressed as infinite series of sine functions. He also developed the Fourier transform, an integral transform that allows signals to be analyzed in both the time and frequency domains, enabling the examination of signal components.

In this chapter, we will systematically explain the method of expanding periodic functions into Fourier series using the theory of orthogonal bases in linear algebra. By manipulating the period, we can approximate non-periodic functions with periodic functions, leading to the derivation of the Fourier integral, which is the precursor to the Fourier transform.

9-1　Orthogonal Functions

In the chapter of linear algebra, we defined vector operations in three-dimensional space and extended them to the infinite-dimensional space of function vectors. The vector space of continuous functions, denoted as $C^0(\mathrm{R})$ (where $C^n(\mathrm{R})$ represents functions with n order they are all continuous derivatives), in fact, it is an inner product space. The inner product between two functions $f(x)$ and $g(x)$ is defined as follows.

$$< f(x), g(x) >= \int_a^b f(x)g(x)dx$$

As a result, this naturally leads to the following definitions: the norm (length, size) of a function f, $\| f \| = \sqrt{< f(x), f(x) >}$; a function f is said to be normalized (unitary) if $\| f \| = 1$; and two functions f and g are defined to be orthogonal if $< f, g > = 0$. These definitions have a similar form to those in finite-dimensional Cartesian coordinate systems for vectors. If the sizes $\| f \|$ and $\| g \|$ of the function set $\{ f(x), g(x) \}$ over the interval $x \in [a, b]$ are both non-zero, then the set is referred to as an orthogonal set. The following example illustrates this:

Example 1

For a function set $S = \{1, (2x - 1)\}$, $\forall x \in [0, 1]$,

(1) find $\| 1 \|$, $\|(2x - 1)\|$,　　(2) prove that S is an orthogonal function set.

Solution

(1)　$\| 1 \| = \sqrt{<1,1>} = \sqrt{\int_0^1 1 dx} = \sqrt{x \big|_0^1} = 1$,

therefore

$$|(2x-1)| = \sqrt{<2x-1, 2x-1>} = \sqrt{\int_0^1 (2x-1)^2 dx} = \sqrt{\int_0^1 (4x^2 - 4x + 1)dx}$$

$$= \sqrt{\left(\frac{4}{3}x^3 - 2x^2 + x\right)\bigg|_0^1} = \sqrt{\frac{1}{3}}\ .$$

(2) $< 1, 2x-1 > = \int_0^1 1 \cdot (2x-1)dx = (x^2 - x)\big|_0^1 = 0$, at $x \in [0, 1]$,

also $\| 1 \|$, $\|(2x - 1)\|$ are both non-zero,

$\therefore \quad s = \{1, 2x - 1\}$ is an orthogonal function set over the interval $x \in [0, 1]$.   Q.E.D.

Next, we introduce a slightly more general inner product:

$$< f(x), g(x) >_w = \int_a^b w(x)f(x)g(x)dx \,,$$

here $w(x)$ is a weight function with its values always greater than zero, known as the weighting function. This weighting function is analogous to the credits of courses we take, as it can emphasize the importance of different x values. Thus, the previous Example 1 corresponds to the special case where $w(x) = 1$. Consequently, the natural norm is defined as: $\|f(x)\| = \sqrt{\int_a^b w(x)f^2(x)dx}$. Two functions $f(x)$ and $g(x)$ are said to be orthogonal with respect to the weighting function $w(x)$ if $< f(x), g(x) >_w = \int_a^b w(x)f(x) \cdot g(x)dx = 0$. In most cases in this chapter, we still assume $w(x) = 1$.

For a function set $S = \{v_1, v_2, \cdots, v_n\}$ can also use the Gram-Schmidt orthogonalization process to transform a linearly independent set of functions into an orthogonal set. Assuming "S" is linearly independent, the corresponding algorithm is as follows:

$$\varphi_1 = v_1(x)$$

$$\varphi_i(x) = v_i(x) - \sum_{j=1}^{i-1} \frac{< v_i(x), \varphi_j(x) >}{\|\varphi_j\|^2} \varphi_j(x) \quad 1 \leq j \leq n$$

Alternatively, one can choose either unitary functions $\dfrac{\varphi_i}{\| \varphi_i \|}$, then

$\{\dfrac{\varphi_1}{\| \varphi_1 \|}, \dfrac{\varphi_2}{\| \varphi_2 \|}, \cdots, \dfrac{\varphi_n}{\| \varphi_n \|}\}$ is a set of unitary functions.

In this way, we can form a basis (coordinate axes) for the function space. Consequently, if there exists a function set $S = \{\phi_1(x), \phi_2(x), \cdots\}$ such that for $x \in [a, b]$ satisfying $< \phi_m(x), \phi_n(x) > = \int_a^b w(x)\phi_m(x) \cdot \phi_m(x)dx = 0$ $(m \neq n)$, we refer to the set S at $x \in [a, b]$ with respect to the weight function $w(x)$ as an orthogonal function set while $S' = \{\dfrac{\phi_1(x)}{\| \phi_1(x) \|}, \dfrac{\phi_2(x)}{\| \phi_2(x) \|}, \cdots\}$ is refer to a set of unitary functions.

Example 2

Convert the linearly independent set $\{1, x, x^2\}$ into an orthogonal set, where $x \in [0, 1]$.

Solution

(1) Let $\varphi_1(x) = 1$.

(2) $\varphi_2(x) = x - \dfrac{<x, 1>}{\|1\|^2} \cdot 1 = x - \dfrac{\displaystyle\int_0^1 x\,dx}{\displaystyle\int_0^1 1^2\,dx} \cdot 1 = x - \dfrac{\dfrac{1}{2}}{1} \cdot 1 = x - \dfrac{1}{2}$.

(3) $\varphi_3(x) = x^2 - \dfrac{<x^2, 1>}{\|1\|^2} \cdot 1 - \dfrac{<x^2, x - \dfrac{1}{2}>}{\|x - \dfrac{1}{2}\|^2} \cdot (x - \dfrac{1}{2})$,

$\therefore \varphi_3(x) = x^2 - \dfrac{1}{3} - \dfrac{\dfrac{1}{12}}{\dfrac{1}{12}} \cdot (x - \dfrac{1}{2}) = x^2 - x + \dfrac{1}{6}$.

(4) $\{1, x - \dfrac{1}{2}, x^2 - x + \dfrac{1}{6}\}$ is an orthogonal set.

Q.E.D.

Example 3

Please prove that $\{1, \cos x, \cos 2x, \cos 3x, \cdots\}$, with the weight function $w(x) = 1$, forms an orthogonal set of functions at $x \in [-\pi, \pi]$.

Solution　Let $\varphi_n(x) = \cos nx$, $n = 0, 1, 2, 3, \cdots$,

$<\varphi_n(x), \varphi_0(x)> = \displaystyle\int_{-\pi}^{\pi} \cos nx \cdot 1\,dx = \dfrac{1}{n}\sin nx \Big|_{-\pi}^{\pi} = 0$, $(n = 1, 2, 3, \cdots)$,

$<\varphi_n(x), \varphi_m(x)> = \displaystyle\int_{-\pi}^{\pi} \cos nx \cdot \cos mx\,dx = \dfrac{1}{2}\int_{-\pi}^{\pi}[\cos(m+n)x + \cos(m-n)x]\,dx$

$= \dfrac{1}{2}[\dfrac{1}{m+n}\sin(m+n)x + \dfrac{1}{m-n}\sin(m-n)x]\Big|_{-\pi}^{\pi} = 0 \quad (m \neq n)$,

therefore, $\{1, \cos x, \cos 2x, \cdots\}$　$x \in [-\pi, \pi]$　relative to the weight function $w(x) = 1$ is an orthogonal set.

Q.E.D.

9-1-1 The Sturm-Liouville Boundary Value Problem

The function space is infinite-dimensional. Therefore, the orthogonal set of functions obtained through the Gram-Schmidt process may not necessarily become a basis for the function space. However, from the perspective of solving ODEs and PDEs, we can at least expect to find a set of orthogonal functions that can be used to expand all solutions. The question is, where should we start to find this set of functions? The Sturm-Liouville problem, proposed by Sturm (1809-1882, France) and Liouville, may shed some light on this issue. Let's start by understanding the outline of this boundary value problem (BVP) with the following question.

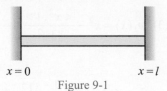

$x = 0$ $\qquad\qquad$ $x = l$

Figure 9-1

For a beam with both ends fixed, its vibration equation in the position direction x can be represented by an ordinary differential equation $y'' + \lambda y = 0$. The boundary conditions are $y(0) = y(l) = 0$, where $y(x)$ represents the vibration displacement. We know that $y(x) = 0$ (the trivial solution) is always a solution, but this is not what we are looking for. The non-trivial solutions of this differential equation are related to the parameter λ, and their analysis is as follows:

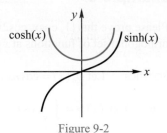

Figure 9-2

1. **$\lambda < 0$**

    Let $\lambda = -k^2$ ($k > 0$), then $y'' - k^2 y = 0$, $y(x) = d_1 e^{kx} + d_2 e^{-kx}$ or
    $y(x) = c_1 \cosh kx + c_2 \sinh kx$, from $y(0) = 0 \Rightarrow c_1 = 0$, let $y(l) = 0 \Rightarrow c_2 \sinh kl = 0$
    $\Rightarrow c_2 = 0$ (because $\sinh kl \neq 0$, $kl > 0$), $\therefore$ $y(x) = 0 \Rightarrow$ Trivial solution.

2. **$\lambda = 0$**

    $y'' = 0 \Rightarrow y(x) = c_1 + c_2 x$, from $y(0) = 0 \Rightarrow c_1 = 0$, $y(l) = 0 \Rightarrow c_2 l = 0 \Rightarrow c_2 = 0$
    $\therefore$ $y(x) = 0 \Rightarrow$ the trivial solution.

3. **$\lambda > 0$**

    Let $\lambda = k^2$ ($k > 0$) $\Rightarrow y'' - k^2 y = 0 \Rightarrow$ the eigenvalue of ODE, $m = \pm ik$
    $\Rightarrow y(x) = c_1 \cos kx + c_2 \sin kx$, from $y(0) = 0 \Rightarrow c_1 = 0$, $y(l) = 0 \Rightarrow c_2 \sin kl = 0$,
    if $c_2 \neq 0$, then $\sin kl = 0$ (also $\sin n\pi = 0$, $n = 1, 2, 3, \cdots$) $\Rightarrow kl = n\pi$, $n = 1, 2, 3, \cdots$

    $\Rightarrow k = \dfrac{n\pi}{l}$, $n = 1, 2, 3, \cdots$.

Based on the above discussion, we can obtain non-trivial solutions at certain values of λ, which we call these values "eigenvalues," and the corresponding non-trivial solutions are referred to as eigenfunctions:

$$\begin{cases} \text{Eigenvalues: } \lambda = k^2 = (\dfrac{n\pi}{l})^2, \ n = 1, 2, 3, \ \cdots \\[3mm] \text{Eigenfunctions: } y(x) = c_2 \sin(\dfrac{n\pi}{l} x), \ n = 1, 2, 3, \ \cdots \end{cases}$$

In general, the Sturm-Liouville boundary value problem refers to a nonlinear ODE of the following form:

$$\text{DE}: \frac{d}{dx}[p(x)\frac{dy}{dx}] + (q(x) + \lambda r(x))y = 0$$

$$\text{BC1}: \begin{cases} \alpha_1 y(a) + \beta_1 y'(a) = 0, \ \alpha_1{}^2 + \beta_1{}^2 \neq 0 \\ \alpha_2 y(b) + \beta_2 y'(b) = 0, \alpha_2{}^2 + \beta_2{}^2 \neq 0 \end{cases}$$

$$\text{BC2}: \begin{cases} y(d) = y(d+T) \\ y'(d) = y'(d+T) \end{cases}$$

$p(x)$, $q(x)$ and $r(x)$ are continuous functions in the interval $[a, b]$, in this interval $p(x) > 0$, $r(x) > 0$, $\forall x$ and $p(x)$ is a differentiable function while $r(x)$ is referred to as the weight function. When the differential equation (DE) is accompanied by the first type of boundary condition (BC1), it is called the regular Sturm-Liouville boundary value problem. On the other hand, when the differential equation (DE) is accompanied by the second type of boundary condition (BC2), it is referred to as the periodic Sturm-Liouville boundary value problem. The values λ that satisfy the equation and its accompanying boundary conditions are called eigenvalues, and the corresponding non-zero functional solutions are called eigenfunctions. Thus, the explanation given earlier pertains to the regular Sturm-Liouville boundary value problem, and below, we will provide an example of a periodic boundary value problem.

Example 4

DE: $y'' + \lambda y = 0$, BC: $\begin{cases} y(0) - y(\pi) = 0 \\ y'(0) - y'(\pi) = 0 \end{cases}$.

Solution

① $\lambda = -k^2 < 0 \ (k > 0)$

$y'' - k^2 y = 0 \Rightarrow y(x) = c_1 \cosh kx + c_2 \sinh kx$,

$\qquad\qquad y'(x) = c_1 k \sinh kx + c_2 k \cosh kx$,

from $y(0) = y(\pi), y'(0) = y'(\pi)$, we obtain $c_1 = c_1 e^{kx}$, then $c_1 = c_2 = 0$,

$\therefore \qquad y(x) = 0$.

② $\lambda = 0$

$y'' = 0 \Rightarrow y(x) = c_1 + c_2\,x,\ y'(x) = c_2,$

from $y(0) = y(\pi) \Rightarrow c_1 - (c_1 + c_2\,\pi) = 0 \Rightarrow c_2 = 0,$

$\quad y'(0) = y'(\pi) = 0 \Rightarrow c_2 = c_2$ (naturally established)

$\therefore \qquad \left.\begin{cases} y(x) = c_1 \\ \lambda = 0 \end{cases}\right\} \ \cdots ①$

③ $\lambda = k^2 > 0\ (k > 0)$

$y'' + k^2 y = 0 \Rightarrow y(x) = c_1 \cos kx + c_2 \sin kx$

$\qquad\qquad y'(x) = -c_1\,k \sin kx + c_2\,k \cos kx,$

from $y(0) - y(\pi) = 0 \Rightarrow c_1 - (c_1 \cos k\pi + c_2 \sin k\pi) = 0$

$y'(0) - y'(\pi) = 0 \Rightarrow c_2\,k - (c_1\,k \sin k\pi + c_2\,k \cos k\pi) = 0$

$\Rightarrow \begin{cases} c_1 \cdot (1 - \cos k\pi) - c_2 \sin k\pi = 0 \\ c_1 k \sin k\pi + c_2 k \cdot (1 - \cos k\pi) = 0 \end{cases} \Rightarrow \begin{bmatrix} 1 - \cos k\pi & -\sin k\pi \\ \sin k\pi & (1 - \cos k\pi) \end{bmatrix}\begin{bmatrix} c_1 \\ c_2 \end{bmatrix} = 0.$

If c_1, c_2 exist non-trivial solution,

then $\begin{vmatrix} 1 - \cos k\pi & -\sin k\pi \\ \sin k\pi & (1 - \cos k\pi) \end{vmatrix} = 0 \Rightarrow (1 - \cos k\pi)^2 + \sin^2 k\pi = 0$

$\Rightarrow 1 - 2\cos k\pi + \cos^2 k\pi + \sin^2 k\pi = 0 \Rightarrow 2 - 2\cos k\pi = 0 \Rightarrow \cos k\pi = 1,\ \therefore k\pi = 2n\pi$

$\therefore k = 2n,\ n = 1, 2, 3, \cdots$

$\therefore \qquad \left.\begin{cases} y(x) = c_1 \cos kx + c_2 \sin kx = c_1 \cos 2nx + c_2 \sin 2nx \\ \lambda = k^2 = (2n)^2 = 4n^2,\ n = 1, 2, 3, \cdots \end{cases}\right\} \cdots ②$

④ By merging ① and ②: $\begin{cases} \text{Eigenvalues: } \lambda = 4n^2,\ n = 0, 1, 2, 3,\ \cdots \\ \text{Eigenfunctions: } y(x) = c_1 \cos 2nx + c_2 \sin 2nx \end{cases},$

that is $y(x) = \begin{cases} c_1,\ \lambda = 0 \\ c_1 \cos 2nx + c_2 \sin 2nx\ ;\ \lambda = 4n^2,\ n = 1, 2, 3, \cdots, (\lambda > 0) \end{cases}.$

Q.E.D.

Upon closer observation, we can write the ODE as $y'' = -\lambda y$. As a result, $-\lambda$ becomes the eigenvalue of the differential operator D^2; and the solution y becomes the eigenfunction. For instance, in the previous explanation, $\lambda = k^2 = (\dfrac{n\pi}{l})^2$ was the eigenvalue, and $y(x) = c_2 \sin(\dfrac{n\pi}{l}x)$ is corresponding to the eigenfunction (which is also a solution of the ODE). Similarly, in Example 4, $\lambda = 4n^2$ was the eigenvalue, and its corresponding eigenfunction was $y(x) = c_1 \cos 2nx + c_2 \sin 2nx$. In fact, for the above boundary value problem, common eigenvalues (and their corresponding eigenfunctions) are shown in Table 1.

Table 1

	Equation	Boundary condition	Eigenfunction and eigenvalue
(1)	$y'' + \lambda y = 0$	$y(0) = y(l) = 0$	$\left\{ \sin \dfrac{n\pi}{l} x \right\}_{n=1}^{\infty}$, $\lambda = (\dfrac{n\pi}{l})^2$, $n = 1, 2, 3, \cdots$
(2)	$y'' + \lambda y = 0$	$y'(0) = y'(l) = 0$	$\left\{ \cos \dfrac{n\pi}{l} x \right\}_{n=0}^{\infty}$, $\lambda = (\dfrac{n\pi}{l})^2$, $n = 0, 1, 2, 3, \cdots$
(3)	$y'' + \lambda y = 0$	$y(0) = y'(l) = 0$	$\left\{ \sin[\dfrac{(n-\frac{1}{2})\pi}{l} \cdot x] \right\}_{n=1}^{\infty}$, $\lambda = \left[\dfrac{(n-\frac{1}{2})\pi}{l} \right]^2$, $n = 1, 2, 3, \cdots$
(4)	$y'' + \lambda y = 0$	$y'(0) = y(l) = 0$	$\left\{ \cos[\dfrac{(n-\frac{1}{2})\pi}{l} \cdot x] \right\}_{n=1}^{\infty}$, $\lambda = \left[\dfrac{(n-\frac{1}{2})\pi}{l} \right]^2$, $n = 1, 2, 3, \cdots$
(5)	$y'' + \lambda y = 0$	$\begin{cases} y(0) = y(T) \\ y'(0) = y'(T) \end{cases}$	$\left\{ 1, \cos \dfrac{2n\pi}{T} \cdot x, \sin \dfrac{2n\pi}{T} x \right\}_{n=1}^{\infty}$, $\lambda = (\dfrac{2n\pi}{T})^2$, $n = 1, 2, 3, \cdots$

9-1-2 The Properties of Sturm-Liouville Boundary Value Problem

We can observe that all eigenfunctions in Table 1 satisfy the following three properties:

Properties Description

(1) There exists an infinite number of eigenvalues $\lambda_1, \lambda_2, \lambda_3, \ldots$, and $\lim\limits_{n \to \infty} \lambda_n \to \infty$ (approach infinity).

(2) All eigenvalues are real numbers.

(3) The eigenfunctions corresponding to different eigenvalues with respect to the weight function $r(x)$ over the interval are orthogonal.[1]

Let's go back to the initial question of finding an orthogonal set. In the process of finding solutions in Example 1, λ played a crucial role, guiding us to the types of solutions that can be expressed as linear combinations of 1. exponential functions, 2. polynomial functions, and 3. sine or cosine functions. In engineering applications, periodic data such as signals are often best represented by sine and cosine functions. This is the origin of the Fourier series, which is widely used in engineering.

---

[1] To verify that the eigenfunction set listed in Table 1 forms an orthogonal set, we can utilize the trigonometric product-to-sum and difference-to-sum formulas.

$\cos A \cos B = \dfrac{1}{2}[\cos(A+B) + \cos(A-B)]$, $\sin A \sin B = \dfrac{1}{2}[\cos(A-B) - \cos(A+B)]$,

$\cos A \sin B = \dfrac{1}{2}[\sin(A+B) - \sin(A-B)]$, $\sin A \cos B = \dfrac{1}{2}[\sin(A+B) + \sin(A-B)]$.

9-1-3 Orthogonal Set of Sine and Cosine Functions

Recalling in a vector space V, if $\beta = \{x_1, \ldots, x_n\}$ is an orthogonal basis, then any vector v in V can be uniquely written as the form $v = \sum_{i=1}^{n} <v, x_i> x_i$, if V is a function inner product space, then consider the following common orthogonal set:

1. The Common Orthogonal Set

(1) $\left\{ \sin\dfrac{n\pi}{l}x \right\}_{n=1}^{\infty}$, $x \in [0, l]$.

(2) $\left\{ \cos\dfrac{n\pi}{l}x \right\}_{n=0}^{\infty} = \left\{ 1, \cos\dfrac{n\pi}{l}x \right\}_{n=1}^{\infty}$, $x \in [0, l]$.

(3) $\left\{ 1, \cos\dfrac{2n\pi}{T}x, \sin\dfrac{2n\pi}{T}x \right\}_{n=1}^{\infty}$, $x \in [0, T]$.

At this point, the function $f(x)$ can be expanded using these three sets of orthogonal functions

2. The Common Function Expansions

(1) $f(x) = \sum_{n=1}^{\infty} b_n \sin\dfrac{n\pi}{l}x$, $b_n = \dfrac{<f(x), \sin\dfrac{n\pi}{l}x>}{<\sin\dfrac{n\pi}{l}x, \sin\dfrac{n\pi}{l}x>} = \dfrac{2}{l}\int_0^l f(x)\sin\dfrac{n\pi}{l}xdx$.

(2) $f(x) = a_0 + \sum_{n=1}^{\infty} a_n \cdot \cos\dfrac{n\pi}{l}x$,

$a_n = \dfrac{<f(x), \cos\dfrac{n\pi}{l}x>}{<\cos\dfrac{n\pi}{l}x, \cos\dfrac{n\pi}{l}x>} = \dfrac{2}{l}\int_0^l f(x)\cos\dfrac{n\pi}{l}xdx$.

(3) $f(x) = a_0 + \sum_{n=1}^{\infty} [a_1 \cos\dfrac{2n\pi}{T}x + b_n \sin\dfrac{2n\pi}{T}x]$,

$a_0 = \dfrac{<f(x), 1>}{<1, 1>} = \dfrac{1}{T}\int_0^T f(x)dx$,

$a_n = \dfrac{<f(x), \cos\dfrac{2n\pi}{T}x>}{<\cos\dfrac{2n\pi}{T}x, \cos\dfrac{2n\pi}{T}x>} = \dfrac{2}{T}\int_0^T f(x)\cos\dfrac{2n\pi}{T}xdx$,

$b_n = \dfrac{<f(x), \sin\dfrac{2n\pi}{T}x>}{<\sin\dfrac{2n\pi}{T}x, \sin\dfrac{2n\pi}{T}x>} = \dfrac{2}{T}\int_0^T f(x)\sin\dfrac{2n\pi}{T}xdx$.

Example 5

If the boundary value problem is $y'' + \lambda y = 0$, $y(0) = y(\ell) = 0$.

(1) Find the eigenfunctions and eigenvalues for this boundary value problem.

(2) Utilize the above eigenfunctions to expand the function $f(x) = 1$ up to its first three non-zero terms.

Solution

(1) This sub-question has been previously explained in the preceding text, and there is no need to repeat it here,

the eigenvalues $\lambda = (\dfrac{n\pi}{\ell})^2$; the eigenfunctions $y(x) = c_2 \sin(\dfrac{n\pi}{\ell}x)$, $n = 1, 2, 3, \cdots$.

(2) Therefore, from the orthogonal expansion of sine functions, we know

$$f(x) = 1 = \sum_{n=1}^{\infty} c_n \cdot \sin \frac{n\pi}{\ell} x,$$

$$c_n = \frac{< f(x), \sin \dfrac{n\pi}{\ell} x >}{< \sin \dfrac{n\pi}{\ell} x, \sin \dfrac{n\pi}{\ell} x >} = \frac{2}{\ell} \int_0^{\ell} 1 \cdot \sin \frac{n\pi}{\ell} x dx = \frac{2}{\ell} \cdot (-\frac{\ell}{n\pi}) \cdot \cos \frac{n\pi}{\ell} x \Big|_0^{\ell}$$

$$= \frac{2}{n\pi}[1 - (-1)^n] = \begin{cases} \dfrac{4}{n\pi}, & n = 1, 3, 5, \cdots \\ 0, & n = 2, 4, 6, \cdots \end{cases},$$

$$f(x) = \sum_{n=1,3,5,\cdots}^{\infty} \frac{4}{n\pi} \sin(\frac{n\pi}{\ell} x) \approx \frac{4}{\pi} \cdot [\sin \frac{\pi}{\ell} x + \frac{1}{3} \sin \frac{3\pi}{\ell} x + \frac{1}{5} \sin \frac{5\pi}{\ell} x].$$ Q.E.D.

9-1 Exercises

Basic questions

Prove that the sets of functions in the following questions 1~3 form an orthogonal set of functions within the specified interval.

1. $\{x, x^2\}, [-1, 1]$.

2. $\{x^2, x^3 + x\}, [-1, 1]$.

3. $\{\cos x, \sin^2 x\}, [0, \pi]$.

4. Consider a linearly independent set $\{1, x\}, -1 \le x \le 1$. Use the Gram-Schmidt orthogonalization process to transform this set into a normalized orthogonal set of functions. $\{P_1, P_2\}, -1 \le x \le 1$.

5. If $f(x) = 2x, g(x) = 3 + cx$ are orthogonal functions within $0 \le x \le 1$ (1) Find the value of c (2) Find the normalized orthogonal set of functions.

6. Find the eigenvalues and eigenfunctions
$$\begin{cases} y'' + \lambda y = 0, 0 \le x \le \pi \\ y(0) = y(\pi) = 0 \end{cases}.$$

7. Find the eigenvalues and eigenfunctions
$$\begin{cases} y'' + \lambda y = 0, \\ y(-\pi) = y(\pi) \\ y'(-\pi) = y'(\pi) \end{cases}.$$

Advanced questions

1. Prove that the sets of functions in the following questions form an orthogonal set of functions within the specified interval.

(1) $\{x, \cos 2x\}, [-\frac{\pi}{2}, \frac{\pi}{2}]$.

(2) $\{\sin \frac{n\pi}{l} x\}\Big|_{n=1,2,3,\cdots}^{\infty}, [0, l]$.

(3) $\{1, \cos \frac{n\pi}{l} x\}\Big|_{n=1,2,3,\cdots}^{\infty}, [0, l]$.

(4) $\{1, \cos \frac{n\pi}{l} x, \sin \frac{m\pi}{l} x\}\Big|_{n,m=1,2,3,\cdots}^{\infty}, [-l, l]$.

2. What is the normalized orthogonal set of the orthogonal set $\{1, \cos x, \cos 2x, \cos 3x, \cdots\}$, $-\pi \le x \le \pi$?

3. What is the normalized orthogonal set of the orthogonal set ?
$$\{1, \cos \frac{n\pi}{l} x, \sin \frac{m\pi}{l} x\}\Big|_{n,m=1,2,3,\cdots}^{\infty}, [-l, l]$$

4. Consider a linearly independent set
$$\{1, x\}, 0 \le x \le \frac{1}{2}, \text{ using the Gram-Schmidt}$$
orthogonalization process to transform this set into a normalized orthogonal set of functions
$$\{P_1, P_2\}, 0 \le x \le \frac{1}{2}.$$

Find the eigenvalues and eigenfunctions of following questions.

5. $\begin{cases} y'' + \lambda y = 0, 0 \le x \le \pi \\ y'(0) = y(\pi) = 0 \end{cases}.$

6. $\begin{cases} y'' + \lambda y = 0, 0 \le x \le \pi \\ y(0) = y'(1) = 0 \end{cases}.$

7. $\begin{cases} y'' + \lambda y = 0, \\ y(-\pi) = y(\pi) \\ y'(-\pi) = y'(\pi) \end{cases}.$

9-2 Fourier Series

In this section, we explain the Fourier expansion of periodic functions and discuss the Gibbs phenomenon that occurs at discontinuity points, which helps us understand the applicability of Fourier expansion—it depends on the number of breakpoints in the graph of the function being expanded. Additionally, if a function is not defined over the entire domain (e. g., R), we can use known waveforms to piece together a function suitable for Fourier expansion. This approach plays a crucial role in processing periodic data (signals), such as analyzing sound waves. Since signals in physics generally exhibit periodic characteristics, we first introduce periodic functions.

9-2-1 Periodic Functions

If the function $f(x)$ satisfies $f(x + T) = f(x)$, then $f(x)$ is called a periodic function, where the smallest positive number T is referred to as the period of $f(x)$. For example, the periods of $\sin kx$ and $\cos kx$ are $\dfrac{2\pi}{|k|}$.

The Properties of Periodic Functions

(1) The period of f is T, then $\displaystyle\int_{d}^{d+T} f(x)\,dx = \int_{-\frac{T}{2}}^{\frac{T}{2}} f(x)\,dx$.[2]

(2) The period of f is T, then $f(kx)$ is the period $\dfrac{T}{|k|}$.

(3) The periods of f_1 and f_2 are T_1 and T_2, respectively, and m, n are the smallest positive integers which cause $mT_1 = nT_2$, then the period of $f_1(x) + f_2(x)$ is either mT_1 or nT_2.

---

[2] If we assume $0 < d < T < T + d$, then we can derive the following from periodicity

$$\int_{0}^{T} f(x)\,dx = \int_{0}^{d} f(x)\,dx + \int_{d}^{T} f(x)\,dx = \int_{0}^{d} f(x)\,dx + \int_{d}^{d+T} f(x)\,dx - \int_{T}^{T+d} f(x)\,dx$$

$$= \int_{0}^{d} f(x)\,dx + \int_{d}^{d+T} f(x)\,dx - \int_{0}^{d} f(x)\,dx = \int_{d}^{d+T} f(x)\,dx .$$

If d exceeds one period, then by shifting back within one period, we can obtain the same result.

Example 1

Determine the periods of the following functions.

(1) $3 \sin x + 2 \sin 3x$ (2) $2 + 5 \sin 4x + 4 \cos 7x$

(3) $2 \sin 3\pi x + 7 \cos \pi x$ (4) $7 \cos(\frac{1}{2}\pi x) + 5 \sin(\frac{1}{3}\pi x)$

Solution

(1) The period of $\sin x$ is 2π, the period of $\sin 3x$ is $\dfrac{2\pi}{3}$,

then the period of $3 \sin x + 2 \sin 3x$ is $2\pi (1 \cdot 2\pi = 3 \cdot \dfrac{2\pi}{3})$.

(2) The period of $\sin 4x$ is $\dfrac{2\pi}{4} = \dfrac{\pi}{2}$, the period of $\cos 7x$ is $\dfrac{2\pi}{7}$.

$\therefore$ the period of $2 + 5 \sin 4x + 4 \cos 7x$ is $2\pi (4 \cdot \dfrac{\pi}{2} = 7 \cdot \dfrac{2\pi}{7})$.

(3) The period of $\sin 3\pi x$ is $\dfrac{2\pi}{3\pi} = \dfrac{2}{3}$, the period of $\cos \pi x$ is $\dfrac{2\pi}{\pi} = 2$,

then the period of $2 \sin 3\pi x + 7 \cos \pi$ is $2 \ (3 \cdot \dfrac{2}{3} = 1 \cdot 2)$.

(4) The period of $\cos(\frac{1}{2}\pi x)$ is $\dfrac{2\pi}{\frac{1}{2}\pi} = 4$, the period of $\sin(\frac{1}{3}\pi x)$ is $\dfrac{2\pi}{\frac{1}{3}\pi} = 6$,

$\therefore$ the period of $7 \cos(\frac{1}{2}\pi x) + 5 \sin(\frac{1}{3}\pi x)$ is $12 \ (3 \cdot 4 = 2 \cdot 6)$.　　Q.E.D.

9-2-2 The Fourier Series of Periodic Functions

We refer to the expansion $f(x) = a_0 + \sum\limits_{n=1}^{\infty}[a_n \cos \dfrac{2n\pi}{T}x + b_n \sin \dfrac{2n\pi}{T}x]$ for $x \in [0, T]$

as the **Fourier series**.

▶ **Theorem 9-2-1**

The Coefficients of Fourier Series

When the period of $f(x)$ is $2l$, the computation of the Fourier series coefficients is given as follows:

(1) $a_0 = \dfrac{1}{2l}\displaystyle\int_{-l}^{l} f(x)dx$,   (2) $a_n = \dfrac{1}{l}\displaystyle\int_{-l}^{l} f(x)\cos \dfrac{n\pi}{l}x dx$,

(3) $b_n = \dfrac{1}{l}\displaystyle\int_{-l}^{l} f(x)\sin \dfrac{n\pi}{l}x dx$.

Proof

From the section 9-1-3, we know:

(1) $a_0 = \dfrac{1}{T}\displaystyle\int_0^T f(x)dx = \dfrac{1}{T}\displaystyle\int_{-\frac{T}{2}}^{\frac{T}{2}} f(x)dx = \dfrac{1}{2l}\displaystyle\int_{-l}^l f(x)dx$.

(2) $a_n = \dfrac{2}{T}\displaystyle\int_0^T f(x)\cos\dfrac{2n\pi}{T}xdx = \dfrac{2}{T}\displaystyle\int_{-\frac{T}{2}}^{\frac{T}{2}} f(x)\cos\dfrac{2n\pi}{T}xdx = \dfrac{1}{l}\displaystyle\int_{-l}^l f(x)\cos\dfrac{n\pi}{l}xdx$.

(3) $b_n = \dfrac{2}{T}\displaystyle\int_0^T f(x)\sin\dfrac{2n\pi}{T}xdx = \dfrac{2}{T}\displaystyle\int_{-\frac{T}{2}}^{\frac{T}{2}} f(x)\sin\dfrac{2n\pi}{T}xdx = \dfrac{1}{l}\displaystyle\int_{-l}^l f(x)\sin\dfrac{n\pi}{l}xdx$. ◀

Example 2

$$f(x) = \begin{cases} 0, -\pi < x \le 0 \\ \pi - x, 0 < x \le \pi \end{cases}, \quad f(x) = f(x+2\pi) \text{ , find the Fourier series of } f(x).$$

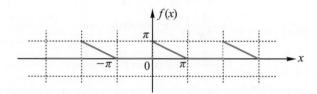

Solution   Let $f(x) = a_0 + \displaystyle\sum_{n=1}^{\infty} a_n \cos nx + b_n \sin nx$,

$$a_0 = \dfrac{1}{2\pi}\int_{-\pi}^{\pi} f(x)dx = \dfrac{1}{2\pi}\int_0^{\pi}(\pi - x)dx = \dfrac{1}{2\pi}\left(\pi x - \dfrac{1}{2}x^2\right)\Big|_0^{\pi} = \dfrac{\pi}{4} \ ;$$

$$a_n = \dfrac{1}{\pi}\int_{-\pi}^{\pi} f(x)\cos nxdx = \dfrac{1}{\pi}\int_0^{\pi}(\pi - x)\cos nxdx$$

$$= \dfrac{1}{\pi}\left[(\pi - x)\dfrac{\sin nx}{n} - \dfrac{1}{n^2}\cos nx\right]\Big|_0^{\pi} = \dfrac{1-(-1)^n}{n^2\pi} \ ;$$

$$b_n = \dfrac{1}{\pi}\int_{-\pi}^{\pi} f(x)\sin nxdx = \dfrac{1}{\pi}\int_0^{\pi}(\pi - x)\sin nxdx$$

$$= \dfrac{1}{\pi}\left[-\dfrac{1}{n}(\pi - x)\cos nx + \dfrac{1}{n^2}\sin nx\right]\Big|_0^{\pi} = \dfrac{1}{n} ,$$

$$\therefore f(x) = \dfrac{\pi}{4} + \sum_{n=1}^{\infty}\left[\dfrac{1-(-1)^n}{n^2\pi}\cos nx + \dfrac{1}{n}\sin nx\right].$$

Q.E.D.

Example 3

Assuming that $f(x) = x^2$, $0 < x < 2\pi$, and $f(x) = (x + 2\pi)$, find the Fourier series of $f(x)$.

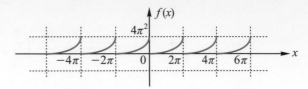

Solution

$$f(x) = a_0 + \sum_{n=1}^{\infty} [a_n \cos nx + b_n \sin nx],$$

$$a_0 = \frac{1}{2\pi} \int_0^{2\pi} x^2 dx = \frac{1}{2\pi} \cdot \frac{1}{3} x^3 \Big|_0^{2\pi} = \frac{4}{3}\pi^2 ;$$

$$a_n = \frac{2}{T} \int_0^{T} f(x) \cos \frac{2n\pi}{T} x dx$$

$$= \frac{2}{2\pi} \int_0^{2\pi} x^2 \cos nx dx$$

$$= \frac{2}{2\pi} \cdot [\frac{2x}{n^2} \cos nx] \Big|_0^{2\pi} = \frac{4}{n^2} ;$$

$$b_n = \frac{2}{T} \int_0^{T} f(x) \sin \frac{2n\pi}{T} x dx = \frac{2}{2\pi} \int_0^{2\pi} x^2 \sin nx dx = -\frac{4\pi}{n},$$

$$\therefore f(x) = \frac{4}{3}\pi^2 + \sum_{n=1}^{\infty} \cdot [\frac{4}{n^2} \cos nx + (-\frac{4}{n}\pi) \sin nx] . \qquad \text{Q.E.D.}$$

9-2-3 Gibbs Phenomenon

The numerical approximate solutions of Example 3 using the Fourier series with $n = 3, 6, 15$ are shown in Figure 9-3. As n increases, the Fourier series approaches $f(x)$ more closely at continuous points. However, at the discontinuity point $x = 0$, there exists a phenomenon known as the Gibbs phenomenon, where a spike occurs. This spike does not disappear with an increase in the number of terms; it only becomes narrower in scope. We refer to this as the Gibbs phenomenon.

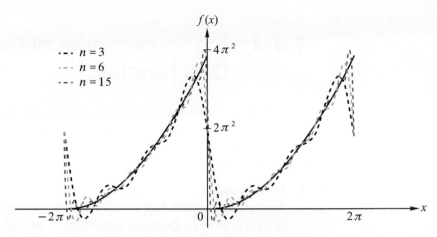

Figure 9-3   Diagram of Gibbs phenomenon

Therefore, it is important to understand under what conditions the magnitude of this spike error can be controlled. The answer is as follows:

▶ Theorem 9-2-2

The Convergence of the Fourier Series

Assuming $f(x) = f(x + T)$ and f is piecewise continuous and the definite integral of f over one period is bounded, then the value of the Fourier series of $f(x)$ at the discontinuity points is $\dfrac{1}{2}[f(x^+) + f(x^-)]$. ◀

Using Example 3 as an example, the Fourier series is

$f(x) = \dfrac{4}{3}\pi^2 + \sum\limits_{n=1}^{\infty}[\dfrac{4}{n^2}\cos nx + (-\dfrac{4}{n}\pi)\sin nx]$. Therefore, at the origin $x = 0$, we have

$\dfrac{4}{3}\pi^2 + \sum\limits_{n=1}^{\infty}[\dfrac{4}{n^2}\cos n\cdot 0 + (-\dfrac{4}{n}\pi)\sin n\cdot 0] = 2\pi^2 = \dfrac{f(0^+) + f(0^-)}{2}$, which is a good

example. As a result, we can obtain $\sum\limits_{n=1}^{\infty}\dfrac{4}{n^2} = \dfrac{2}{3}\pi^2$, that is

$\sum\limits_{n=1}^{\infty}\dfrac{1}{n^2} = \dfrac{1}{1^2} + \dfrac{1}{2^2} + \dfrac{1}{3^2} + \cdots = \dfrac{\pi^2}{6}$. Hence, the Fourier series can also be used to

compute some challenging series sums.

9-2-4 The Fourier Series of Even Function and Odd Function

Next, we will discuss the properties of odd and even functions when expanded into Fourier series. Let's briefly review odd and even functions and their related properties, paying special attention to their behavior when encountering definite integrals. This is directly related to the Fourier series.

1. **Even Function**

As shown in Figure 9-4(a), a function that is symmetric about the y-axis, like $f(x) = x^2$, is called an even function. Generally, it satisfies the equation $f(-x) = f(x)$. Additionally, as $f(x) = \cos x$, it is also an example.

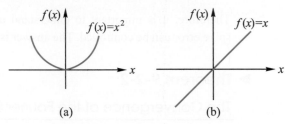

(a)   (b)

Figure 9-4   The diagram of even and odd function

2. **Odd Function**

As shown in Figure 9-4(b), a function that is symmetric about the origin, like $f(x) = x$, is called an odd function. Generally, it satisfies the equation $f(-x) = -f(x)$. Additionally, as $f(x) = \sin x$, it is also an example.

When performing addition, subtraction, and multiplication between odd and even functions, the following rules apply. Please note that the reciprocal of an odd or even function remains its original odd or even nature. Therefore, after division, the odd or even nature remains the same as multiplication:

(1) even function $\pm$ even function = even function;
 odd function $\pm$ odd function = odd function;
 odd function $\pm$ even function = Neither odd nor even function.

(2) even function $\times$ even function = even function;
 odd function $\times$ odd function = even function;
 odd function $\times$ even function = odd function.

In terms of integration, even and odd functions have the following properties:

(3) $f(x)$ is an even function over $x \in (-a, a)$, then $\int_{-a}^{a} f(x)dx = 2\int_{0}^{a} f(x)dx$.

(4) $f(x)$ is an odd function over $x \in (-a, a)$, then $\int_{-a}^{a} f(x)dx = 0$.

From the above properties (1) to (2), we can directly deduce the following two points for powers of polynomials and trigonometric functions:

(5) The power function $f(x) = x^n$, if $n = 0, 2, 4, 6, \cdots$ is even, then $f(x)$ is an even function. If $n = 1, 3, 5, 7, \cdots$ is odd, then $f(x)$ is an odd function.

(6) $\sin^k x$, $\cos^k x$, if k is even, then the period is $\dfrac{2\pi}{2} = \pi$, if k is odd, then the period is 2π.

3. **The Fourier Series of Even Function**

When $f(x)$ is an even function, then $f(x) \sin x$ is odd function, so from property (3), we have $\displaystyle\int_{-a}^{a} f(x) \sin x = 0$; on the other hand, $f(x) \cos x$ is an even function, therefore from property 4, we obtain $\displaystyle\int_{-a}^{a} f(x) \cos x\, dx = 2\int_{0}^{a} f(x) \cos x\, dx$. Thus, the Fourier series of $f(x)$ is

$$f(x) = a_0 + \sum_{n=1}^{\infty} a_n \cos \frac{n\pi}{l} x ,$$

where $a_0 = \dfrac{1}{2l} \displaystyle\int_{-l}^{l} f(x)\, dx = \dfrac{1}{l} \int_{0}^{l} f(x)\, dx$, $a_n = \dfrac{2}{l} \displaystyle\int_{0}^{l} f(x) \cos \frac{n\pi}{l} x\, dx$.

4. **The Fourier Series of Odd Function**

If $f(x)$ is an odd function, then the situation is reversed compared to the even function case. Consequently, we obtain the Fourier series of $f(x)$ as

$$f(x) = \sum_{n=1}^{\infty} b_n \sin \frac{n\pi}{l} x ,$$

where $b_n = \dfrac{2}{l} \displaystyle\int_{0}^{l} f(x) \sin \frac{n\pi}{l} x\, dx$.

Below, we will use several examples to illustrate the Fourier series expansion of odd and even functions.

Example 4

Find the Fourier series of $f(x)$, where $f(x) = x^2(-\pi \le x < \pi)$, and $f(x) = f(x + 2\pi)$.

Solution

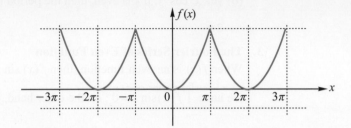

$f(x)$ is on even function, and the period $T = 2\pi$, $l = \pi$, $\therefore f(x) = a_0 + \sum_{n=1}^{\infty} a_n \cos nx$,

where $a_0 = \dfrac{2}{2\pi} \int_0^{\pi} x^2 \cdot dx = \dfrac{\pi^2}{3}$,

$a_n = \dfrac{2}{\pi} \int_0^{\pi} x^2 \cos nx\, dx = \dfrac{2}{\pi} \left[\dfrac{2x}{n^2} \cos nx \right]_0^{\pi} = \dfrac{4}{n^2} \cdot (-1)^n$,

$\therefore f(x) = \dfrac{1}{3}\pi^2 + \sum_{n=1}^{\infty} (-1)^n \cdot \dfrac{4}{n^2} \cdot \cos nx$. Q.E.D.

Example 5

Find the Fourier series of $f(x) = x(-\pi < x < \pi)$, $f(x) = f(x + 2\pi)$.

Solution $f(x) = x$ at $-\pi < x < \pi$ is odd function, $\therefore f(x) = \sum_{n=1}^{\infty} b_n \sin nx$,

$b_n = \dfrac{2}{\pi} \int_0^{\pi} x \cdot \sin nx\, dx = \dfrac{2}{\pi} \cdot \left[-\dfrac{x}{n} \cos nx \right]_0^{\pi} = (-1)^{n+1}$,

$\therefore f(x) = \sum_{n=1}^{\infty} (-1)^{n+1} \cdot \dfrac{2}{n} \sin nx$. Q.E.D.

Another interesting point is that if the given function is already a Fourier series, there is no need to expand it further. We can directly compare the coefficients. Let's use an example to illustrate this.

Example 6

$f(x) = 1 + \sin^2 x$, find the Fourier series of $f(x)$.

Solution $f(x)$ is an even function with the period $T = \pi$ (The period of $\sin x = 2\pi$ $\Rightarrow \sin^2 x$'s period is π) also $f^{(n)}(x) \in C_p$

$$f(x) = 1 + \sin^2 x = 1 + \frac{1 - \cos 2x}{2} = \frac{3}{2} \cdot 1 + (-\frac{1}{2}) \cos 2x,$$

and the Fourier series of $f(x)$ is:

$$f(x) = 1 + \sum_{n=1}^{\infty} a_n \cos nx = \frac{3}{2} \cdot 1 + (-\frac{1}{2}) \cos 2x,$$

so comparing the coefficients we can obtain:

$$\therefore \begin{cases} a_0 = \dfrac{3}{2} \\ a_n = \begin{cases} -\dfrac{1}{2}, n = 1 \\ 0, n = 2, 3, \cdots \end{cases} \end{cases}, \quad \therefore \text{ the Fourier series of } f(x) \text{ is } \frac{3}{2} - \frac{1}{2} \cos 2x,$$

which means $f(x)$ is already a Fourier series.   Q.E.D.

9-2-5   Half-Range Expansions

If $f(x)$ is only defined on $(0, l)$, then pass through

$$F(x) = \begin{cases} f(x), 0 < x < l \\ g(x), -l < x < 0 \end{cases}$$

We extend the function to $-l < x < l$, and then replicate the graph of F to the left (or right), resulting in a periodic function defined on R. This process is illustrated in Figure 9-5.

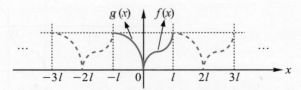

Figure 9-5   The schematic diagram of half-range expansions

In this case, the Fourier series of $F(x)$ is called the half-range Fourier series expansion of $f(x)$. Naturally, the choice of $g(x)$ is made to facilitate the computation of Fourier coefficients. Below are two typical approaches:

1. **Half-Range Even Expansion (Fourier Cosine Series)**

    By taking $g(x) = f(-x)$ on $-\ell < x < 0$, which means $F(x) = \begin{cases} f(x), 0 < x < l \\ f(-x), -l < x < 0 \end{cases}$, then

    here $F(x)$ becomes an even function, thus from the Fourier expansion for even functions (on $0 < x < l$):

    $$f(x) = F(x) = a_0 + \sum_{n=1}^{\infty} a_n \cos\frac{n\pi x}{l},$$

    where $a_0 = \frac{1}{l}\int_0^l f(x)dx$, $a_n = \frac{2}{l}\int_0^l f(x)\cos\frac{n\pi}{l}xdx$, as shown in Figure 9-6.

    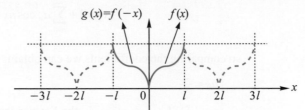

    Figure 9-6   Half-range even function expansion

2. **Half-Range Odd Expansion (Fourier Sine Series)**

    By taking $g(x) = -f(-x)$ on $-\ell < x < 0$, that is $F(x) = \begin{cases} f(x), 0 < x < l \\ -f(-x), -l < x < 0 \end{cases}$, now

    $F(x)$ is an odd function, thus from the Fourier expansion for odd functions (on $0 < x < l$):

    $$f(x) = F(x) = \sum_{n=1}^{\infty} b_n \sin\frac{n\pi}{l}x,$$

    where $b_n = \frac{2}{l}\int_0^l f(x)\sin\frac{n\pi}{l}xdx$, as shown in Figure 9-7.

    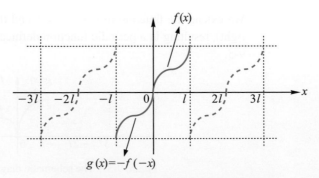

    Figure 9-7   Half-range odd function expansion

Example 7

Let $f(t) = t^2$, $0 < t < 1$, find the half-range Fourier cosine series and Fourier sine series of $f(t)$.

Solution

(1) Half-range even expansion $f(t) = a_0 + \sum\limits_{n=1}^{\infty} a_n \cos n\pi t$, $0 < t < 1$,

$$a_0 = \frac{1}{1}\int_0^1 t^2 dt = \frac{1}{3}t^3\Big|_0^1 = \frac{1}{3},\quad a_n = \frac{2}{1}\int_0^1 t^2 \cos n\pi t dt = (-1)^n \cdot \frac{4}{n^2\pi^2},$$

$$\therefore f(t) = \frac{1}{3} + \sum_{n=1}^{\infty} (-1)^n \frac{4}{n^2\pi^2}\cos n\pi t, \ 0 < t < 1.$$

(2) Half-range odd expansion $f(t) = \sum\limits_{n=1}^{\infty} b_n \sin n\pi t$, $0 < t < 1$,

$$b_n = \frac{2}{1}\int_0^1 t^2 \sin n\pi t dt = -\frac{4}{n^3\pi^3} + \frac{2\cdot(2-n^2\pi^2)}{n^3\pi^3}(-1)^n,$$

$$\therefore f(t) = \sum_{n=1}^{\infty}[-\frac{4}{n^3\pi^3} + \frac{2\cdot(2-n^2\pi^2)}{n^3\pi^3}(-1)^n]\cdot\sin n\pi t, \ 0 < t < 1.$$ Q.E.D.

9-2-6 Find Infinite Series by Using Fourier Series

In a vector space, if the vector u is represented as $u = \sum\limits_{i=1}^{n} <u, v_i> v_i$ by an orthogonal basis, $\beta = \{v_1, \cdots, v_n\}$, then

$$\|u\|^2 = <\sum_{i=1}^{n}<u,v_i>v_i, \sum_{i=1}^{n}<u,v_i>v_i> = \sum_{i=1}^{n}<u,v_i>^2.$$

Taking $\beta = \left\{1, \cos\frac{2n\pi}{T}x, \sin\frac{2n\pi}{T}x\right\}_{n=1}^{\infty}$ as a set of orthogonal function basis on $x \in [0, T]$, then the Fourier expansion $f(x) = a_0 + \sum\limits_{n=1}^{\infty}[a_n\cos\frac{2n\pi}{T}x + b_n\sin\frac{2n\pi}{T}x]$.

Through $<f(x), f(x)> = a_0^2 + \sum\limits_{n=1}^{\infty} a_n^2\cdot\|\cos\frac{2n\pi}{T}x\|^2 + b_n^2\|\sin\frac{2n\pi}{T}x\|^2$, then we

get $\frac{1}{T}\|f(x)\|^2 = a_0^2 + \sum\limits_{n=1}^{\infty}\frac{1}{2}(a_n^2 + b_n^2)$, thus, according to the definition of the inner product of functions:

$$\frac{1}{T}\int_0^T f^2(x)dx = a_0^2 + \sum_{n=1}^{\infty}\frac{1}{2}(a_n^2 + b_n^2)$$

where $\| \cos \frac{2n\pi}{T} x \|^2 = \frac{T}{2}$, $\| \sin \frac{2n\pi}{T} x \|^2 = \frac{T}{2}$, this is known as the **Parseval's** equality.

If we take a finite number of terms, then $\frac{1}{T} \int_0^T f^2(x)dx \geq a_0^2 + \sum_{n=1}^{M} \frac{1}{2}(a_n^2 + b_n^2)$ is called

the **Bessel's** inequality. This concept can be used to compute some infinite series that were previously difficult to solve in calculus.

Example 8

Please prove that following identities.

(1) $1 + \frac{1}{2^4} + \frac{1}{3^4} + \frac{1}{4^4} + \cdots = \frac{1}{90} \pi^4$.

(2) $\frac{\pi^2}{12} = \sum_{n=1}^{\infty} (-1)^{n+1} \frac{1}{n^2}$.

Solution

(1) In the Fourier expansion of Example 4 $f(x) = \frac{1}{3}\pi^2 + \sum_{n=1}^{\infty} (-1)^n \times \frac{4}{n^2} \times \cos nx$, then

using the Parseval's equality, we have

$$\frac{2}{2\pi} \int_0^\pi x^4 dx = \frac{1}{\pi} \times \frac{1}{5} x^5 \Big|_0^\pi = \frac{1}{9}\pi^4 + 8\sum_{n=1}^{\infty} \frac{1}{n^4},$$

then $\frac{4}{45}\pi^4 = 8\sum_{n=1}^{\infty} \frac{1}{n^4}$, $\sum_{n=1}^{\infty} \frac{1}{n^4} = \frac{1}{1^4} + \frac{1}{2^4} + \frac{1}{3^4} + \cdots = \frac{1}{90}\pi^4$ can be certified.

(2) From the Example 5, we get $f(x) = \sum_{n=1}^{\infty} (-1)^{n+1} \cdot \frac{2}{n} \sin nx$, then using the Parseval's

equality, we know

$$\frac{1}{2\pi} \int_{-\pi}^\pi x^2 dx = \frac{1}{2} \sum_{n=1}^{\infty} (\frac{2}{n})^2 .$$

Now we can get $\sum_{n=1}^{\infty} \frac{1}{n^2} = \frac{1}{6}\pi^2$, then

$$\sum_{n=1}^{\infty} (-1)^{n+1} \frac{1}{n} = (\frac{1}{1^2} + \frac{1}{2^2} + \frac{1}{3^2} + \cdots) - 2(\frac{1}{2^2} + \frac{1}{4^2} + \frac{1}{6^2} + \cdots)$$

$$= \sum_{n=1}^{\infty} \frac{1}{n^2} - 2 \cdot \frac{1}{4} \sum_{n=1}^{\infty} \frac{1}{n^2} = \frac{\pi^2}{12} .$$

Q.E.D.

9-2   Exercises

Basic questions

1. Let the periodic function
$$f(x)=\begin{cases}0,-\pi<x<0\\4,0\le x<\pi\end{cases}\ (f(x)=f(x+2\pi)),$$
and find its Fourier series.

Using odd (or even) functions, find the Fourier series of the periodic functions $f(x)$ in the following 2~4 questions.

2. $f(x)=\begin{cases}-2\pi,\ 0\le x<1\\2\pi,\ -1\le x<0\end{cases},f(x)=f(x+2).$

3. $f(x)=|x|,-\pi\le x\le\pi,f(x)=f(x+2\pi).$

4. $f(x)=x,-\pi\le x\le\pi,f(x)=f(x+2\pi).$

5. Expand $f(x)=x^2,0<x<L$ into
(1) Fourier cosine series.
(2) Fourier sine series.
(3) Fourier series.

6. If $f(x+2\pi)=f(x)$, and
$$f(x)=\begin{cases}-1,\ -\pi\le x\le 0\\1,\ \ \ 0<x<\pi\end{cases},$$
find its Fourier series.

Advanced questions

1. Find the Fourier series for the following periodic functions.

(1) $f(x)=\begin{cases}-3,\ -\pi<x<0\\2,\ 0\le x<\pi\end{cases},f(x)=f(x+2\pi).$

(2) $f(x)=\begin{cases}2,\ -1<x<0\\2x,\ 0\le x<1\end{cases},f(x)=f(x+2).$

(3) $f(x)=\begin{cases}0,\ -1<x<0\\3x,\ 0\le x<1\end{cases},f(x)=f(x+2).$

2. Let the periodic function $f(x)$ is
$f(x)=x+\pi,-\pi\le x\le\pi,$
and $f(x)=f(x+2\pi).$
(1) Find the Fourier series of $f(x)$.
(2) Prove that $1-\dfrac{1}{3}+\dfrac{1}{5}-\dfrac{1}{7}+\cdots=\dfrac{\pi}{4}.$

3. Let the periodic function $f(x)$ is
$$f(x)=\begin{cases}0,-\pi<x<0\\\sin x,0\le x<\pi\end{cases},\ f(x)=f(x+2\pi).$$
(1) Find the Fourier series of $f(x)$.
(2) Prove that
$$\frac{1}{2}+\frac{1}{1\cdot3}-\frac{1}{3\cdot5}+\frac{1}{5\cdot7}-\frac{1}{7\cdot9}+\cdots=\frac{\pi}{4}.$$

4. Using the properties of odd and even functions, find the Fourier series for the following periodic functions.

(1) $f(x)=\begin{cases}2,1\le x<2\\0,-1\le x<1\\2,-2\le x<-1\end{cases},\ f(x)=f(x+4).$

(2) $f(x)=x^2,-1\le x\le1,\ f(x)=f(x+2).$

(3) $f(x)=\begin{cases}x+2,0\le x<\pi\\x-2,-\pi\le x<0\end{cases},\ f(x)=f(x+2\pi).$

5. $f(x)=3+x^2,0<x<3$, if expanded using Fourier series, Fourier cosine series, and Fourier sine series, which of the following statements is correct?
(1) $f(6)=3$ in the Fourier sine series expansion.
(2) $f(3)=12$ in the Fourier cosine series expansion.
(3) $f(0)=3$ in the Fourier series expansion.
(4) $f(-1)=4$ in the Fourier series expansion.
(5) $f(-3)=12$ in the Fourier cosine series expansion.

6. Let $f(x)=\begin{cases}-x+3,1\le x<4\\0,0\le x<1\end{cases},0<x<4,$

if $g(x)=a_0+\sum_{n=1}^{\infty}a_n\cos\dfrac{n\pi}{2}x,$

$a_0=\dfrac{1}{2}\int_0^? f(x)dx,$

which of the following statements is correct?
(1) $g(x)=g(-x)$ (2) $g(1)=1$
(3) $g(x)=g(x+2)$ (4) $g(-3/2)=0$
(5) $g(7/2)=0.$

7. Find the half-range cosine and half-range sine
   Fourier expansions for the following functions.

   (1) $f(x) = \begin{cases} 0, \dfrac{1}{2} \le x < 1 \\ 1, 0 < x < \dfrac{1}{2} \end{cases}$.

   (2) $f(x) = \begin{cases} 1, \dfrac{1}{2} \le x < 1 \\ 0, 0 < x < \dfrac{1}{2} \end{cases}$.

   (3) $f(x) = \begin{cases} \pi - x, \dfrac{\pi}{2} < x < \pi \\ x, 0 < x \le \dfrac{\pi}{2} \end{cases}$.

   (4) $f(x) = \begin{cases} x - \pi, \pi < x < 2\pi \\ 0, 0 < x \le \pi \end{cases}$.

8. Given the periodic function
   $$f(x) = \begin{cases} \pi x + x^2, -\pi < x < 0 \\ \pi x - x^2, 0 < x < \pi \end{cases}$$

   (1) Find the Fourier series of $f(x)$.

   (2) Use Parseval's identity to verify
   $$1 + \frac{1}{3^6} + \frac{1}{5^6} + \frac{1}{7^6} + \cdots = \frac{\pi^6}{960}.$$

9. Find the Fourier series of
   $$f(x) = \frac{x^2}{2}(-\pi < x < \pi), \ f(x) = f(x + 2\pi).$$

   Using above results to calculate
   $$1 - \frac{1}{4} + \frac{1}{9} - \frac{1}{16} + \frac{1}{25} - \cdots.$$

10. $f(x) = \cos^3 x$, $g(x) = \sin^3 x$
    (1) Are $f(x)$ and $g(x)$ odd functions or even
        functions?
    (2) Find the Fourier series of $f(x)$ and $g(x)$.

11. $f(x) = 3\sin\dfrac{\pi}{2}x + 5\sin 3\pi x$, where

    $-2 < x < 2$, find the Fourier series of $f(x)$.

9-3 Complex Fourier Series and Fourier Integral

In general, Fourier series are often represented in the form of trigonometric functions, such as sine and cosine waves. However, in some applications like signal processing, it is more convenient to use exponential form to represent the series as it helps to visualize the magnitude and phase of the signal, making it easier to understand the signal. Therefore, in this section, we will learn how to convert the trigonometric form of Fourier series into complex form.

9-3-1 Complex Fourier Series

From Euler's formula, we know that $\begin{cases} \cos\theta = \dfrac{e^{i\theta} + e^{-i\theta}}{2} \\ \sin\theta = \dfrac{e^{i\theta} - e^{-i\theta}}{2i} \end{cases}$. Therefore, in the general

Fourier expansion, the sine and cosine functions can be replaced with complex

exponential functions. By substituting $\theta = \dfrac{2n\pi}{T}$ into the general Fourier expansion

equations, $f(x) = a_0 + \sum\limits_{n=1}^{\infty}[a_n \cos\dfrac{2n\pi}{T}x + b_n \sin\dfrac{2n\pi}{T}x]$, after rearranging, we get:

$f(x) = \sum\limits_{n=-\infty}^{\infty} c_n e^{i\frac{2n\pi}{T}x}$, where $c_n = \dfrac{1}{T}\int_0^T f(x)e^{-i\frac{2n\pi}{T}x}dx$, $n = 0, \pm 1, \pm 2, \pm 3, \cdots$ is called

the complex form of Fourier series for $f(x)$, let's illustrate how to represent a periodic function as a complex form of Fourier series with an example.

Example 1

Let the periodic function $f(t) = \begin{cases} 1, 0 \le t \le 1 \\ -1, 1 \le t \le 2 \end{cases}$, $f(t) = f(t+2)$, find the complex Fourier

series of $f(t)$.

Solution

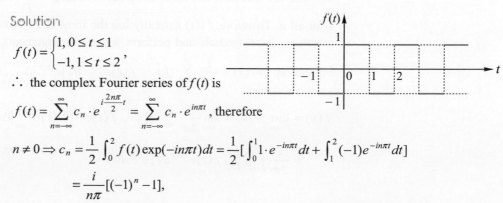

$f(t) = \begin{cases} 1, 0 \le t \le 1 \\ -1, 1 \le t \le 2 \end{cases}$,

$\therefore$ the complex Fourier series of $f(t)$ is

$f(t) = \sum\limits_{n=-\infty}^{\infty} c_n \cdot e^{i\frac{2n\pi}{2}t} = \sum\limits_{n=-\infty}^{\infty} c_n \cdot e^{in\pi t}$, therefore

$n \ne 0 \Rightarrow c_n = \dfrac{1}{2}\int_0^2 f(t)\exp(-in\pi t)dt = \dfrac{1}{2}[\int_0^1 1 \cdot e^{-in\pi t}dt + \int_1^2 (-1)e^{-in\pi t}dt]$

$= \dfrac{i}{n\pi}[(-1)^n - 1]$,

$$n = 0 \Rightarrow c_0 = \frac{1}{2} \int_0^2 f(t)dt = \frac{1}{2} [\int_0^1 1 dt + \int_1^2 (-1)dt] = 0,$$

$$\therefore f(t) = \sum_{n=-\infty, n\neq 0}^{\infty} \frac{i}{n\pi}[(-1)^n - 1]e^{in\pi t}$$

$$= \sum_{n=-\infty}^{-1} \frac{i}{n\pi}[(-1)^n - 1]e^{in\pi t} + \sum_{n=1}^{\infty} \frac{i}{n\pi}[(-1)^n - 1]e^{in\pi t}.$$

Q.E.D.

9-3-2 Fourier Integral

In the previous discussions, we focused on the Fourier series for periodic functions. However, not all signals in engineering systems are periodic. Therefore, for non-periodic signals, we first assume that they are periodic functions and then represent them using the complex Fourier series. We then consider the period approaching infinity, as shown in Figure 9-8.

By doing so, the original Fourier series becomes an integral form, known as the Fourier integral. The derivation is as follows.

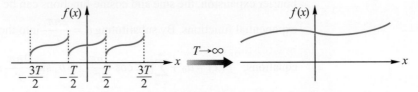

Figure 9-8   The schemaric diagram of Fourier integral

Let $f_T(x)$ be a periodic function with period T, and its complex Fourier series is

$$f_T(x) = \sum_{n=-\infty}^{\infty} [\frac{1}{T} \int_{-\frac{T}{2}}^{\frac{T}{2}} f(\tau)e^{-i\frac{2n\pi}{T}\tau} d\tau]e^{i\frac{2n\pi}{T}x}$$

In our approach, we consider the period T to tend towards infinity, therefore $w_n = \frac{2n\pi}{T}$ will tend towards zero, and the spacing between $\Delta w_n = \frac{2\pi}{T}$ values will also tend towards 0 for all n. Therefore, $f_T(x)$ naturally has the form of a Riemann integral. In fact, if we substitute these symbols and perform some rearrangements, we obtain the following expression: $f_T(x) = \frac{1}{2\pi} \sum_{n=-\infty}^{\infty} \{[\int_{-\frac{T}{2}}^{\frac{T}{2}} f(\tau)e^{-iw_n\tau} d\tau]e^{iw_n x}\}\Delta w_n$, then

$$f(x) = \lim_{T\to\infty} f_T(x) = \lim_{T\to\infty} \frac{1}{2\pi} \sum_{n=-\infty}^{\infty} \{[\int_{-\frac{T}{2}}^{\frac{T}{2}} f(\tau)e^{-iw_n\tau} d\tau] e^{iw_n x}\}\Delta w_n$$

$$= \frac{1}{2\pi} \int_{-\infty}^{\infty} [\int_{-\infty}^{\infty} f(\tau)e^{-iw\tau} d\tau] e^{iwx} dw.$$

▶ **Definition 9-3-1**

The Complex Fourier Integral

We refer to the expression $\dfrac{1}{2\pi}\displaystyle\int_{-\infty}^{\infty}\int_{-\infty}^{\infty}f(\tau)\,e^{-iw(\tau-x)}d\tau dw$ as the **complex Fourier integral** of $f(x)$. ◀

The Fourier integral is an improper integral, thus raising concerns about its existence. The following theorem ensures its convergence.

▶ **Theorem 9-3-1**

The Convergence of Fourier Integral

If a function $f(x)$ satisfies: $\begin{cases}(1)\ \displaystyle\int_{-\infty}^{\infty}|f(x)|\,dx \text{ exists}\\[2mm](2)\ f(x),\,f'(x) \text{ are piecewise continuous functions}\end{cases}$, then

$$\frac{1}{2\pi}\int_{-\infty}^{\infty}\int_{-\infty}^{\infty}f(\tau)e^{-iw(\tau-x)}d\tau dw=\frac{1}{2}[f(x^{+})+f(x^{-})],$$

the Fourier integral converges to $f(x)$ at continuous points while converges to the average at discontinuity points. ◀

Even with the help of the above theorem, complex integrals remain relatively challenging to compute. In fact, following the derivation of the Fourier integral, by letting the period T tend to infinity in the general form of the Fourier series, we obtain the following formula:

$$f(x)=\int_{0}^{\infty}A(w)\cos wxdw+\int_{0}^{\infty}B(w)\sin wxdw$$

where $A(w)=\dfrac{1}{\pi}\displaystyle\int_{-\infty}^{\infty}f(\tau)\cos w\tau d\tau$, $B(w)=\dfrac{1}{\pi}\displaystyle\int_{-\infty}^{\infty}f(\tau)\sin w\tau d\tau$, is called the **full trigonometric Fourier integral**. This provides a more friendly algorithm for our calculations. The full trigonometric Fourier integral is typically used when dealing with non odd and non-even functions.

Example 2

$$f(x)=\begin{cases}e^{-x},\ x>0\\0,\ x\le 0\end{cases}$$

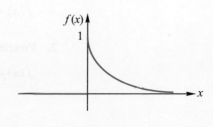

(1) Find the Fourier integral of $f(x)$.

(2) Use the result of (1) to find $\displaystyle\int_{0}^{\infty}\frac{\cos x}{1+x^{2}}dx$.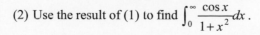

Solution

(1) Find the Fourier integral of $f(x)$. Since $f(x)$ is a non-odd and non-even function, we use the full trigonometric Fourier integral.

$$f(x) = \int_0^\infty [A(\omega)\cos\omega x + B(\omega)\sin\omega x]\,d\omega,$$

where $A(\omega) = \dfrac{1}{\pi}\int_{-\infty}^\infty f(x)\cos\omega x\,dx = \dfrac{1}{\pi}\int_0^\infty e^{-x}\cos\omega x\,dx$

$$= \frac{1}{\pi}\cdot\frac{1}{1+\omega^2}[-(-1\cdot 1)] = \frac{1}{\pi(1+\omega^2)}\,;$$

$$B(\omega) = \frac{1}{\pi}\int_{-\infty}^\infty f(x)\sin\omega x\,dx = \frac{1}{\pi}\int_0^\infty e^{-x}\sin\omega x\,dx$$

$$= \frac{1}{\pi}\cdot\frac{1}{1+\omega^2}[-e^{-x}\sin\omega x - e^{-x}\cdot\omega\cos\omega x]\Big|_0^\infty = \frac{\omega}{\pi(1+\omega^2)},$$

$$\therefore f(x) = \int_0^\infty [A(\omega)\cos\omega x + B(\omega)\sin\omega x]\,d\omega$$

$$= \int_0^\infty [\frac{1}{\pi}\cdot\frac{1}{1+\omega^2}\cos\omega x + \frac{1}{\pi}\cdot\frac{\omega}{1+\omega^2}\sin\omega x]\,d\omega.$$

(2) From (1), we know $f(x) = \dfrac{1}{\pi}\displaystyle\int_0^\infty \dfrac{1}{1+\omega^2}\cdot[\cos\omega x + \omega\sin\omega x]\,d\omega$

$$x = 1 \Rightarrow f(x) = e^{-1} = \frac{1}{\pi}\int_0^\infty \frac{1}{1+\omega^2}[\cos\omega + \omega\sin\omega]\,d\omega\cdots①$$

$$x = -1 \Rightarrow f(x) = 0 = \frac{1}{\pi}\int_0^\infty \frac{1}{1+\omega^2}[\cos\omega - \omega\sin\omega]\,d\omega\cdots②$$

① + ② obtain

$$e^{-1} = \frac{1}{\pi}\int_0^\infty \frac{2\cos\omega}{1+\omega^2}\,d\omega \Rightarrow \int_0^\infty \frac{\cos\omega}{1+\omega^2}\,d\omega = \frac{\pi}{2}e^{-1} = \frac{\pi}{2e}$$

$$\Rightarrow \int_0^\infty \frac{\cos x}{1+x^2}\,dx = \frac{\pi}{2e}.$$

Q.E.D.

If f is an odd function, then $A(w) = \dfrac{1}{\pi}\displaystyle\int_{-\infty}^\infty f(\tau)\cos w\tau\,d\tau = 0$; conversely, if f is an even function, then $B(w) = \dfrac{1}{\pi}\displaystyle\int_{-\infty}^\infty f(x)\sin wx\,dx = 0$. In these cases, the full trigonometric Fourier integral reduces to the following two situations:

1. **Fourier Cosine Integral (When $f(x)$ Is Even)**

$$f(x) = \frac{2}{\pi}\int_0^\infty [\int_0^\infty f(x)\cos wx\,dx]\cos wx\,dw$$

2. **Fourier Sine Integral (When $f(x)$ Is Odd)**

$$f(x) = \frac{2}{\pi}\int_0^\infty [\int_0^\infty f(x)\sin wx\,dx]\sin wx\,dw$$

Example 3

$$f(x) = \begin{cases} 1+x, -1 \le x \le 0 \\ -(x-1), 0 < x \le 1 \\ 0, \text{others} \end{cases}$$

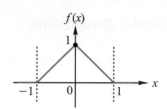

(1) Find the Fourier integral of $f(x)$.

(2) Use the above result to calculate $\int_0^\infty \dfrac{1}{\omega^2}[\cos\dfrac{\omega}{2} - \cos\dfrac{\omega}{2}\cos\omega]d\omega$.

Solution

(1) $f(x)$ is even function, $\therefore$ the Fourier integral of $f(x)$ is the Fourier cosine integral is

$$f(x) = \int_0^\infty A(\omega)\cos\omega x d\omega,$$

where $A(\omega) = \dfrac{2}{\pi}\int_0^\infty f(x)\cos\omega x dx = \dfrac{2}{\pi}\int_0^1 (-x+1)\cos\omega x dx$

$$= \dfrac{2}{\pi}\cdot[\dfrac{1}{\omega}(-x+1)\sin\omega x - \dfrac{1}{\omega^2}\cos\omega x]\Big|_0^1 = \dfrac{2}{\pi}(\dfrac{1-\cos\omega}{\omega^2}),$$

$$\therefore f(x) = \int_0^\infty \dfrac{2}{\pi}\cdot(\dfrac{1-\cos\omega}{\omega^2})\cos\omega x d\omega\cdots \text{①}$$

(2) $x = \dfrac{1}{2}$ substitute into ①

$$\Rightarrow \dfrac{1}{2} = \int_0^\infty \dfrac{2}{\pi}\cdot(\dfrac{1-\cos\omega}{\omega^2})\cos\dfrac{\omega}{2}d\omega$$

$$\Rightarrow \int_0^\infty \dfrac{1}{\omega^2}\cdot[\cos\dfrac{\omega}{2} - \cos\dfrac{\omega}{2}\cos\omega]d\omega = \dfrac{\pi}{4}.$$

Q.E.D.

9-3 Exercises

Basic questions

1. Let the periodic function be $f(x) = \begin{cases} 0, -\pi < x < 0 \\ 1, 0 < x < \pi \end{cases}$

($f(x) = f(x + 2\pi)$), find the complex Fourier series of $f(x)$.

2. Let $f(x) = \begin{cases} 0, 2\pi < x \\ 4, \pi < x < 2\pi \\ 0, x < \pi \end{cases}$, find its Fourier integral

expression (Full trigonometric Fourier integral).

3. Let $f(x) = \begin{cases} 0, |x| > 2 \\ \pi, |x| < 2 \end{cases}$, find its Fourier cosine or

Fourier sine integral (by first determining if $f(x)$ is an odd or even function).

4. Let $f(x) = \begin{cases} 0, |x| > \pi \\ x, |x| < \pi \end{cases}$, find its Fourier cosine or

Fourier sine integral (by first determining if $f(x)$ is an odd or even function).

5. Let $f(x) = e^{-kx}$ ($k > 0, x > 0$), find its Fourier cosine or Fourier sine integral.

6. (1) Let $f(x) = \begin{cases} 0, |x| > 1 \\ 1, |x| < 1 \end{cases}$, find the Fourier

integral expression of $f(x)$.

(2) Calculate $\int_0^\infty \dfrac{\sin \omega}{\omega} d\omega$ using the results from the previous question.

Advanced questions

1. Find the complex Fourier series of the following functions over the given intervals.

$$f(x) = \sum_{n=-\infty}^{\infty} c_n e^{i\frac{2n\pi}{T}x}$$

(1) $f(x) = \begin{cases} 1, 0 < x < 2 \\ -1, -2 < x < 0 \end{cases}$.

(2) $f(x) = \begin{cases} 1, 1 < x < 2 \\ 0, 0 < x < 1 \end{cases}$.

(3) $f(x) = \begin{cases} 0, \dfrac{1}{4} < x < \dfrac{1}{2} \\ 1, 0 < x < \dfrac{1}{4} \\ 0, -\dfrac{1}{2} < x < 0 \end{cases}$.

2. Find the Fourier integral expression of following functions (Full Trigonometric Fourier Integral).

(1) $f(x) = \begin{cases} 0, x > 3 \\ x, 0 < x < 3 \\ 0, x < 0 \end{cases}$.

(2) $f(x) = \begin{cases} 0, x > \pi \\ \sin x, 0 < x < \pi \\ 0, x < 0 \end{cases}$.

(3) $f(x) = \begin{cases} e^{-x}, 0 < x \\ 0, x < 0 \end{cases}$.

3. Find the Fourier cosine or Fourier sine integral of following functions.

(1) $f(x) = \begin{cases} 0, x > 1 \\ 5, 0 < x < 1 \\ -5, -1 < x < 0 \\ 0, x < -1 \end{cases}$.

(2) $f(x) = \begin{cases} 0, |x| > \pi \\ |x|, |x| < \pi \end{cases}$.

4. Find the Fourier cosine or Fourier sine integral of $f(x)$: $f(x) = e^{-x} \cos x, x > 0$.

5. (1) Let $f(x) = e^{-a|x|}, a > 0$, find the Fourier integral expression of $f(x)$.

(2) Calculate $\int_0^\infty \dfrac{\cos 2x}{x^2 + 4} dx$ using the results from the previous question.

9-4 **Fourier Transform**

In many signal processing applications, the Fourier Transform is frequently utilized. It enables the transformation of originally challenging signals (such as discontinuous or signals with abrupt features) into a superposition of numerous sine and cosine waves. By accumulating these transformed signals, one can calculate the frequencies, amplitudes, and phases of different sinusoidal components present in the original signal. This understanding allows us to delve deeper into the true essence of complex original signals.

Mathematically, there are some relationships between the Fourier Transform and Laplace Transform, especially in many of their transformation formulas, which will be explained in detail in this section.

9-4-1 Complex Form of the Fourier Transform

Looking back at the Fourier integral formula

$$f(x) = \frac{1}{2\pi} \int_{-\infty}^{\infty} \int_{-\infty}^{\infty} f(\tau) e^{-iw(\tau-x)} d\tau dw,$$

the integral on the inside provides a function transformation: $f \mapsto \int_{-\infty}^{\infty} f(x) e^{-iwx} dx$, it is known as **the complex Fourier transform**

$$\mathscr{F}\{f(x)\} = F(w) = \int_{-\infty}^{\infty} f(x) e^{-iwx} dx$$

represented by the symbol (or can be defined as $\frac{1}{\sqrt{2\pi}} \int_{-\infty}^{\infty} f(x) e^{-iwx} dx$). Therefore, the Fourier integral can be rewritten as: $f(x) = \frac{1}{2\pi} \int_{-\infty}^{\infty} F(w) e^{iwx} dw$ then obtaining the complex inverse Fourier transform $\mathscr{F}^{-1}\{F(w)\} = \frac{1}{2\pi} \int_{-\infty}^{\infty} F(w) e^{iwx} dw$ (or denoted as $\frac{1}{\sqrt{2\pi}} \int_{-\infty}^{\infty} f(x) e^{iwx} dx$), the parity (odd or even) of $f(x)$ also affects the Fourier transform, due to Euler's formula:

$$\int_{-\infty}^{\infty} f(x) e^{-iwx} dx = \int_{-\infty}^{\infty} f(x) \cos wx dx - i \int_{-\infty}^{\infty} f(x) \sin wx dx.$$

Consequently, we have:

1. $f(x)$ is even function, then

$$\mathscr{F}\{f(x)\} = F(w) = 2\int_0^\infty f(x)\cos wx\,dx$$

2. $f(x)$ is odd function, then

$$\mathscr{F}\{f(x)\} = F(w) = -2i\int_0^\infty f(x)\sin wx\,dx$$

▶ **Theorem 9-4-1**

The Existence Theorem of Fourier Transform (Sufficient but not

Necessary Conditions)

If $\int_{-\infty}^\infty |f(x)|\,dx$ exists, then the Fourier transform of $f(x)$ exists. ◀

Example 1

Find the Fourier transform of following functions

$$f(x) = \begin{cases} -1, & -1 < x < 0 \\ 1, & 0 < x < 1 \\ 0, & \text{other} \end{cases}.$$

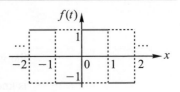

Solution The Fourier (inverse) transform are respectively:

$$\mathscr{F}\{f(x)\} = \int_{-\infty}^\infty f(x)\,e^{-iwx}\,dx = F(w),$$

$$\mathscr{F}^{-1}\{F(w)\} = \frac{1}{2\pi}\int_{-\infty}^\infty F(w)\,e^{iwx}\,dw = f(x).$$

Therefore $\mathscr{F}\{f(x)\} = \int_{-\infty}^\infty f(x)\,e^{-iwx}\,dx = \int_{-1}^0 (-1)\,e^{-iwx}\,dx + \int_0^1 1\cdot e^{-iwx}\,dx$

$$= (-1)\cdot\frac{1}{(-iw)}e^{-iwx}\bigg|_{-1}^0 + \frac{1}{(-iw)}e^{-iwx}\bigg|_0^1$$

$$= \frac{1}{iw}\cdot[1-e^{iw}] - \frac{1}{iw}\cdot[e^{-iw}-1] = \frac{1}{iw}[1-e^{iw}-e^{-iw}+1]$$

$$= \frac{1}{iw}[2-(e^{iw}+e^{-iw})] = \frac{1}{iw}\cdot(2-2\cos w).^3 \qquad \text{Q.E.D.}$$

---

[3] Since $f(x)$ is an odd function, the Fourier transform of Example 1 can also be written as

$$\mathscr{F}\{f(x)\} = -2i\int_0^\infty f(x)\sin wx\,dx = -2i\int_0^1 1\cdot\sin wx\,dx = \frac{2i}{w}\cos wx\bigg|_0^1 = \frac{2i}{w}(\cos w-1) = \frac{1}{iw}(2-2\cos w).$$

Example 2

Find the Fourier transform of following functions.

(1) $f(t) = e^{-a|t|}, a > 0$.   (2) $f(t) = \begin{cases} e^{-2t}, t > 0 \\ e^{3t}, t < 0 \end{cases}$.

Solution

(1) $f(t) = e^{-a|t|}$, $a < 0$ is even function,

$$\therefore \mathscr{F}\{f(t)\} = \int_{-\infty}^{\infty} f(t) e^{-iwt} dt = \int_{-\infty}^{\infty} e^{-a|t|}[\cos wt - i\sin wt] dt = 2\int_{0}^{\infty} e^{-at} \cos wt dt$$

$$= \frac{2}{a^2 + w^2} \cdot [(-a)e^{-at} \cos wt - e^{-at}(-w)\sin wt]\Big|_{0}^{\infty}$$

$$= \frac{2}{a^2 + w^2} \cdot [-(-a)] = \frac{2a}{a^2 + w^2} \,^4.$$

The original function and the Fourier transform plot at a = 0.5 are depicted below.

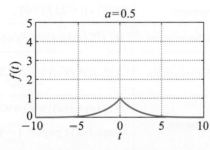

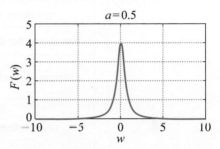

(2) $f(t) = \begin{cases} e^{-2t}, t > 0 \\ e^{3t}, t < 0 \end{cases}$, the function is neither odd nor even,

$$\therefore \mathscr{F}\{f(t)\} = \int_{-\infty}^{\infty} f(t) \cdot e^{-iwt} dt = \int_{-\infty}^{0} e^{3t} \cdot e^{-iwt} dt + \int_{0}^{\infty} e^{-2t} \cdot e^{-iwt} dt$$

$$= \int_{-\infty}^{0} e^{(3-iw)t} dt + \int_{0}^{\infty} e^{-(2+iw)t} dt$$

$$= \frac{1}{3-iw} e^{(3-iw)t}\Big|_{-\infty}^{0} + \frac{1}{-(2+iw)} e^{-(2+iw)t}\Big|_{0}^{\infty}$$

$$= \frac{1}{3-iw}[1-0] - \frac{1}{2+iw}[0-1] = \frac{1}{3-iw} + \frac{1}{2+iw}.$$ Q.E.D.

4
Since $f(x)$ is even function, so

$$\mathscr{F}\{f(t)\} = 2\int_{0}^{\infty} f(t)\cos wt dt = 2\int_{0}^{\infty} e^{-at} \cos wt dt = 2\mathscr{L}\{\cos wt\}\Big|_{s \to a} = 2\frac{a}{a^2 + w^2}.$$

When dealing with integrals that span from 0 to infinity and involve exponential terms, the thought of applying Laplace transforms for solving should come to mind.

9-4-2 The Properties of Fourier Transform

The important formulas of the Fourier transform are listed in Table 2. For each formula, we will first examine practical examples and then proceed to derive the general symbols for the formulas.

Table 2

Properties		Formulas	Corresponding example
linear operation		$\mathscr{F}\{cf(x)+dg(x)\}=cF(w)+dG(w)$	3
scale transformation		$\mathscr{F}\{f(ax)\}=\dfrac{1}{|a|}F(\dfrac{w}{a})$	
translation property	variable x	$\mathscr{F}\{f(x-\xi)\}=e^{-iw\xi}F(w)$	4
	variable w	$\mathscr{F}\{e^{iw_0x}f(x)\}=F(w-w_0)$	5
symmetry property		$\mathscr{F}\{F(x)\}=2\pi f(-w)$	
transform after differentiation		$\mathscr{F}\{f^{(n)}(x)\}=(iw)^n F(w)$	5
differentiation after transform		$\mathscr{F}\{(-ix)^n f(x)\}=\dfrac{d^n F(w)}{dw^n}$	
convolution theorem		$\mathscr{F}\{f(x)*g(x)\}=F(w)G(w)$, where $f(x)*g(g)=\displaystyle\int_{-\infty}^{\infty}f(\tau)g(x-\tau)d\tau$	7

Example 3

(1) Let $f(x)=\begin{cases}1-|x|, & |x|\le 1 \\ 0, & |x|>1\end{cases}$, find $\mathscr{F}\{f(x)\}$.

(2) If $g(x)=\begin{cases}1-|7x|, & |x|\le 1/7 \\ 0, & |x|>1/7\end{cases}$, find $\mathscr{F}\{g(x)\}$.

Solution

(1) $\mathscr{F}\{f(x)\} = 2\int_0^1 (1-x)\cos\omega x\,dx = 2[(1-x)\dfrac{\sin\omega x}{\omega} - \dfrac{\cos\omega x}{\omega^2}]\Big|_0^1$

$\qquad = \dfrac{2(1-\cos\omega)}{\omega^2} = \mathscr{F}(\omega)$.

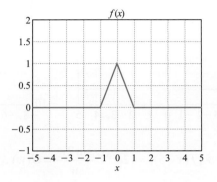
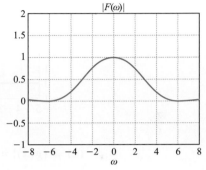

(2) $\mathscr{F}\{g(x)\} = \mathscr{F}\{f(7x)\} = \dfrac{1}{7}\mathscr{F}(\dfrac{\omega}{7}) = \dfrac{14(1-\cos\dfrac{\omega}{7})}{\omega^2} = G(\omega)$. [5]

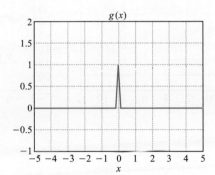
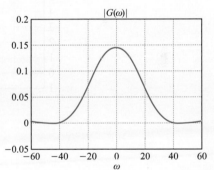

Q.E.D.

Example 4

(1) If $f(x) = \begin{cases} 6, & -2 \le x \le 2 \\ 0, & |x| > 2 \end{cases}$, find $\mathscr{F}\{f(x)\}$.

(2) If $g(x) = \begin{cases} 6, & 3 \le x \le 7 \\ 0, & x < 3 \text{ and } x > 7 \end{cases}$, find $\mathscr{F}\{g(x)\}$.

---

[5] From Example 3, it can be observed that the scale transformation formula holds when $a > 0$. This implies that if the graph of $f(x)$ is compressed along the x-axis by a factor of a, its Fourier transform graph will stretch along the w-axis by a factor of a, while the height becomes $1/a$ times the original. Conversely, if $a < 0$, the transformed graph will be a mirror reflection of the graph which we obtain when $a > 0$ with respect to the vertical axis.

Solution

(1) $\mathscr{F}\{f(x)\} = 2\int_0^2 6 \cdot \cos wx\, dx = 12\dfrac{\sin 2w}{w}$.

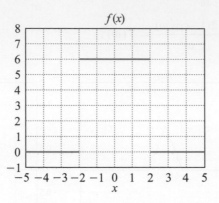
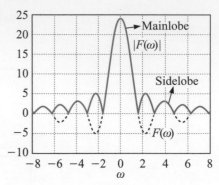

(2) $\mathscr{F}\{g(x)\} = \mathscr{F}\{f(x-5)\} = 12e^{-5iw}\dfrac{\sin 2w}{w} = G(w)$.

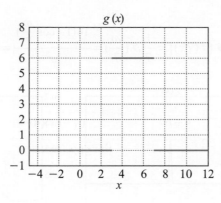
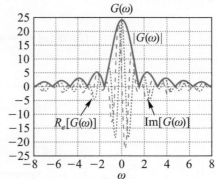

Q.E.D.

1. **Proof of the x-Translation Formula $\mathscr{F}\{f(x-\xi)\} = e^{-iw\xi}\mathscr{F}(w)$**

   Based on the definition $\mathscr{F}\{f(x-\zeta)\} = \int_{-\infty}^{\infty} f(x-\zeta)\,e^{-iwx}dx$, perform variables transform $x - \zeta = \tau$, $dx = d\tau$, then

   $$\mathscr{F}\{f(\tau)\} = \int_{-\infty}^{\infty} f(\tau)e^{-iw(\tau+\zeta)}d\tau = \int_{-\infty}^{\infty} f(\tau)\,e^{-iw\tau}e^{-iw\zeta}\,d\tau = e^{-iw\zeta}\int_{-\infty}^{\infty} f(\tau)e^{-iw\tau}d\tau$$

   $$= e^{-iw\zeta}\mathscr{F}\{f(x)\} = e^{-iw\zeta}\mathscr{F}(w).$$

2. **Proof of the w-Translation Formula $\mathscr{F}\{e^{iw_0 x}f(x)\} = F(w-w_0)$**

   According to the definition

   $$\mathscr{F}\{e^{iw_0 x}f(x)\} = \int_{-\infty}^{\infty} f(x)e^{iw_0 x}e^{-iwx}dx = \int_{-\infty}^{\infty} f(x)e^{-i(w-w_0)x}dx = F(w-w_0).$$

3. **Proof of Symmetry Property** $\mathscr{F}\{F(x)\} = 2\pi f(-w)$

By interchanging x and w in the Fourier integral $f(x) = \dfrac{1}{2\pi} \displaystyle\int_{-\infty}^{\infty} F(w)e^{iwx}dw$, then

$f(w) = \dfrac{1}{2\pi} \displaystyle\int_{-\infty}^{\infty} F(x)e^{iwx}dx \Rightarrow \displaystyle\int_{-\infty}^{\infty} F(x)e^{iwx}dx = 2\pi f(w)$. Let $w = -w$

$\Rightarrow \displaystyle\int_{-\infty}^{\infty} F(x)e^{-iwx}dx = 2\pi f(-w) \Rightarrow \mathscr{F}\{F(x)\} = 2\pi f(-w)$. Let's illustrate the

symmetry property with an example. Through $\mathscr{F}\{e^{-|x|}\} = \dfrac{2}{1+w^2}$, the symmetry

property implies $\mathscr{F}\{\dfrac{2}{1+x^2}\} = 2\pi e^{-|-w|} = 2\pi e^{-|w|}$. Now, let's explore how to utilize

the Fourier transform to solve ODE.

Example 5

Use Fourier transform to solve: $y'(x) - 4y(x) = e^{-4x}u(x), \quad -\infty < x < \infty$.

Solution   Applying the Fourier transform to ODE
$\mathscr{F}\{y'(x)\} - 4\mathscr{F}\{y(x)\} = \mathscr{F}\{e^{-4x} \cdot u(x)\}$,

because $\mathscr{F}(e^{-4x} \cdot u(x)) = \displaystyle\int_{-\infty}^{\infty} e^{-4x} \cdot 1 \cdot e^{-ixw}dx = \displaystyle\int_{-\infty}^{\infty} e^{-(4x+iw)x}dx = \dfrac{1}{4+iw}$,

$\therefore iwY(w) - 4Y(w) = \dfrac{1}{4+iw} \Rightarrow Y(w) = -\dfrac{1}{(4-iw)(4+iw)} = -\dfrac{1}{16+w^2}$,

also $\mathscr{F}\{e^{-a|x|}\} = \dfrac{2a}{a^2+w^2}$, therefore $\mathscr{F}^{-1}\{\dfrac{1}{a^2+w^2}\} = \dfrac{1}{2a}e^{-a|x|}$,

then $y(x) = \mathscr{F}^{-1}\{Y(w)\} = -\mathscr{F}^{-1}\{\dfrac{1}{4^2+w^2}\} \Rightarrow y(x) = -\dfrac{1}{8}e^{-4|x|}$.   Q.E.D.

4. **Derivation of the Formula for Differentiating First and Then Taking the Transform:**
$\mathscr{F}\{f^{(n)}(x)\} = (iw)^n F(w)$
If $f(x) \in C$, $f'(x) \in C_p$ (piecewise continuous), and $f(\pm\infty)$ is bounded $\to 0$, then

$$\mathscr{F}\{f'(x)\} = \int_{-\infty}^{\infty} f'(x)e^{-iwx}dx = f(x)e^{-iwx}\Big|_{-\infty}^{\infty} + iw\int_{-\infty}^{\infty} f(x)e^{-iwx}dx$$

$$= iw\int_{-\infty}^{\infty} f(x)e^{-iwx}dx = iwF(w) .$$

If $f(x), f'(x), f''(x), \cdots f^{(n-1)}(x) \in C$, $f^{(n)}(x) \in C_p$, and $f(\pm\infty), f'(\pm\infty), f''(\pm\infty), \cdots$,
is bounded $\to 0$, then repeatedly applying the result for $n = 1$ will lead to

$$\mathscr{F}\{f^{(n)}(x)\} = (iw)^n F(w)$$

Next, let's explore how to utilize the formula for differentiation after transform to solve the Fourier transform of complex functions.

5. **Derivation of the Formula for Differentiation After Transform:**

$$\mathcal{F}\{(-ix)^n f(x)\} = \frac{d^n F(w)}{dw^n}$$

$$\mathcal{F}\{f(x)\} = F(w) = \int_{-\infty}^{\infty} f(x) e^{-iwx} dx,$$

$$\frac{dF(w)}{dw} = \int_{-\infty}^{\infty} \frac{\partial}{\partial w} f(x)(-ix) e^{-iwx} dx = \int_{-\infty}^{\infty} f(x)(-ix) e^{-iwx} dx$$

$$= \int_{-\infty}^{\infty} [(-ix) f(x)] e^{-iwx} dx = \mathcal{F}\{(-ix) f(x)\}$$

$$\Rightarrow \mathcal{F}\{(-ix) f(x)\} = \frac{dF(w)}{dw} \text{ or } \mathcal{F}\{x(f(x))\} = i \frac{dF(w)}{dw}, \text{ repeatedly applying this}$$

result leads to the general formula

$$\mathcal{F}\{(-ix)^n f(x)\} = \frac{d^n F(w)}{dw^n}$$

Example 6

$f(x) = 4x^2 e^{-3|x|}$ try to find $\mathcal{F}\{f(x)\}$.

Solution

$$\mathcal{F}\{e^{-3|x|}\} = 2 \int_0^{\infty} e^{-3x} \cos wx \, dx = \frac{6}{w^2 + 9},$$

so $\mathcal{F}\{4x^2 e^{-3|x|}\} = -4 \frac{d^2}{dw^2}(\frac{6}{w^2 + 9}) = -24[\frac{8w^2}{(w^2 + 9)^3} - \frac{2}{(w^2 + 9)^2}]$.   Q.E.D.

6. **Proof of the Convolution Theorem**

According to the definition of convolution $f(x) * g(x) = \int_{-\infty}^{\infty} f(\tau) g(x - \tau) d\tau$, thus,

based on the definition of the Fourier transform:

$$\mathcal{F}\{[f(x) * g(x)]\} = \int_{-\infty}^{\infty} [f(x) * g(x)] e^{-iwx} dx = \int_{-\infty}^{\infty} [\int_{-\infty}^{\infty} f(\tau) g(x - \tau) d\tau] e^{-iwx} dx .$$

Performing variables transform, let $x - \tau = u \Rightarrow x = u + \tau$, $dx = du$, then the upper integral becomes

$$\int_{-\infty}^{\infty} [\int_{-\infty}^{\infty} f(\tau) g(u) d\tau] e^{-iw(u+\tau)} du = \int_{-\infty}^{\infty} f(\tau) e^{-iw\tau} d\tau \int_{-\infty}^{\infty} g(u) e^{-iwu} du$$

$$= \mathcal{F}\{f(x)\} \mathcal{F}\{g(x)\}$$

$$= F(w) G(w).$$

Proof is complete. Next, we will discuss the applications of the convolution theorem $\mathcal{F}\{f(x) * g(x)\} = \mathcal{F}(w) \mathcal{G}(w)$.

Example 7

Let $f(t) = e^{-|t|}$, $g(t) = \begin{cases} 1, & -1 \le t \le 1 \\ 0, & \text{others} \end{cases}$, if $y(t) = f(t) * g(t)$, find $\mathscr{F}\{y(t)\}$.

Solution

$$\mathscr{F}\{f(t)\} = 2\int_0^{\infty} e^{-t} \cos wt\, dt = \frac{2}{w^2+1}, \quad \mathscr{F}\{g(t)\} = 2\int_0^1 1\cos wt\, dt = \frac{2}{w}\sin w,$$

$$\mathscr{F}\{y(t)\} = \mathscr{F}\{f(t) * g(t)\} = \mathscr{F}\{f(t)\}\mathscr{F}\{g(t)\}$$

$$= \frac{2}{w^2+1} \times \frac{2}{w}\sin w = \frac{4}{w(1+w^2)}\sin w.$$

Q.E.D.

9-4-3   Parserval's Theorem and Spectral Energy

We use the Fourier transform to analyze a signal, but regardless of how we proceed, the energy conservation property should be maintained before and after the transformation. This concept can be illustrated by Parseval's theorem. Parseval's theorem (equation) states that the total energy accumulated by the function $f(x)$ along the x-axis is equal to the total energy accumulated by its Fourier transform in the frequency domain, specifically, if $\mathscr{F}\{f(x)\} = F(w)$, then

$$\int_{-\infty}^{\infty} |f(x)|^2\, dx = \frac{1}{2\pi}\int_{-\infty}^{\infty} |F(w)|^2\, dw$$

Utilizing Parseval's theorem can be employed to solve some initially difficult improper integrals, as shown in the following example.

Example 8

(1) Let $f(t) = \begin{cases} 1, & -1 \le t \le 1 \\ 0, & \text{others} \end{cases}$, find $\mathscr{F}\{f(t)\}$.

(2) Use Parseval's theorem to calculate $\int_{-\infty}^{\infty} \dfrac{\sin^2 x}{x^2}\, dx$.

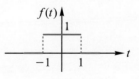

Solution

(1) $\mathscr{F}\{f(t)\} = 2\int_0^1 1\cos wt dt = \dfrac{2}{w}\sin w$.

(2) From Parseval's theorem, we know $\int_{-\infty}^{\infty} |f(x)|^2 \, dx = \dfrac{1}{2\pi} \int_{-\infty}^{\infty} |F(w)|^2 \, dw$.

Therefore

$$\int_{-1}^{1} 1^2 \, dx = \frac{1}{2\pi} \int_{-\infty}^{\infty} \left| \frac{2 \sin w}{w} \right|^2 \, dw ,$$

$$2 = \frac{2}{\pi} \int_{-\infty}^{\infty} \frac{\sin^2 w}{w^2} \, dw \Rightarrow \int_{-\infty}^{\infty} \frac{\sin^2 w}{w^2} \, dw = \pi ,$$

thus

$$\int_{-\infty}^{\infty} \frac{\sin^2 x}{x^2} \, dx = \pi .$$

Q.E.D.

Table 3 presents common Fourier transforms.

Table 3

$f(x)$	$\delta(x)$	$\delta(x - x_0)$	1	$u(x)$	sgn (x)	$\begin{cases} 1, \, |x| < a \\ 0, \, |x| > a \end{cases}$
$F(w)$ $= \mathscr{F}\{f(x)\}$	1	e^{-iwx_0}	$2\pi\delta(w)$	$\pi\delta(w) + \dfrac{1}{iw}$	$\dfrac{2}{iw}$	$\dfrac{2\sin aw}{w}$
$f(x)$	$e^{iw_0 x}$	$\cos w_0 x$	$\sin w_0 x$	e^{-ax^2}	$e^{-a|x|}$	$e^{-ax} u(x)$
$F(w)$ $= \mathscr{F}\{f(x)\}$	$2\pi\delta(w - w_0)$	$\pi[\delta(w - w_0)$ $+ \delta(w + w_0)]$	$i\pi[\delta(w + w_0)$ $- \delta(w - w_0)]$	$\sqrt{\dfrac{\pi}{a}}\, e^{-\frac{w^2}{4a}}$; $a > 0$	$\dfrac{2a}{a^2 + w^2}$; $a > 0$	$\dfrac{1}{a + iw}$; $a > 0$

where $\delta(x)$ is the Dirac delta function, and $u(x)$ is the unit step function while sgn(x) is the sign function.

Example 9

Find $\mathscr{F}\{f(x)\}$, which $f(x) = e^{-ax} \cos(w_0 x) u(x)$, $a > 0$, and $u(x)$ is the unit step function.

Solution

$$\mathscr{F}\{e^{-ax}u(x)\} = \int_0^\infty e^{-ax}e^{-iwx}dx = \frac{1}{a+iw}; a>0,$$

$$\mathscr{F}\{e^{-ax}\cos(w_0x)u(x)\} = \mathscr{F}\{e^{-ax}u(x)\frac{1}{2}(e^{iw_0x}+e^{-iw_0x})\}$$

$$= \frac{1}{2}[\frac{1}{a+i(w-w_0)}+\frac{1}{a+i(w+w_0)}]$$

$$= \frac{1}{2}[\frac{1}{(a+iw)-iw_0}+\frac{1}{(a+iw)+iw_0}]$$

$$= \frac{a+iw}{(a+iw)^2+w_0^2}.$$

Q.E.D.

9-4-4 The Fourier Transform and the Fourier Integral of a Periodic Function

The upper limit of the Fourier integral is $\int_{-\infty}^\infty |f(x)|dx$, which also serves as a condition to check the existence of the function $f(x)$. General periodic functions can have constant positive or negative values, resulting in the divergence of the aforementioned integral. Hence, the existence of their Fourier transforms is not clear. Therefore, by initially representing $f(x)$ as a complex Fourier series $f(x) = \sum_{n=-\infty}^\infty c_n e^{i\frac{2n\pi}{T}x}$, we can subsequently subject it to the Fourier transform. Since the Fourier transform is linear, therefore

$$\mathscr{F}\{f(x)\} = \mathscr{F}\{\sum_{n=-\infty}^\infty c_n e^{i\frac{2n\pi}{T}x}\} = \sum_{n=-\infty}^\infty c_n \mathscr{F}\{e^{i\frac{2n\pi}{T}x}\} = \sum_{n=-\infty}^\infty c_n 2\pi\delta(w-\frac{2n\pi}{T}).$$

Subsequently, applying an inverse transformation once again results in **the Fourier integral of the periodic function**

$$f(x) = \frac{1}{2\pi}\int_{-\infty}^\infty \sum_{n=-\infty}^\infty c_n 2\pi\delta(w-\frac{2n\pi}{T})e^{iwx}dw$$

Example 10

Let $f(x) = \begin{cases} 1, & 0 \le x \le 1 \\ 0, & 1 \le x \le 2 \end{cases}$, also

$f(x) = f(x+2), \quad -\infty < x < \infty$.

(1) Find the Fourier transform of $f(x)$. (2) Find the Fourier integral of $f(x)$.

Solution First, express $f(x)$ as a complex Fourier series.

$$f(x) = \sum_{n=-\infty}^{\infty} c_n e^{i\frac{2n\pi}{2}x} = \sum_{n=-\infty}^{\infty} c_n e^{in\pi x},$$

where $n \neq 0 \Rightarrow c_n = \frac{1}{2}\int_0^2 f(x)e^{-in\pi x}dx = \frac{1}{2}\int_0^1 1\cdot e^{-in\pi x}dx = \frac{1}{2in\pi}\cdot(1-e^{-in\pi x})$,

$$n = 0 \Rightarrow c_0 = \frac{1}{2}\int_0^2 f(x)\cdot 1 dx = \frac{1}{2}\int_0^1 1\cdot 1 dx = \frac{1}{2}.$$

(1) $\mathscr{F}\{f(x)\} = \mathscr{F}\{\sum_{n=-\infty}^{\infty} c_n\cdot e^{in\pi x}\} = \sum_{n=-\infty}^{\infty} c_n\cdot\mathscr{F}\{e^{in\pi x}\} = \sum_{n=-\infty}^{\infty} 2\pi c_n\cdot\delta(w-n\pi)$.

(2) $f(x) = \frac{1}{2\pi}\int_{-\infty}^{\infty} F(w)\cdot e^{iwx}dw = \frac{1}{2\pi}\int_{-\infty}^{\infty}\sum_{n=-\infty}^{\infty} 2\pi\cdot c_n\cdot\delta(w-n\pi)e^{iwx}dw$

$$= \int_{-\infty}^{\infty}\sum_{n=-\infty}^{\infty} c_n\cdot\delta(w-n\pi)e^{iwx}dw.$$

Q.E.D.

9-4-5 Fourier Cosine and Fourier Sine Transforms

Recalling how we derived the Fourier Transform by observing the forms of the inner and outer layers of the Fourier integral

$$f(x) = \frac{1}{2\pi}\int_{-\infty}^{\infty}[\int_{-\infty}^{\infty} f(x)e^{-wx}dx]e^{iwx}dw;$$

a similar approach can be employed for the Fourier cosine and Fourier sine transforms. Specifically, by referring to the Fourier cosine integral

$f(x) = \frac{2}{\pi}\int_0^{\infty}[\int_0^{\infty} f(x)\cos wx dx]\cos wx dw$ and the Fourier sine integral

$f(x) = \frac{2}{\pi}\int_0^{\infty}[\int_0^{\infty} f(x)\sin wx dx]\sin wx dw$, we can deduce the Fourier sine (cosine) transform.

1. **Fourier Cosine Transform (Inverse Transform)**

$\mathscr{F}_c\{f(x)\} = \int_0^{\infty} f(x)\cos wx dx = F_c(w)$ (or $\mathscr{F}_c\{f(x)\} = \sqrt{\frac{2}{\pi}}\int_0^{\infty} f(x)\cos wx dx$)

$\mathscr{F}_c^{-1}\{F_c(x)\} = f(x) = \frac{2}{\pi}\int_0^{\infty} F_c(w)\cos wx dw$

(or $\mathscr{F}_c^{-1}\{F_c(w)\} = \sqrt{\frac{2}{\pi}}\int_0^{\infty} F_c(w)\cos wx dw$)

2. Fourier Sine Transform (Inverse Transform)

$$\mathscr{F}_s\{f(x)\} = \int_0^\infty f(x)\sin wx dx = F_s(w) \quad (\text{or } \mathscr{F}_s\{f(x)\} = \sqrt{\frac{2}{\pi}} \int_0^\infty f(x)\sin wx dx)$$

$$\mathscr{F}_s^{-1}\{F_s(w)\} = f(x) = \frac{2}{\pi} \int_0^\infty F_s(w)\sin wx dw$$

$$(\text{or } \mathscr{F}_s^{-1}\{F_s(w)\} = \sqrt{\frac{2}{\pi}} \int_0^\infty F_s(w)\sin wx dw)$$

Example 11

$$f(x) = \begin{cases} 1, 0 < x \le a \\ 0, x > a \end{cases}$$

(1) Find the Fourier cosine transform of $f(x)$.

(2) Calculate $\int_0^\infty \dfrac{\sin 2ax}{x} dx$.

Solution

(1) $\mathscr{F}_c\{f(x)\} = \int_0^\infty f(x)\cos wx dx = F_c(w)$,

$$\mathscr{F}_c^{-1}\{F_c(w)\} = \frac{2}{\pi} \int_0^\infty F_c(w)\cos wx dw = f(x),$$

$$\therefore \mathscr{F}_c\{f(x)\} = \int_0^a 1\cdot\cos wx dx = \frac{1}{w}\sin wx \Big|_0^a = \frac{\sin wa}{w}.$$

(2) $f(x) = \mathscr{F}_c^{-1}\{F_c(w)\} = \mathscr{F}_c^{-1}\{\dfrac{\sin wa}{w}\} = \dfrac{2}{\pi}\displaystyle\int_0^\infty \dfrac{\sin wa}{w}\cdot\cos wx dw$, furthermore, since

$f(x)$ has a discontinuity at a, thus, the convergence of $f(x)$ at $x = a$ leads to

$\dfrac{1}{2}[f(x^+) + f(x^-)] = \dfrac{1}{2}$. Substituting $x = a$, we obtain $\dfrac{2}{\pi}\displaystyle\int_0^\infty \dfrac{\sin wa}{w}\cos wa dw = \dfrac{1}{2}$,

that is $\displaystyle\int_0^\infty \dfrac{\sin 2wa}{w} dw = \dfrac{\pi}{2}$, therefore $\displaystyle\int_0^\infty \dfrac{\sin 2ax}{x} dx = \dfrac{\pi}{2}$.   **Q.E.D.**

Next, let's examine the Fourier sine (cosine) transforms of the first and second order derivatives. When applied to regularly differentiable functions, these transforms prove to be advantageous. Their common formulas are presented in Table 4:

Table 4

Properties	Formulas	Corresponding Example
Transforming the first derivative	$\mathscr{F}_c\{f'(x)\} = w\mathscr{F}_s\{f(x)\} - f(0)$ $\mathscr{F}_s\{f'(x)\} = -w\mathscr{F}_c\{f(x)\}$	12
Transforming the second derivative	$\mathscr{F}_c\{f''(x)\} = -f'(0) - w^2\mathscr{F}_c\{f(x)\}$ $\mathscr{F}_s\{f''(x)\} = +wf(0) - w^2\mathscr{F}_s\{f(x)\}$	

Example 12

If $f(x) = e^{-ax}$, $a > 0$, using the transformation of the second derivative, find the Fourier sine and cosine transforms of $f(x)$.

Solution $f(x) = e^{-ax}$, $f''(x) = a^2 e^{-ax} = a^2 f(x)$, also $f(0) = 1$, $f'(0) = -a$, then from $\mathscr{F}\{kf(x)\} = k\mathscr{F}\{f(x)\}$.

(1) $\mathscr{F}_s\{f''(x)\} = -w^2 \mathscr{F}_s\{f(x)\} + wf(0) = a^2 \mathscr{F}_s\{f(x)\}$,

   obtaining $\mathscr{F}_s\{f(x)\} = \dfrac{w}{a^2 + w^2}$.

(2) $\mathscr{F}_c\{f''(x)\} = -w^2 \mathscr{F}_c\{f(x)\} - f'(0) = a^2 \mathscr{F}_c\{f(x)\}$,

   obtaining $\mathscr{F}_c\{f(x)\} = \dfrac{a}{a^2 + w^2}$.   Q.E.D.

Regarding the applications of the Fourier cosine and sine transforms, they will be utilized in Chapter 10 to solve partial differential equations.

3. **Proof of the Formula for Transforming the First Derivative**

   Supposing that $f(x) \in C$, $f'(x) \in C_p$, and $f(\pm\infty)$ is bounded $\to 0$, then from the definition of Fourier transform, we obtain

   $$\mathscr{F}_c\{f'(x)\} = \int_0^\infty f'(x)\cos wx\,dx = \cos wx \cdot f(x)\Big|_0^\infty + \int_0^\infty f(x)w \cdot \sin wx\,dx$$

   $$= -f(0) + w\int_0^\infty f(x)\sin wx\,dx = w\mathscr{F}_s\{f(x)\} - f(0),$$

   $$\mathscr{F}_s\{f'(x)\} = \int_0^\infty f'(x)\sin wx\,dx = \sin wx \cdot f(x)\Big|_0^\infty - \int_0^\infty w \cdot f(x)\cos wx\,dx$$

   $$= -w\int_0^\infty f(x)\cos wx\,dx = -w\mathscr{F}_c\{f(x)\}.$$

4. **Proof of the Formula for Transforming the Second Derivative**

   Supposing that $f(x)$, $f'(x) \in C$, and $f''(x) \in C_p$, $f(\infty) = 0$, $f'(\infty) = 0$

   $$\mathscr{F}_c\{f''(x)\} = \int_0^\infty f''(x)\cos wx\,dx$$

   $$= \cos wx \cdot f'(x)\Big|_0^\infty + w\sin wx \cdot f(x)\Big|_0^\infty - w^2\int_0^\infty f(x)\cos wx\,dx$$

   $$= -f'(0) - w^2\mathscr{F}_c\{f(x)\},$$

   $$\mathscr{F}_s\{f''(x)\} = \int_0^\infty f''(x)\cdot\sin wx\,dx$$

   $$= \sin wx \cdot f'(x)\Big|_0^\infty - w\cos wx \cdot f(x)\Big|_0^\infty - w^2 f(x)\sin wx\,dx$$

   $$= wf(0) - w^2\mathscr{F}_s\{f(x)\}.$$

9-4 Exercises

Basic questions

In the following questions1~2, find the Fourier transform of the given function $f(x)$.

1. $f(t) = \begin{cases} e^{-t}, t > 0 \\ 0, t < 0 \end{cases}$.

2. $f(t) = \begin{cases} 1, & |t| < 1 \\ 0, & \text{others} \end{cases}$.

3. Find the Fourier sine transform and the Fourier cosine transform of $f(t) = e^{-t}$.

4. Find the Fourier sine transform and the Fourier cosine transform of following functions:
$f(t) = \begin{cases} k, 0 < t < a \\ 0, t > a \end{cases}$.

Advanced questions

1. If the Fourier transform is defined as
$\mathscr{F}\{f(t)\} = \int_{-\infty}^{\infty} f(t) \cdot e^{-iwt} dt$, find the Fourier transforms of the following functions.
(1) $f(t) = e^{-a|t|}$.
(2) $f(t) = e^{-at} \cdot u(t)$.
(3) $f(t) = te^{-at} \cdot u(t)$.
(4) $f(t) = e^{-at} \sin(w_0 t) \cdot u(t)$.

2. If the Fourier transform is defined as
$\mathscr{F}\{f(t)\} = \int_{-\infty}^{\infty} f(t) \cdot e^{-iwt} dt$, find the Fourier transforms of the following functions.
(1) $f(t) = 10 \cdot e^{-3|t+1|}$.
(2) $f(t) = \begin{cases} k \cos w_0 t, & -a < t < a \\ 0, & \text{others} \end{cases}$.
(3) $f(t) = \begin{cases} 1 + e^t, & 0 < t < 1 \\ 0, & \text{others} \end{cases}$.
(4) $\begin{cases} e^{-at}, t \geq 0 \\ 0, t < 0 \end{cases}$, $a > 0$.
(5) $f(t) = 3e^{-4|t+2|}$.

(6) $f(t) = \dfrac{1}{a^2 + t^2}$.

(7) $f(t) = \dfrac{t}{4 + t^2}$.

3. Calculate the Fourier transform of following functions
$f(x) = \begin{cases} 1+x, & 0 \leq x \leq 1 \\ 1-x, & -1 \leq x \leq 0 \\ 0, & \text{others} \end{cases}$.

4. Calculate the Fourier transform of following functions
$f(x) = \begin{cases} x, & -1 \leq x \leq 1 \\ 0, & \text{others} \end{cases}$.

5. Given that $\mathscr{F}[f(x)] = \dfrac{1}{\sqrt{2\pi}} \int_{-\infty}^{\infty} f(x) e^{-iwx} dx$;
$\mathscr{F}[e^{-ax^2}] = \dfrac{1}{\sqrt{2a}} e^{-w^2/4a}$, and
$\mathscr{F}[\dfrac{1}{x^2+a^2}] = \sqrt{\dfrac{\pi}{2}} \dfrac{e^{-a|w|}}{a}$;
$\mathscr{F}[e^{-ax}u(x)] = \dfrac{1}{\sqrt{2\pi}(a+iw)}$, $a > 0$,
find
(1) $\int_{-\infty}^{\infty} \dfrac{\cos 6x}{x^2+9}$,
(2) $\int_{-\infty}^{\infty} \exp[-4x - x^2] dx$.

6. The definition of the periodic function $f(t)$ within one period is provided below. Calculate its Fourier transform.
$f(t) = \begin{cases} 0, -\pi < t < 0 \\ \sin t, 0 \leq t < \pi \end{cases}$

7. Utilize the Fourier transform to solve ODE $y' - 2y = e^{-2t} \cdot u(t)$, $-\infty < t < \infty$ which $u(t)$ is the unit step function.

8. Utilize the Fourier transform to solve ODE $\ddot{x} + 2\dot{x} + 2x = f(t)$,
$f(t) = \begin{cases} 1, & 0 < t < 1 \\ 0, & \text{others} \end{cases}$.

9. Utilize the Fourier transform to solve
ODE $y'' + 6y' + 5y = \delta(t-3), \ -\infty < t < \infty$,
which $\delta(t)$ is the Dirac delta function.

10. Try to verify Parseval's Theorem

$$\int_{-\infty}^{\infty} |f(x)|^2 \, dx = \frac{1}{2\pi} \int_{-\infty}^{\infty} |F(w)|^2 \, dw,$$

where $\mathcal{F}[f(x)] = F(w)$.

11. $f(x) = \displaystyle\sum_{n=-\infty}^{\infty} \delta(x - kT)$, find the Fourier transform

and Fourier integral of $f(x)$ as shown in the Figure.

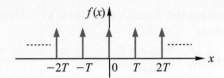

10 Partial Differential Equation

D'Alembert
(1717~1783,
France)

There are many mathematicians who have made significant contributions to partial differential equations, besides Fourier and Leibniz mentioned in Chapter 9. One of them is the French mathematician Jean d'Alembert, who in 1747 published the paper "Recherches sur la courbe que forme une corde tendue mise en vibration" (Research on the curve formed by a vibrating stretched string). In this paper, he introduced the concept of partial derivatives for the mathematical description of string vibrations, which is regarded as the beginning of the study of partial differential equations.

Learning Objectives

10-1
Introduction to Partial Differential Equation (PDE)

10-1-1 Understanding the derivation of the wave equation from string vibrations

10-1-2 Understanding the classification of second-order PDEs: hyperbolic, parabolic, elliptic; Dirichlet, Neumann, Robin

10-1-3 Mastering the solution of one-dimensional second-order PDEs: homogeneous solutions, particular solutions, method of successive integration

10-1-4 Mastering the solution of one-dimensional second-order Laplace PDE: utilizing the combination of variable method

10-2
Solving Second-Order PDE Using the Method of Separation of Variables

10-2-1 Mastering the solution method for one-dimensional second-order wave equation: separation of variables

10-2-2 Further mastering the process of separation of variables

10-2-3 Mastering the solution of one-dimensional second-order heat conduction PDE: separation of variables

10-2-4 Mastering the solution of two-dimensional second-order Laplace PDE: separation of variables

10-2-5 Organizing the general solution of the Laplace equation

10-3
Solving Non-Homogeneous Partial Differential Equation

10-3-1 Mastering the solution method for non-homogeneous PDE: method of eigenfunction expansion

10-3-2 Mastering the solution method for non-homogeneous PDE: transient solutions, steady-state solutions

10-3-3 Mastering the correction method for non-homogeneous boundary conditions: method of interpolation functions

10-4
Solving PDE Using Integral Transformations

10-4-1 Mastering integral transform solutions for second-order PDE: Laplace transform

10-4-2 Mastering integral transform solutions for second-order PDEs: Fourier transform

10-5
Partial Differential Equations in Non-Cartesian Coordinate Systems (Selected Reading)

10-5-1 Mastering the solution methods and properties of the Bessel equation

10-5-2 Understanding the modified Bessel equation and its solutions

10-5-3 Understanding the graphs of Bessel functions

10-5-4 Understanding the boundary value problem associated with the Bessel equation: the Sturm-Liouville boundary value problem.

10-5-5 Mastering the solution methods for partial differential equations in polar coordinates

ExampleVideo

Taking the example of heat flow, to describe the indoor temperature perceived by students in different seats within a classroom, one can observe that when the air conditioning is initially turned on, everyone feels hot. After a while, people start feeling cool, and those sitting near the air conditioning vents even put on sweaters as they begin to feel cold. This indicates that the temperature distribution within the classroom depends on both time and position, represented as $T(x, t)$. To capture this behavior, an equation naturally needs to incorporate both $\dfrac{\partial}{\partial x}$ and $\dfrac{\partial}{\partial t}$ (as seen in the heat conduction equation). This is the intuitive origin of PDE. The heat conduction equation was first introduced by Fourier in 1822, and he obtained its series solution using the method of separation of variables.

Starting from physical problems, this chapter gradually introduces different types of PDE and presents corresponding solution methods. Throughout the process, there will be occasional references to how ODE were solved initially, such as power series solutions, integral transforms, Fourier analysis, and so on. Devising ways to reduce PDE to ODE is also a significant underlying approach. PDEs find extensive use in engineering, especially in fluid mechanics, unit operations, heat conduction, and electromagnetics, where they play a particularly crucial role.

10-1   Introduction to Partial Differential Equation (PDE)

The type of solution for ODE is solely determined by the coefficient functions and the source functions. However, since PDE involve two or more variables, the appearance of mixed terms of partial derivatives in the equations becomes crucial. This section starts by describing wave equations and introduces the general form of second-order PDE. It then presents methods for solving the first class of constant coefficient second-order PDE: by a straight line that reducing the PDE to a constant coefficient ODE on a restricted domain. Exceptions to this method are also discussed.

The second class of solutions involves transforming the PDE into a first-order ODE with an integrating factor through a transition function. This part is explained through examples.

10-1-1 Wave Equation

Consider the wave motion of a string, as shown in Figure 10-1, while making the following assumptions:

1. The unit mass of the string is ρ, and the displacement function is represented by $u(x, t)$.

2. The string is perfectly elastic (bending).

3. Gravity's effect is negligible.

4. Only longitudinal motion (in the u-direction) is considered.

Now, let's perform a mechanics analysis: For a small segment of the string depicted in Figure 10-2(b), the force equilibrium in the x-direction is $T_2 \cos \beta = T_1 \cos \alpha = T$ (constant), while the resultant force in the u-direction is $T_2 \cos \beta - T_1 \cos \alpha$, therefore, according to Newton's second law,

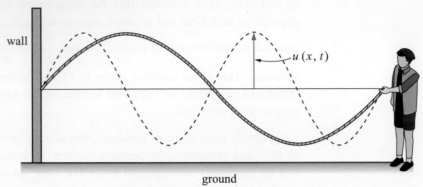

wall

$u(x, t)$

ground

Figure 10-1　Schematic Diagram of Wave Motion Equilibrium in a String

$\Sigma F = ma$, we know $T_2 \sin \beta - T_1 \sin \alpha = (\rho \Delta x) \dfrac{\partial^2 u}{\partial t^2}$ then dividing the two equations

yields $\dfrac{T_2 \sin \beta}{T_2 \cos \beta} - \dfrac{T_1 \sin \alpha}{T_1 \cos \alpha} = (\dfrac{\rho \Delta x}{T}) \dfrac{\partial^2 u}{\partial t^2} \Rightarrow \tan \beta = \tan \alpha = (\dfrac{\rho \Delta x}{T}) \dfrac{\partial^2 u}{\partial t^2}$, so

$\displaystyle \lim_{\Delta x \to 0} \dfrac{\left. \dfrac{\partial u}{\partial x} \right|_{x+\Delta x} - \left. \dfrac{\partial u}{\partial x} \right|_{x}}{\Delta x} = \dfrac{\rho}{T} \dfrac{\partial^2 u}{\partial t^2}$ (since $\tan \beta$, $\tan \alpha$ respectively represent the tangent

slopes at $x + \Delta x$ and x) , then we can get

$$\dfrac{\partial^2 u}{\partial x^2} = \dfrac{1}{\alpha^2} \dfrac{\partial^2 u}{\partial t^2},$$

where $\alpha^2 = \dfrac{T}{\rho}$. This equation describing one-dimensional wave motion is a second-order partial differential equation. Such equations are quite common in physical systems. The following will introduce their classification and solutions.

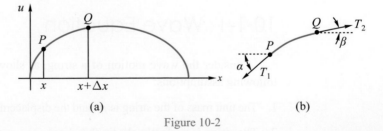

u

Q

P

x　　$x+\Delta x$

(a)

Q　T_2

β

P

α

T_1

(b)

Figure 10-2

10-1-2 The Classification of Second-Order PDE

The common form of a second-order linear partial differential equation is:

$$A\frac{\partial^2 u}{\partial x^2} + B\frac{\partial^2 u}{\partial x \partial y} + C\frac{\partial^2 u}{\partial y^2} = D = f(x, y, u, \frac{\partial u}{\partial x}, \frac{\partial u}{\partial y}) \cdots ①$$

where A, B, C, D are functions of x, y, u, $\frac{\partial u}{\partial x}$, $\frac{\partial u}{\partial y}$. If we let $\Delta = B^2 - 4AC$, then second-order partial differential equations can be classified as

1. $\Delta > 0 \Rightarrow$ Hyperbolic PDE

2. $\Delta = 0 \Rightarrow$ Parabolic PDE

3. $\Delta < 0 \Rightarrow$ Elliptic PDE

Table 1 presents examples that will be encountered in this chapter:

Table 1

Wave equation	$\dfrac{\partial^2 u}{\partial x^2} = \dfrac{1}{c^2}\dfrac{\partial^2 u}{\partial t^2}$	$A = 1, B = 0, \ C = -\dfrac{1}{c^2}$; $\Delta = 0^2 - 4\cdot 1 \cdot \dfrac{(-1)}{c^2} = \dfrac{4}{c^2} > 0$; Hyperbolic
Heat equation	$\dfrac{\partial^2 u}{\partial x^2} = \dfrac{1}{\alpha^2}\dfrac{\partial u}{\partial t}$	$A = 1, B = 0, C = 0;$ $\Delta = 0^2 - 4 \times 1 \times 0 = 0;$ Parabolic
Laplace equation	$\dfrac{\partial^2 u}{\partial x^2} + \dfrac{\partial^2 u}{\partial y^2} = 0$	$A = 1, B = 0, C = 1;$ $\Delta = 0^2 - 4 \times 1 \times 1 = -4 < 0;$ Elliptic

Second-order PDE can also be classified into three categories based on different boundary conditions (BC). Let u be the sought solution in the PDE. Depending on the manner in which boundary values are specified, it can be categorized into:

1. $u\big|_{\text{on the border}} = u_0$ is called Dirichlet condition

2. $\dfrac{\partial u}{\partial n}\bigg|_{\text{on the border}} = u_1$ is called Neumann condition

3. $(\dfrac{\partial u}{\partial n} + hu)\bigg|_{\text{on the border}} = u_3$ is called Robin (mixed) condition (which h is a constant)

where $\dfrac{\partial u}{\partial n}$ is the directional derivative in the direction of the boundary normal vector.

When $B \neq 0$, the above equation ① becomes a variable coefficient second-order partial differential equation. Its solution is quite challenging and often requires numerical methods using computers, as analytical solutions are seldom attainable. However, if all coefficients are constants, the solution process becomes significantly easier. Next, we will introduce how to solve second-order constant coefficient PDE.

10-1-3 Solve Second-Order Constant Coefficient PDE

For the general form $A\dfrac{\partial^2 u}{\partial x^2} + B\dfrac{\partial^2 u}{\partial x \partial y} + C\dfrac{\partial^2 u}{\partial y^2} = f(x, y)$, it is referred to as a homogeneous equation when $f = 0$ and as a non-homogeneous equation when $f \neq 0$. The solution approach for these types of PDE is similar to that of second-order constant coefficient ODE. It involves first finding the homogeneous solution u_h and then determining a particular solution u_p. The general solution is $u(x, y) = u_h + u_p$.

1. **Homogeneous Solution**

   Now $f(x, y) = 0$, that is $A\dfrac{\partial^2 u}{\partial x^2} + B\dfrac{\partial^2 u}{\partial x \partial y} + C\dfrac{\partial^2 u}{\partial y^2} = 0$. Considering that u is restricted

   to a straight line, which means let $u = \varphi(y + mx) = \varphi(g)$, where

   $g(x, y) = y + mx$. Then, by the chain rule, we have:

   $$\begin{cases} \dfrac{\partial u}{\partial x} = \dfrac{d\varphi}{dg}\dfrac{\partial g}{\partial x} = m \cdot \dfrac{d\varphi}{dg}, \text{ similarly } \dfrac{\partial^2 u}{\partial x^2} = m^2\dfrac{d^2\varphi}{dg^2} \\[3mm] \dfrac{\partial u}{\partial y} = \dfrac{d\varphi}{dg}\dfrac{\partial g}{\partial y} = \dfrac{d\varphi}{dg}, \text{ similarly } \dfrac{\partial^2 u}{\partial y^2} = \dfrac{d^2\varphi}{dg^2} \\[3mm] \dfrac{\partial^2 u}{\partial x \partial y} = m \cdot \dfrac{d^2\varphi}{dg^2} \end{cases}$$

   substituting into the original PDE and simplifying,

   $$Am^2\dfrac{d^2\varphi}{dg^2} + Bm\dfrac{d^2\varphi}{dg^2} + C\dfrac{d^2\varphi}{dg^2} = (Am^2 + Bm + C)\dfrac{d^2\varphi}{dg^2} = 0, \text{ but } \dfrac{d^2\varphi}{dg^2} \neq 0, \text{ so}$$

   $Am^2 + Bm + C = 0$ is referred to as the **auxiliary equation**, also $\Delta = b^2 - 4ac$ known as **the discriminant**. Consequently, based on the solutions of the auxiliary equation, the following two classifications emerge:

   (1) $\Delta \neq 0$, $m = m_1, m_2$ (two distinct roots)
       then the homogeneous solution is $u_h(x, y) = \phi(y + m_1 x) + \varphi(y + m_2 x)$.

   (2) $\Delta = 0$, $m = m_0, m_0$ (repeated root)
       then the homogeneous solution is $u_h(x, y) = \varphi(y + m_0 x) + x\varphi(y + m_0 x)$.

   The method described above for finding homogeneous solutions of this type of PDE is known as the D'Alembert solution. An example will be used below to illustrate this.

Example 1

Find a set of wave equations using the D'Alembert solution for the following partial differential equation $\dfrac{\partial^2 u}{\partial x^2} = \dfrac{1}{c^2}\dfrac{\partial^2 u}{\partial t^2}$ with initial conditions are $u(x, 0) = F(x)$, $u_t(x, 0) = G(x)$, and $-\infty < x < \infty$.

Solution

(1) Let $u = f(x + mt)$. $\therefore \begin{cases} \dfrac{\partial^2 u}{\partial x^2} = f''(x + mt) \\ \dfrac{\partial^2 u}{\partial t^2} = m^2 f''(x + mt) \end{cases}$, and substitute into differential equation,

we obtain $f''(x + mt) = \dfrac{m^2}{c^2}f''(x + mt) \Rightarrow (1 - \dfrac{m^2}{c^2})f'' = 0$,

its auxiliary equation: $1 - \dfrac{m^2}{c^2} = 0$, $\therefore m = \pm c$

$\therefore u(x, t) = f(x + ct) + g(x - ct)$.

(2) $u(x, t) = f(x + ct) + g(x - ct)$, $u_t(x, t) = cf'(x + ct) - cg'(x - ct)$, by $u(x, 0) = F(x) \Rightarrow f(x) + g(x) = F(x)$.

By $u_t(x, 0) = G(x) \Rightarrow cf'(x) - cg'(x) = G(x) \Rightarrow f'(x) - g'(x) = \dfrac{1}{c}G(x)$

$\Rightarrow \displaystyle\int_a^x [f'(\tau) - g'(\tau)]d\tau = \int_a^x \dfrac{1}{c}G(\tau)d\tau$

$\Rightarrow f(x) - f(a) - [g(x) - g(a)] = \dfrac{1}{c}\displaystyle\int_a^x G(\tau)d\tau$,

thus, we get $f(x) - g(x) = \dfrac{1}{c}\displaystyle\int_a^x G(\tau)d\tau + f(a) - g(a)$.

(3) $\begin{cases} f(x) + g(x) = F(x) \\ f(x) - g(x) = \dfrac{1}{c}\displaystyle\int_a^x G(\tau)d\tau + f(a) - g(a) \end{cases}$

$\Rightarrow f(x) = \dfrac{1}{2}F(x) + \dfrac{1}{2c}\displaystyle\int_a^x G(\tau)d\tau + \dfrac{1}{2}[f(a) - g(a)]$

$\Rightarrow g(x) = \dfrac{1}{2}F(x) - \dfrac{1}{2c}\displaystyle\int_a^x G(\tau)d\tau - \dfrac{1}{2}[f(a) - g(a)]$,

$\therefore f(x + ct) = \dfrac{1}{2}F(x + ct) + \dfrac{1}{2c}\displaystyle\int_a^{x+ct} G(\tau)d\tau + \dfrac{1}{2}[f(a) - g(a)]$,

$g(x - ct) = \dfrac{1}{2}F(x - ct) - \dfrac{1}{2c}\displaystyle\int_a^{x-ct} G(\tau)d\tau - \dfrac{1}{2}[f(a) - g(a)]$.

$$\therefore u(x,t) = f(x+ct) + g(x-ct) = \frac{1}{2}F(x+ct) + \frac{1}{2}F(x-ct) + \frac{1}{2c}\int_{x-ct}^{x+ct} G(\tau)d\tau$$

$$= \frac{1}{2}[F(x+ct) + F(x-ct)] + \frac{1}{2c}\int_{x-ct}^{x+ct} G(\tau)d\tau$$

The above equation is referred to as the D'Alembert solution of the wave equation.

Q.E.D.

2. **Particular Solution**

For $f(x,t) = $ constant, e^{mx+ny}, $\cos(mx+ny)$, $\sin(mx+ny)$, $x^m y^n$, we can refer to the assumptions provided in Table 2 and use the method of undetermined coefficients to find particular solutions. If the form of u_p is the same as the homogeneous solution, it must be multiplied by x^k for correction, where k is the smallest positive integer that makes u_p distinct from u_h.

Table 2

$f(x,y)$	k	e^{mx+ny}	$\cos(mx+ny)$	$\sin(mx+ny)$	$x^m y^n$
Assumption of $u_p(x,y)$	$ax^2 + bxy + cy^2$	$A \cdot e^{mx+ny}$	$A \cdot \cos(mx+ny)$	$A \cdot \sin(mx+ny)$	A homogeneous polynomial of degree $(m+n+2)$ of x, y

Example 2

Solve this partial differential equation $\dfrac{\partial^2 u}{\partial x^2} - \dfrac{\partial^2 u}{\partial y^2} = x - y$.

Solution

(1) Find the homogeneous solution: because $\dfrac{\partial^2 u}{\partial x^2} - \dfrac{\partial^2 u}{\partial y^2} = 0$, the auxiliary equation is $m^2 - 1 = 0$, we obtain $m = \pm 1$, therefore

$$u_h(x,y) = \phi(y+x) + \varphi(y-x).$$

(2) Find the particular solution: $f(x,y) = x - y$ is a polynomial, from assumption of Table 2, we can let $u_p = ax^3 + bx^2 y + cxy^2 + dy^3$, and substitute it into the PDE, we obtain $\quad 6ax + 2by - 2cx - 6dy = x - y \Rightarrow (6a - 2c)x + (2b - 6d)y = x - y$,

$$\therefore \begin{cases} 6a - 2c = 1 \\ 2b - 6d = -1 \end{cases}, \text{ let } \begin{cases} c = 0 \\ b = 0 \end{cases} \Rightarrow \begin{cases} a = \dfrac{1}{6} \\ d = \dfrac{1}{6} \end{cases},$$

$$\therefore u_p = \frac{1}{6}(x^3 + y^3).$$

(3) Thus, $u(x,y) = u_h(x,y) + u_p(x,y) = \phi(y+x) + \varphi(y-x) + \dfrac{1}{6}x^3 + \dfrac{1}{6}y^3$.   Q.E.D.

3. Method of Successive Integration

If the particular solution cannot be directly assumed using the above method, an alternative approach is to attempt solving second-order constant coefficient linear PDE directly through successive integration. An example is provided below:

Example 3

Solving the following partial differential equation:

$$\frac{\partial^2 u}{\partial x^2} = xe^y, \quad \begin{cases} u(0, y) = y^2 \\ u(1, y) = \sin y \end{cases}.$$

Solution

(1) $\dfrac{\partial}{\partial x}(\dfrac{\partial u}{\partial x}) = xe^y$, let $\dfrac{\partial u}{\partial x} = v$, $\therefore \dfrac{\partial v}{\partial x} = xe^y \Rightarrow v(x, y) = \dfrac{1}{2}x^2 e^y + f(y)$,

$\therefore \dfrac{\partial u}{\partial x} = \dfrac{1}{2}x^2 e^y + f(y) \Rightarrow u = \dfrac{1}{6}x^3 e^y + x \cdot f(y) + g(y)$.

(2) By $u(0, y) = y^2 \Rightarrow g(y) = y^2$,

$u(1, y) = \sin y \Rightarrow \dfrac{1}{6}e^y + f(y) + y^2 = \sin y \Rightarrow f(y) = -\dfrac{1}{6}e^y - y^2 + \sin y$,

$\therefore u(x, y) = \dfrac{1}{6}x^3 e^y + x \cdot (-\dfrac{1}{6}e^y - y^2 + \sin y) + y^2$.   Q.E.D.

10-1-4 Solving Second-Order Heat Conduction PDE Using the Combination of Variable Method

Mathematicians have explored a relatively straightforward solution method for second-order heat conduction PDE, known as the combination of variable method. The following example will illustrate this approach.

Example 4

Solve the following heat conduction partial differential equation $\dfrac{\partial u}{\partial t} = D\dfrac{\partial^2 u}{\partial x^2}$,

boundary conditions are $u(0, t) = c_0$, $u(\infty, t) = c_\infty$; initial conditions are $u(x, 0) = c_\infty$, $0 \le x < \infty$, $0 \le t < \infty$.

Solution

(1) Let $u(x,t) = V(g) = V(\dfrac{x}{2\sqrt{Dt}})$, where $g = \dfrac{x}{2\sqrt{Dt}}$ (combing the variables

transformation of variables) , then, using the chain rule, it is observed that

$$\begin{cases} \dfrac{\partial u}{\partial t} = \dfrac{dV}{dg}\cdot\dfrac{\partial g}{\partial t} = -\dfrac{x}{4\sqrt{D}}t^{-\frac{3}{2}}\cdot\dfrac{dV}{dg} = -\dfrac{x}{4\sqrt{Dt}}\cdot\dfrac{1}{t}\cdot\dfrac{dV}{dg} = -\dfrac{x}{4t\sqrt{Dt}}\cdot\dfrac{dV}{dg} \\[2mm] \dfrac{\partial u}{\partial x} = \dfrac{dV}{dg}\dfrac{\partial g}{\partial x} = \dfrac{1}{2\sqrt{Dt}}\cdot\dfrac{dV}{dg} \\[2mm] \dfrac{\partial^2 u}{\partial x^2} = \dfrac{1}{4Dt}\dfrac{d^2V}{dg^2} \end{cases}$$

substituting this into the PDE and simplifying yields $\dfrac{d^2V}{dg^2} + 2g\dfrac{dV}{dg} = 0$ (through a

change of variables, the PDE is reduced to an ODE). The boundary conditions

become $u(0,t) = c_0 \Rightarrow V(g=0) = c_0$, and $\begin{cases} u(\infty,t) = c_\infty \\ u(x,0) = c_\infty \end{cases} \Rightarrow V(g=\infty) = c_\infty$.

(2) From $\begin{cases} \dfrac{d^2V}{dg^2} + 2g\dfrac{dV}{dg} = 0 \\ V(0) = c_0 \, , V(\infty) = c_\infty \end{cases}$ (the second-order variable coefficient ODE),

we let $\dfrac{dV}{dg} = H \;\Rightarrow\; \dfrac{dH}{dg} + 2gH = 0$ (reduce order),

let $I(g) = e^{\int 2g\,dg} = e^{g^2}$ (integrating factor)

$\Rightarrow (H\cdot e^{g^2})' = 0 \Rightarrow H\cdot e^{g^2} = c_1 \Rightarrow H = c_1 e^{-g^2}$,

$\therefore \dfrac{dV}{dg} = H = c_1 e^{-g^2} \Rightarrow V = c_1 \displaystyle\int_0^g e^{-z^2}\,dz + c_2$,

$\therefore V = c_3\cdot\dfrac{2}{\sqrt{\pi}}\displaystyle\int_0^g e^{-z^2}\,dz + c_2 \Rightarrow V = c_3\,erf(g) + c_2$

(where the error function is defined as $erf(g) = \dfrac{2}{\sqrt{\pi}}\displaystyle\int_0^g e^{-\mu^2}\,d\mu$).

Next, substitute in the boundary conditions: since $erf(0) = 0$, $erf(\infty) = 1$, so
$V(0) = c_0 \Rightarrow c_2 = c_0$,
$V(\infty) = c_\infty \Rightarrow c_3\cdot 1 + c_0 = c_\infty \Rightarrow c_3 = c_\infty - c_0$,
$\therefore V(g) = (c_\infty - c_0)\,erf(g) + c_0$.

(3) substitute into $g = \dfrac{x}{2\sqrt{Dt}}$ and return to the original variables,

we obtain the solution $\quad u(x,t) = V(g) = (c_\infty - c_0)\,erf(\dfrac{x}{2\sqrt{Dt}}) + c_0$.　　**Q.E.D.**

10-1   Exercises

Basic questions

Attempt to determine the type of the following partial differential equation as hyperbolic, parabolic, or elliptic.

1. $\dfrac{\partial^2 u}{\partial x^2} + \dfrac{\partial^2 u}{\partial x \partial y} + 3\dfrac{\partial^2 u}{\partial y^2} = 0$.

2. $\dfrac{\partial^2 u}{\partial x^2} + 4\dfrac{\partial^2 u}{\partial x \partial y} + 3\dfrac{\partial^2 u}{\partial y^2} = 0$.

3. $\dfrac{\partial^2 u}{\partial x^2} + 4\dfrac{\partial^2 u}{\partial x \partial y} + 4\dfrac{\partial^2 u}{\partial y^2} = 0$.

4. $\dfrac{\partial^2 u}{\partial x^2} = 4\dfrac{\partial^2 u}{\partial y^2}$.

5. $\dfrac{\partial^2 u}{\partial x \partial y} - 4\dfrac{\partial^2 u}{\partial y^2} + 3\dfrac{\partial^2 u}{\partial x^2} = 0$.

6. $\dfrac{\partial^2 u}{\partial x^2} + \dfrac{\partial^2 u}{\partial y^2} = u$.

7. $\dfrac{\partial^2 u}{\partial x^2} = \dfrac{\partial u}{\partial t}$.

Advanced questions

1. Solve the following partial differential equation using the method of direct integration.
$$\dfrac{\partial^2 u}{\partial x \partial y} = x^2 y , \quad \begin{cases} u(x, 0) = x^2 \\ u(1, y) = \cos y \end{cases} .$$

2. The partial differential equation is $\dfrac{\partial^2 u}{\partial t^2} = \dfrac{\partial^2 u}{\partial x^2}$, initial condition is
$$u(x, 0) = F(x) = \begin{cases} \cos x, & -\pi \le x \le \pi \\ 0, & \text{others} \end{cases} ,$$
$\dfrac{\partial u}{\partial t}(x, 0) = 0$ and $-\infty < x < \infty$.

(1) Find a set of wave equations using the D'Alembert solution for the following partial differential equation.

(2) Find the solution when $u(x, t)$ at $t = 3$.

3. Solve the following constant coefficient PDE:
$$\dfrac{\partial^4 u}{\partial x^4} + 2\dfrac{\partial^4 u}{\partial x^2 \partial y^2} + \dfrac{\partial^4 u}{\partial y^4} = 0 .$$

4. Solve the following constant coefficient PDE: $\dfrac{\partial^2 u}{\partial x^2} = \dfrac{\partial^2 u}{\partial t^2} + 12t^2$.

5. Solve the following heat conduction partial differential equation
$$\dfrac{\partial T}{\partial t} = \alpha \dfrac{\partial^2 T}{\partial x^2} , \text{ initial condition is } T(x, 0) = 0,$$
boundary conditions are is $T(0, t) = 1$, $T(\infty, t) = 0$, $0 \le x < \infty$, $0 \le t < \infty$.

10-2 Solving Second-Order PDE Using the Method of Separation of Variables

The variables in the unknown function can be decoupled under appropriate boundary conditions, with the most common condition being homogeneous boundary conditions (usually set to 0). The PDE to be addressed in this section involve the operator $\Delta = \nabla^2$, as shown in Table 3.

Table 3

Wave equation	$\nabla^2 u = \dfrac{1}{c^2}\dfrac{\partial^2 u}{\partial t^2}$	One-dimensional $\dfrac{\partial^2 u}{\partial x^2} = \dfrac{1}{c^2}\dfrac{\partial^2 u}{\partial t^2}$ Three-dimensional $\dfrac{\partial^2 u}{\partial x^2} + \dfrac{\partial^2 u}{\partial y^2} + \dfrac{\partial^2 u}{\partial z^2} = \dfrac{1}{c^2}\dfrac{\partial^2 u}{\partial t^2}$
Heat equation	$\nabla^2 u = \dfrac{1}{\alpha^2}\dfrac{\partial u}{\partial t}$	One-dimensional $\dfrac{\partial^2 u}{\partial x^2} = \dfrac{1}{\alpha^2}\dfrac{\partial u}{\partial t}$ Three-dimensional $\dfrac{\partial^2 u}{\partial x^2} + \dfrac{\partial^2 u}{\partial y^2} + \dfrac{\partial^2 u}{\partial z^2} = \dfrac{1}{\alpha^2}\dfrac{\partial u}{\partial t}$
Laplace equation	$\nabla^2 u = 0$	Two-dimensional $\nabla^2 u = \dfrac{\partial^2 u}{\partial x^2} + \dfrac{\partial^2 u}{\partial y^2} = 0$

10-2-1 Solve the Wave Equation Using the Method of Separation of Variables

In the previous section, we derived the wave equation for a string, which can be used to describe the wave state of a violin string. Now, we will explain how to use the method of separation of variables to solve it. The unknown function $u(x, t)$ of this wave equation represents the vibration displacement (deflection) of the violin string at different positions and times. In this case, we are only considering one-dimensional (x-direction) positions, and the equation is $\dfrac{\partial^2 u}{\partial x^2} = \dfrac{1}{c^2}\dfrac{\partial^2 u}{\partial t^2}$, $c^2 = \dfrac{T}{\rho}$. If the ends of the string are fixed (tied) at $x = 0$ and $x = 1$, then the boundary conditions are $u(0, t) = u(l, t) = 0$. Additionally, the motion of the string depends on its initial displacement and velocity. We can assume that its initial displacement ($t = 0$) is $f(x)$, and its initial velocity is $\dfrac{\partial u}{\partial t}(x, 0) = g(x)$. Thus, the wave equation describing the vibration of the violin string is:

$$\text{DE}：\frac{\partial^2 u}{\partial x^2} = \frac{1}{c^2}\frac{\partial^2 u}{\partial t^2}(0 < x < l, t > 0)$$

$$\text{BC}：u(0, t) = u(l, t) = 0$$

$$\text{IC}：u(x, 0) = f(x), u_t(x, 0) = g(x)$$

Next, we will utilize the method of separation of variables to solve it. Let's assume that the variables x and t in $u(x, t)$ can be separated into $F(x)$ and $H(t)$, respectively, so we let $u(x, t) = F(x)H(t)$. Substituting this into differential equation (DE), the original equation becomes $F''(x)H(t) = \frac{1}{c^2}F(x)H''(t)$. Dividing both sides by $F(x)H(t)$ yields

$\frac{F''(x)}{F(x)} = \frac{1}{c^2}\frac{H''(t)}{H(t)}$. Since the left and right sides of the equation are functions of x and

t separately, in order for them to be equal, they must be constant functions, that is $\frac{F''(x)}{F(x)} = \frac{1}{c^2}\frac{H''(t)}{H(t)} = -\lambda$ (constant). Therefore, we obtain two ODEs

$\begin{cases} F''(x) + \lambda F(x) = 0 \\ H''(t) + c^2\lambda H(t) = 0 \end{cases}$, furthermore, the BC of the PDE tell us that $u(0, t) = F(0)H(t) = 0$,

then $F(0) = 0$, $u(l, t) = F(l)H(t) = 0$, implying that $F(l) = 0$. Therefore, from the variable x, we can deduce that the ODE forms a boundary value problem (BVP): $\begin{cases} F''(x) + \lambda F(x) = 0 \\ F(0) = F(l) = 0 \end{cases}$. Regarding this BVP, three cases can be discussed based on the assumed eigenvalues λ:

1. $\lambda < 0$

   $\lambda = -p^2$ $(p > 0)$, $F'' - p^2 F = 0$, $\therefore F(x) = c_1 \cosh px + c_2 \sinh px$, also

   $F(0) = 0 \Rightarrow c_1 = 0$,

   $F(l) = 0 \Rightarrow c_2 \sinh px = 0 \Rightarrow c_2 = 0$,

   then $F(x) = 0$.

   This kind of solution can be observed directly from DE, and it is not the solution we are seeking. This type of solution is also known as a trivial solution.

2. $\lambda = 0$

   $F''(x) = 0$, $\therefore F(x) = c_1 + c_2 x$, also $F(0) = 0 \Rightarrow c_1 = 0$, $F(l) = 0 \Rightarrow c_2 = 0$,

   then $F(x) = 0$.

   → The zero solution is also considered a trivial solution.

3. $\lambda > 0$

   $\lambda = p^2 (P > 0) \Rightarrow F''(x) + p^2 F = 0$, $F(x) = c_1 \cos px + c_2 \sin px$, also

   $F(0) = 0 \Rightarrow c_1 = 0$,

   $F(l) = 0 \Rightarrow c_2 \sin pl = 0$, if $c_2 \neq 0$, $\therefore \sin pl = 0$, $\therefore p = \frac{n\pi}{l}$, $n = 1, 2, 3, \cdots$,

   then $F(x) = c_2 \sin\frac{n\pi}{l}x$, $\lambda = (\frac{n\pi}{l})^2$, $n = 1, 2, 3, \cdots$.

   This is a non-trivial solution since it depends on n. We can express it as

$$F_n(x) = c_2 \sin\frac{n\pi}{l}x \,, n = 1, 2, 3, \cdots.$$

Therefore, we substitute $\lambda = (\frac{n\pi}{l})^2$ of 3. into $H''(t) + c^2\lambda H(t) = 0$, we obtain

$$H'' + (\frac{cn\pi}{l})^2 H = 0 \,,$$

$\therefore H_n(t) = d_1 \cos\frac{cn\pi}{l}t + d_2 \sin\frac{cn\pi}{l}t$, so the solutions of $F(x)$ and $H(t)$ are

$$\begin{cases} F_n(x) = c_2 \sin\dfrac{n\pi}{l}x, n = 1, 2, 3, \cdots \\ H_n(t) = d_1 \cos(\dfrac{cn\pi}{l}t) + d_2 \sin(\dfrac{cn\pi}{l}t), \ n = 1, 2, 3, \cdots \end{cases}$$, thus, for natural numbers n, we

have $u_n(x, t) = F_n(x) \cdot H_n(t) = [A_n \cos(\frac{cn\pi}{l}t) + B_n \sin(\frac{cn\pi}{l}t)]\sin\frac{n\pi}{l}x \,, n = 1, 2, 3, \cdots$,

these functions are known as the **eigenfunctions** of the vibrating string, while $\lambda_n = (\frac{n\lambda}{l})^2$ is the **eigenvalue**, and $\{\lambda_1, \lambda_2, \lambda_3, \cdots\}$ is the **vibration spectrum** of the string. Since the corresponding eigenfunctions for $n = 1, 2, 3, \cdots$ are all solutions, the principle of superposition implies that the general solution is

$$u(x, t) = \sum_{n=1}^{\infty}[A_n \cos(\frac{cn\pi}{l}t) + B_n \sin(\frac{cn\pi}{l}t)] \cdot \sin(\frac{n\pi}{l}x) \,, \text{ calculating } \frac{\partial}{\partial t} \text{ yields}$$

$$u_t(x, t) = \sum_{n=1}^{\infty}[-A_n \frac{cn\pi}{l}\sin(\frac{cn\pi}{l}t) + B_n \frac{cn\pi}{l}\cos(\frac{cn\pi}{l}t)]\sin(\frac{n\pi}{l}x) \,, \text{ substituting the}$$

initial conditions, we have:

$$u(x, 0) = f(x) \Rightarrow f(\dot{x}) = \sum_{n=1}^{\infty}A_n \sin(\frac{n\pi}{l}x);$$

$$u_t(x, 0) = g(x) \Rightarrow g(x) = \sum_{n=1}^{\infty}B_n \frac{cn\pi}{l}\sin(\frac{n\pi}{l}x),$$

hence, using the orthogonality of the eigenfunctions, the coefficients in the general solution are determined as follows:

$$\begin{cases} A_n = \dfrac{<f(x), \sin\dfrac{n\pi}{l}x>}{<\sin\dfrac{n\pi}{l}x, \sin\dfrac{n\pi}{l}x>} = \dfrac{2}{l}\displaystyle\int_0^l f(x)\sin\dfrac{n\pi}{l}xdx \\ B_n \dfrac{cn\pi}{l} = \dfrac{2}{l}\displaystyle\int_0^l g(x)\sin\dfrac{n\pi}{l}xdx \Rightarrow B_n = \dfrac{2}{cn\pi}\displaystyle\int_0^l g(x)\sin\dfrac{n\pi}{l}xdx \end{cases} .$$

▶ **Theorem 10-2-1**

Solution of the Wave Equation

Subject to the constraint at BC: $u(0, t) = u(l, t) = 0$ and IC: $u(x, 0) = f(x)$, $u_t(x, 0) = g(x)$,

the general solution of wave equation $\dfrac{\partial^2 u}{\partial x^2} = \dfrac{1}{c^2}\dfrac{\partial^2 u}{\partial t^2}$ $(0 < x < l, t > 0)$ is

$$u(x, t) = \sum_{n=1}^{\infty}[A_n \cos(\frac{cn\pi}{l}t) + B_n \sin(\frac{cn\pi}{l}t)] \cdot \sin(\frac{n\pi}{l}x),$$

where $A_n = \dfrac{2}{l}\displaystyle\int_0^l f(x)\sin\dfrac{n\pi}{l}x\,dx$, $\quad B_n = \dfrac{2}{cn\pi}\displaystyle\int_0^l g(x)\sin\dfrac{n\pi}{l}x\,dx$. ◀

For instance, when $\ell = \pi$, $c = 5$, $f(x) = \sin 2x$, when $g(x) = \pi - x$, substituting the above formulas of A_n, B_n, we obtain $A_2 = 1$, $B_n = \dfrac{2}{5n^2}$, thus

$u(x, t) = \cos(10t)\sin(2x) + \displaystyle\sum_{n=1}^{\infty}\dfrac{2}{5n^2}\sin(5nt)\sin(nx)$. Through the above calculation process of solving the wave equation using the method of separation of variables, we can summarize the approach and procedure for solving PDE using separation of variables in the following section.

10-2-2 Method and Procedure for Solving PDE Using Separation of Variables

From the above solving process, the following steps can be obtained.

Step 1 By assuming a separable variable form, the PDE is transformed into a system of coupled ODEs. For example, let $u(x, t) = F(x)H(t)$ be substituted into the PDE, leading to the separation into two categories of ODEs:

$$\begin{cases} \text{The ODE of the position function } F(x) \\ \text{The ODE of the temporal function } H(t) \end{cases}$$

Step 2 Utilize the integrating factor of the position function $F(x)$, coupled with the boundary value problem, to determine the eigenfunctions and eigenvalues.

Step 3 Substitute the eigenvalues of $F(x)$ into the ODE for the temporal function $H(x)$, and solve for the temporal function.

Step 4 Express the general solution of the PDE using the principle of superposition.

Step 5 Determine the undetermined parameters in $u(x, t)$ using the initial conditions in conjunction with Fourier series (for bounded domains) or Fourier integral (for unbounded domains).

Next, we will discuss when the length of the string is extremely long, such as in the case of a suspension bridge's cable. In such cases, the wave equation can be treated as a problem in a semi-infinite domain, and the solution process involves the use of Fourier integral. We can illustrate this with the following example.

Example 1

Solve following PDE.

DE: $\dfrac{\partial^2 u}{\partial t^2} = c^2 \dfrac{\partial^2 u}{\partial x^2}$, $0 < x < \infty$ (this is a semi-infinite domain), $t > 0$.

BC: $u(0, t) = 0$, $u(\infty, t)$ is bounded. (Because the displacement of the physical system's vibration, while possibly unmeasurable at infinity, but can be assumed to have a finite value)

IC: $u(x, 0) = f(x)$, $u_t(x, 0) = g(x)$.

Solution

(1) Let $u(x, t) = F(x)H(t)$ substitute into DE $\Rightarrow FH'' = c^2 F''H$ dividing both sides equally

$$c^2 FH \Rightarrow \frac{H''}{c^2 H} = \frac{F''}{F} = -\lambda \Rightarrow \begin{cases} F'' + \lambda F = 0 \\ H'' + \lambda c^2 H = 0 \cdots \text{①} \end{cases}$$

by BC: $u(0, t) = F(0)H(t) = 0 \Rightarrow F(0) = 0$; $u(\infty, t) = F(\infty)H(t)$ is bounded, $\therefore F(\infty)$ is bounded.

(2) By $F'' + \lambda F = 0$, $F(0) = 0$, $F(\infty)$ is bounded, it forms a boundary value problem in a semi-infinite domain. The solution is discussed as follows:

① $\lambda < 0$,

$\lambda = -w^2\, (w > 0) \Rightarrow F'' - w^2 F = 0$, $\therefore F(x) = c_1 e^{wx} + c_2 e^{-wx}$ when solving, exponential functions are more convenient for the infinite domain, while $\sinh x$, $\cosh x$ are more suitable for the finite domain, by $F(0) = 0 \Rightarrow c_1 + c_2 = 0$, $F(\infty)$ is bounded $\Rightarrow c_1 = 0$, $\therefore c_2 = 0$ so $F(x) = 0 \Rightarrow$ is zero solution (a trivial solution, not what we are looking for).

② $\lambda = 0$,

$F'' = 0$, $\therefore F(x) = c_1 + c_2 x$ by $F(0) = 0 \Rightarrow c_1 = 0$, $F(\infty)$ is bounded $\Rightarrow c_2 = 0$ $\therefore F(x) = 0$ is the trivial solution, which is also a trivial solution and not what we are seeking.

③ $\lambda = w^2 > 0$ $(0 < w < \infty)$,

$F'' + w^2 F = 0$, $\therefore F(x) = c_1 \cos wx + c_2 \sin wx$, by $F(0) = 0 \Rightarrow c_1 = 0$, $F(\infty)$ is bounded $\Rightarrow F(\infty) = c_2 \sin wx$ is bounded, $\therefore F(x) = c_2 \sin wx$, $0 < w < \infty$,

$\lambda = w^2$, $0 < w < \infty$,

$\therefore$ Let $\lambda = w^2$ substitute into ① $\Rightarrow H'' + c^2 w^2 H = 0$,

$\therefore H_w(t) = d_1 \cos cwt + d_2 \sin cwt$, the general solution of $F(x)$ and $H(t)$

are $\begin{cases} F_w(x) = c_2 \sin wx, 0 < w < \infty \\ H_w(t) = d_1 \cos cwt + d_2 \sin cwt, 0 < w < \infty \end{cases}$.

(3) $u_w(x, t) = F_w(x) \cdot H_w(t) = (A_w \cos cwt + B_w \sin cwt) \cdot \sin wx$, where $0 < x < \infty$,

because it is a continuous system, by the principle of superposition, we know

$$u(x, t) = \int_0^\infty u_w(x, t) dw = \int_0^\infty [A_w \cos cwt + B_w \sin cwt] \sin wx dw \cdots ②$$

$$u_t(x, t) = \int_0^\infty [-cwA_w \sin cwt + cwB_w \cos cwt] \sin wx dw .$$

(4) From the initial condition, we can determine

$$u(x, 0) = f(x) \Rightarrow f(x) = \int_0^\infty A_w \cdot \sin wx dw \quad \text{(the Fourier sine integral)}$$

$$\therefore A_w = \frac{2}{\pi} \int_0^\infty f(x) \sin wx dx ,$$

$$u_t(x, 0) = g(x) \Rightarrow g(x) = \int_0^\infty cw \cdot B_w \cdot \sin wx dw \Rightarrow cw \cdot B_w = \frac{2}{\pi} \int_0^\infty g(x) \sin wx dx$$

$$\therefore B_w = \frac{2}{cw\pi} \int_0^\infty g(x) \sin wx dx ,$$

let A_w and B_w substitute into ②, we obtain $u(x, t)$.　Q.E.D.

From the solution of this problem, we obtain a boundary value problem in a semi-infinite domain. Common types of boundary value problems in a semi-infinite domain are presented in Table 4. In the example, we have already introduced the first type, and the reader can practice solving the second type on their own.

Table 4

	DE	BC	Eigenfunctions and Eigenvalues
(1)	$y'' + \lambda y = 0$, $0 < x < \infty$	$y(0) = 0$, $y(\infty)$ is bounded	$\sin wx$, $\lambda = w^2$, $0 < w < \infty$
(2)	$y'' + \lambda y = 0$, $0 < x < \infty$	$y'(0) = 0$, $y(\infty)$ is bounded	$\cos wx$, $\lambda = w^2$, $0 \le w < \infty$

10-2-3 Separation of Variables Method for Solving the Heat Conduction Equation

In the previous section, we derived the wave equation for a string. Another common partial differential equation in physical systems is the heat conduction equation, or the concentration diffusion equation. Consider a cylindrical iron rod as shown in Figure 10-3, with a cross-sectional area of A and a uniform length along interval $x \in [0, l]$. For the sake of deriving the physical system conveniently, we assume

(1) Heat conduction occurs only in the x-direction within the iron rod.

(2) The cylindrical surface around the iron rod is adiabatic, meaning heat does not escape from the curved surface.

(3) There is no internal heat source within the iron rod.

(4) The iron rod is homogeneous, with a density (mass per unit volume) denoted as ρ (constant) .

(5) The specific heat of the material of the iron rod is r, and the thermal conductivity coefficient is k, which are constants.

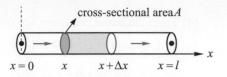

Figure 10-3

The heat Q within the mass m of an element is $Q = rmu$, where $u(x, t)$ represents the temperature distribution along the iron rod. Because the heat flux Q_t across the cross-section is proportional to the cross-sectional area A and the partial derivative of temperature $u(x, t)$ with respect to x, then $Q_t = -kA\dfrac{\partial u}{\partial x}$ (referred to as equation ①). The negative sign indicates that heat conduction occurs in the direction of decreasing temperature (from higher to lower temperatures). Considering two cross-sections at x and $x + \Delta x$ that are very close to each other, the temperature can be approximated as $u(x,t)$ for both sections, then $Q = r\rho A(u \cdot \Delta x)$, where $m = \rho(A\Delta x)$ represents the mass within this small change Δx of the iron rod. Therefore, whether heat conduction occurs along the positive x-direction, the change on the right-hand side of equation ① within Δx can be written as:

$$-kAu_x(x, t) - [-kAu_x(x + \Delta x, t)] = kA[u_x(x + \Delta x, t) - u_x(x, t)]$$

By taking a partial derivative of $Q = r\rho A(u \cdot \Delta x)$ to t, we obtain: $Q_t = r\rho A(\Delta x) \cdot \dfrac{\partial u}{\partial t}$

$\therefore r\rho A(\Delta x)\dfrac{\partial u}{\partial t} = kA[u_x(x + \Delta x, t) - u_x(x, t)]$, that is

$$\frac{k}{r\rho}\ \frac{u_x(x + \Delta x, t) - u_x(x, t)}{\Delta x} = \frac{\partial u}{\partial t},$$

we take $\Delta x \to 0$, then we get: $\dfrac{k}{r\rho}\dfrac{\partial^2 u}{\partial x^2} = \dfrac{\partial u}{\partial t}$. Let $\dfrac{k}{r\rho} = c^2$ be a constant, which is called

the thermal diffusivity, the heat conduction equation can be written as $\dfrac{\partial^2 u}{\partial x^2} = \dfrac{1}{c^2}\dfrac{\partial u}{\partial t}$, if

the temperatures at both ends of the iron rod are 0, and the initial temperature of the rod is $f(x)$, then we can describe it with the following PDE, and proceed to solve it using the method of separation of variables.

$$\text{DE :} \quad \frac{\partial^2 u}{\partial x^2} = \frac{1}{c^2}\frac{\partial u}{\partial t},\, 0 < x < l, t > 0$$

$$\text{BC :} \quad u(0, t) = u(l, t) = 0$$

$$\text{IC :} \quad u(x, 0) = f(x)$$

To Solve

We assume that the temperature distribution function $u(x, t)$ has separable variables x and t, then

(1) Let $u(x, t) = F(x) H(t)$ substitute into DE $\Rightarrow F''H = \dfrac{1}{c^2} F\dot{H}$, dividing both sides by

$FH \Rightarrow \dfrac{F''}{F} = \dfrac{1}{c^2} \dfrac{\dot{H}}{H} = -\lambda$ (constants) $\Rightarrow \begin{cases} F'' + \lambda F = 0 \\ \dot{H} + c^2 \lambda H = 0 \cdots \text{①} \end{cases}$ now by BC, we can

obtain BVP $\begin{cases} F'' + \lambda F = 0 \\ F(0) = F(l) = 0 \end{cases}$.

(2) Following the approach of Example 1, we first obtain the eigenfunctions and then write down $F_n(x)$, $H_n(t)$ separately:

$$F_n(x) = c_2 \sin(\dfrac{n\pi}{l} x) , \quad \lambda = (\dfrac{n\pi}{l})^2 , n = 1, 2, 3, \cdots .$$

Let $\lambda = (\dfrac{n\pi}{l})^2$ substitute into ①

$$\Rightarrow \dot{H} + (\dfrac{cn\pi}{l})^2 H = 0, \quad H_n(t) = d_1 \cdot \exp[-(\dfrac{cn\pi}{l})^2] t$$

$$\therefore \begin{cases} F_n(x) = c_2 \sin(\dfrac{n\pi}{l} x), n = 1, 2, 3, \cdots \\ H_n(t) = d_1 e^{-(\dfrac{n\pi c}{l})^2 t}, \quad n = 1, 2, 3, \cdots \end{cases}$$

(3) The eigenfunctions are $u_n(x, t) = F_n(x) \cdot H_n(t) = A_n \cdot e^{-(\frac{cn\pi}{l})^2 t} \cdot \sin(\dfrac{n\pi}{l} x)$,

$n = 1, 2, 3, \cdots$, by the principle of superposition, we know:

$$u(x, t) = \sum_{n=1}^{\infty} A_n \cdot e^{-(\frac{cn\pi}{l})^2 t} \cdot \sin(\dfrac{n\pi}{l} x) \cdots \text{②}$$

(4) By initial conditions: $u(x, 0) = f(x) \Rightarrow f(x) = \sum_{n=1}^{\infty} A_n \cdot \sin(\dfrac{n\pi}{l} x)$

$$\therefore A_n = \dfrac{2}{l} \int_0^l f(x) \sin(\dfrac{n\pi}{l} x) dx ,$$

let A_n substitute into ②, we obtain $u(x, t)$.

▶ **Theorem 10-2-2**

The Solution of the Heat Conduction Equation

Restrict to BC: $y(0, t) = u(l, t) = 0$ and IC: $u(x, 0) = f(x)$, the general solution of the heat conduction equation $\dfrac{\partial^2 u}{\partial x^2} = \dfrac{1}{c^2}\dfrac{\partial u}{\partial t}$ $(0 < x < l, t > 0)$ is

$$u(x, t) = \sum_{n=1}^{\infty} A_n e^{-(\frac{cn\pi}{\ell})^2 t}\sin(\frac{n\pi}{l}x)\,, \text{ where } A_n = \frac{2}{l}\int_0^l f(x)\sin(\frac{n\pi}{l}x)dx\,. \quad ◀$$

In practical applications, if $l = 80$, the initial temperature

$$u(x, 0) = f(x) = 100\sin(\frac{\pi}{80}x)\,, \text{ then } u(x, t) = \sum_{n=1}^{\infty} A_n \cdot e^{-(\frac{cn\pi}{80})^2 t}\cdot\sin(\frac{n\pi}{80}x)\,,$$

by $u(x, 0) = f(x) = 100\cdot\sin(\dfrac{\pi}{80}x) = \displaystyle\sum_{n=1}^{\infty} A_n \cdot \sin(\dfrac{n\pi}{80}x)$

(which has a value only when $n = 1$) $\Rightarrow$ when $n = 1 \Rightarrow A_1 = 100$; when $n \neq 1 \Rightarrow A_n = 0$,

$$\therefore u(x, t) = A_1 \cdot e^{-(\frac{c\pi}{80})^2 t}\cdot\sin(\frac{\pi}{80}x) = 100\cdot e^{-(\frac{c\pi}{80})^2 t}\cdot\sin(\frac{\pi}{80}x)\,.$$

In the previous section, we assumed that the temperature at both ends ($x = 0$ and $x = \ell$) of the iron rod is 0. However, in the insulation devices commonly used, there can be a situation where both ends are insulated, meaning the presence of $\dfrac{\partial u}{\partial x}(0, t) = \dfrac{\partial u}{\partial x}(l, t) = 0$ at $x = 0$ and $x = l$. The solution is illustrated in the following example.

Example 2

Solve following PDE.

DE: $\dfrac{\partial^2 u}{\partial x^2} = \dfrac{1}{c^2}\dfrac{\partial u}{\partial t}$, $0 < x < l$, $t > 0$;

BC: $u_x(0, t) = u_x(l, t) = 0$ (both ends are insulated) ;

IC: $u_x(x, 0) = f(x)$.

Solution

(1) Let $u_x(x, t) = F(x)H(t)$ substitute into DE $\Rightarrow F''H = \dfrac{1}{c^2}F\cdot\dot{H}$, dividing both sides by

$$FH \Rightarrow \frac{F''}{F} = \frac{1}{c^2}\frac{\dot{H}}{H} = -\lambda \Rightarrow \begin{cases} F'' + \lambda F = 0 \\ \dot{H} + c^2\lambda H = 0 \end{cases}, \text{ by}$$

BC, we obtain

$$u_x(0, t) = F'(0)H(t) = 0 \Rightarrow F'(0) = 0\,,$$
$$u_x(l, t) = F'(l)H(t) = 0 \Rightarrow F'(l) = 0\,.$$

(2) the solution $\begin{cases} F'' + \lambda F = 0 \\ F'(0) = F'(l) = 0 \end{cases}$:

① let $\lambda < 0$, $\lambda = -p^2 \ (p > 0) \Rightarrow F'' - p^2 F = 0$,

$\therefore F(x) = c_1 \cosh px + c_2 \sinh px$

$\Rightarrow F'(x) = c_1 p \sinh px + c_2 p \cosh px$,

by $F'(0) = 0 \Rightarrow c_2 = 0$, $F'(l) = 0 \Rightarrow c_1 = 0$, $\therefore F(x) = 0$ is trivial solution.

② $\lambda = 0$, $\therefore F(x) = c_1 + c_2 x$, $F'(x) = c_2$, also $F'(0) = 0 = F'(l) \Rightarrow c_2 = 0$,

$\therefore F(x) = c_1$, the eigenfunctions $F_0(x) = c_1 \cdot 1$.

③ $\lambda > 0$, $\lambda = p^2 \ (p > 0) \Rightarrow F'' + p^2 F = 0 \Rightarrow F(x) = c_1 \cos px + c_2 \sin px$

$\Rightarrow F'(x) = -c_1 p \sin px + c_2 p \cos px$,

also $F'(0) = 0 \Rightarrow c_2 = 0$, $F'(l) = 0 \Rightarrow c_1 p \sin pl = 0$, if $c_1 \neq 0$, $\therefore \sin pl = 0$,

$\therefore pl = n\pi$, $p = \dfrac{n\pi}{l}$, $n = 1, 2, 3, \cdots$, $\therefore F(x) = c_1 \cos(\dfrac{n\pi}{l} x)$, $n = 1, 2, 3, \cdots$,

$\therefore F_n(x) = \begin{cases} c_1, \lambda = 0, n = 0 \\ c_1 \cos(\dfrac{n\pi}{l} x), \lambda = (\dfrac{n\pi}{l})^2, n = 1, 2, 3, \cdots \end{cases}$, let $\lambda = (\dfrac{n\pi}{l})^2$ substitute

into $\dot{H} + c^2 \lambda H = 0 \ \Rightarrow \dot{H} + (\dfrac{cn\pi}{l})^2 H = 0$,

$\therefore H_n(t) = d_1 e^{-(\frac{cn\pi}{l})^2 t}$, $n = 1, 2, \cdots$, and $F_n(x) = \begin{cases} c_1, n = 0 \\ c_1 \cos(\dfrac{n\pi}{l} x), n = 1, 2, 3, \cdots \end{cases}$

(3) $u_n(x, t) = F_n(x) H_n(t) = A_0 + A_n e^{-(\frac{cn\pi}{l})^2 t} \cdot \cos(\dfrac{n\pi}{l} x)$, by the principle of

superposition, we know:

$$u(x, t) = A_0 + \sum_{n=1}^{\infty} A_n e^{-(\frac{cn\pi}{l})^2 t} \cdot \cos(\dfrac{n\pi}{l} x) \cdots ①$$

(4) By initial conditions: $u(x, 0) = f(x) \Rightarrow f(x) = A_0 + \sum_{n=1}^{\infty} A_n \cos(\dfrac{n\pi}{l} x)$ (Fourier

cosine series), thus, we have

$$\begin{cases} A_0 = \dfrac{1}{l} \int_0^l f(x) dx \\ A_n = \dfrac{2}{l} \int_0^l f(x) \cdot \cos(\dfrac{n\pi}{l} x) dx \end{cases}$$

let A_0 and A_n substitute into ①, then we obtain $u(x, t)$.

Q.E.D.

10-2-4 Solve Steady-State Two-Dimensional Heat Conduction Problems Using the Method of Separation of Variables.

Next, we discuss the steady-state two-dimensional heat conduction problem. For the heat conduction equation $\nabla^2 u = \dfrac{1}{c^2}\dfrac{\partial u}{\partial t}$, if the heat transfer within the system reaches a steady state, the temperature u becomes independent of time t. Consequently, we can obtain $\dfrac{\partial u}{\partial t} = 0$, that is $\nabla^2 u = 0$. If considering the two-dimensional heat conduction equation, we can derive $\nabla^2 u = \dfrac{\partial^2 u}{\partial x^2} + \dfrac{\partial^2 u}{\partial y^2} = 0$, this known as the Laplace equation.

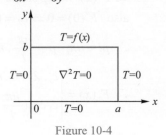

Figure 10-4

Assuming the temperature distribution is denoted as $u = T(x, t)$, and it exists in the region R ($0 \le x \le a$, $0 \le y \le b$) within the xy-plane, with boundary conditions $T(0, y) = T(a, y) = 0$, $T(x, 0) = 0$, and $T(x, b) = f(x)$, this can be described by the following PDE. We can also employ the method of separation of variables to solve this Laplace equation.

$$\text{DE:}\ \frac{\partial^2 T}{\partial x^2} + \frac{\partial^2 T}{\partial y^2} = 0$$

$$\text{BC:}\ T(0, y) = T(a, y) = 0 = T(x, 0) = 0$$
$$T(x, b) = f(x)$$

The solving process is shown in the following.

STEP 1 Let $T(x, y) = F(x)Q(y)$ substitute into $\text{DE} \Rightarrow F''Q + FQ'' = 0 \Rightarrow F''Q = -FQ''$, dividing both sides by FQ

$$\Rightarrow \frac{F''(x)}{F(x)} = -\frac{Q''(y)}{Q(y)} = -\lambda \Rightarrow \begin{cases} F''(x) + \lambda F(x) = 0 \\ Q''(y) - \lambda Q(y) = 0 \end{cases}.$$

Substituting into BC , we find that:

$$T(0, y) = F(0)Q(y) = 0 \Rightarrow F(0) = 0\ ;$$
$$T(a, y) = F(a)Q(y) = 0 \Rightarrow F(a) = 0\ ;$$
$$T(x, 0) = F(x)Q(0) = 0 \Rightarrow Q(0) = 0,$$

hence, we obtain the homogeneous BVP $\begin{cases} F'' + \lambda F = 0 \\ F(0) = F(a) = 0 \end{cases}$.

In an abstract sense, the homogeneity lies in the x-direction, allowing us to expand in the x-direction.

STEP 2 Next, let's proceed to find the eigenfunctions:

(1) $\lambda = -p^2 \, (p > 0) \Rightarrow$ the trivial solution;

(2) $\lambda = 0 \Rightarrow$ the trivial solution;

(3) $\lambda = p^2 \, (p > 0) \Rightarrow F'' + p^2 F = 0$, $F(x) = c_1 \cos px + c_2 \sin px$,

by $F(0) = 0 \Rightarrow c_1 = 0$; $F(a) = 0 \Rightarrow c_2 \sin pa = 0$, also $c_2 \neq 0$,

$\therefore \sin pa = 0 \Rightarrow pa = n\pi$, thus $p = \dfrac{n\pi}{a}$,

$\therefore F_n(x) = c_2 \sin(\dfrac{n\pi}{a} x)$, $n = 1, 2, 3, \cdots$, $p^2 = \lambda = (\dfrac{n\pi}{a})^2$,

let λ substitute into

$$Q'' - \lambda Q = 0 \Rightarrow Q''(y) - (\dfrac{n\pi}{a})^2 Q(y) = 0,$$

$$\therefore Q(y) = d_1 \sinh(\dfrac{n\pi}{a} y) + d_2 \cosh(\dfrac{n\pi}{a} y).$$

By $Q(0) = 0 \Rightarrow d_2 = 0$,

$$\therefore Q_n(y) = d_1 \sinh(\dfrac{n\pi}{a} y) = d_3 \dfrac{\sinh(\dfrac{n\pi}{a} y)}{\sinh(\dfrac{n\pi}{a} b)}.$$

STEP 3 $T_n(x, y) = F_n(x) Q_n(y) = A_n \cdot \dfrac{\sinh(\dfrac{n\pi}{a} y)}{\sinh(\dfrac{n\pi}{a} b)} \cdot \sin(\dfrac{n\pi}{a} x)$, $n = 1, 2, 3, \cdots$, by the

principle of superposition, we know

$$T(x, y) = \sum_{n=1}^{\infty} A_n \cdot \dfrac{\sinh(\dfrac{n\pi}{a} y)}{\sinh(\dfrac{n\pi}{a} b)} \cdot \sin(\dfrac{n\pi}{a} x) \cdots \text{①}$$

STEP 4 By $T(x, b) = f(x) = \sum_{n=1}^{\infty} A_n \cdot \sin(\dfrac{n\pi}{a} x)$,

$$A_n = \dfrac{2}{a} \int_0^a f(x) \sin(\dfrac{n\pi}{a} x) dx,$$

let A_n substitute into ①, we obtain $T(x, y)$.

▶ **Theorem 10-2-3**

The Solution of the Laplace Equation

Restrict to BC: $T(0, y) = T(a, y) = T(x, 0) = 0$, $T(x, b) = f(x)$, the general solution of the

Laplace equation $\dfrac{\partial^2 T}{\partial x^2} + \dfrac{\partial^2 T}{\partial y^2} = 0$ is:

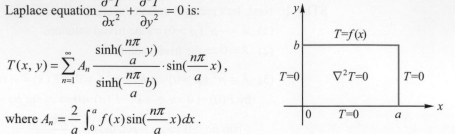

$$T(x, y) = \sum_{n=1}^{\infty} A_n \frac{\sinh(\dfrac{n\pi}{a} y)}{\sinh(\dfrac{n\pi}{a} b)} \cdot \sin(\dfrac{n\pi}{a} x),$$

where $A_n = \dfrac{2}{a} \displaystyle\int_0^a f(x) \sin(\dfrac{n\pi}{a} x) dx$.

10-2-5 Organization of the Solution Methods for the Laplace Equation

The common types of Laplace's equation can be categorized into the following four classes:

1. DE: $\dfrac{\partial^2 u}{\partial x^2} + \dfrac{\partial^2 u}{\partial y^2} = 0$; BC: $u(0, y) = u(a, y) = u(x, 0) = 0$, $u(x, b) = f(x)$

   Fourier expansion of the boundary function $f(x)$:

   $$\begin{cases} u(x, y) = \displaystyle\sum_{n=1}^{\infty} A_n \cdot \dfrac{\sinh(\dfrac{n\pi}{a} y)}{\sinh(\dfrac{n\pi}{a} b)} \cdot \sin(\dfrac{n\pi}{a} x) \\[4mm] A_n = \dfrac{2}{a} \displaystyle\int_0^a f(x) \sin(\dfrac{n\pi}{a} x) dx \end{cases}.$$

2. DE: $\dfrac{\partial^2 u}{\partial x^2} + \dfrac{\partial^2 u}{\partial y^2} = 0$; BC: $u(0, y) = u(a, y) = u(x, b) = 0$, $u(x, 0) = f(x)$

   Fourier expansion of the boundary function $f(x)$:

   $$\begin{cases} u(x) = \displaystyle\sum_{n=1}^{\infty} B_n \cdot \dfrac{\sinh[\dfrac{n\pi}{a} (b - y)]}{\sinh(\dfrac{n\pi}{a} b)} \cdot \sin(\dfrac{n\pi}{a} x) \\[4mm] f(x) = \displaystyle\sum_{n=1}^{\infty} B_n \sin(\dfrac{n\pi}{a} x) \\[4mm] B_n = \dfrac{2}{a} \displaystyle\int_0^a f(x) \sin(\dfrac{n\pi}{a} x) dx \end{cases}.$$

3. DE: $\dfrac{\partial^2 u}{\partial x^2} + \dfrac{\partial^2 u}{\partial y^2} = 0$; BC: $u(0, y) = 0$, $u(a, y) = g(y)$, $u(x, 0) = u(x, b) = 0$

Fourier expansion of the boundary function $g(y)$:

$$
\begin{cases}
u(x, y) = \displaystyle\sum_{n=1}^{\infty} C_n \cdot \dfrac{\sinh(\dfrac{n\pi}{b}x)}{\sinh(\dfrac{n\pi}{b}a)} \sin(\dfrac{n\pi}{b}y) \\[4mm]
g(y) = \displaystyle\sum_{n=1}^{\infty} C_n \cdot \sin(\dfrac{n\pi}{b}y) \\[4mm]
C_n = \dfrac{2}{b} \displaystyle\int_0^b g(y)\sin(\dfrac{n\pi}{b}y)\,dy
\end{cases}
$$

4. DE: $\dfrac{\partial^2 u}{\partial x^2} + \dfrac{\partial^2 u}{\partial y^2} = 0$; BC: $u(0, y) = g(y)$, $u(a, y) = u(x, 0) = u(x, b) = 0$

Fourier expansion of the boundary function $g(y)$:

$$
\begin{cases}
u(x, y) = \displaystyle\sum_{n=1}^{\infty} D_n \dfrac{\sinh[\dfrac{n\pi}{b}(a-x)]}{\sinh(\dfrac{n\pi}{b}a)} \sin(\dfrac{n\pi}{b}y) \\[4mm]
g(y) = \displaystyle\sum_{n=1}^{\infty} D_n \sin(\dfrac{n\pi}{b}y) \\[4mm]
D_n = \dfrac{2}{b} \displaystyle\int_0^b g(y)\sin(\dfrac{n\pi}{b}y)\,dy
\end{cases}
$$

The following example will illustrate how to solve a Laplace's equation with non-homogeneous boundary conditions (often denoted as zero) using the superposition principle when there is more than one such condition.

Example 3

Solve following PDE:

DE: $\dfrac{\partial^2 u}{\partial x^2} + \dfrac{\partial^2 u}{\partial y^2} = 0$,

BC: $u(0, y) = 0$, $u(a, y) = g(y)$,
$u(x, 0) = 0$, $u(x, b) = f(x)$.

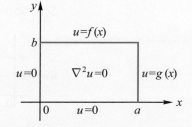

Solution

As both the x and y directions are non-homogeneous, direct separation of variables cannot be applied. However, since the Laplace's equation is linear, we can decompose BC as follows. The right-hand side diagrams both exhibit homogeneous BC. Then, following the principle of superposition, these solutions can be added together to yield the solution for the original non-homogeneous BVP.

(1) By Theorem 10-2-3, it is known that u_1 (the separation of variables method) :

$$u_1(x, y) = \sum_{n=1}^{\infty} A_n \frac{\sinh(\frac{n\pi}{a}y)}{\sinh(\frac{n\pi}{a}b)} \sin(\frac{n\pi}{a}x) , \quad A_n = \frac{2}{a} \int_0^a f(x)\sin(\frac{n\pi}{a}x)dx .$$

(2) Exchanging x and y to obtain $u_2(x, y)$:

$$\begin{cases} u_2(x, y) = \sum_{n=1}^{\infty} B_n \dfrac{\sinh(\frac{n\pi}{b}x)}{\sinh(\frac{n\pi}{b}a)} \cdot \sin(\frac{n\pi}{b}y) \\[3mm] g(y) = \sum_{n=1}^{\infty} B_n \sin(\frac{n\pi}{b}y) \\[3mm] B_n = \dfrac{2}{b} \int_0^b g(y)\sin(\frac{n\pi}{b}y)dy \end{cases}.$$

(3) By the principle of superposition, we know: $u(x, y) = u_1(x, y) + u_2(x, y)$.   Q.E.D.

Example 4

Given that BVP is linear, attempt to decompose the non-homogeneous BC represented in the attached diagram into a sum of homogeneous BC, and indicate the general form of the solution.

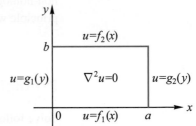

Solution

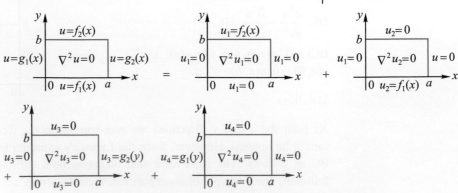

Therefore, based on the principle of superposition, the solution is $u = u_1 + u_2 + u_3 + u_4$.
Q.E.D.

We have previously introduced the Laplace's equation in the rectangular coordinate system (xy coordinates). Now, we will discuss the solution of the Laplace's equation in cylindrical coordinates (polar coordinates), where the Laplace's equation in polar coordinates (r, θ) is $\nabla^2 u = \dfrac{\partial^2 u}{\partial r^2} + \dfrac{1}{r}\dfrac{\partial u}{\partial r} + \dfrac{1}{r^2}\dfrac{\partial^2 u}{\partial \theta^2} = 0$. The following example demonstrates how to solve it using the separation of variables method.

Example 5

The Laplace's PDE in polar coordinate form, with the domain of u lying within a circle. Solve following PDE:

DE: $\nabla^2 u = \dfrac{\partial^2 u}{\partial r^2} + \dfrac{1}{r}\dfrac{\partial u}{\partial r} + \dfrac{1}{r^2}\dfrac{\partial^2 u}{\partial \theta^2} = 0$,

BC: $\begin{cases} u(r, \theta) = u(r, \theta + 2\pi) \\ u_\theta(r, \theta) = u_\theta(r, \theta + 2\pi) \end{cases}$,

$u(\rho, \theta) = f(\theta)$, $u(0, \theta)$ is bounded, $0 \le r \le \rho$.

Solution

(1) Let $u(r, \theta) = R(r)Q(\theta)$ substitute into DE $\Rightarrow R''Q + \dfrac{1}{r}R'Q + \dfrac{1}{r^2}RQ'' = 0$

$\Rightarrow r^2 R''Q + rR'Q + RQ'' = 0$, dividing both sides by RQ:

$\Rightarrow -\dfrac{r^2 R'' + rR'}{R} = \dfrac{Q''}{Q} = -\lambda$

$\Rightarrow \begin{cases} Q'' + \lambda Q = 0 \\ r^2 R'' + rR' - \lambda R = 0 \text{ (called equation ①)} \end{cases}$,

thus we obtain

$u(r, \theta) = u(r, \theta + 2\pi) \Rightarrow Q(\theta) = Q(\theta + 2\pi)$,

$u_\theta(r, \theta) = u_\theta(r, \theta + 2\pi) \Rightarrow Q'(\theta) = Q'(\theta + 2\pi)$.

(2) By $\begin{cases} Q'' + \lambda Q = 0 \\ Q(\theta) = Q(\theta + 2\pi),\, Q'(\theta) = Q'(\theta + 2\pi) \end{cases}$,

① $\lambda = -p^2 < 0$, $p > 0 \Rightarrow$ the trivial solution.

② $\lambda = 0 \Rightarrow Q'' = 0 \Rightarrow Q(\theta) = c_1$, let $\lambda = 0$ substitute into ① $r^2 R'' + rR' = 0$ (ODE of equal dimension), let

$r = e^t$, $t = \ln r$, $D \triangleq \dfrac{d}{dt} \Rightarrow [D(D-1)+D]R = 0 \Rightarrow D^2 R = 0 \Rightarrow m = 0, 0$,

$\therefore R = d_1 + d_2 t$, $\therefore R_0(r) = d_1 + d_2 \ln r$

$\therefore u_0 = R_0(r) \cdot Q_0(\theta) = a_0 + b_0 \ln r$ (where $\lambda = 0$).

(3) $\lambda = p^2 > 0 (p > 0)$,　$Q'' + p^2 Q = 0$,　$Q(\theta) = c_1 \cos p\theta + c_2 \sin p\theta$,

by $\begin{cases} Q(\theta) = Q(\theta + 2\pi) \\ Q'(\theta) = Q'(\theta + 2\pi) \end{cases} \Rightarrow p = n$, $n = 1, 2, 3, \cdots$.

Let $\lambda = p^2 = n^2$ substitute into ①

$\Rightarrow r^2 R'' + rR' - n^2 R = 0$ (ODE of equal dimension), let $r = e^t$, $t = \ln r$, $D \overset{\Delta}{=} \dfrac{d}{dt}$

$\Rightarrow [D \cdot (D-1) + D - n^2]R = 0 \Rightarrow [D^2 - n^2]R = 0$,

$\therefore m = \pm n$, $\therefore R_n(r) = d_3 r^n + d_4 r^{-n}$, we obtain

$$\theta_n(\theta) = c_1 \cos p\theta + c_2 \sin p\theta, n = 1, 2, 3, \cdots.$$

(4) $u_n(r, \theta) = R_n(r) \cdot Q_n(\theta) = (a_n r^n + b_n r^{-n}) \cos n\theta + (c_n r^n + d_n r^{-n}) \sin n\theta$, by the principle of superposition, we know

$$u(r, \theta) = a_0 + b_0 \ln r + \sum_{n=1}^{\infty} (a_n r^n + b_n r^{-n}) \cos n\theta + (c_n r^n + d_n r^{-n}) \sin n\theta.$$

(5) By $u(0, \theta)$ is bounded, we obtain $b_0 = 0$, $b_n = 0$, $d_n = 0$

$\Rightarrow u(r, \theta) = a_0 + \sum_{n=1}^{\infty} a_n r^n \cos n\theta + c_n r^n \sin n\theta$　(called equation ②),

by $u(\rho, \theta) = f(\theta)$

$\Rightarrow f(\theta) = a_0 + \sum_{n=1}^{\infty} a_n \rho^n \cos n\theta + c_n \rho^n \sin n\theta$　(Fourier series)

$$\therefore a_0 = \frac{1}{2\pi} \int_0^{2\pi} f(\theta) d\theta,$$

$$a_n \rho^n = \frac{1}{\pi} \int_0^{2\pi} f(\theta) \cos n\theta d\theta,$$

$$c_n \rho^n = \frac{1}{\pi} \int_0^{2\pi} f(\theta) \sin n\theta d\theta.$$

find a_0, a_n, c_n then substitute into ②, we obtain $u(r, \theta)$.　**Q.E.D.**

10-2   Exercises

Basic questions

1. Solve the wave equation:

DE: $\dfrac{\partial^2 u}{\partial x^2} = \dfrac{\partial^2 u}{\partial t^2}$ $(0 < x < l, t > 0)$,

IC: $u(x, 0) = \begin{cases} \dfrac{2}{l} x, & 0 < x < \dfrac{l}{2} \\ \dfrac{2}{l}(l-x), & \dfrac{l}{2} < x < l \end{cases}$,

$\dfrac{\partial u}{\partial t}(x, 0) = 0$,

BC: $u(0, t) = u(l, t) = 0$.

2. Solve the heat conduction equation:

DE: $\dfrac{\partial^2 T}{\partial x^2} = \dfrac{1}{\alpha^2} \dfrac{\partial T}{\partial t}$

$(0 < x < l, t > 0, \alpha > 0)$,

IC: $T(x, 0) = 100 \cdot \sin(\dfrac{\pi x}{l})$,

BC: $T(0, t) = T(l, t) = 0$.

3. Solve the Laplace's equation:

DE: $\dfrac{\partial^2 u}{\partial x^2} + \dfrac{\partial^2 u}{\partial y^2} = 0 \, (1 < x < 3, \, 1 < y < 3, t > 0)$,

BC: $u(x, 1) = u(x, 3) = 0$,

$u(1, y) = u(3, y) = 0$.

4. Suppose there is a bivariate function $u(x, y)$ defined outside a certain circle (as shown in the attached figure). Additionally, it is known to satisfy the following DE and BC:

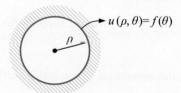
$u(\rho, \theta) = f(\theta)$

DE: $\nabla^2 u = \dfrac{\partial^2 u}{\partial r^2} + \dfrac{1}{r} \dfrac{\partial u}{\partial r} + \dfrac{1}{r^2} \dfrac{\partial^2 u}{\partial \theta} = 0$;

$\rho \le r \le \infty$.

BC: $\begin{cases} u(r, \theta) = u(r, \theta + 2\pi) \\ u_\theta(r, \theta) = u_\theta(r, \theta + 2\pi) \end{cases}$;

$u(\rho, \theta) = f(\theta) \, ; u(\infty, \theta)$ is bounded.

Please solve this partial differential equation.

Advanced questions

1. Solve the following wave equation.

DE: $4\dfrac{\partial^2 u}{\partial x^2} = \dfrac{\partial^2 u}{\partial t^2}$ $(-\infty < x < \infty, t > 0)$.

IC: $u(x, 0) = 0$, $\dfrac{\partial u}{\partial t}(x, 0) = \delta(x)$,

where $\delta(x)$ is the Dirac delta function.

BC: $u(\pm\infty, t)$ is bounded.

2. Solve the following heat conduction equation

(1) DE: $k\dfrac{\partial^2 u}{\partial x^2} = \dfrac{\partial u}{\partial t}$, $k > 0$

$(0 < x < 2, t > 0)$.

IC: $u(x, 0) = \begin{cases} x, 0 < x < 1 \\ 0, 1 < x < 2 \end{cases}$.

BC: $u_x(0, t) - u_x(2, t) = 0$.

(2) DE: $\dfrac{\partial^2 u}{\partial x^2} = \dfrac{\partial u}{\partial t}$, $-\pi < x < \pi, t > 0$.

IC: $u(x, 0) = f(x) = x + x^2$.

BC: $u(-\pi, t) = u(\pi, t) = 0$.

3. Solve the following Laplace's equation.

DE: $\dfrac{\partial^2 u}{\partial x^2} = \dfrac{\partial^2 u}{\partial y^2} = 0$

$(0 < x < a, 0 < y < b, t > 0)$.

BC: $u(x, 0) = 0$,

$u(x, b) = (a - x)\sin(x)$,

$u(0, y) = u(a, y) = 0$.

10-3 Solving Non-Homogeneous Partial Differential Equation

When the PDE includes a nonzero source function, the simple assumption of separable variables may not lead to an ODE problem. However, thanks to the introduction of Fourier series (as seen in Chapter 9) and drawing from the experience of solving ODE power series solutions (from Chapter 4), this forms the first method we are about to introduce: the method of eigenfunction expansion.

In the second part, we encounter second-order PDE with non-homogeneous boundary conditions. At this point, we reference the concept of system states from circuit theory. We assume the solution to be a combination of transient and steady-state components. From a mathematical perspective, we are using externally applied functions to manage the boundary conditions. As a result, the original PDE transforms into a joint equation of a constant-coefficient ODE and homogeneous boundary PDE, both of which we already know how to solve.

The third part then goes on to discuss situations where both the PDE and its BC are non-homogeneous. In such cases, we refer to the concept of interpolation from numerical analysis to modify the original PDE into a problem with homogeneous BC. This allows us to once again utilize the eigenfunction expansion method to tackle the problem.

10-3-1 Eigenfunction Expansion

The eigenfunctions obtained from the Sturm-Liouville boundary value problem form an orthogonal basis that can be used to represent any solution function within that interval. The table below lists common orthogonal coordinate functions and their corresponding expansion forms. However, it's essential to note that when using eigenfunction expansion to solve a PDE, the boundary conditions must satisfy the "Sturm-Liouville" homogeneous boundary conditions.

From the recommended orthogonal basis in Table 5, we can simplify the principles of solving the PDE as follows:

1. $\left.\begin{array}{l}\text{DE homogeneous} \\ \text{BC homogeneous}\end{array}\right\}$ can be solved using the separation of variables method or the eigenfunction expansion method.

2. $\left.\begin{array}{l}\text{DE non-homogeneous} \\ \text{BC homogeneous}\end{array}\right\}$ separation of variables cannot be used, but the eigenfunction expansion method is applicable ($\because$ BC is homogeneous).

The following examples will illustrate how to utilize the eigenfunction expansion method to solve the PDE.

Table 5

Boundary conditions	Eigenfunction	Expansion form
$u(0, t) = u(l, t) = 0$	$\{\sin\frac{n\pi}{l}x\}_{n=1}^{\infty}$	$\begin{cases} u(x, t) = \sum_{n=1}^{\infty} a_n(t)\sin(\frac{n\pi}{l}x) \\ a_n(t) = \frac{2}{l}\int_0^l u(x, t)\sin(\frac{n\pi}{l}x)dx \end{cases}$
$u_x(0, t) = u_x(l, t) = 0$	$\{\cos\frac{n\pi}{l}x\}_{n=0}^{\infty}$ or $\{1, \cos\frac{n\pi}{l}x\}_{n=1}^{\infty}$	$\begin{cases} u(x, t) = a_0 1 + \sum_{n=1}^{\infty} a_n(t)\cos(\frac{n\pi}{l}x) \\ a_0 = \frac{1}{l}\int_0^l u(x, t)dx \\ a_n(t) = \frac{2}{l}\int_0^l u(x, t)\cos\frac{n\pi}{l}xdx \end{cases}$
$u(0, t) = u_x(l, t) = 0$	$\{\sin\frac{(n-\frac{1}{2})\pi}{l}x\}_{n=1}^{\infty}$	$\begin{cases} u(x, t) = \sum_{n=1}^{\infty} a_n(t)\sin\frac{(n-\frac{1}{2})\pi}{l}x \\ a_n(t) = \frac{2}{l}\int_0^l u(x, t)\sin\frac{(n-\frac{1}{2})\pi}{l}xdx \end{cases}$
$u_x(0, t) = u(l, t) = 0$	$\{\cos\frac{(n-\frac{1}{2})\pi}{l}x\}_{n=1}^{\infty}$	$\begin{cases} u(x, t) = \sum_{n=1}^{\infty} a_n(t)\cdot\cos\frac{(n-\frac{1}{2})\pi}{l}x \\ a_n(t) = \frac{2}{l}\int_0^l u(x, t)\cos\frac{(n-\frac{1}{2})\pi}{l}xdx \end{cases}$
$u(x, t) = u(x + T, t)$ $u_x(x, t) = u_x(x + T, t)$ (BC are periodic)	$\{1, \sin\frac{2n\pi}{T}x, \cos\frac{2n\pi}{T}x\}_{n=1}^{\infty}$	$u(x, t)$ $= a_0 1 + \sum_{n=1}^{\infty}[a_n(t)\cos\frac{2n\pi}{T}x + b_n(t)\sin\frac{2n\pi}{T}x]$ $\begin{cases} a_0 = \frac{1}{T}\int_0^T u(x, t)dx \\ a_n(t) = \frac{2}{T}\int_0^T u(x, t)\cos\frac{2n\pi}{T}xdx \\ b_n(t) = \frac{2}{T}\int_0^T u(x, t)\sin\frac{2n\pi}{T}xdx \end{cases}$

Example 1

Utilize the eigenfunction expansion method to solve the following PDE.

(1) DE: $\dfrac{\partial^2 u}{\partial t^2} = c^2 \dfrac{\partial^2 u}{\partial x^2}$, $0 < x < l, t > 0$.

    BC: $u(0, t) = u(l, t) = 0$.

    IC: $u(x, 0) = f(x)$, $u_t(x, 0) = g(x)$.

(2) DE: $\dfrac{\partial^2 u}{\partial t^2} = c^2 \dfrac{\partial^2 u}{\partial x^2} + p(x, t)$, $0 < x < l, t > 0$.

    BC: $u(0, t) = u(l, t) = 0$.

    IC: $u(x, 0) = f(x)$, $u_t(x, 0) = g(x)$.

Solution

(1) ① By $u(0, t) = u(l, t) = 0$, we obtain that the eigenfunction $\{\sin \dfrac{n\pi}{l} x\}_{n=1}^{\infty}$ is a functional coordinate basis.

Let $u(x, t) = \displaystyle\sum_{n=1}^{\infty} a_n(t) \cdot \sin \dfrac{n\pi}{l} x$ (equation ①) substitute into the ordinary DE

$\Rightarrow \displaystyle\sum_{n=1}^{\infty} a_n''(t) \sin(\dfrac{n\pi}{l} x) = c^2 \cdot \sum_{n=1}^{\infty} a_n(t) \cdot [-(\dfrac{n\pi}{l})^2] \cdot \sin(\dfrac{n\pi}{l} x)$

$\Rightarrow \displaystyle\sum_{n=1}^{\infty} [a_n''(t) + (\dfrac{cn\pi}{l})^2 a_n(t)] \sin(\dfrac{n\pi}{l} x) = 0 \Rightarrow a_n''(t) + (\dfrac{cn\pi}{l})^2 a_n(t) = 0$

$\Rightarrow a_n(t) = A_n \cos(\dfrac{cn\pi}{l} t) + B_n \sin(\dfrac{cn\pi}{l} t)$ substitute into ①,

$\therefore u(x, t) = \displaystyle\sum_{n=1}^{\infty} [A_n \cos(\dfrac{cn\pi}{l} t) + B_n \sin(\dfrac{cn\pi}{l} t)] \cdot \sin \dfrac{n\pi}{l} x \cdots$ ②,

$u_t(x, t) = \displaystyle\sum_{n=1}^{\infty} [-\dfrac{cn\pi}{l} A_n \sin(\dfrac{cn\pi}{l} t) + \dfrac{cn\pi}{l} B_n \cos(\dfrac{cn\pi}{l} t)] \cdot \sin \dfrac{n\pi}{l} x$.

② By $u(x, 0) = f(x) \Rightarrow f(x) = \displaystyle\sum_{n=1}^{\infty} A_n \cdot \sin(\dfrac{n\pi}{l} x)$,

where $A_n = \dfrac{2}{l} \displaystyle\int_0^l f(x) \sin(\dfrac{n\pi}{l} x) dx$,

by $u_t(x, 0) = g(x) \Rightarrow g(x) = \displaystyle\sum_{n=1}^{\infty} \dfrac{cn\pi}{l} B_n \sin \dfrac{n\pi}{l} x$,

$\dfrac{cn\pi}{l} B_n = \dfrac{2}{l} \displaystyle\int_0^l g(x) \cdot \sin \dfrac{n\pi}{l} x dx \Rightarrow B_n = \dfrac{2}{cn\pi} \int_0^l g(x) \cdot \sin(\dfrac{n\pi}{l} x) dx$,

let A_n, B_n substitute into ②, we obtain $u(x, t)$.

(2) ① By $u(0, t) = u(l, t) = 0$, we obtain the eigenfunction $\{\sin\frac{n\pi}{l}x\}_{n=1}^{\infty}$,

let $u(x, t) = \sum_{n=1}^{\infty} a_n(t) \cdot \sin\frac{n\pi}{l}x$, and $p(x, t) = \sum_{n=1}^{\infty} q_n(t)\sin\frac{n\pi}{l}x$, where

$q_n(t) = \frac{2}{l}\int_0^l p(x, t)\cdot\sin\frac{n\pi}{l}xdx$,

let $u(x, t)$ and $p(x, t)$ substitute into ordinary DE

$\Rightarrow \sum_{n=1}^{\infty} a_n''(t)\sin\frac{n\pi}{l}x = c^2 \cdot \sum_{n=1}^{\infty} a_n(t)\cdot[-(\frac{n\pi}{l})^2 \cdot\sin\frac{n\pi}{l}x] + \sum_{n=1}^{\infty} q_n(t)\sin\frac{n\pi}{l}x$

$\Rightarrow \sum_{n=1}^{\infty} [a_n''(t) + (\frac{cn\pi}{l})^2 a_n - q_n(t)]\sin\frac{n\pi}{l}x = 0$,

$\therefore a_n''(t) + (\frac{cn\pi}{l})^2 a_n(t) - q_n(t) = 0$

$\Rightarrow a_n''(t) + (\frac{cn\pi}{l})^2 a_n(t) = q_n(t)$, $a_{nh}(t) = A_n\cos\frac{n\pi}{l}t + B_n\sin\frac{cn\pi}{l}t$;

$a_{np}(t) = \frac{1}{D^2 + (\frac{cn\pi}{l})^2}\cdot q_n(t)$, let $\zeta(t) = a_{np}(t)$,

$\therefore a_n(t) = a_{nh}(t) + a_{np}(t) = A_n\cos\frac{cn\pi}{l}t + B_n\sin\frac{cn\pi}{l}t + \zeta(t)$

$\therefore u(x, t) = \sum_{n=1}^{\infty} [A_n\cos\frac{cn\pi}{l}t + B_n\sin\frac{cn\pi}{l}t + \zeta(t)]\cdot\sin(\frac{n\pi}{l}x)\cdots$ ③

we obtain

$u_t(x, t) = \sum_{n=1}^{\infty} [-\frac{cn\pi}{l}A_n\sin\frac{cn\pi}{l}t + \frac{cn\pi}{l}B_n\cos\frac{cn\pi}{l}t + \zeta'(t)]\cdot\sin\frac{n\pi}{l}x$.

② By $u(x, 0) = f(x) \Rightarrow f(x) = \sum_{n=1}^{\infty} [A_n + \zeta(0)]\sin\frac{n\pi}{l}x$,

$\therefore A_n + \zeta(0) = \frac{2}{l}\int_0^l f(x)\cdot\sin\frac{n\pi}{l}xdx$

$\Rightarrow A_n = -\zeta(0) + \frac{2}{l}\int_0^l f(x)\sin\frac{n\pi}{l}xdx$,

by $u_t(x, 0) = g(x) \Rightarrow g(x) = \sum_{n=1}^{\infty} [\frac{cn\pi}{l}B_n + \zeta'(0)]\sin\frac{n\pi}{l}x$,

$\therefore \frac{cn\pi}{l}B_n + \zeta'(0) = \frac{2}{l}\int_0^l g(x)\cdot\sin\frac{n\pi}{l}xdx$

$\Rightarrow B_n = \frac{l}{cn\pi}[-\zeta'(0) + \frac{2}{l}\int_0^l g(x)\sin\frac{n\pi}{l}xdx]$,

let A_n, B_n substitute into ③, we obtain $u(x, t)$.

Q.E.D.

Example 2

Utilize the eigenfunction expansion method to solve the following PDE.

DE: $\dfrac{\partial^2 u}{\partial t^2} = \dfrac{\partial^2 u}{\partial x^2} - 6x$, $t > 0, 0 < x < 1$.

BC: $u(0, t) = u(1, t) = 0$.
IC: $u(x, 0) = u_t(x, 0) = 0$.

Solution

(1) By $u(0, t) = u(1, t) = 0$, we get the eigenfunction $\{\sin n\pi x\}_{n=1}^{\infty}$, let

$$u(x, t) = \sum_{n=1}^{\infty} a_n(t) \cdot \sin n\pi x \,,$$

and $-6x = \sum_{n=1}^{\infty} q_n(t) \sin n\pi x \,,$

$$\therefore q_n(t) = \frac{2}{1} \int_0^1 (-6x) \cdot \sin n\pi x \, dx = 2 \cdot [\frac{6x}{n\pi} \cos n\pi x]\Big|_0^1 = (-1)^n \cdot \frac{12}{n\pi} \,.$$

Let the eigenfunction expansion of $u(x, t)$ and $-6x$ substitute into the ordinary DE

$$\Rightarrow \sum_{n=1}^{\infty} a_n''(t) \sin n\pi x = \sum_{n=1}^{\infty} a_n(t) [-(n\pi)^2] \cdot \sin n\pi x + \sum_{n=1}^{\infty} (-1)^n \frac{12}{n\pi} \sin n\pi x$$

$$\Rightarrow \sum_{n=1}^{\infty} [a_n''(t) + (n\pi)^2 a_n - (-1)^n \frac{12}{n\pi}] \sin n\pi x = 0 \,, \quad \therefore a_n'' + (n\pi)^2 a_n - (-1)^n \frac{12}{n\pi} = 0 \,,$$

$$a_n(t) = A_n \cos n\pi t + B_n \sin n\pi t + \frac{1}{D^2 + (n\pi)^2} \cdot (-1)^n \cdot \frac{12}{n\pi}$$

$$\Rightarrow a_n(t) = A_n \cos n\pi t + B_n \sin n\pi t + (-1)^n \cdot \frac{12}{(n\pi)^3} \,,$$

$$\therefore u(x, t) = \sum_{n=1}^{\infty} [A_n \cos n\pi t + B_n \sin n\pi t + (-1)^n \cdot \frac{12}{(n\pi)^3}] \cdot \sin n\pi x \cdots ①$$

$$u_t(x, t) = \sum_{n=1}^{\infty} [-n\pi A_n \sin n\pi t + n\pi B_n \cos n\pi t] \cdot \sin n\pi x \,.$$

(2) By $u(x, 0) = 0 \Rightarrow \sum_{n=1}^{\infty} [A_n + (-1)^n \cdot \frac{12}{(n\pi)^3}] \sin n\pi x = 0 \Rightarrow A_n + (-1)^n \cdot \frac{12}{n^3 \pi^3} = 0 \,,$

therefore, $A_n = -(-1)^n \cdot \frac{12}{n^3 \pi^3}$, by $u_t(x, 0) = 0 \Rightarrow [\sum_{n=1}^{\infty} n\pi B_n] \sin n\pi x = 0 \,, \quad \therefore B_n = 0,$

$$\therefore u(x, t) = \sum_{n=1}^{\infty} [-(-1)^n \cdot \frac{12}{n^3 \pi^3} \cos n\pi t + (-1)^n \cdot \frac{12}{n^3 \pi^3}] \cdot \sin n\pi x$$

$$= \sum_{n=1}^{\infty} (-1)^n \cdot \frac{12}{n^3 \pi^3} [1 - \cos n\pi t] \sin n\pi x \,. \qquad \text{Q.E.D.}$$

Example 3

Solve following PDE:

DE: $\dfrac{\partial u}{\partial t} = \dfrac{\partial^2 u}{\partial x^2} + \sin \pi x$,

BC: $u(0, t) = 0$, $u(1, t) = 0$,

IC: $u(x, 0) = \sin 2\pi x$, $0 \le x \le 1$, $t > 0$.

Solution

(1) By $u(0, t) = u(1, t)$ we have the eigenfunction $\{\sin n\pi x\}_{n=1}^{\infty}$, let

$$u(x, t) = \sum_{n=1}^{\infty} a_n(t) \cdot \sin n\pi x \text{ (called equation ①) };$$

$$\sin \pi x = \sum_{n=1}^{\infty} q_n(t) \cdot \sin n\pi x \text{ (called equation ②) , where}$$

$$q_n(t) = \begin{cases} 1, n=1 \\ 0, \text{ others} \end{cases} \Rightarrow \begin{cases} q_1 = 1 \\ q_n = 0, n \ne 1 \end{cases},$$

let equation ① and ② substitute into DE

$$\Rightarrow \sum_{n=1}^{\infty} a_n'(t) \sin n\pi x = \sum_{n=1}^{\infty} a_n(t) \cdot [-(n\pi)^2] \sin n\pi x + \sum_{n=1}^{\infty} q_n(t) \cdot \sin n\pi x$$

$$\Rightarrow \sum_{n=1}^{\infty} [a_n'(t) + (n\pi)^2 \cdot a_n(t) - q_n(t)] \sin n\pi x = 0 \Rightarrow \begin{cases} a_1'(t) + \pi^2 a_1(t) - 1 = 0 \\ a_n'(t) + (n\pi)^2 a_n(t) = 0 , n \ne 1 \end{cases},$$

let $a_1(t)$, $a_n(t)$ substitute into the orthogonal function expansion of $u(x, t)$

$$\Rightarrow \begin{cases} a_1(t) = c_1 e^{-\pi^2 t} + \dfrac{1}{\pi^2}, n = 1 \\ a_n(t) = c_n e^{-(n\pi)^2 t} , n \ne 1 \end{cases},$$

we obtain $u(x, t) = (c_1 e^{-\pi^2 t} + \dfrac{1}{\pi^2}) \sin \pi x + \sum_{n=2}^{\infty} c_n \cdot e^{-(n\pi)^2 t} \cdot \sin n\pi x$.

(2) By $u(x, 0) = \sin 2\pi x \Rightarrow 1 \cdot \sin 2\pi x = (c_1 + \dfrac{1}{\pi^2}) \sin \pi x + \sum_{n=2}^{\infty} c_n \cdot \sin n\pi x$

$$\Rightarrow \begin{cases} c_1 + \dfrac{1}{\pi^2} = 0 \to c_1 = -\dfrac{1}{\pi^2} \\ c_2 = 1 \\ c_n = 0 , n = 3, 4, 5, \cdots \end{cases},$$

$$\therefore u(x, t) = [-\dfrac{1}{\pi^2} e^{-\pi^2 t} + \dfrac{1}{\pi^2}] \sin \pi x + e^{-4\pi^2 t} \cdot \sin 2\pi x.$$

Q.E.D.

10-3-2 Assuming Transient and Steady-State Solutions to Solve PDE

As mentioned earlier, for non-homogeneous BC that are not time-dependent, we can assume that the general solution is a sum of transient state solution and steady state solution:

$$u(x,t) = \underbrace{\phi(x,t)}_{\text{transient state solution}} + \underbrace{v(x)}_{\text{steady state solution}}.$$

The goal is to transform the non-homogeneous BVP into a homogeneous one for ease of solving. The following example illustrates this process.

Example 4

Solve the following PDE using the separation of variables method.

DE: $\dfrac{\partial u}{\partial t} = c^2 \dfrac{\partial^2 u}{\partial x^2}$.

BC: $u(0, t) = u_1$, $u(l, t) = u_2$, u_1, u_2 are constants.
IC: $u(x, 0) = f(x)$.

Solution

(1) Let $u(x, t) = \phi(x, t) + v(x)$ substitute into DE, we obtain $\dfrac{\partial \phi}{\partial t} = c^2 (\dfrac{\partial^2 \phi}{\partial x^2} + v'')$,

let $v'' = 0$, also

$$u(0, t) = \phi(0, t) + v(0) = u_1 \Rightarrow \begin{cases} \phi(0, t) = 0 \\ v(0) = u_1 \end{cases};$$

$$u(l, t) = \phi(l, t) + v(l) = u_2 \Rightarrow \begin{cases} \phi(l, t) = 0 \\ v(l) = u_2 \end{cases};$$

Then, the original system can be transformed into a second-order

constant-coefficient linear ODE $\begin{cases} v'' = 0 \\ v(0) = u_1 \,,\, v(l) = u_2 \end{cases}$, and a second-order PDE with

homogeneous boundary conditions that can be solved using the separation of

variables method $\begin{cases} \dfrac{\partial \phi}{\partial t} = c^2 \dfrac{\partial^2 \phi}{\partial x^2} \\ \phi(0, t) = \phi(l, t) = 0 \end{cases}$.

(2) To solve the steady state solution: $v'' = 0$, $v(x) = c_1 + c_2 x$, by $v(0) = u_1 \Rightarrow c_1 = u_1$,

$$v(l) = c_1 + c_2 l = u_2 \Rightarrow c_2 = \frac{u_2 - u_1}{l} , \quad \therefore v(x) = u_1 + \frac{u_2 - u_1}{l} x .$$

(3) To solve the transient state solution $\phi(x, t)$: $\begin{cases} \dfrac{\partial \phi}{\partial t} = c^2 \dfrac{\partial^2 \phi}{\partial x^2} \\ \phi(0, t) = \phi(l, t) = 0 \end{cases}$,

by the separation of variables method , we can find:

$$\phi(x, t) = \sum_{n=1}^{\infty} [A_n \cdot e^{-(\frac{cn\pi}{l})^2 t}] \cdot \sin \frac{n\pi}{l} x \cdots ①$$

(4) The general solution $u(x, t) = \phi(x, t) + v(x)$:

by IC: $u(x, 0) = f(x) \Rightarrow f(x) = \phi(x, 0) + v(x)$

$$\Rightarrow \phi(x, 0) = f(x) - v(x) = \sum_{n=1}^{\infty} [A_n] \cdot \sin \frac{n\pi}{l} x ,$$

where $A_n = \dfrac{2}{l} \int_0^l [f(x) - v(x)] \sin(\dfrac{n\pi}{l} x) dx$.

Substituting A_n into ①, we have $\phi(x, t)$, then the solution for the non-homogeneous
PDE is $\Rightarrow u(x, t) = \phi(x, t) + v(x)$.   Q.E.D.

Example 5

DE: $\dfrac{\partial^2 u}{\partial x^2} = a^2 \dfrac{\partial u}{\partial t}$, a is a constant.

BC: $u(0, t) = 0$, $u(l, t) = 100$.

IC: $u(x, 0) = 100$, $0 \le x \le l$, $t > 0$.

Solution

(1) Let $u(x, t) = \phi(x, t) + v(x)$ substitute into DE $\Rightarrow \dfrac{\partial^2 \phi}{\partial x^2} + v''(x) = a^2 \dfrac{\partial \phi}{\partial t}$,

let $v'' = 0 \Rightarrow \dfrac{\partial^2 \phi}{\partial x^2} = a^2 \dfrac{\partial \phi}{\partial t}$,

by $u(0, t) = \phi(0, t) + v(0) = 0 \Rightarrow \begin{cases} \phi(0, t) = 0 \\ v(0) - 0 \end{cases}$;

by $u(l, t) = \phi(l, t) + v(l) = 100 \Rightarrow \begin{cases} \phi(l, t) = 0 \\ v(l) = 100 \end{cases}$,

also $u(x, 0) = \phi(x, 0) + v(x) = 100 \Rightarrow \phi(x, 0) = 100 - v(x)$,

$\therefore \begin{cases} v'' = 0 \\ v(0) = 0 , v(l) = 100 \end{cases} \Rightarrow v(x) = \dfrac{100}{l} x$.

$$(2) \begin{cases} \text{DE} : \dfrac{\partial^2 \phi}{\partial x^2} = a^2 \dfrac{\partial \phi}{\partial t} \\[2mm] \text{BC} : \phi(0,t) = \phi(l,t) = 0 \\[2mm] \text{IC} : \phi(x,0) = 100 - v(x) \end{cases}, \text{ by the separation of variables method, we can find}$$

$$\phi(x,t) = \sum_{n=1}^{\infty} [A_n \cdot e^{-(\frac{n\pi}{al})^2 t}] \cdot \sin \frac{n\pi}{l} x \Rightarrow \phi(x,0) = 100 - \frac{100}{l} x = \sum_{n=1}^{\infty} [A_n] \cdot \sin \frac{n\pi}{l} x,$$

$$A_n = \frac{2}{l} \int_0^l (100 - \frac{100}{l} x) \sin(\frac{n\pi}{l} x) dx = \frac{200}{n\pi},$$

$$\phi(x,t) = \sum_{n=1}^{\infty} [\frac{200}{n\pi} \cdot e^{-(\frac{n\pi}{al})^2 t}] \cdot \sin \frac{n\pi}{l} x.$$

$$\therefore u(x,t) = \phi(x,t) + v(x).$$

Q.E.D.

10-3-3 Non-Homogeneous BVP

Following we consider a BVP where both the DE and BC are non-homogeneous, that is DE: $\dfrac{\partial u}{\partial t} = \alpha^2 \dfrac{\partial^2 u}{\partial x^2} + p(x,t)$; BC: $u(0,t) = A(t)$, $u(l,t) = B(t)$; IC: $u(x,0) = f(x)$, $0 \le x \le l$, $t > 0$. First, we consider the interpolation functions for $A(t)$ and $B(t)$, $A(t)(1 - \dfrac{x}{l}) + B(t)\dfrac{x}{l}$; simultaneously, we contemplate the idea of modifying the BC in a manner analogous to the transient solution $\varphi(x,t)$, let the general solution

$$u(x,t) = A(t)(1 - \frac{x}{l}) + B(t)\frac{x}{l} + \varphi(x,t)$$

substitute into DE, we obtain $A'(t)(1 - \dfrac{x}{l}) + B'(t)\dfrac{x}{l} + \dfrac{\partial \varphi}{\partial t} = \alpha^2 \dfrac{\partial^2 \varphi}{\partial x^2} + p(x,t)$.

On the other hand, the original BC substitute into $u(x, t)$ results in a new BC: $\varphi(0,t) = 0$; $\varphi(l,t) = 0$, while IC then transforms into $\varphi(x,0) = g(x)$. Consequently, the original BVP becomes:

$$\begin{cases} \dfrac{\partial \varphi}{\partial t} = \alpha^2 \dfrac{\partial^2 \varphi}{\partial x^2} + p(x,t) - A'(t)(1 - \dfrac{x}{l}) - B'(t)\dfrac{x}{l} \\[2mm] \varphi(0,t) = 0, \varphi(l,t) = 0 \qquad\qquad\qquad \cdots(A) \\[2mm] \varphi(x,0) = g(x) = f(x) - A(0)(1 - \dfrac{x}{l}) - B(0)\dfrac{x}{l} \end{cases}$$

Next, we can proceed to solve the above PDE using the "eigenfunction expansion method." The following example will illustrate this approach.

Example 6

Solve PDE: $\dfrac{\partial u(x,t)}{\partial t} = \dfrac{\partial^2 u(x,t)}{\partial x^2} + W(x,t),\ 0 < x < 1,\ 0 < t$,

$u(0, t) = a(t),\ 0 < t;\ u(1, t) = b(t),\ 0 < t;\ u(x, 0) = f(x),\ 0 \le x \le 1.$

Solution   Substituting the corresponding information from the DE into the equation (A) above, we obtain:

$$\begin{cases} \dfrac{\partial \phi}{\partial t} = \dfrac{\partial^2 \phi}{\partial x^2} + W(x,t) - \{b'(t) - a'(t)\}x - a'(t) \cdots ① \\ \text{BC}\ :\ \phi(0, t) = 0,\ \ \phi(1, t) = 0 \\ \text{IC}\ :\ \phi(x, 0) = f(x) - \{b(0) - a(0)\}x - a(0) \end{cases},$$

because $\phi(0, t) = \phi(1, t) = 0$, the corresponding eigenfunction is $\{\sin(n\pi x)\}_{n=1}^{\infty}$, by

eigenfunction expansion method, we know $\phi(x, t) = \displaystyle\sum_{n=1}^{\infty} a_n(t) \sin n\pi x$ (called equation ②)

and $W(x, t) - \{b'(t) - a'(t)\}x - a'(t) = \displaystyle\sum_{n=1}^{\infty} b_n \sin n\pi x$ (called equation ③) where

$b_n(t) = 2\displaystyle\int_0^1 \{W(x, t) - [b'(t) - a'(t)]x - a'(t)\} \sin n\pi x\, dx$, let equation ② and ③

substitute into equation ①, we have

$$\sum_{n=1}^{\infty} a'_n(t) \sin n\pi x = \sum_{n=1}^{\infty} a_n(t)(-n^2\pi^2) \sin n\pi x + \sum_{n=1}^{\infty} b_n \sin n\pi x,$$

that is $\displaystyle\sum_{n=1}^{\infty} \{a'_n(t) + n^2\pi^2 a_n(t) - b_n(t)\} \sin n\pi x = 0$, so $a'_n(t) + n^2\pi^2 a_n(t) = b_n(t)$,

we find that $a_n(t) = [A_n + \displaystyle\int_0^t e^{n^2\pi^2\tau} b_n(\tau)d\tau]e^{-n^2\pi^2 t}$,

therefore, $\phi(x, t) = \displaystyle\sum_{n=1}^{\infty} \{[A_n + \int_0^t e^{n^2\pi^2\tau} b_n(\tau)d\tau]e^{-n^2\pi^2 t}\} \sin n\pi x$,

then by $\phi(x, 0) = f(x) - \{b(0) - a(0)\}x - a(0) = \displaystyle\sum_{n=1}^{\infty} A_n \sin n\pi x$,

we obtain $A_n = 2\displaystyle\int_0^1 \{f(x) - \{b(0) - a(0)\}x - a(0)\} \sin n\pi x\, dx$,

the solution is $u(x, t) = \phi(x, t) + \{b(t) - a(t)\}x + a(t)$.   Q.E.D.

10-3　Exercises

Basic questions

1. Use the eigenfunction expansion method to solve following PDE.

$$\frac{\partial^2 u}{\partial t^2} = c^2 \frac{\partial^2 u}{\partial x^2}, t > 0, 0 < x < l,$$

$$u(0, t) = u_x(l, t) = u_t(x, 0) = 0,$$

$$u(x, 0) = \frac{x}{l}.$$

2. Use the eigenfunction expansion to find the solution.

DE: $3\frac{\partial^2 u}{\partial x^2} = \frac{\partial u}{\partial t}$; $0 \leq x \leq 2, t > 0.$

IC:　$u(x, 0) = 2[1 - \cos(\pi x)].$

BC: $u(0, t) = 0, u(2, t) = 0.$

3. Use the eigenfunction expansion to find the solution.

DE: $\frac{\partial u}{\partial t} = 4\frac{\partial^2 u}{\partial x^2} + 1.$

BC: $u(0, t) = 0, u(\pi, t) = 0.$
IC: $u(x, 0) = 0, 0 \leq x \leq \pi, t > 0.$

Advanced questions

1. Use the eigenfunction expansion method to solve following PDE.

$$4\frac{\partial^2 u}{\partial t^2} = \frac{\partial^2 u}{\partial x^2}, t > 0, 0 < x < \pi,$$

$$u(0, t) = u(\pi, t) = u(x, 0) = 0,$$

$$u_t(x, 0) = \sin 2x - \sin 3x.$$

2. Use the eigenfunction expansion to find the solution.

DE: $\frac{\partial^2 u}{\partial x^2} = \frac{\partial u}{\partial t}$; $0 \leq x \leq 2, t > 0.$

IC:　$u(x, 0) = 8\cos\frac{3\pi x}{4} - 6\cos\frac{9\pi x}{4}.$

BC: $u_x(0, t) = 0, u(2, t) = 0.$

3. Use the eigenfunction expansion to find the solution.

DE: $c^2 \frac{\partial^2 u}{\partial x^2} = \frac{\partial u}{\partial t}$; $a =$ constant.

IC: $u(x, 0) = 100, 0 \leq x \leq l, t > 0.$
BC: $u(0, t) = 100, u(l, t) = 0.$

4. Use the eigenfunction expansion to find the solution.

DE: $\frac{\partial u}{\partial t} = \frac{\partial^2 u}{\partial x^2} + \sin x.$

BC: $u(0, t) = 0, u(\pi, t) = \pi.$
IC: $u(x, 0) = \sin x, 0 \leq x \leq \pi, t > 0.$

10-4 Solving PDE Using Integral Transformations

In ordinary differential equations, we utilize Laplace transforms to convert ODEs into algebraic equations for solving. Similarly, in PDE, apart from using Laplace transforms, we can also use Fourier integral transforms for solving. Both of these transforms involve integrating with respect to one of the variables, effectively eliminating one variable. This not only transforms the problem into a familiar ODE form but also enables us to handle algebraic equations through inverse transforms. The combination of these approaches provides the general solution to the original equation. This underscores the power of integral transforms once again.

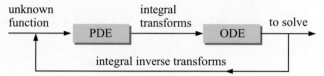

Figure 10-5   The flowchart of using integral transformations to solve PDE

10-4-1   Solve PDE Using Laplace Transform

Let $\hat{u}(x, s) = \mathscr{L}\{u(x, t)\} = \int_0^\infty u(x, t)\, e^{-st} dt$, after the transformation, $\hat{u}(x, s)$ only has the unknown function of a single variable x remains, then

$$\mathscr{L}\{\frac{\partial u}{\partial t}\} = s\hat{u}(x, s) - u(x, 0)$$

$$\mathscr{L}\{\frac{\partial^2 u}{\partial t^2}\} = s^2 \hat{u}(x, s) - su(x, 0) - u_t(x, 0)$$

$$\mathscr{L}\{\frac{\partial u}{\partial x}\} = \int_0^\infty \frac{\partial u}{\partial x} e^{-st} dt = \frac{\partial}{\partial x} \int_0^\infty u \cdot e^{-st} dt = \frac{\partial \hat{u}}{\partial x} = \frac{d\hat{u}}{dx}$$

$$\mathscr{L}\{\frac{\partial^2 u}{\partial x^x}\} = \frac{\partial^2 \hat{u}}{\partial x^2} = \frac{d^2 \hat{u}}{dx^2}$$

Through these transformations, we can convert a PDE with two variables, x and t, into an ODE with only a single variable x. The following examples will illustrate how to use the Laplace transform to solve PDE.

Example 1

Use Laplace Transform to solve following PDE.

DE: $\dfrac{\partial^2 u(x, t)}{\partial x^2} = \dfrac{\partial u(x, t)}{\partial t}$.

BC: $u(0, t) = 1, u(1, t) = 1$.

IC: $u(x, 0) = 1 + \sin \pi x, 0 < x < 1, t > 0$.

Solution

(1) Let $\mathcal{L}\{u(x,t)\} = \hat{u}(x,s)$, taking the Laplace transform of the DE and BC results in

an ODE $\dfrac{d^2\hat{u}}{dx^2} - s\hat{u}(x,s) = -(1+\sin\pi x)$, $\hat{u}(0,s) = \dfrac{1}{s}$, $\hat{u}(1,s) = \dfrac{1}{s}$

(2) From the method of undetermined coefficients for the second-order ODE, we obtain

the general solution $\hat{u}(x,s) = c_1 e^{\sqrt{s}x} + c_2 e^{-\sqrt{s}x} + \dfrac{1}{s} + \dfrac{1}{s+\pi^2}\sin\pi x$

then by the boundary conditions: $\hat{u}(0,s) = \dfrac{1}{s} \Rightarrow c_1 + c_2 + \dfrac{1}{s} = \dfrac{1}{s}$;

$\hat{u}(1,s) = \dfrac{1}{s} \Rightarrow c_1 e^{\sqrt{s}} + c_2 e^{-\sqrt{s}} + \dfrac{1}{s} = \dfrac{1}{s}$, we obtain $\begin{cases} c_1 = 0 \\ c_2 = 0 \end{cases}$, thus

$\hat{u}(x,s) = \dfrac{1}{s} + \dfrac{1}{s+\pi^2}\sin\pi x$

(3) The final step involves using the inverse Laplace transform to transform the solution back to the original function:

$u(x,t) = \mathcal{L}^{-1}\{\hat{u}(x,s)\} = \mathcal{L}^{-1}\{\dfrac{1}{s} + \dfrac{1}{s+\pi^2}\sin\pi x\} = 1 + e^{-\pi^2 t}\sin\pi x$.   Q.E.D.

Example 2

Use Laplace Transform to solve following PDE.

DE: $\dfrac{\partial G(x,t)}{\partial t} = \dfrac{\partial^2 G(x,t)}{\partial x^2}$, $x > 0, t > 0$.

BC: $G(0,t) = \begin{cases} 1, 0 < t \le 1 \\ 0, t > 1 \end{cases}$, $G(\infty, t)$ is bounded.

IC: $G(x, 0) = 0$.

【Hint】 $\mathcal{L}\{erf_c(\dfrac{a}{2\sqrt{t}})\} = \dfrac{1}{s}e^{-\sqrt{s}a}$.

Solution

(1) Let $\mathcal{L}\{G(x,t)\} = \hat{G}(x,s)$, taking the Laplace transform of the DE and BC then obtain

$$\dfrac{d^2\hat{G}}{dx^2} - s\hat{G}(x,s) = 0; \quad \hat{G}(0,s) = \dfrac{1}{s} - \dfrac{1}{s}e^{-s}.$$

(2) First, from the homogeneous solution of the second-order ODE, we can assume
$\hat{G}(x,s) = c_1 e^{\sqrt{s}x} + c_2 e^{-\sqrt{s}x}$, also $G(\infty, t)$ is bounded, that is $\hat{G}(\infty, s)$ is also bounded,
$\therefore c_1 = 0 \Rightarrow \hat{G}(x,s) = c_2 e^{-\sqrt{s}x}$. And by boundary conditions

$$\hat{G}(0, s) = \frac{1}{s} - \frac{1}{s}e^{-s} \Rightarrow c_2 = \frac{1}{s} - \frac{1}{s}e^{-s},$$

$$\therefore \hat{G}(x, s) = (\frac{1}{s} - \frac{1}{s}e^{-s})e^{-\sqrt{s}x} = \frac{1}{s}e^{-\sqrt{s}x} - \frac{1}{s}e^{-s}e^{-\sqrt{s}x}.$$

(3) Now because $\mathscr{L}^{-1}\{\frac{1}{s}e^{-\sqrt{s}a}\} = erf_c(\frac{a}{2\sqrt{t}})$,

$$\therefore G(x, t) = \mathscr{L}^{-1}\{\hat{G}(x, s)\} = \mathscr{L}^{-1}\{\frac{1}{s}e^{-\sqrt{s}x} - \frac{1}{s} \cdot e^{-\sqrt{s}x} \cdot e^{-s}\}$$

$$= erf_c(\frac{x}{2\sqrt{t}}) - erf_c(\frac{x}{2\sqrt{t-1}}) \cdot H(t-1),$$

where $H(t-1) = \begin{cases} 1, t \geq 1 \\ 0, \text{others} \end{cases}$, is the unit step function.

Q.E.D.

Example 3   First-Order PDE

Use Laplace transform to solve following PDE:

DE: $\dfrac{\partial u}{\partial x} + x\dfrac{\partial u}{\partial t} = 0$,

IC: $u(x, 0) = 0$,

BC: $u(0, t) = 4t$.

Solution

(1) Let $\mathscr{L}\{u(x, t)\} = \hat{u}(x, s)$, taking the Laplace transform of the DE and BC,

we obtain $\dfrac{d\hat{u}}{dx} + sx\hat{u} = 0$; $\hat{u}(0, s) = \dfrac{4}{s^2}$,

by the integrating factor $I = e^{\int sx dx} = e^{\frac{1}{2}sx^2}$ we obtain $\dfrac{d}{dx}(\hat{u}e^{\frac{1}{2}sx^2}) = 0$, which implies

the existence of a constant c_1 causes $\hat{u}(x, s) = c_1 e^{-\frac{1}{2}sx^2}$.

Then by boundary conditions, we obtain $c_1 = \dfrac{4}{s^2}$, thus $\hat{u}(x, s) = \dfrac{4}{s^2}e^{-\frac{1}{2}x^2 s}$.

(2) Next, we perform the inverse Laplace transform to obtain the original general solution:

$$u(x, t) = \mathscr{L}^{-1}\{\hat{u}(x, s)\} = \mathscr{L}^{-1}\{\frac{4}{s^2}e^{-\frac{1}{2}x^2 s}\} = 4(t - \frac{1}{2}x^2)H(t - \frac{1}{2}x^2),$$

where $H(t - \dfrac{1}{2}x^2) = \begin{cases} 1, t \geq \dfrac{1}{2}x^2 \\ 0, t < \dfrac{1}{2}x^2 \end{cases}$.

Q.E.D.

10-4-2 Solve PDE Using Fourier Transform

In the earlier discussion, when we used the Laplace transform to solve PDE, we transformed the variable t into s. This transformed the original bivariate function $u(x, t)$ into a univariate function $\hat{u}(x, s)$, simplifying the problem to a single variable. Now, when using the Fourier transform to solve PDEs, we transform the bivariate function $u(x, t)$ into a univariate function $u^*(w, t)$, resulting in a unknown function $u^*(w, t)$ with single variable t. This allows us, similar to the Laplace transform, to simplify the PDE into an ODE for solving. In the following discussion, we will address two types: the semi-infinite domain and the all-infinite domain cases.

1. **Semi-Infinite Domain: $0 \leq x \leq \infty$**

    Divided into the following two types of BC

    (1) $u(0, t) = A(t)$, and $u(\infty, t)$ is bounded: at this point, we perform a Fourier sine transform on the general solution.

    $$\mathscr{F}_s\{u(x, t)\} = \int_0^\infty u(x, t)\sin wx\, dx = u^*(w, t) ;$$

    $$\mathscr{F}_s^{-1}\{u^*(w, t)\} = \frac{2}{\pi}\int_0^\infty u^*(w, t)\cdot \sin wx\, dw = u(x, t).$$

    For a second-order equation, by the common formula $\mathscr{F}_s\{f''(x)\} = -w^2 F(w) + wf(0)$, we obtain

    $$\mathscr{F}_s\{\frac{\partial^2 u}{\partial x^2}\} = -w^2 u^*(w, t) + wu(0, t),$$

    therefore we have $\begin{cases} \mathscr{F}_s\{\dfrac{\partial u}{\partial t}\} = \dfrac{du^*(w, t)}{dt} \\ \mathscr{F}_s\{\dfrac{\partial^2 u}{\partial t^2}\} = \dfrac{d^2 u^*(w, t)}{dt^2} \end{cases}$, then substituting it back into the

    original equation to solve using ODE techniques.

    (2) $\dfrac{\partial u}{\partial x}(0, t) = B(t)$ and $u(\infty, t)$ is bounded: switching to Fourier cosine transform

    $$\mathscr{F}_c\{u(x, t)\} = \int_0^\infty u(x, t)\cos wx\, dx = u^*(w, t) ;$$

    $$\mathscr{F}_c^{-1}\{u^*(w, t)\} = \frac{2}{\pi}\int_0^\infty u^*(w, t)\cos wx\, dw = u(x, t).$$

    Hence, we have $\begin{cases} u(x,t) \text{ Laplace transform to } \hat{u}(x,s) \Rightarrow t \to s \\ u(x,t) \text{ Fourier transform to } u^*(w,t) \Rightarrow x \to w \end{cases}$.

    Below is an example to illustrate this.

Example 4

Use Fourier transform to solve following PDE:

DE: $\dfrac{\partial u}{\partial t} = c^2 \dfrac{\partial^2 u}{\partial x^2}$, $0 \le x < \infty$, $t > 0$,

BC: $u(0, t) = 0$, $u(\infty, t)$ is bounded,

IC: $u(x, 0) = f(x) = \begin{cases} \pi - x, 0 \le x \le \pi \\ 0, x > \pi \end{cases}$.

Solution

(1) Let $\mathscr{F}_s \{u(x, t)\} = \displaystyle\int_0^\infty u(x, t) \cdot \sin wx\, dx = u^*(w, t)$, taking the Fourier transform of the DE and IC

$\Rightarrow \dfrac{du^*}{dt} = c^2 \cdot [-w^2 \cdot u^* + wu(0, t)]$, $\therefore \dfrac{du^*}{dt} + c^2 w^2 u^* = 0 \cdots ①$

also $\mathscr{F}_s \{u(x, 0)\} = u^*(w, 0) = \displaystyle\int_0^\infty u(x, 0) \sin wx\, dx$

$\Rightarrow u^*(w, 0) = \displaystyle\int_0^\pi (\pi - x) \sin wx\, dx = [-\dfrac{\pi - x}{w} \cos wx - \dfrac{1}{w^2} \sin wx]\Big|_0^\pi = \dfrac{\pi}{w} - \dfrac{1}{w^2} \sin w\pi$.

(2) By $① \Rightarrow u^*(w, t) = ke^{-c^2 w^2 t}$, also $u^*(w, 0) = \dfrac{\pi}{w} - \dfrac{1}{w^2} \sin w\pi$,

$\therefore k = \dfrac{\pi}{w} - \dfrac{1}{w^2} \sin w\pi$,

$\therefore u^*(w, t) = (\dfrac{\pi}{w} - \dfrac{1}{w^2} \sin w\pi) \cdot e^{-c^2 w^2 t}$.

(3) $\mathscr{F}_s^{-1} \{u^*(w, t)\} = \dfrac{2}{\pi} \displaystyle\int_0^\infty u^*(w, t) \sin wx\, dw = u(x, t)$,

$\therefore u(x, t) = \dfrac{2}{\pi} \displaystyle\int_0^\infty (\dfrac{\pi}{w} - \dfrac{1}{w^2} \sin w\pi) e^{-c^2 w^2 t} \sin wx\, dw$. 　Q.E.D.

2. **All-Infinite Domain: $-\infty < x < \infty$**

Referring to practical applied problems, assuming the general solution is bounded, then $u(\pm\infty, t) \to 0$, taking the Fourier exponential transform of the general solution:

$\begin{cases} \mathscr{F}\{u(x, t)\} = \displaystyle\int_{-\infty}^\infty u(x, t) e^{-iwx}\, dx = u^*(w, t) \\ \mathscr{F}^{-1}\{u^*(w, t)\} = \dfrac{1}{2\pi} \displaystyle\int_{-\infty}^\infty u^*(w, t) e^{iwx}\, dw = u(x, t) \end{cases}$,

the common formula: $\mathscr{F}\{\dfrac{\partial^2 u}{\partial x^2}\} = -w^2 u^*(w, t)$.

Example 5

Use Fourier transform to solve following PDE:

DE: $k \dfrac{\partial^2 u}{\partial x^2} = \dfrac{\partial u}{\partial t}$, $-\infty < x < \infty$, $t > 0$,

BC: $u(\pm\infty, t)$ is bounded,

IC: $u(x, 0) = f(x) = e^{-x^2}$.

【Hint】 $\mathcal{F}\{e^{-\frac{x^2}{4\bar{p}^2}}\} = 2\sqrt{\pi}\, p e^{-p^2 w^2}$.

Solution

(1) Let $\mathcal{F}\{u(x, t)\} = \displaystyle\int_{-\infty}^{\infty} u(x, t) \cdot e^{-iwx} dx = u^*(w, t)$,

taking the Fourier transform of the DE and BC results in a second-order ODE

$$
\begin{cases}
\dfrac{du^*}{dt} + kw^2 u^* = 0 \\[2mm]
u^*(w, 0) = f^*(w) = \sqrt{\pi}\, e^{-\frac{w^2}{4}}
\end{cases}.
$$

(2) From the homogeneous solution of the second-order ODE, we obtain

$u^*(w, t) = c_1 e^{-kw^2 t}$, and by substituting this into the boundary conditions, we find

$c_1 = f^*(w)$, the general solution is $u^*(w, t) = f^*(w) e^{-kw^2 t}$.

(3) Finally, performing the inverse Fourier transform to obtain the general solution function

$$
u(x, t) = \mathcal{F}^{-1}\{u^*(w, t)\} = \frac{1}{2\pi} \int_{-\infty}^{\infty} f^*(w) e^{-kw^2 t} e^{iwx} dw
$$

$$
= \frac{1}{2\pi} \int_{-\infty}^{\infty} \sqrt{\pi}\, e^{-\frac{w^2}{4}} e^{-kw^2 t} \cdot e^{iwx} dw.
$$

Q.E.D.

10-4   Exercises

Basic questions

1. Use Laplace Transform to solve following PDE:

$$\frac{\partial u}{\partial x} + 2x\frac{\partial u}{\partial t} = 2x \; ;$$

$u(x,0) = 1, u(0,t) = 1.$

2. Use Laplace Transform to solve following PDE:

$$c^2\frac{\partial^2 u(x,t)}{\partial x^2} = \frac{\partial u(x,t)}{\partial t}, 0 < x < \infty, t > 0,$$

$u(x,0) = 0, \; \dfrac{\partial u(x,0)}{\partial t} = 0,$

$u(0,t) = f(t), u(\infty, t) = $ is bounded.

Advanced questions

1. Use Laplace Transform to solve following PDE:

$$\frac{\partial u(x,t)}{\partial x} = \frac{\partial^3 u(x,t)}{\partial t^3}, 0 < x < \infty, t > 0,$$

$u(x,0) = 0, \dfrac{\partial u(x,0)}{\partial t} = 0, \; \dfrac{\partial^2 u(x,0)}{\partial t^2} = e^{8x},$

$u(\infty, t) = $ is bounded.

2. Use Laplace Transform to solve following PDE:

$$a^2\frac{\partial^2 u(x,t)}{\partial x^2} - g = \frac{\partial^2 u(x,t)}{\partial t^2}, 0 < x < \infty, t > 0,$$

$u(x,0) = \dfrac{\partial u(x,0)}{\partial t} = 0,$

$u(0,t) = 0, \; \lim\limits_{x\to\infty}\dfrac{\partial u(x,t)}{\partial x} = 0.$

3. Use Fourier transform to solve following PDE:

$$\frac{\partial u(x,t)}{\partial t} = \frac{\partial^2 u(x,t)}{\partial x^2}, 0 < x < \infty, t > 0,$$

$u(0,t) = g(t),$

$u(x,0) = 0, u(x,t)$ is bounded.

4. Use Fourier transform to solve following PDE:

$$\frac{\partial^2 u(x,y)}{\partial x^2} + \frac{\partial^2 u(x,y)}{\partial y^2} = 0,$$

$0 < x < \infty, 0 < y < \infty, u(0,y) = 0,$

$u(x,0) = F(x) = \begin{cases} x, 0 \le x \le 2 \\ 0, x > 0 \end{cases},$

$u(x,y)$ is bounded.

5. Use Fourier transform to solve following PDE:

$$\frac{\partial u(x,t)}{\partial t} - \frac{\partial^2 u(x,t)}{\partial x^2} + tu(x,t) = 0, 0 < x < \infty, t > 0,$$

$\dfrac{\partial u}{\partial x}(0,t) = 0, u(x,0) = xe^{-x},$

$\dfrac{\partial u}{\partial t}(0,t) = 0.$

6. Use Fourier transform to solve following PDE:

$$\frac{\partial^2 u(x,t)}{\partial t^2} = 4\frac{\partial^2 u(x,t)}{\partial x^2}, -\infty < x < \infty, t > 0,$$

$u(x,0) = 0,$

$\dfrac{\partial u}{\partial t}(x,0) = \delta(x)$ is an impulse function.

10-5 Partial Differential Equations in Non-Cartesian Coordinate Systems (Selected Reading)

We have primarily focused on solving partial differential equation problems in Cartesian coordinate systems in the preceding chapters. However, in real physical systems, there are numerous problems formulated in non-Cartesian coordinate systems. One of the most typical and frequently encountered examples is the polar coordinate system and cylindrical coordinate system. For instance, wave propagation or heat conduction problems formulated in polar or cylindrical coordinates are common in physics. As indicated in Section 10-1, the general form of the PDE for two-dimensional wave equation in polar coordinates is

$$\nabla^2 u = \frac{\partial^2 u}{\partial r^2} + \frac{1}{r}\frac{\partial u}{\partial r} + \frac{1}{r^2}\frac{\partial^2 u}{\partial \theta^2} = \frac{1}{c^2}\frac{\partial^2 u}{\partial t^2},$$

if it is axially symmetric, then it is independent of θ, and its PDE can be expressed as follows:

DE: $\dfrac{\partial^2 \varphi}{\partial r^2} + \dfrac{1}{r}\dfrac{\partial \varphi}{\partial r} = \dfrac{1}{c^2}\dfrac{\partial^2 \varphi}{\partial t^2}$ $(0 < r < a, t > 0)$.

Assuming the following boundary conditions:
$\varphi(a, t) = 0$, $\varphi(0, t)$ is bounded, and initial

conditions: $\varphi(r, 0) = f(r)$, $\dfrac{\partial \varphi}{\partial t}(r, 0) = g(r)$.

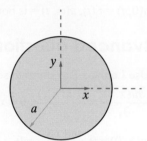

Figure 10-6 A schematic diagram of the vibration problem for a circular membrane

we can solve the PDE using the method of separation of variables as follows: let $\varphi(r, t) = F(r)H(t)$ and substitute into DE

$$F''H + \frac{1}{r}F'H = \frac{1}{c^2}FH''$$

dividing both sides of the above equation by FH yields $\dfrac{F'' + \dfrac{1}{r}F'}{F} = \dfrac{\dfrac{1}{c^2}H''}{H} = -\lambda$.

After simplification, the ordinary differential equation with respect to position and time variables is as follows

$$\begin{cases} F'' + \dfrac{1}{r}F' + \lambda F = 0 \cdots\cdots(*) \\ H'' + \lambda c^2 H = 0 \end{cases}$$

Take $(*) \times r^2$, and we get

$$\begin{cases} r^2 F'' + rF' + \lambda r^2 F = 0 \\ H'' + \lambda c^2 H = 0 \end{cases}$$

At this point, the differential equation with respect to the radial variable r, based on the boundary conditions, leads to the following boundary value problem

$$\begin{cases} r^2 F'' + rF' + \lambda r^2 F = 0 \\ F(a) = 0,\ F(0) \text{ is bounded} \end{cases} \quad \cdots\cdots(**)$$

This forms a boundary value problem, sharing many concepts with Section 9-1 of Chapter 9. However, the solution to this differential equation involves the use of Bessel functions. Therefore, let's first introduce how to solve this differential equation using Bessel functions.

10-5-1 Bessel Equations: Solution and Properties

We express the equation (**) as the common Bessel equation in the following form

$$x^2 y'' + xy' + (x^2 - \mu^2) y = 0 \cdots (***)$$

From the power series solution in Chapter 4, it is known that $x = 0$ is a regular singular point Therefore, the Frobenius series solution of (***) can be expressed as

$$\begin{cases} y(x) = \sum_{n=0}^{\infty} a_n x^{n+r} \\ y'(x) = \sum_{n=0}^{\infty} (n+r) a_n x^{n+r-1} \\ y''(x) = \sum_{n=0}^{\infty} (n+r)(n+r-1) a_n x^{n+r-2} \end{cases}$$

substituting the above expression into (***) yields

$$\Rightarrow x^2 \sum_{n=0}^{\infty} (n+r)(n+r-1) a_n x^{n+r-2} + x \sum_{n=0}^{\infty} (n+r) a_n x^{n+r-1} + (x^2 - \mu^2) \sum_{n=0}^{\infty} a_n x^{n+r} = 0$$

$$\Rightarrow \sum_{n=0}^{\infty} (n+r)(n+r-1) a_n x^{n+r} + \sum_{n=0}^{\infty} (n+r) a_n x^{n+r} - \sum_{n=0}^{\infty} \mu^2 a_n x^{n+r} + \sum_{n=0}^{\infty} a_n x^{n+r+2} = 0$$

$$\Rightarrow \sum_{n=0}^{\infty} [(n+r)^2 - \mu^2] a_n x^{n+r} + \sum_{n=0}^{\infty} a_n x^{n+r+2} = 0$$

$$\Rightarrow \sum_{n=0}^{\infty} [(n+r+\mu)(n+r-\mu)] a_n x^{n+r} + \sum_{n=2}^{\infty} a_{n-2} x^{n+r} = 0$$

$$\Rightarrow (r+\mu)(r-\mu) a_0 x^r + (1+r+\mu)(1+r-\mu) a_1 x^{1+r}$$

$$+ \sum_{n=2}^{\infty} [a_n \cdot (n+r+\mu)(n+r-\mu) + a_{n-2}] x^{n+r} = 0$$

By comparing coefficients, we obtain

$$\Rightarrow \begin{cases} (r+u)(r-u)a_0 = 0 \\ (1+r+u)(1+r-u)a_1 = 0 \\ (n+r+u)(n+r-u)a_n + a_{n-2} = 0 \end{cases}$$

$$\Rightarrow \begin{cases} a_0 \neq 0 \Rightarrow r = \mu(\text{big indicial root}), -\mu(\text{small indicial root}) \\ a_1 = 0 \\ a_n = -\dfrac{a_{n-2}}{(n+r+\mu)(n+r-\mu)}; n = 2, 3, 4, \ldots \end{cases}$$

From the power series solution in Chapter 4, it is known that $r = \mu$ is the larger root, and the corresponding solution must exist, also $a_1 = 0$, $a_0 \neq 0$,

$a_n = -\dfrac{1}{(n+r+\mu)(n+r-\mu)} a_{n-2}$; $n = 2, 3, 4 \cdots$, let $r = \mu$ substitute into above equation

and we get

$$\Rightarrow a_n = -\frac{1}{n \cdot (n+2\mu)} a_{n-2} ; n = 2, 3, 4 \cdots$$

$$n = 2 \rightarrow a_2 = -\frac{1}{(2+2\mu) \cdot 2} a_0$$

$$n = 3 \rightarrow a_3 = -\frac{1}{(3+2\mu) \cdot 3} a_1 = 0 \quad \therefore a_5 = 0$$

$$\Rightarrow a_{2n+1} = 0 , n = 0, 1, 2, 3, \cdots$$

$$n = 4 \rightarrow a_4 = -\frac{1}{(4+2\mu) \cdot 4} a_2 = (-1)^2 \cdot \frac{1}{(4+2\mu)(2+2\mu) \cdot 2 \cdot 4} a_0$$

$$\therefore a_{2n} = (-1)^n \cdot \frac{1}{(2n+2\mu)\cdots(4+2\mu)(2+2\mu) \cdot 2 \cdot 4 \cdots (2n)} a_0$$

$$= (-1)^n \cdot \frac{1}{2^n \cdot (n+\mu)\cdots(2+\mu)(1+\mu) \cdot 2^n \cdot n!} \cdot a_0$$

$$a_{2n+1} = 0$$

$$y_1(x) = y(x)\Big|_{r=\mu} = \sum_{n=0}^{\infty} a_n x^{n+r}\Big|_{r=\mu} = x^u \cdot \sum_{n=0}^{\infty} a_n x^n\Big|_{r=\mu} = x^\mu [\sum_{n=0}^{\infty} a_{2n} x^{2n} + \sum_{n=0}^{\infty} a_{2n+1} x^{2n+1}]$$

$$= x^\mu \sum_{n=0}^{\infty} (-1)^n \cdot \frac{1}{2^{2n} \cdot (n+\mu)\cdots(2+\mu)(1+\mu) \cdot n!} a_0 x^{2n} \text{ must exist by Chapter 4}$$

$$= a_0 \cdot \sum_{n=0}^{\infty} \frac{(-1)^n \cdot 1}{2^{2n} \cdot (n+\mu)\cdots(2+\mu)(1+\mu) \cdot n!} x^{2n+u}$$

Multiply both the numerator and denominator by $\Gamma(\mu+1)\cdot 2^{\mu}$

$$\Rightarrow y_1(x) = a_0 \sum_{n=0}^{\infty} (-1)^n \cdot \frac{1\cdot\Gamma(\mu+1)\cdot 2^{\mu}}{2^{2n}(n+\mu)\cdots(2+\mu)(1+\mu)\cdot n!\,\Gamma(\mu+1)\cdot 2^{\mu}} x^{2n+\mu}$$

$$= a_0\Gamma(1+\mu)2^{\mu}\cdot \sum_{n=0}^{\infty} (-1)^n \frac{1}{\Gamma(n+\mu+1)\cdot n!}(\frac{x}{2})^{2n+\mu}$$

$$= A_0 \sum_{n=0}^{\infty} (-1)^n \cdot \frac{1}{\Gamma(n+\mu+1)n!}(\frac{x}{2})^{2n+\mu}$$

With the relevant power series solutions in hand, for the sake of convenience in subsequent representations, we now introduce the definitions of Bessel functions.

▶ Definition 10-5-1

Bessel Functions of the First Type

(1) The first type Bessel function of order μ:

$$J_{\mu}(x) = \sum_{n=0}^{\infty} \frac{(-1)^n}{\Gamma(n+1+\mu)n!}(\frac{x}{2})^{2n+\mu}\,.$$

(2) The first type Bessel function of order $-\mu$:

$$J_{-\mu}(x) = \sum_{n=0}^{\infty} \frac{(-1)^n}{\Gamma(n+1-\mu)n!}(\frac{x}{2})^{2n-\mu}\,.$$

◀

▶ Theorem 10-5-1

The Properties of the First Type Bessel Function

The Bessel equation of order μ: $x^2 y'' + xy' + (x^2 - \mu^2)y = 0$.

At $\mu \neq$ integer $\Rightarrow J_{\mu}(x)$ and $J_{-\mu}(x)$ are linearly independent (L.I.),

then the solution of the original ODE: $y(x) = c_1 J_{\mu}(x) + c_2 J_{-\mu}(x)$.

◀

▶ Theorem 10-5-2

The Properties of the First Type Bessel Function

If $\mu = m$ is integer, then $J_m(x)$ and $J_{-m}(x)$ are linearly dependent (L.D.),

and $J_{-m}(x) = (-1)^m J_{-m}(x)$.

◀

Proof

$$J_{-m}(x) = \sum_{n=0}^{\infty} (-1)^n \frac{1}{\Gamma(n+1-m) \cdot n!} (\frac{x}{2})^{2n-m}$$

$$= \sum_{n=0}^{m-1} \frac{(-1)^n}{\Gamma(n+1-m) \cdot n!} (\frac{x}{2})^{2n-m} + \sum_{n=m}^{\infty} \frac{(-1)^n}{\Gamma(n+1-m) \cdot n!} (\frac{x}{2})^{2n-m} \quad {}^1$$

$$= \sum_{n=m}^{\infty} \frac{(-1)^n}{\Gamma(n+1-m) \cdot n!} (\frac{x}{2})^{2n-m} \; ; \text{ let } k = n-m \begin{pmatrix} n : m \to \infty \\ k : 0 \to \infty \end{pmatrix}$$

$$= \sum_{k=0}^{\infty} \frac{(-1)^{k+m}}{\Gamma(k+1)(k+m)!} \cdot (\frac{x}{2})^{2k+m}$$

$$= (-1)^m \sum_{k=0}^{\infty} \frac{(-1)^r}{\Gamma(k+1+m) \cdot k!} (\frac{x}{2})^{2k+m}$$

$$= (-1)^m \sum_{n=0}^{\infty} \frac{(-1)^n}{\Gamma(n+1+m) \cdot n!} (\frac{x}{2})^{2n+m} = (-1)^m J_m(x),$$

$$\therefore J_{-m}(x) = (-1)^m J_m(x),$$

$$\therefore J_m(x) \text{ and } J_{-m}(x) \text{ are L.D.} \qquad \blacktriangleleft$$

When $\mu = m$ is an integer, only one linearly independent solution can be obtained, and the other solution is rather complex. Therefore, it is necessary to define another Bessel function, known as the Bessel function of the second type Y_μ.

▶ **Definition 10-5-2**

Bessel Functions of the Second Type

$$Y_\mu(x) \Rightarrow \begin{cases} \mu \neq \text{integers} \Rightarrow Y_\mu(x) = \dfrac{\cos u\pi J_\mu(x) - J_{-\mu}(x)}{\sin \mu\pi} \quad {}^2 \\ \mu = \text{integers} \Rightarrow Y_m(x) = \lim_{\mu \to m} Y_\mu(x) \end{cases} \qquad \blacktriangleleft$$

▶ **Theorem 10-5-3**

The Bessel Equation of Order μ

$x^2 y'' + xy' + (x^2 - \mu^2) y = 0$, at $\mu \geq 0$, its solution can expresses as:

$y(x) = c_1 J_\mu(x) + c_2 Y_\mu(x)$. ^{3} $\qquad \blacktriangleleft$

---

^{1} $\Gamma(1) = 1, \Gamma(0), \Gamma(-1), \Gamma(-2)\cdots = \infty$.

^{2} Since Y_μ contains the linear combination of $J_\mu(x)$ and $J_{-\mu}(x)$, Y_μ is still a solution of μ order Bessel function.

^{3} $\mu \neq$ integers, in $Y_\mu(x)$, include $J_\mu(x)$ and $J_{-\mu}(x)$, $\begin{cases} \therefore Y_\mu(x) \text{ and } J_\mu(x) \text{ are L.D.} \\ \mu = \text{integer} \\ Y_\mu(x) \text{ and } J_\mu(x) \text{ are L.I.} \end{cases}$.

Example 1

Solve the following differential equation

$x^2 y'' + xy' + (x^2 - 16)y = 0$.

Solution

$x^2 y'' + xy' + (x^2 - 4^2)y = 0$ is the fourth-order Bessel equation.

∴ Its solution is:

$$y(x) = c_1 J_4(x) + c_2 Y_4(x).$$

$J_4(x)$ is the 4th-order Bessel function of the first type, $Y_4(x)$ is the 4th-order Bessel function of the second type.   Q.E.D.

Example 2

Solve the following differential equation

$x^2 y'' + xy' + (x^2 - 0.25)y = 0$.

Solution

$x^2 y'' + xy' + (x^2 - (\frac{1}{2})^2)y = 0$ is an $\frac{1}{2}$-order Bessel equation,

$$\therefore y(x) = c_1 J_{\frac{1}{2}}(x) + c_2 Y_{\frac{1}{2}}(x).$$

where $J_{\frac{1}{2}}(x)$ is an $\frac{1}{2}$-order Bessel function of the first type, $Y_{\frac{1}{2}}(x)$ is an $\frac{1}{2}$-order Bessel function of the second type.   Q.E.D.

10-5-2 The Modified Bessel Equation

We have previously introduced the solution to Bessel equations, and now we will discuss the modified Bessel equations and their solutions.

1. **Definition**

   $x^2 y'' + xy' + (-x^2 - \mu^2)y = 0 \Rightarrow \mu$ order the modified Bessel equation,

   or $x^2 y'' + xy' - (x^2 + \mu^2)y = 0$.

2. **Solution**

   $x^2 y'' + xy' + (-x^2 - \mu^2)y = 0$

   $\Rightarrow x^2 y'' + xy' + (i^2 x^2 - \mu^2)y = 0$

   $\Rightarrow y(x)$ the solutions, there exists one linearly independent solution $J_\mu(ix)$, and another independent solution $Y_\mu(ix)$,

$$y(x) = c_1 J_\mu(ix) + c_2 Y_\mu(ix),$$

where $J_\mu(ix) = \sum_{n=0}^{\infty} \dfrac{(-1)^n}{\Gamma(n+1+\mu)n!} (\dfrac{ix}{2})^{2n+\mu} = \sum_{n=0}^{\infty} \dfrac{(-1)^n}{\Gamma(n+1+\mu)n!} (i)^{2n+\mu} \cdot (\dfrac{x}{2})^{2n+\mu}$

$$= \sum_{n=0}^{\infty} \dfrac{(-1)^n}{\Gamma(n+1+\mu)n!} (-1)^n \cdot i^\mu \cdot (\dfrac{x}{2})^{2n+\mu}$$

$$= (i)^\mu \cdot \sum_{n=0}^{\infty} \dfrac{1}{\Gamma(n+1+\mu)n!} (\dfrac{x}{2})^{2n+\mu},$$

$$\therefore I_\mu(x) = \sum_{n=0}^{\infty} \dfrac{1}{\Gamma(n+1+\mu)n!} (\dfrac{x}{2})^{2n+\mu}.$$

3. The Modified Bessel Function

(1) The modified Bessel function of first type $I_\mu(x)$:

$\mu \ne$ integer, $I_\mu(x)$ and $I_{-\mu}(x)$ are L.I.,

$\mu =$ integer $= m$, $I_\mu(x)$ and $I_{-\mu}(x)$ are L.D.,

$$\begin{cases} I_\mu(x) = \sum_{n=0}^{\infty} \dfrac{1}{\Gamma(n+1+\mu)n!} (\dfrac{x}{2})^{2n+\mu} \\ \quad \rightarrow \text{The modified Bessel function of first type } \mu \text{ order} \\ I_{-\mu}(x) = \sum_{n=0}^{\infty} \dfrac{1}{\Gamma(n+1-\mu)n!} (\dfrac{x}{2})^{2n-\mu} \\ \quad \rightarrow \text{The modified Bessel function of first type } -\mu \text{ order} \end{cases}$$

(2) The modified Bessel function of second type $K_\mu(x)$:

$$K_\mu(x) = \begin{cases} \mu \ne \text{integer}, \ K_\mu(x) = \dfrac{\pi}{2} \cdot \dfrac{I_{-\mu}(x) - I_\mu(x)}{\sin \mu\pi} \\ \mu = m = \text{integer}, \ K_m(x) = \lim\limits_{\mu \to m} K_\mu(x) \end{cases}.$$

4. The Solution of Modified Bessel Equation

$x^2 y'' + xy' + (-x^2 - \mu^2)y = 0$, at $\mu \ge 0$, its solution can expresses as:

$y(x) = c_1 I_\mu(x) + c_2 K_\mu(x).$[4]

Example 3

Solve the following differential equation

$x^2 y'' + xy' - (x^2 + 3)y = 0$.

---

[4] When $\mu > 0$ does not have to be an integer $\Rightarrow y(x) = c_1 I_\mu(x) + c_2 I_{-\mu}(x)$.

Solution

$x^2 y'' + xy' - \left[-x^2 - (\sqrt{3})^2 \right] y = 0$ is the modified Bessel equation of the first type with order $\sqrt{3}$,

its solution is: $y(x) = c_1 I_{\sqrt{3}}(x) + c_2 I_{-\sqrt{3}}(x) \Rightarrow$ (This solution is valid when μ is not an

$$\text{integer})$$
$$= d_1 I_{\sqrt{3}}(x) + d_2 K_{\sqrt{3}}(x)$$

where $I_{\sqrt{3}}(x)$ is the modified Bessel function of the first type with order $\sqrt{3}$,

$K_{\sqrt{3}}(x)$ is the modified Bessel function of the second type with order $\sqrt{3}$. `Q.E.D.`

Example 4

$I_\mu(x) = \displaystyle\sum_{n=0}^{\infty} \frac{1}{\Gamma(n+1+\mu)n!} (\frac{x}{2})^{2n+\mu}$,

when μ is an integer, $\mu = m$, please verify $\Rightarrow I_{-m}(x) = I_m(x)$, $I_m(x)$ and $I_{-m}(x)$ are L.D..

Solution

$$I_{-m}(x) = \sum_{n=0}^{\infty} \frac{1}{\Gamma(n+1-m)n!} (\frac{x}{2})^{2n-m}$$

$$= \sum_{n=0}^{m-1} \frac{1}{\Gamma(n+1-m)n!} (\frac{x}{2})^{2n-m} + \sum_{n=m}^{\infty} \frac{1}{\Gamma(n+1-m)n!} (\frac{x}{2})^{2n-m}$$

$$= \sum_{n=m}^{\infty} \frac{1}{\Gamma(n+1-m)n!} (\frac{x}{2})^{2n-m} \text{ ; let } n = n + m$$

$$= \sum_{n=0}^{\infty} \frac{1}{\Gamma(n+1)(n+m)!} (\frac{x}{2})^{2n+m}$$

$$= \sum_{n=0}^{\infty} \frac{1}{\Gamma(n+m+1)n!} (\frac{x}{2})^{2n+m} = I_m(x) .$$ `Q.E.D.`

10-5-3　The Graph of Bessel Function

It is difficult to find the exact solution of Bessel function in related applications, so it is often necessary to describe its solution through graphics. The Bessel function and modified Bessel function graphics will be introduced below.

1. The Graph of Bessel Function

$$J_\mu(x) = \sum_{n=0}^{\infty} \frac{(-1)^n}{\Gamma(n+1+\mu)n!}(\frac{x}{2})^{2n+\mu},$$

$$\mu = 0 \Rightarrow J_0(x) = \sum_{n=0}^{\infty} \frac{(-1)^n}{\Gamma(n+1)n!}(\frac{x}{2})^{2n}$$

$$= \sum_{n=0}^{\infty} \frac{(-1)^n}{(n!)^2}(\frac{x}{2})^{2n} = 1 - (\frac{x}{2})^2 + \frac{1}{(2!)^2}(\frac{x}{2})^4 - \frac{1}{(3!)^2}(\frac{x}{2})^6 + \cdots,$$

$$\therefore J_0(0) = 1 : \text{zero point},$$

$$\mu > 0 \Rightarrow J_\mu(x) = \sum_{n=0}^{\infty} \frac{(-1)^n}{\Gamma(n+1+\mu)n!}(\frac{x}{2})^{2n+\mu}$$

$$= \frac{1}{\Gamma(1+\mu)}(\frac{x}{2})^\mu - \frac{1}{\Gamma(2+\mu)}(\frac{x}{2})^{2+\mu} + \cdots,$$

$$\therefore J_\mu(0) = 0 \; ; \mu > 0,$$

$$J_\mu(0) \to \infty \; ; \mu < 0, \mu \neq -1, -2, -3, \cdots.$$

When x is large enough, the function period is $T = 2\pi$ and $Y_\mu(x) \Rightarrow \mu > 0 \Rightarrow x \to 0$, $\lim_{x \to 0} Y_\mu(x) \to -\infty$, its graph can be obtained as follows:

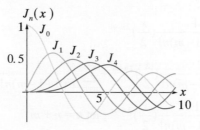

Figure 10-7

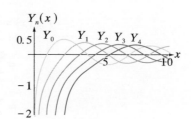

Figure 10-8

2. The Graph of Modified Bessel Function

$$I_\mu(x) = \sum_{n=0}^{\infty} \frac{1}{\Gamma(n+1+\mu)n!}(\frac{x}{2})^{2n+\mu},$$

$$I_0(x) = \sum_{n=0}^{\infty} \frac{1}{\Gamma(n+1)n!}(\frac{x}{2})^{2n} = \sum_{n=0}^{\infty} \frac{1}{(n!)^2}(\frac{x}{2})^{2n}$$

$$= 1 + \left(\frac{x}{2}\right)^2 + \frac{1}{(2!)^2}\left(\frac{x}{2}\right)^4 + \cdots$$

(if $x \to \infty, I \to \infty$),

$$I_0(0) = 1, \quad I_0'(0) = 0.$$

When $\mu > 0$,

Figure 10-9

$$I_\mu(x) = \sum_{n=0}^{\infty} \frac{1}{\Gamma(n+1+\mu)n!}(\frac{x}{2})^{2n+\mu} = \frac{1}{\Gamma(1+\mu)}(\frac{x}{2})^\mu + \cdots,$$

$I_\mu(0) = 0$, so when $K_\mu(x)$ at $x \to 0, K_\mu(x) \to \infty$,

when $x \to \infty, K_\mu(x) \to 0.$

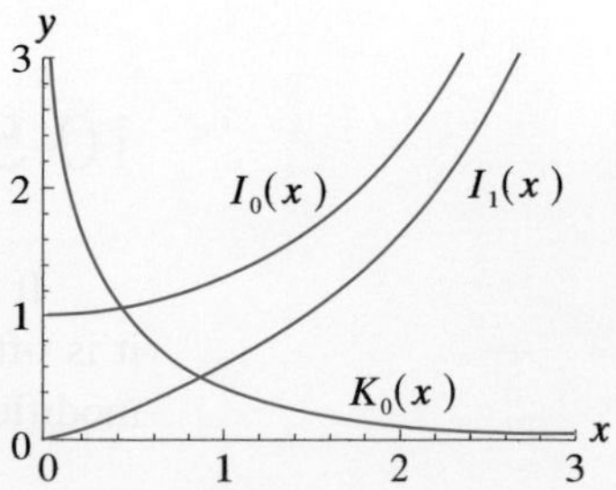

Its graph can be obtained as Figure 10-9:

10-5-4 Bessel Equation Corresponds to the Boundary Value Problem

Then we use the boundary value problem in Chapter 9 to analyze Bessel equation. The form can be transformed into

$$(xy')' + (-\frac{\mu^2}{x} + \lambda x)y = 0$$

It can be seen from the first section of Chapter 9 that $p(x) = x$, $q(x) = -\dfrac{\mu^2}{x}$, weight function $r(x) = x$. Therefore, it can form a standard Sturm—Liouville boundary value problem, and then solve this type of boundary value problem as follows,

DE: $x^2 y'' + xy' + (\lambda x^2 - \mu^2)y = 0$, $\mu > 0$, $0 \le x \le \ell$,

B.C: $y(\ell) = 0$, $y(0)$ is bounded.

(1) $\lambda = -k^2 < 0$ ($k > 0$)

$x^2 y'' + xy' + (-k^2 x^2 - \mu^2)y = 0$ ......The modified Bessel

$\Rightarrow \therefore y(x) = c_1 I_\mu(kx) + c_2 K_\mu(kx)$,

$y(0)$ is bounded $\Rightarrow c_2 = 0$,

$y(\ell) = 0 \Rightarrow c_1 I_\mu(k\ell) = 0 \Rightarrow c_1 = 0$,

$\therefore y(x) = 0 \Rightarrow$ this is the trivial solution, not the solution we want.

(2) $\lambda = 0$

$x^2 y'' + xy' - \mu^2 y = 0$ (Cauchy equal dimensions)

Let $x = e^t$, $t = \ln x$, $dt = \dfrac{1}{x}dx$, $D \triangleq \dfrac{d}{dt}$

$\Rightarrow xy' = Dy$, $x^2 y'' = D \cdot (D-1)y$ substitute into above DE

$\Rightarrow [D(D-1) + D - \mu]y = 0$

$\Rightarrow (D^2 - \mu^2)y = 0$, $n = \pm\mu$,

$\therefore y = c_1 e^{\mu t} + c_2 e^{-\mu t}$,

$y(x) = c_1 x^\mu + c_2 x^{-\mu}$; $\mu > 0$,

$y(0)$ is bounded $\Rightarrow c_2 = 0$,

$y(\ell) = 0 \Rightarrow c_1 \ell^\mu = 0 \Rightarrow c_1 = 0$,

$\therefore y(x) = 0 \Rightarrow$ this is the trivial solution, not the solution we want.

(3) $\lambda = k^2 > 0$ $(k > 0)$

$$x^2 y'' + xy' + (k^2 x^2 - \mu^2) y = 0$$
$$\Rightarrow y(x) = c_1 J_\mu (kx) + c_2 Y_\mu (kx)$$

$y(0)$ is bounded $\Rightarrow c_2 = 0$,
$y(\ell) = 0 \Rightarrow c_1 J_\mu (k\ell) = 0$,
if $c_1 \neq 0$, then $J_\mu (k\ell) = 0$.
Let $J_\mu (\alpha_n) = 0$, $n = 1, 2, 3, \cdots$,

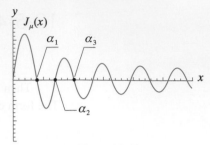

Figure 10-10

then $k\ell = \alpha_n$, $k = \dfrac{\alpha_n}{\ell}$, $n = 1, 2, 3, \cdots$, therefore, the solution corresponding to

Bessel equation can be obtained as

$$\therefore \begin{cases} y(x) = c_1 J_\mu (\dfrac{\alpha_n}{\ell} \cdot x) \\ \lambda = k^2 = (\dfrac{\alpha_n}{\ell})^2, n = 1, 2, 3, \cdots \end{cases}.$$

From the boundary value problem theory in Chapter 9, Section 1, we can know the

characteristic function $\left\{ J_\mu (\dfrac{\alpha_n}{l} x) \right\}_{n=1}^{\infty}$ at $x \in [0, l]$ is a set of orthogonal functions, that

is $< J_\mu (\dfrac{\alpha_n}{l} x), J_\mu (\dfrac{\alpha_n}{l} x) >_w = \int_0^l x J_\mu (\dfrac{\alpha_n}{l} x) \cdot J_\mu (\dfrac{\alpha_n}{l} x) dx = \begin{cases} 0; \ m \neq n \\ \dfrac{1}{2} l^2 J_{\mu+1}(\alpha_n); \ m = n \end{cases}$

Then $\left\{ J_\mu (\dfrac{\alpha_n}{l} x) \right\}_{n=1}^{\infty}$ at $x \in [0, l]$ is a Complete orthogonal set.

$\left\{ J_\mu (\dfrac{\alpha_n}{l} x) \right\}_{n=1}^{\infty}$ at $x \in [0, l]$ is a Complete orthogonal set, it can be known from the

generalized Fourier series that any function $f(x)$ can be expressed as follows

$\therefore f(x) = \displaystyle\sum_{n=1}^{\infty} a_n J_\mu (\dfrac{\alpha_n}{\rho} x) dx$, we call it the Fourier Bessel series and its coefficients are

$$a_n = \frac{< f(x), J_\mu (\dfrac{\alpha_n}{l} x) >}{< J_\mu (\dfrac{\alpha_n}{l} x), J_\mu (\dfrac{\alpha_n}{l} x) >} = \frac{2}{l^2 J_{\mu+1}^2 (\alpha_n)} \int_0^l x \cdot f(x) \cdot J_\mu (\dfrac{\alpha_n}{l} x) dx$$

10-5-5   Solving Polar Coordinate Partial Differential Equation

With the above basic concepts in mind, we can return to the problem at the beginning of this chapter and then solve the following PDE.

DE: $\dfrac{\partial^2 \varphi}{\partial r^2} + \dfrac{1}{r} \dfrac{\partial \varphi}{\partial r} = \dfrac{1}{c^2} \dfrac{\partial^2 \varphi}{\partial t^2}$ $(0 < r < a, t > 0)$, the boundary conditions are $\varphi(a,t) = 0$, $\varphi(0,t)$ is bounded,

the initial conditions are $\varphi(r,0) = f(r)$, $\dfrac{\partial \varphi}{\partial t}(r,0) = g(r)$.

(1) By the method of separation of variables, we let $\varphi(r,t) = F(r)H(t)$ substitute into DE

$$F''H + \frac{1}{r}F'H = \frac{1}{c^2}FH''$$

dividing by FH

$$\Rightarrow \frac{F'' + \dfrac{1}{r}F'}{F} = \frac{\dfrac{1}{c^2}H''}{H} = -\lambda$$

$$\Rightarrow \begin{cases} r^2F'' + rF' + \lambda r^2 F = 0 \\ H'' + \lambda c^2 H = 0 \end{cases}$$

(2) From $\begin{cases} r^2F'' + rF' + \lambda r^2 F = 0 \\ F(a) = 0, F(0) \text{ is bounded} \end{cases}$,

　(i)　$\lambda < 0$, $\lambda = -p^2 \,(p > 0)$

　　　$\Rightarrow r^2F'' + rF' + (-p^2r^2 - 0)F = 0$ (is 0 order modified Bessel),

　　　$\therefore F(r) = c_1 I_0(pr) + c_2 k_0(pr)$,

　　　by $F(0)$ is bounded $\Rightarrow c_2 = 0$,

　　　$F(a) = 0 \Rightarrow c_1 I_0(pr) = 0 \Rightarrow c_1 = 0$,

　　　$\therefore F(r) = 0 \Rightarrow$ trivial solution, not the solution we want.

　(ii)　$\lambda = 0$

　　　$\Rightarrow r^2F'' + rF' = 0$ (Cauchy equal dimensions),

　　　let $r = e^t$, $t = \ln r$

　　　$\Rightarrow [D \cdot (D-1) + D]F = 0$,

　　　$D^2F - 0 \cdot m^2 = 0$,

　　　$\therefore F(r) = c_1 + c_2 t = c_1 + c_2 \ln r$,

　　　by $F(0)$ is bounded $\Rightarrow c_2 = 0$,

　　　$F(a) = 0 \Rightarrow c_1 = 0$, $\therefore F(r) = 0 \Rightarrow$ trivial solution, not the solution we want.

　(iii)　$\lambda = p^2 > 0$, $(p > 0)$

　　　$\Rightarrow r^2F'' + rF + (p^2r^2 - 0)F = 0$ (0 order Bessel),

　　　$\therefore F(r) = c_1 J_0(pr) + c_2 Y_0(pr)$,

　　　by $F(0)$ is bounded $\Rightarrow c_2 = 0$,

　　　$F(a) = 0 \Rightarrow c_1 J_0(pa) = 0$,

　　　let $J_0(\alpha_n) = 0 \Rightarrow pa = \alpha_n$, $p = \dfrac{\alpha_n}{a}$,

$$\therefore F(r) = c_1 J_0(\frac{\alpha_n}{a} r), \quad \lambda = (\frac{\alpha_n}{a})^2, \, n = 1, 2, \cdots,$$

let $\lambda = (\frac{\alpha_n}{a})^2$ substitute into $H'' + \lambda c^2 H = 0$,

$$\therefore H'' + (\frac{c\alpha_n}{a})^2 H = 0$$

$$\Rightarrow H(t) = d_1 \cos\frac{c\alpha_n}{a} t + d_2 \sin\frac{c\alpha_n}{a} t$$

$$\Rightarrow \begin{cases} F_n(r) = c_1 J_0(\frac{\alpha_n}{a} r), \, n = 1, 2, 3, \cdots \\ \\ H_n(r) = d_1 \cos\frac{c\alpha_n}{a} t + d_2 \sin\frac{c\alpha_n}{a} t, \, n = 1, 2, 3, \cdots \end{cases}.$$

(3) $\varphi_n(r,t) = F_n(r) H_n(t) = [A_n \cos\frac{c\alpha_n}{a} t + B_n \sin\frac{c\alpha_n}{a} t] J_0(\frac{\alpha_n}{a} r)$,

It can be seen from the superposition principle:

$$\varphi(r,t) = \sum_{n=1}^{\infty} [A_n \cos\frac{c\alpha_n}{a} t + B_n \sin\frac{c\alpha_n}{a} t] J_0(\frac{\alpha_n}{a} r) \cdots\cdots (**)$$

$$\varphi_t(r,t) = \sum_{n=1}^{\infty} [-\frac{c\alpha_n}{a} A_n \sin\frac{\alpha_n c}{a} t + \frac{c\alpha_n}{a} B_n \cos\frac{c\alpha_n}{a} t] J_0(\frac{\alpha_n}{a} r).$$

(4) From IC , we know:

$$\varphi(r,0) = f(r) = \sum_{n=1}^{\infty} A_n J_0(\frac{\alpha_n}{a} r) \text{ (Fourier Bessel series)}$$

$$\Rightarrow A_n = \frac{< f(r), J_0(\frac{\alpha_n}{a} r) >}{< J_0(\frac{\alpha_n}{a} r), J_0(\frac{\alpha_n}{a} r) >},$$

$$\varphi_t(r,0) = g(r) \Rightarrow g(r) = \sum_{n=1}^{\infty} \frac{c\alpha_n}{a} B_n J_0(\frac{\alpha_n}{a} r)$$

$$\Rightarrow B_n = (\frac{a}{c\alpha_n}) \frac{< g(r), J_0(\frac{\alpha_n}{a} r) >}{< J_0(\frac{\alpha_n}{a} r), J_0(\frac{\alpha_n}{a} r) >},$$

let A_n, B_n substitute into (**) and we get $\varphi(r,t)$.

Next, let's illustrate the solution of the polar coordinate form of the heat conduction PDE with an example.

Example 5

Solve following PDE:

$$\frac{\partial \phi}{\partial t} = k(\frac{\partial^2 \phi}{\partial r^2} + \frac{1}{r} \frac{\partial \phi}{\partial r}),$$

the initial condition is $\phi(r,0) = F(r) \, (0 < r < 1)$,

the boundary condition is $\phi(1,t) = 0$, $\phi(0,t)$ is bounded.

Solution

(1) Let $\phi(r, t) = R(r)T(t)$

substitute into PDE and we get

$$R\dot{T} = k(R''T + \frac{1}{r}R'T),$$

that is

$$\frac{R'' + \frac{1}{r}R'}{R} = \frac{1}{k}\frac{\dot{T}}{T} = -\lambda \text{ (constant)},$$

therefore

$$\begin{cases} r^2R''(r) + rR'(r) + \lambda r^2 R(r) = 0 \\ \dot{T}(t) + \lambda k T(t) = 0 \end{cases} \cdots \text{①}$$

(2) By boundary condition,

$\phi(1, t) = R(1)T(t) = 0,$

we know $R(1) = 0$, through

$\phi(0, t) = R(0)T(t) = $ finite value,

we get $R(0) = $ finite value.

(i) $\lambda < 0$ let $\lambda = -p^2$ $(0 < p < \infty)$, substitute into equation ①, then we get

$$r^2R''(r) + rR'(r) - p^2r^2R(r) = 0.$$

The above formula is the 0th order modified Bessel equation, and its solution is

$$R(r) = c_1 I_0(pr) + c_2 K_0(pr).$$

From $\lim_{r \to 0^+} R(r) = $ finite value, we know $c_2 = 0$ $(\because K_0(0^+) \to \infty)$

$$R(a) = c_1 I_0(pa) = 0 \Rightarrow c_1 = 0.$$

So $R(r) = 0$ has no non-zero solution.

(ii) $\lambda = 0$ substitute into equation ①, then we get

$$r^2R''(r) + rR'(r) = 0.$$

The above formula is Cauchy's equal-dimensional equation, and its solution is

$$R(r) = c_1 + c_2 \ln r.$$

From $\lim_{r \to 0^+} R(r) = $ finite value, we know $c_2 = 0$ $(\because \ln 0^+ \to -\infty)$

$$R(a) = 0 \Rightarrow c_1 = 0,$$

thus, $R(r) = 0$ has no non-zero solution.

(iii) $\lambda > 0$ let $\lambda = p^2$ $(0 < p < \infty)$, substitute into equation ①, then we get

$$r^2 R''(r) + rR'(r) + p^2 r^2 R(r) = 0.$$

The above formula is the 0th order Bessel equation, and its solution is

$$R(r) = c_1 J_0(pr) + c_2 Y_0(pr).$$

From $\lim\limits_{r \to 0^+} R(r) = $ finite value, we know $c_2 = 0$ $(\because Y(0^+) \to -\infty)$

$$R(a) = c_1 J_0(pa) = 0,$$

let $c_1 \neq 0$, then $J_0(pa) = 0$, and let $J_0(\alpha_n) = 0$, that is, α_n is the root of $J_0(x) = 0$, so

$$p = \alpha_n \ (n = 1, 2, 3...),$$

therefore,

$$R(r) = c_1 J_0(\alpha_n r) = R_n(r),$$

let $\lambda = p^2 = \alpha_n^2$ substitute into equation ①, then we get

$$\dot{T} + \alpha_n^2 kT = 0 .$$

The above formula is a first-order ODE, and its solution is

$$T(t) = de^{-\alpha_n^2 kt} = T_n .$$

Thus

$$\phi_n(r,t) = T_n(t) R_n(r) = A_n e^{-\alpha_n^2 kt} J_0(\alpha_n r) .$$

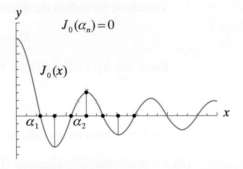

(3) We acquire by superposition principle

$$\phi(r,t) = \sum_{n=1}^{\infty} \phi_n(r,t) = \sum_{n=1}^{\infty} A_n e^{-\alpha_n^2 kt} J_0(\alpha_n r) .$$

Through

$$\phi(r,0) = F(r) = \sum_{n=1}^{\infty} A_n J_0(\alpha_n r) ,$$

then through Fourier—Bessel series, we get

$$A_n = \frac{< F(r), J_0(\alpha_n r) >}{< J_0(\alpha_n r), J_0(\alpha_n r) >} ,$$

substitute A_n into $\phi(r, t)$, that is the solution.

Q.E.D.

10-5 Exercises

Basic questions

Solve the following differential equations.

1. $x^2 y'' + xy' + (x^2 - 36)y = 0$.

2. $x^2 y'' + xy' + (x^2 - 0.81)y = 0$.

3. $x^2 y'' + xy' - (x^2 + 25)y = 0$.

4. $x^2 y'' + xy' + (-x^2 - 0.25)y = 0$.

Advanced questions

1. Solve the following PDF.
$$\frac{\partial \phi}{\partial t} = \frac{\partial^2 \phi}{\partial r^2} + \frac{1}{r} \frac{\partial \phi}{\partial r},$$
$(0 \leq r \leq 1, t \geq 0)$ the initial condition is
$\phi(r, t = 0) = 1 - r^2$, and the boundary condition is
that $\phi(r = 1, t) = 0$, $\phi(r = 0, t)$ is bounded.

2. Starting from separation of variables, solve the
boundary-value problem
$$\frac{\partial^2 u}{\partial r^2} + \frac{1}{r} \frac{\partial u}{\partial r} + \frac{\partial^2 u}{\partial z^2} = 0, 0 \leq r \leq 2, 0 \leq z \leq 1,$$
$u(r, 0) = 0, u(r, 1) = 1, u(2, z) = 0.$

11 Complex Analysis

Cauchy (1789~1857, France)

Cauchy (1789-1857) was a French mathematician, physicist, and astronomer whose research spanned a wide range of fields. His significant contribution, proposed in 1821, was the definition of the limit $\varepsilon - \delta$, which greatly advanced the theoretical study of calculus. This concept was also extended to complex functions, leading to the establishment of calculus for complex variables. Cauchy's groundbreaking work laid a solid foundation for the later applications of complex functions in fluid dynamics, thermodynamics, electromagnetics, communication systems, and more.

Learning Objectives

11-1
Basic Concepts of Complex Number

11-1-1 Understanding complex numbers: basic operations, norm, conjugate
11-1-2 Understanding complex numbers in polar form: definition, calculation of argument
11-1-3 Familiarity with De Moivre's theorem
11-1-4 Proficiency in calculating complex n-th roots

11-2
Complex Functions

11-2-1 Understanding complex variable functions and classification: multivalued functions, single-valued functions
11-2-2 Mastering the decomposition of multivalued functions into single-valued functions using branch cuts: logarithmic functions
11-2-3 Familiarity with common elementary functions: polynomials, rational functions, exponential, (inverse) trigonometric, (inverse) hyperbolic

11-3
Differentiation of Complex Functions

11-3-1 Understanding limits and continuity of complex functions
11-3-2 Familiarity with derivatives of complex functions: operations, L'Hôpital's rule
11-3-3 Familiarity with analyticity conditions: Cauchy-Riemann equations; and applications: holomorphic functions
11-3-4 Understanding singular points and classification: poles, zeros, branch points, removable singularities, essential singularities

11-4
Integration of Complex Functions

11-4-1 Understanding integration of complex functions: basic properties, ML theorem
11-4-2 Understanding the complex version of the Green's theorem
11-4-3 Mastering the Cauchy integral theorem: fundamental inferences and contour deformation principle
11-4-4 Familiarity with the Cauchy integral formula
11-4-5 Mastering of the maximum (minimum) modulus theorem, argument principle

11-5
Taylor Series Expansion and Laurent Series Expansion

11-5-1 Understanding power series of complex functions: Taylor series expansion
11-5-2 Understanding power series of complex functions: Laurent series expansion; and its relation to isolated singularities

11-6
Residue Theorem

No
section-- Getting acquainted with residue calculation through Laurent series expansion: residue theorem

11-7
Definite Integral of Real Variable Functions

11-7-1 Mastering residue theorem for computing real variable improper integrals: trigonometric function type
11-7-2 Mastering residue theorem for computing real variable improper integrals: rational function type
11-7-3 Mastering residue theorem for computing real variable improper integrals: complex Fourier transform (integral) type
11-7-4 Mastering residue theorem for computing real variable improper integrals: Laplace inverse transform type

ExampleVideo

The theoretical origin of complex functions dates back to the late 18th century. In 1774, the mathematician Euler was the first to consider the integration of complex functions. By the 19th century, physicists Cauchy and Riemann, while studying fluid mechanics, delved deeper into the study of complex functions, giving rise to the Cauchy-Riemann equations. Subsequently, complex functions underwent comprehensive development, becoming a major branch of mathematics beyond calculus. Since the 20th century, complex functions have found extensive application in fields such as aerodynamics, electromagnetics, and engineering probability, becoming a highly significant mathematical analytical tool.

This chapter will start with an introduction to complex number operations, then delve into calculus of complex functions, and introduce the concept of residues. This concept will then be employed to compute certain definite integrals that are difficult to solve using traditional calculus methods.

11-1　Basic Concepts of Complex Number

When the discriminant $\Delta = b^2 - 4ac < 0$ of the quadratic equation $ax^2 + bx + c = 0$, the formulaic solutions $\dfrac{-b \pm \sqrt{\Delta}}{2a}$ become meaningless in the realm of real numbers. Mathematicians introduced the concept of imaginary numbers $\sqrt{-1} = i$, and by incorporating them into the original real number system, they formed the set $\mathbf{C} = \{x + yi \mid x, y \in \mathbf{R}\}$ now known as the complex number system or the complex plane, as depicted in Figure 11-1. When the **imaginary part** y in the set is 0, $\mathbf{C}$ reduces to the real number system $\mathbf{R}$ composed solely of **real parts** x. Geometrically, from the perspective of the complex plane, the algebraic structure of the complex number system naturally parallels that of $\mathbf{R}^2$.

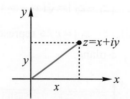

Figure 11-1　Complex plane

11-1-1 Algebraic Operations in the Complex Plane

1. **Addition and Subtraction**

   $z_1 \pm z_2 = (a_1 \pm a_2) + i(b_1 \pm b_2)$.

2. **Multiplication**

   $z_1 \cdot z_2 = (a_1 + ib_1) \cdot (a_2 + ib_2) = (a_1 a_2 - b_1 b_2) + i(a_1 b_2 + b_1 a_2)$.

3. Division

$$\frac{z_1}{z_2} = \frac{a_1 + ib_1}{a_2 + ib_2} = \frac{(a_1 + ib_1)(a_2 - ib_2)}{(a_2 + ib_2)(a_2 - ib_2)} = \frac{(a_1 a_2 + b_1 b_2) + i(a_2 b_1 - a_1 b_2)}{a_2^2 + b_2^2}.$$

4. Equality

$z_1 = a_1 + ib_1$, $z_2 = a_2 + ib_2$, if $z_1 = z_2$, then $a_1 = a_2$, $b_1 = b_2$.

Example 1

If $z_1 = 4 + 3i$, $z_2 = 2 - 5i$, please calculate: (1) $z_1 \cdot z_2$, (2) $\dfrac{z_1}{z_2}$.

Solution

(1) $z_1 \cdot z_2 = (4 + 3i) \cdot (2 - 5i) = (8 + 15) + i(-20 + 6) = 23 - 14i$.

(2) $\dfrac{z_1}{z_2} = \dfrac{4 + 3i}{2 - 5i} = \dfrac{(4 + 3i)(2 + 5i)}{(2 - 5i)(2 + 5i)} = \dfrac{-7 + 26i}{29} = -\dfrac{7}{29} + \dfrac{26}{29}i$.

Q.E.D.

5. Norm

For $z = x + iy$, the norm is defined as $|z| = \sqrt{x^2 + y^2}$. Sometimes, we also refer to $|z|$ as the absolute value of the complex number. This can be seen as the complex version of the distance concept on the real number plane. As a result, the distance properties commonly used on the real number plane can also be extended to the complex number plane. Readers can derive this based on the definition.

(1) $|z|$ it represents on the z-plane, the distance between z and the origin point O.

(2) If $z_1 = x_1 + iy_1$, $z_2 = x_2 + iy_2$, then

$|z_1 - z_2| = \sqrt{(x_1 - x_2)^2 + (y_1 - y_2)^2}$ represents the distance between z_1 to z_2.

(3) $|z - z_1| = r \Rightarrow$ represents a circle with z_1 as its center and r as its radius on the z-plane.

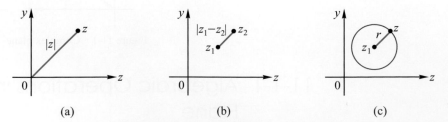

(a)　　　　　　　　(b)　　　　　　　　(c)

Figure 11-2　The distance of complex plane

(4) $|z_1 z_2 \cdots z_n| = |z_1||z_2|\cdots|z_n|$.

(5) $\left|\dfrac{z_1}{z_2}\right| = \dfrac{|z_1|}{|z_2|}$.

(6) $|z_1| - |z_2| \le |z_1| \le |z_1| + |z_2|$.

6. **Complex Conjugate Numbers**

The conjugate of $z = x + iy$ is $\bar{z} = x - iy$. The conjugate of a complex number is the result of reflecting it across the real axis, as illustrated in Figure 11-3. The conjugate satisfies the following properties:

$$(1)\begin{cases} x = \operatorname{Re}(z) = \dfrac{z + \bar{z}}{2} \\ y = \operatorname{Im}(z) = \dfrac{z - \bar{z}}{2i} \end{cases} \qquad (2)\,\overline{(\bar{z})} = z \qquad (3)\,\overline{z_1 z_2} = \overline{z_1} \cdot \overline{z_2}$$

$$(4)\,\overline{\left(\dfrac{z_1}{z_2}\right)} = \dfrac{\overline{z_1}}{\overline{z_2}} \qquad (5)\,|z| = |\bar{z}| \qquad (6)\,z \cdot \bar{z} = |z|^2$$

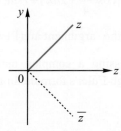

Figure 11-3   Complex conjugate diagram

Example 2

If $z_1 = 7 + 24i$, $z_2 = 3 - 4i$, please calculate: $(1)\,|z_1|$, $(2)\,|z_1 \cdot z_2{}^2|$.

Solution

(1) $|z_1| = |7 + 24i| = \sqrt{7^2 + 24^2} = 25$.

(2) $|z_1 \cdot z_2{}^2| = |(7 + 24i)(3 - 4i)^2| = |7 + 24i| \cdot |3 - 4i|^2 = 25 \times 5^2 = 25 \times 25 = 625$. **Q.E.D.**

11-1-2 Polar Form of a Complex Number

Through polar coordinates, $\begin{cases} x = r\cos\theta \\ y = r\sin\theta \end{cases}$, we have $z = x + iy = r[\cos\theta + i\sin\theta]$,

and it is known that $r = |z|$.

$\theta = \tan^{-1}\left(\dfrac{y}{x}\right)$ is referred to as the argument of z,

usually denoted by the symbol $\theta = \arg(z)$.

From the definition of the tangent function, we know that

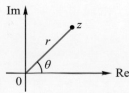

Figure 11-4

$$\theta = \arg(z) = \tan^{-1}(\frac{y}{x}) = \begin{cases} 2n\pi + \text{Tan}^{-1}(\frac{y}{x}), \ x > 0 \\ \\ (2n+1)\pi + \text{Tan}^{-1}(\frac{y}{x}), \ x < 0 \end{cases} , \text{ where } n = 0, \pm1, \pm2, \cdots$$

and $-\frac{\pi}{2} < \text{Tan}^{-1}(\frac{y}{x}) < \frac{\pi}{2}$. If $-\pi < \theta \le \pi$, $\theta = \text{Arg}(z)$ is referred to as the principal argument of z; the principal argument can also be taken in the range $0 \le \theta \le 2\pi$.

For instance, the complex $1 + i = \sqrt{2}[\cos\frac{\pi}{4} + i\sin\frac{\pi}{4}] = \sqrt{2}[\cos(\frac{9}{4}\pi) + i\sin\frac{9}{4}\pi]$

$$= \sqrt{2}[\cos(\frac{\pi}{4} + 2n\pi) + i\sin(\frac{\pi}{4} + 2n\pi)]; \ n = 0, \pm1, \pm2, \cdots,$$

then the argument $\arg(1+i) = \frac{\pi}{4} + 2n\pi$; the principal argument $\text{Arg}(1+i) = \frac{\pi}{4}$.

Expressing a complex number using its argument is most famously demonstrated through Euler's formula. In fact, this formula plays a crucial role in many calculations:

$$e^{i\theta} = \cos\theta + i\sin\theta .$$

In terms of proofs, this is derived from the Taylor series expansions of exponential and trigonometric functions:

$$e^{i\theta} = \sum_{n=0}^{\infty} \frac{(i\theta)^n}{n!} = \sum_{n=0}^{\infty} \frac{(i\theta)^{2n}}{(2n)!} + \sum_{n=0}^{\infty} \frac{(i\theta)^{2n+1}}{(2n+1)!} = \sum_{n=0}^{\infty} \frac{(-1)^n \theta^{2n}}{(2n)!} + i \sum_{n=0}^{\infty} \frac{(-1)^n \theta^{2n+1}}{(2n+1)!}$$
$$= \cos\theta + i\sin\theta,$$

and $|e^{i\theta}| = \sqrt{\cos^2\theta + \sin^2\theta} = 1$; further rewriting yields $\begin{cases} \cos\theta = \dfrac{e^{i\theta} + e^{-i\theta}}{2} \\ \\ \sin\theta = \dfrac{e^{i\theta} - e^{-i\theta}}{2i} \end{cases}$, thus, from

polar coordinates, we obtain $z = r(\cos\theta + i\sin\theta) = re^{i\theta}$, which is called the complex polar form of z.

▶ **Theorem 11-1-1**

The Polar Form of a Complex Number

If the argument of the complex $z = x + iy$ is θ,
and $r = |z|$, then $z = r(\cos\theta + i\sin\theta) = re^{i\theta}$. ◀

The complex polar form implies the following Theorem 11-1-2.

▶ **Theorem 11-1-2**

Properties of Arguments

(1) $\arg(z_1 z_2) = \arg(z_1) + \arg(z_2)$, but $\mathrm{Arg}(z_1 z_2) \neq \mathrm{Arg}(z_1) + \mathrm{Arg}(z_2)$.

(2) $\arg(z_1/z_2) = \arg(z_1) - \arg(z_2)$.

(3) $\mathrm{Arg}(z_1/z_2) \neq \mathrm{Arg}(z_1) - \mathrm{Arg}(z_2)$. ◀

Important Properties

(1) ① $e^{i2k\pi} = \cos 2k\pi + i \sin 2k\pi = 1$; $k = 0, \pm 1, \pm 2, \cdots$.

② $e^{i(2k+\frac{1}{2})\pi} = i$, $e^{i(2k+1)\pi} = -1$, $e^{i(2k+\frac{3}{2})\pi} = -i$.

(2) $z = (x_0 + r\cos\theta) + i(y_0 + r\sin\theta) = z_0 + re^{i\theta}$, where
$r = |z - z_0|$, $\theta = \arg(z - z_0)$.

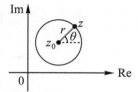

(3) $z_1 = r_1 e^{i\theta_1}$, $z_2 = r_2 e^{i\theta_2}$ we have (as shown in Figure 11-5)

Figure 11-5

① $z_1 z_2 = r_1 r_2\, e^{i(\theta_1 + \theta_2)}$, ② $\dfrac{z_1}{z_2} = \dfrac{r_1}{r_2} e^{i(\theta_1 - \theta_2)}$.

The following three examples illustrate various applications of Euler's formula:

Example 3

If $z_1 = i$, $z_2 = 1 - \sqrt{3}\,i$, find (1) $\arg(\dfrac{z_1}{z_2})$, (2) $\arg(z_1 \cdot z_2)$.

Solution

$z_1 = i = e^{i(\frac{\pi}{2} + 2m\pi)} \Rightarrow \arg(z_1) = \dfrac{\pi}{2} + 2m\pi$, $m = 0, \pm 1, \pm 2, \ldots$;

$z_2 = 1 - \sqrt{3}\,i = 2e^{i(-\frac{\pi}{3} + 2n\pi)} \Rightarrow \arg(z_2) = -\dfrac{\pi}{3} + 2n\pi$, $n = 0, \pm 1, \pm 2, \ldots$,

(1) $\arg(\dfrac{z_1}{z_2}) = \arg(z_1) - \arg(z_2) = \dfrac{5}{6}\pi + 2k\pi$, $k = 0, \pm 1, \pm 2, \ldots$.

(2) $\arg(z_1 \cdot z_2) = \arg(z_1) + \arg(z_2) = \dfrac{1}{6}\pi + 2l\pi$, $l = 0, \pm 1, \pm 2, \ldots$.

Q.E.D.

Example 4

What is $(\sqrt{2i} - 1)^{1001}$?

(1) 0   (2) 1   (3) i (4) $1 + i$ (5) $1000 - 1000i$.

Solution

$$\sqrt{2i} = (2e^{\frac{\pi}{2}i})^{\frac{1}{2}} = \sqrt{2}e^{\frac{\pi}{4}i} = \sqrt{2}(\cos\frac{\pi}{4} + i\sin\frac{\pi}{4}) = (1+i),$$

so $(\sqrt{2i} - 1) = (1 + i - 1) = i$, therefore $(\sqrt{2i} - 1)^{1001} = i^{1001} = i^{1000}i = (i^2)^{500}i = i$.   Q.E.D.

Example 5

If $z = x + iy = re^{i\theta}$, $i = \sqrt{-1}$, please verify $\sin^2 z + \cos^2 z = 1$.

Solution

Because $\sin z = \dfrac{1}{2i}(e^{iz} - e^{-iz})$, $\cos z = \dfrac{1}{2}(e^{iz} + e^{-iz})$,

$$\sin^2 z + \cos^2 z = \{\frac{1}{2i}(e^{iz} - e^{-iz})\}^2 + \{\frac{1}{2}(e^{iz} + e^{-iz})\}^2$$

$$= -\frac{1}{4}(e^{2iz} - 2 + e^{-2iz}) + \frac{1}{4}(e^{2iz} + 2 + e^{-2iz}) = 1.$$ Q.E.D.

11-1-3 De Moiver's Theorem

1. If $z_1 = r_1(\cos\theta_1 + i\sin\theta_1)$, $z_2 = r_2(\cos\theta_2 + i\sin\theta_2)$,

    then ① $z_1 z_2 = r_1 r_2 [\cos(\theta_1 + \theta_2) + i\sin(\theta_1 + \theta_2)]$,

    ② $\dfrac{z_1}{z_2} = \dfrac{r_1}{r_2}[\cos(\theta_1 - \theta_2) + i\sin(\theta_1 - \theta_2)]$.

2. If $z = r[\cos\theta + i\sin\theta]$,

    then $z^n = r^n[\cos\theta + i\sin\theta]^n = r^n[\cos n\theta + i\sin n\theta]$

    $\Rightarrow [\cos\theta + i\sin\theta]^n = \cos n\theta + i\sin n\theta$.

By directly substituting $z = r(\cos\theta + i\sin\theta) = re^{i\theta}$, we can verify De Moivre's theorem. Here leave this for the readers to practice on their own. Now, let's take a look at the examples.

Example 6

Express $z = 1 + \sqrt{3}i$ in complex polar form and find z^3.

Solution

(1) $z = 1 + \sqrt{3}i = x + iy$, $x = 1$, $y = \sqrt{3}$, $r = \sqrt{x^2 + y^2} = 2$,

$\theta = \tan^{-1}(\dfrac{y}{x}) = \tan^{-1}(\sqrt{3}) = \dfrac{\pi}{3}$, thus, $z = re^{(2n\pi + \theta)i} = 2e^{(2n\pi + \frac{\pi}{3})i}$, $n = 0, \pm 1, \pm 2, \cdots$.

(2) $z^3 = (2e^{(2n\pi + \frac{\pi}{3})i})^3 = 2^3 e^{(6n\pi + \pi)i} = -8$.

Q.E.D.

Example 7

Verify $\cos 5\theta = 16\cos^5 \theta - 20\cos^3 \theta + 5\cos\theta$.

Solution

$$[\cos 5\theta + i\sin 5\theta] = (\cos\theta + i\sin\theta)^5$$
$$= C_0^5 \cos^5 \theta + C_1^5 \cos^4 \theta(i\sin\theta) + C_2^5 \cos^3 \theta(i\sin\theta)^2$$
$$+ C_3^5 \cos^2 \theta(i\sin\theta)^3 + C_4^5 \cos\theta(i\sin\theta)^4 + C_5^5 (i\sin\theta)^5$$
$$= \cos^5 \theta - 10\cos^3 \theta\sin^2 \theta + 5\cos\theta\sin^4 \theta$$
$$+ i(5\cos^4 \theta\sin\theta - 10\cos^2 \theta\sin^3 \theta + \sin^5 \theta),$$
$$\text{so } \cos^5\theta = \cos^5\theta - 10\cos^3\theta\sin^2\theta + 5\cos\theta\sin^4\theta$$
$$= \cos^5 \theta - 10\cos^3 \theta(1 - \cos^2 \theta) + 5\cos\theta(1 - 2\cos^2 \theta + \cos^4 \theta)$$
$$= 16\cos^5 \theta - 20\cos^3 \theta + 5\cos\theta.$$

Q.E.D.

11-1-4 Complex Roots

In Section 11-1, we defined addition, subtraction, multiplication, and division of complex numbers, and we further simplified these operations using the unique concept of the complex argument. As a unified entity, complex numbers naturally allow us to consider equations involving them. This leads us to the problem of finding roots of complex number equations (which is different from inventing the complex number i to solve the equation $x^2 + 1 = 0$ with no real solutions—please do not confuse these cases). Now, let's discuss how to find complex roots, which fundamentally relies on the application of Euler's formula.

The *n*-th Roots

Assuming that $w^n = z$, then w is called the *n*-th root of z, denoted as $w = z^{\frac{1}{n}}$. If $z = r(\cos\theta + i\sin\theta) = re^{i\theta} = re^{i(\theta+2k\pi)}$; θ is the principal argument (defined on $[-\pi, \pi]$), then

$$w = z^{\frac{1}{n}} = [re^{i(\theta+2k\pi)}]^{\frac{1}{n}} = r^{\frac{1}{n}} \cdot e^{i(\frac{\theta}{n}+\frac{2k}{n}\pi)} = r^{\frac{1}{n}}[\cos(\frac{\theta+2k\pi}{n}) + i\sin(\frac{\theta+2k\pi}{n})],$$

$k = 0, 1, 2, \cdots, (n-1)$, represents that when z in the original complex plane is mapped to another complex plane w through $w = f(z) = z^{\frac{1}{n}}$, it forms n corresponding points (w_1, w_2, $\cdots$, w_n), as illustrated in Figure 11-6.

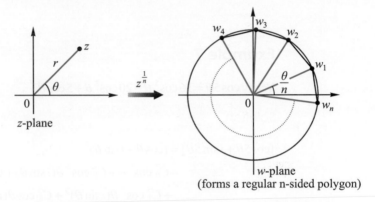

Figure 11-6　The schematics diagram of complex roots

If $w^n = 1$, then w is referred to as the *n*-th root of 1, to be more specific.:

(1) $k = 0$; $w_1 = r^{\frac{1}{n}}(\cos\frac{\theta}{n} + i\sin\frac{\theta}{n}) = r^{\frac{1}{n}}e^{i(\frac{\theta}{n})}$.

(2) $k = 1$; $w_2 = r^{\frac{1}{n}}(\cos\frac{\theta+2\pi}{n} + i\sin\frac{\theta+2\pi}{n}) = r^{\frac{1}{n}}e^{i(\frac{\theta+2\pi}{n})}$.

⋮

(3) $k = n-1$; $w_n = r^{\frac{1}{n}}(\cos\frac{\theta+2(n-1)\pi}{n} + i\sin\frac{\theta+2(n-1)\pi}{n}) = r^{\frac{1}{n}}e^{i[\frac{\theta+2(n-1)\pi}{n}]}$.

Referring to Figure 11-6, you'll notice that $w_1, w_2, \cdots, w_n$ on the circle with radius $r^{\frac{1}{n}}$ form a regular *n*-sided polygon.

Example 8

Find all the roots of $(8-8\sqrt{2}i)^{\frac{1}{4}}$.

Solution

Let

$$w = (8-8\sqrt{2}i)^{\frac{1}{4}} \Rightarrow w^4 = 8-8\sqrt{2}i = 8\sqrt{3}(\frac{1}{\sqrt{3}} - \frac{\sqrt{2}}{\sqrt{3}}i) = 8\sqrt{3} \cdot e^{i\theta} \text{ where}$$

$\theta = -\tan^{-1}(\sqrt{2})$, taking $-\pi \leq \theta \leq \pi$, then $w = (8\sqrt{3})^{\frac{1}{4}} \cdot e^{i(\frac{\theta+2k\pi}{4})}$; $k = 0, 1, 2, 3$, thus
we obtain

$$w_1 = (8\sqrt{3})^{\frac{1}{4}} \cdot e^{i\frac{\theta}{4}}, \quad w_2 = (8\sqrt{3})^{\frac{1}{4}} \cdot e^{i(\frac{\theta}{4}+\frac{\pi}{2})},$$
$$w_3 = (8\sqrt{3})^{\frac{1}{4}} \cdot e^{i(\frac{\theta}{4}+\pi)}, \quad w_4 = (8\sqrt{3})^{\frac{1}{4}} \cdot e^{i(\frac{\theta}{4}+\frac{3}{2}\pi)}.$$

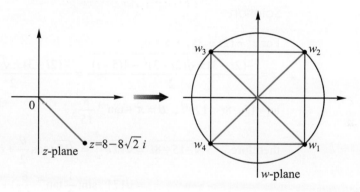

Q.E.D.

Example 9

If $z \neq 0$, (1)find n distinct roots of $w = z^{\frac{1}{n}}$,   (2)find the cube roots of 27.

Solution

(1) $w = z^{\frac{1}{n}} = [re^{i(\theta+2k\pi)}]^{\frac{1}{n}} = r^{\frac{1}{n}}e^{i(\frac{\theta+2k\pi}{n})}$, where we take $-\pi \leq \theta \leq \pi$, $k = 0, 1, 2, \cdots$
$n - 1$.

(2) the cube roots of $27 \Rightarrow w = (3^3 \cdot e^{i2k\pi})^{\frac{1}{3}} = 3e^{i\frac{2k\pi}{3}}$, $k = 0, 1, 2$.

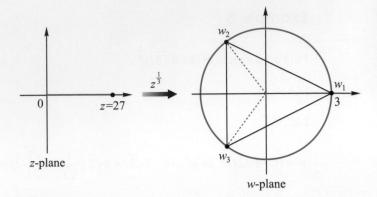

z-plane

w-plane

Example 10

$z = x + iy, i = \sqrt{-1}$, solve complex equation $z^2 + (2i - 3)\,z + 5 = 0$.

Solution

For $z^2 + (2i - 3)\,z + 5 = 0$,

$$z = \frac{-(2i-3) \pm \sqrt{(2i-3)^2 - 4(5-i)}}{2} = \frac{-(2i-3) \pm \sqrt{-15-8i}}{2},$$

let $-15 - 8i = 17e^{i\theta}$, $\quad \theta = \pi + \tan^{-1}\dfrac{8}{15}$,

therefore $\quad \sqrt{-15-8i} = \sqrt{17}\,e^{i\frac{\theta}{2}} = \sqrt{17}\,(\cos\dfrac{\theta}{2} + i\sin\dfrac{\theta}{2})$

$$= \sqrt{17}\,[-\sin(\dfrac{1}{2}\tan^{-1}\dfrac{8}{15}) + i\cos(\dfrac{1}{2}\tan^{-1}\dfrac{8}{15})]$$

$$= -1 + 4i,$$

thus $\quad z = \dfrac{-(2i-3) \pm \sqrt{-15-8i}}{2} = \dfrac{1}{2}[(3-2i) \pm (-1+4i)] = 1+i,\ 2-3i$.   Q.E.D.

11-1   Exercises

Basic questions

1. Let $z = \dfrac{2+i}{1-i}$, $i = \sqrt{-1}$, the complex conjugate of z, $\bar{z} = ?$

2. Let $i = \sqrt{-1}$, if $z = \dfrac{2+3i^{13}}{2i^{15} - i^{20}}$, what is the complex conjugate of z?

3. Calculate $(2+3i)(4-5i) = ?$

4. Express $\dfrac{(4+3i)}{(3+4i)}$ in the form of $a + bi$, where a and b are real numbers, and i is the imaginary unit.

5. $\left| \dfrac{(4-3i)^2 (1-i)^3}{(3+4i)^2 (1+i)} \right| = ?$

Advanced questions

1. $\left[\dfrac{1}{2} - \dfrac{\sqrt{3}}{2} i \right]^8 = ?$

2. Find all the roots of $(1-i)^{\frac{1}{3}}$?

3. Find the five roots of $z^5 + 32 = 0$?

4. Find all the sixth roots of $(-1 + \sqrt{3}i)$?

5. $z = x + iy$, $i = \sqrt{-1}$, solve complex equation $z^2 - (i+5)z + (8+i) = 0$.

6. $z = x + iy$, $i = \sqrt{-1}$, solve complex equation $z^6 + 8z^3 - 9 = 0$.

7. $(\sqrt{2i} - 1)^{2013}$ is which of the following?
   (1) 0   (2) 1   (3) i (4) π (5) $2013 - 2013i$.

11-2 Complex Functions

11-2-1 The Definition and Classification

In the realm of real numbers, a function cannot have a one-to-many relationship. However, in the world of complex numbers, this is possible. For example, in Section 11-1, when considering the square root function $w = f(z) = z^{\frac{1}{2}}$, it asks for what squares to give the complex number z. When restricting f to the unit circle $|e^{i\theta}| = 1$, it is discovered that there's not just one complex number on the unit circle that satisfies $z^2 = i$; in fact, $e^{i(\frac{\pi}{4}+2k\pi)}$ are all solutions, as shown in Figure 11-7. Complex functions that exhibit this kind of multi-valued behavior are called **multivalued functions**, whereas those that do not are called **single-valued functions**, such as $f(z) = z^2$. Naturally, this contradicts geometric intuition, and we will later address this issue using the concept of branch lines.

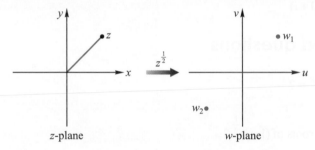

z-plane   w-plane

Figure 11-7   Diagram of Complex Function Square Roots

Inverse functions f^{-1} remain defined similarly to the real case, satisfying $f^{-1}f = f f^{-1} = 1$, it's important to note that the existence of an inverse function for a complex function doesn't necessarily require f to be one-to-one. For instance, $w = f(z) = z^2$, $z = f^{-1}(w) = w^{1/2}$ are good counterexamples. Additionally, it's crucial to understand that a single-valued complex function can be formally represented by the symbol $u + iv = f(x + iy)$, in other words, $\begin{cases} u = u(x, y) \\ v = v(x, y) \end{cases}$, it can be interpreted geometrically as mapping a region R in the z-plane (xy-plane) through $w = f(z)$ to a region R^* in the w-plane, as depicted in Figure 11-8. This notation will prove useful when discussing analytic functions later on.

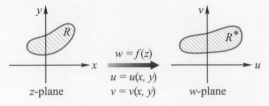

z-plane $u = u(x, y)$ w-plane
$v = v(x, y)$

Figure 11-8   Mapping of Figures on the Complex Plane

11-2-2 Branch Point & Branch Line of Multivalued Function

We start with multivalued functions, $f(z) = (z-\alpha)^{1/3}$, beginning with Euler's formula: $z - \alpha = r \cdot e^{i\theta}$; $0 < r$. Consider point A in the diagram below,

$f(z) = (z-\alpha)^{\frac{1}{3}} = r^{\frac{1}{3}} \cdot e^{i\frac{\theta}{3}}$, this can be divided into two cases:

1. Counterclockwise winding around $z = \alpha$
   for one loop leads to point A', at point A'
   $$f(z) = r^{\frac{1}{3}} e^{i\frac{\theta+2\pi}{3}} = r^{\frac{1}{3}} \cdot e^{i\frac{\theta}{3}} \cdot e^{i\frac{2\pi}{3}}, \text{ where } e^{i\frac{2\pi}{3}} \neq 1,$$
   then the function values $f(A) \neq f(A')$ at points
   A and A', resulting in multivalued behavior.

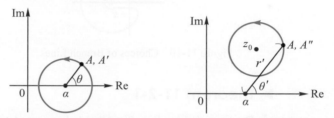

Figure 11-9   Mapping of Figures on the Complex Plane

2. Counterclockwise winding around $z = z_0$ for one loop, excluding the point α is A''.
   At point A'', $f(z) = r'^{\frac{1}{3}} e^{i\frac{\theta'}{3}}$, that is at points A and A'', the function values
   $f(A) = f(A'')$, leading to a single-valued behavior, as shown in Figure 11-9.

From this, it can be seen that even though we are circling around for one loop, the different trajectory leads to either multivalued or single-valued behavior. Thus, for multivalued functions, choosing the appropriate path of circling can restrict them into single-valued functions. In order to rationally interpret multivalued functions geometrically, we "cut" the function at the points where multiple values are generated:

▶ Definition 11-2-1

Branch Point and Branch Line

(1) **Branch Point**
   If the continuous function $f(z)$ changes its value after circling around a fixed point z_0, then $z - z_0$ is referred to as a branch point.

(2) **Branch Line**
   A deliberately chosen straight line or segment, the purpose of which is to transform the multivalued function $f(z)$ into a single-valued function. When the argument of z takes $-\pi < \theta < \pi$ (or $0 < \theta < 2\pi$), the value of $f(z)$ is referred to as the "principal value." By appropriately selecting a branch line, a multivalued function can be converted into a single-valued one.   ◀

1. **Branch Cut**

   Through branch lines, a multivalued function becomes single-valued (with infinitely many possible choices, as there are infinitely many ways to select branch lines). Two commonly used choices are as follows:

   (1) $\begin{cases} z - \alpha = re^{i\theta} \\ 0 < r < \infty \\ -\pi < \theta < \pi \end{cases}$, as shown in Figure11-10.

   (2) $\begin{cases} z - \alpha = re^{i\theta} \\ 0 < r < \infty \\ 0 < \theta < 2\pi \end{cases}$, as shown in Figure 11-11.

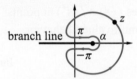

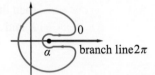

Figure 11-10　Choices of Branch Lines　　　Figure 11-11　Choices of Branch Lines

▶ **Theorem 11-2-1**

Determining the Single Value and Multivaluedness of Polynomial

Functions

During the process of $f(z) = (z - \alpha)^{\beta}$ circling $z = \alpha$, different values of β lead to either single-valued or multivalued behaviors, as discussed below.

$f(z) = (z - \alpha)^{\beta} = r^{\beta} e^{i\beta\theta}$, circling $z = \alpha$ counterclockwise for one loop results in $f(z) = r^{\beta} \cdot e^{i2\pi\beta}$:

(1) $\beta =$ integer, $e^{i2\pi\beta} = 1$, then $f(z)$ is single-valued.

(2) $\beta \neq$ integer, $e^{i2\pi\beta} \neq 1$, then $f(z)$ is multivalued, α is a branch point. ◀

2. **Complex Logarithm Function**

   The complex version of the logarithmic function is one of the most typical examples of branch cutting. Unlike the concept of the real logarithmic function, we employ exponentiation to extract logarithmic values. If $z = re^{i(\theta + 2k\pi)}$, $\theta = \text{Arg}(z)$ is the principal argument, $k = 0, \pm 1, \pm 2, \cdots$, then define

   $$\ln z = \ln r + i(\theta + 2k\pi).$$

   If the principal argument is taken, the resulting value $\ln z = \ln |z| + i\text{Arg}(z)$ is referred to as the principal value of $\ln z$, denoted as $\text{Ln}(z)$. In this case, $\ln(z)$ becomes a single-valued function.

3. Properties

(1) $z^a = e^{a \ln z}$; $\ln(z)^\alpha = \alpha \ln(z)$.

(2) $\ln(z_1 \cdot z_2) = \ln z_1 + \ln z_2$, $\ln(z_1 / z_2) = \ln z_1 - \ln z_2$, but
$\mathrm{Ln}(z_1 \cdot z_2) \neq \mathrm{Ln}(z_1) + \mathrm{Ln}(z_2)$.

(3) $e^{\ln z} = z$, $\ln(e^z) = \ln(e^{z+i2k\pi}) = z + i2k\pi \neq z$.

Note that in Property 3, the multivalued nature of the logarithm function restricts us to a one-way inverse function, which is a significant departure from the real logarithmic function. Now, let's revisit the issue of branch lines. If taking $z - \alpha = re^{i\theta}$, then taking the logarithm $w = \ln(z - \alpha) = \ln(r) + i\theta$. If we circle counterclockwise around $z = \alpha$ for one loop, we obtain $w = \ln(r) + i(\theta + 2\pi)$. Therefore, the choice of branch lines for the logarithm function is illustrated in Figure 10-12. This leads to $w = \ln(z - \alpha) = \ln(r) + i\theta$. More examples are provided in the following examples.

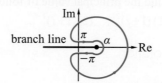

Figure 11-12   ln(z) Branch Diagram

Example 1

If $(z^2 - 1)^{\frac{1}{3}} (z-3)^{\frac{1}{3}}$, find the branch point and branch line.

Solution

(1) The branch point $z = \pm 1$, 3.

(2) The branch line: the rays drawn outward from the branch point are the branch lines. An example of this is shown in the following diagram.

Q.E.D.

Example 2

Try to make the following derivations:
(1) Converting i^i into complex polar form.   (2) Expanding $\ln(1 + i)$ using its definition.

Solution

(1) $i^i = e^{i \ln i} = e^{i(2n\pi + \frac{\pi}{2})i} = e^{-(2n\pi + \frac{\pi}{2})}$ $(n = 0, \ \pm 1, \ \pm 2 \cdots)$.

(2) $\ln(1 + i) = \ln\{\sqrt{2} e^{(2n\pi + \frac{\pi}{4})i}\} = \ln\sqrt{2} + (2n\pi + \frac{\pi}{4})i$ $(n = 0, \pm 1, \pm 2 \cdots)$.   Q.E.D.

Example 3

Find the principal value of $\ln(1-i\sqrt{3})$, and express it in the form of $a+ib$.

Solution

Because $1-i\sqrt{3} = 2e^{(-\frac{\pi}{3}+2n\pi)i}$, the principal value of $\ln(1-i\sqrt{3})$ is (taking $n=0$):

$$\operatorname{Ln}(1-i\sqrt{3}) = \operatorname{Ln}(2e^{-\frac{\pi}{3}i}) = \ln 2 - \frac{\pi}{3}i.$$ Q.E.D.

Example 4

Find the principal value of following complex:
(1) $(1+i)^{2i}$,   (2) $(1+i)^{1-i}$.

Solution

(1) $w = (1+i)^{2i} = e^{2i\ln(1+i)} = e^{2i[\ln\sqrt{2}+i\frac{\pi}{4}]} = e^{2i\ln\sqrt{2}} \cdot e^{-2\cdot\frac{\pi}{4}} = e^{i\ln 2} \cdot e^{-\frac{\pi}{2}}$

$= e^{-\frac{\pi}{2}}[\cos(\ln 2) + i\sin(\ln 2)]$ is the principal value of $(1+i)^{2i}$.

(2) $w = (1+i)^{1-i} = e^{(1-i)\ln(1+i)} = e^{(1-i)[\ln\sqrt{2}+i\frac{\pi}{4}]} = e^{(\ln\sqrt{2}+\frac{\pi}{4})+i(\frac{\pi}{4}-\ln\sqrt{2})}$

$= \sqrt{2} \cdot e^{\frac{\pi}{4}} \cdot [\cos(\frac{\pi}{4}-\ln\sqrt{2}) + i\sin(\frac{\pi}{4}-\ln\sqrt{2})]$

is the principal value of $(1+i)^{1-i}$.   Q.E.D.

11-2-3 Other Common Elementary Functions

1. **Polynomial Functions**

    $P(z) = a_n z^n + a_{n-1}z^{n-1} + \cdots + a_1 z + a_0$, where $a_n \neq 0$, $a_{n-1}, \cdots, a_1, a_0$ are complex.

2. **Rational Functions**

    $f(z) = \dfrac{P(z)}{Q(z)}$, ($P(z)$, $Q(z)$ all of them are polynomials of z).

    For example: $f(z) = \dfrac{z-1}{z^2+1}$.

3. **Exponential Function**

$$f(z) = e^z = e^{x+iy} = e^x \cdot [\cos y + i \sin y].^{1}$$

4. **Trigonometric Function**

$$\sin z = \frac{e^{iz} - e^{-iz}}{2i}, \quad \cos z = \frac{e^{iz} + e^{-iz}}{2}, \quad \tan z = \frac{\sin z}{\cos z},$$

$$\cot z = \frac{\cos z}{\sin z}, \quad \sec z = \frac{1}{\cos z}, \quad \csc z = \frac{1}{\sin z}.$$

▶ **Theorem 11-2-2**

The Properties of Elementary Functions

We can verify the following properties from the above definition:

(1) $\sin(-z) = -\sin z$, $\cos(-z) = \cos z$, $\tan(-z) = -\tan z$.

(2) $\sin^2 z + \cos^2 z = 1$, $1 + \tan^2 z = \sec^2 z$.

(3) $\sin(z_1 \pm z_2) = \sin z_1 \cos z_2 \pm \cos z_1 \sin z_2$, $\cos(z_1 \pm z_2) = \cos z_1 \cos z_2 \pm \sin z_1 \sin z_2$.

(4) $\sin(iz) = i \sinh z$, $\sinh(iz) = i \sin z$, $\cos(iz) = \cosh z$, $\cosh(iz) = \cos z$.

(5) $\begin{cases} \sin(z) = 0 \Leftrightarrow z = 0, \pm\pi, \pm 2\pi, \cdots \\ \cos(z) = 0 \Leftrightarrow z = \pm\dfrac{\pi}{2}, \pm\dfrac{3}{2}\pi, \cdots \end{cases}$, the roots are located on the real axis. ◀

5. **Hyperbolic Function**

$$\sinh(z) = \frac{e^z - e^{-z}}{2}, \quad \cosh(z) = \frac{e^z + e^{-z}}{2}, \quad \tanh(z) = \frac{\sinh(z)}{\cosh(z)},$$

$$\coth(z) = \frac{\cosh(z)}{\sinh(z)}, \quad \text{sech}(z) = \frac{1}{\cosh(z)}, \quad \text{csch}(z) = \frac{1}{\sinh(z)}.$$

---

[1] From the definition, it is known that for all complex numbers z, we have $e^z \neq 0$.

▶ **Theorem 11-2-3**

The Roots of Hyperbolic Function

$$\begin{cases} \sinh(z) = 0 \Leftrightarrow z = 0, \pm i\pi, \pm 2i\pi, \cdots \\ \cosh(z) = 0 \Leftrightarrow z = \pm i\dfrac{\pi}{2}, \pm i\dfrac{3\pi}{2}, \cdots \end{cases}$$, that is the roots of the hyperbolic functions are

located on the imaginary axis, as shown in Figures 11-13 and 11-14. Additionally, there exist angle sum and difference identities between hyperbolic functions and trigonometric functions: let $z = x + iy$, then $w = \sin z = \sin x \cosh y + i \cos x \sinh y$,
$\cos z = \cos (x + iy) = \cos x \cosh y - i \sin x \sinh y$.

Figure 11-13   The root of $\sinh(z) = 0$ Figure 11-14   The root of $\cosh(z) = 0$ Q.E.D.

6. Inverse Trigonometric and Hyperbolic Functions

Similar to solving equations, by expanding trigonometric functions according to their definitions to solve exponential functions (taking the principal value), we obtain the inverse sine function. If $z = \sin w$, then the inverse sine function is expressed as $w = \sin^{-1} z$. Below, we illustrate the process with a detailed example.[2]

Example 5

Please verify $\sin^{-1} z = \dfrac{1}{i} \ln(iz + \sqrt{1-z^2})$.

---

[2] For $\sqrt{1-z^2} = (1-z^2)^{1/2}$; let $1 - z^2 = r \cdot e^{i(\theta + 2k\pi)}$, then $\sqrt{1-z^2} = \sqrt{r} \cdot e^{i(\frac{\theta}{2} + k\pi)} = (-1)^k \sqrt{r} \cdot e^{i\frac{\theta}{2}}$.

Taking $k = 0$ as the principal value, then $\sqrt{1-z^2} = \sqrt{r} \cdot e^{i\frac{\theta}{2}}$, where $(-1)^k$ results in the cancellation of $(\pm)$ as well as the $(\pm)$ in front of the square root.

Solution

By $z = \sin w = \dfrac{e^{iw} - e^{-iw}}{2i} \Rightarrow e^{iw} - 2iz - e^{-iw} = 0 \Rightarrow e^{2iw} - 2ize^{iw} - 1 = 0$

$\Rightarrow e^{iw} = \dfrac{2iz \pm \sqrt{(2iz)^2 + 4}}{2} = iz \pm \sqrt{1 - z^2} \Rightarrow iw = \ln(iz + \sqrt{1 - z^2})$,

taking the principal argument without adding $2k\pi$, and taking 「+」 in front of root,

$w = \sin^{-1} z = \dfrac{1}{i} \ln(iz + \sqrt{1 - z^2})$.   Q.E.D.

Similarly, we can derive the inverse functions of other trigonometric and hyperbolic functions:

(1) $\cos^{-1} z = \dfrac{1}{i} \operatorname{Ln}(z + \sqrt{z^2 - 1})$,   (2) $\sinh^{-1} z = \operatorname{Ln}(z + \sqrt{z^2 + 1})$,

(3) $\cosh^{-1} z = \operatorname{Ln}(z + \sqrt{z^2 - 1})$.

Example 6

Find all the values of $\sin^{-1} 5$.

Solution

Let $z = \sin^{-1} 5$, so $\sin z = 5$, $\dfrac{e^{iz} - e^{-iz}}{2i} = 5$, then $e^{2iz} - 10ie^{iz} - 1 = 0$,

we obtain $e^{iz} = \dfrac{10i \pm \sqrt{-100 + 4}}{2} = 5i \pm 2\sqrt{6}i = (5 \pm 2\sqrt{6})e^{(2n\pi + \frac{\pi}{2})i}$ $(n = 0, \pm 1, \pm 2, \cdots)$,

taking the logarithm of both sides, we have $iz = \ln(5 \pm 2\sqrt{6}) + (2n\pi + \dfrac{\pi}{2})i$,

thus, $z = (2n\pi + \dfrac{\pi}{2}) - i\ln(5 \pm 2\sqrt{6})$ $(n = 0, \pm 1, \pm 2, \cdots)$.   Q.E.D.

Example 7

(1) Verify
$\sin z = \sin x \cosh(y) + i\cos x \sinh(y)$.
(2) Find the solution of $\sin z = \cosh(4)$.

Solution

(1) $\sin z = \sin(x+iy) = \sin x \cos(iy) + \sin(iy)\cos x = \sin x \cosh(y) + i \cos x \sinh(y)$.

(2) $\sin z = \cosh(4) \Rightarrow \sin x \cosh(y) + i \cos x \sinh(y) = \cosh(4)$

$$\Rightarrow \begin{cases} \sin x \cosh(y) = \cosh(4) \\ \cos x \sinh(y) = 0 \Rightarrow \sinh(y) = 0 \text{ or } \cos x = 0 \end{cases}.$$

①   when $\sinh(y) = 0 \Rightarrow y = 0 \Rightarrow \sin x = \cosh(4) > 1$ unsuited.

②   when $\cos x = 0 \Rightarrow x = (n+\dfrac{1}{2})\pi$,

$n = 0, \pm 1, \pm 2, \cdots$,

then $(-1)^n \cosh y = \cosh 4 \Rightarrow n = 0, \pm 2, \pm 4, \cdots$,

$y = \pm 4$, thus,

$$z = x + iy = (n+\dfrac{1}{2})\pi + i(\pm 4) \,; n = 0, \pm 2, \pm 4, \cdots. \quad \boxed{\text{Q.E.D.}}$$

Example 8

Let $z = a + ib$, and the equation is $e^z = i$.
(1) Solve for all values of a, b that satisfy the above equation.
(2) Determine the magnitude and argument arg (z) of z.

Solution

(1) Because $e^z = I$, $z = \ln(i) = \ln(\exp\{(2k\pi + \dfrac{\pi}{2})i\}) = (2k\pi + \dfrac{\pi}{2})i$,

$(k = 0, \pm 1, \pm 2, \cdots)$.

(2) Let $z = re^{\theta i}$, so $r = |z| = (2k\pi + \dfrac{\pi}{2})$, $(k = 0, \pm 1, \pm 2, \cdots)$ is the magnitude of z, and the

argument of z is $\theta = \arg z = 2m\pi + \dfrac{\pi}{2}$, $(m = 0, \pm 1, \pm 2, \cdots)$. $\boxed{\text{Q.E.D.}}$

11-2   Exercises

Basic questions

1. Solve the complex equation $z^2 = 1 + i$.

2. Find the principal value of $(3 + 4i)^{1/3}$.

3. Find all the values of $(1 - i)^{1+i}$.

4. Find all the values of $(2i)^{3i}$.

5. Find all the values of $(1 + i)^{2-i}$.

Advanced questions

1. Find all values of z that satisfy $\sin z = \sqrt{2}$.

2. Find all values of z that satisfy $\cos z = 20$.

3. Find all values of z that satisfy $e^z = 1$.

4. Express $\sin(i\sin i)$ in the form of $a + bi$.

5. Find $\sin(\dfrac{\pi}{2} + \sqrt{2}\,i)$.

6. $f(z) = \cos z$, $z = x + iy$, then $|f(z)| = ?$

7. Find $\sin^{-1} 3$.

8. If $\cos z = 2$, find $\cos 3z$.

11-3 Differentiation of Complex Functions

Next, we will introduce the differentiation of complex functions. As operations are conducted on the complex plane, the concepts and definitions are similar to the differentiation of functions on the real number plane. Therefore, the issue of existence is frequently encountered in complex function differentiation, and readers should pay special attention to it.

11-3-1 Limit

▶ **Definition 11-3-1**

Limit

Let $f(z)$ be a single-valued function defined in a neighborhood δ (which may not include z_0) of $z = z_0$. If for any $\varepsilon > 0$, there exists a real number $\delta > 0$, causes when $|z - z_0| < \delta$ satisfying $|f(z) - l| < \varepsilon$, then we say that when $z \to z_0$, $f(z)$ has a limit l. This is denoted as $\lim_{z \to z_0} f(z) = l$. ◀

From an intuitive perspective on the complex plane, the definition essentially states that no matter how z approaches $z = z_0$ in any manner or direction, $f(z)$ will always get closer to l. Hence, if the existence of the limit is contingent upon the path of approach, then the limit is considered nonexistent. This is illustrated in Figure 11-15. When dealing with functions that exhibit multivalued behavior (as discussed in the previous section), directly taking limit poses the well-defined problem. Therefore, to discuss the limit of multivalued functions, it is necessary to restrict their phase angles to a certain range within 2π, transforming them into single-valued functions. For instance: let

$$f(z) = z^{\frac{1}{2}}, \text{ and } -\pi < \arg z \le \pi, \text{ then } \lim_{z \to i} z^{\frac{1}{2}} = \lim_{z \to e^{i\pi/2}} z^{\frac{1}{2}} = e^{i\frac{\pi}{4}} = \frac{\sqrt{2}}{2} + i\frac{\sqrt{2}}{2}.$$

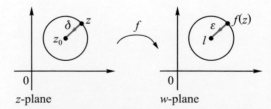

Figure 11-15   Illustration of Complex Limits

The next two theorems are direct consequences of the definitions.

▶ **Theorem 11-3-1**

The Limit Theorem

Let $f(z) = u(x, y) + iv(x, y)$, $z_0 = x_0 + iy_0$, and $w_0 = u_0 + iv_0$, then $\lim\limits_{z \to z_0} f(z) = w_0$ if and only if $\lim\limits_{(x, y) \to (x_0, y_0)} u(x, y) = u_0$, $\lim\limits_{(x, y) \to (x_0, y_0)} v(x, y) = v_0$. ◀

▶ **Theorem 11-3-2**

Rules of Limit Operations

Assuming that $\lim\limits_{z \to z_0} f(z) = A$, $\lim\limits_{z \to z_0} g(z) = B$, then

(1) $\lim\limits_{z \to z_0} [f(z) \pm g(z)] = A \pm B$.

(2) $\lim\limits_{z \to z_0} f(z)g(z) = AB$.

(3) $\lim\limits_{z \to z_0} \dfrac{f(z)}{g(z)} = \dfrac{A}{B}$, $(B \neq 0)$. ◀

11-3-2 Continuity

Once we have the definition of limits, similar to calculus of real functions, we also need to discuss the continuity of complex functions. For a complex function $f(z)$, if the following three conditions hold, then it is said to be continuous at $z = z_0$.

1. $f(z)$ is defined at $z = z_0 \Rightarrow f(z_0)$ exists.

2. $\lim\limits_{z \to z_0} f(z) = l$ exists.

3. $\lim\limits_{z \to z_0} f(z) = f(z_0)$.

▶ **Theorem 11-3-3**

Continuity Theorem

$f(z) = u + iv$ is continuous within R, then u, v must be continuous within R. ◀

11-3-3 Derivative

Having understood the limits and continuity of complex functions, we now proceed to discuss the derivatives and differentials of complex functions. Let $f(z)$ be a single-valued function defined in a neighborhood δ centered at $z = z_0$. Then, the derivative of $f(z)$ at $z = z_0$, $f'(z_0)$, is defined as:

$$f'(z_0) = \lim_{z \to z_0} \frac{f(z) - f(z_0)}{z - z_0} = \lim_{\Delta z \to 0} \frac{f(z_0 + \Delta z) - f(z_0)}{\Delta z}$$

If the limit $f'(z_0)$ exists, then $f(z)$ is said to be differentiable at z_0.

In general functions of z, if they involve $\bar{z}$ (the complex conjugate of z), their derivatives typically do not exist. Functions that are not differentiable in the real number system are also not differentiable in the complex function context, as seen in examples 1 and 2. The properties of complex function derivatives can be summarized as follows.

1. **The Operations of Derivatives**

    Let $f(z)$, $g(z)$ both be differentiable, then

    (1) $\dfrac{d}{dz}[af(z) \pm bg(z)] = a\dfrac{df(z)}{dz} \pm b\dfrac{dg(z)}{dz}$.

    (2) $\dfrac{d}{dz}[f(z)g(z)] = f(z)\dfrac{dg(z)}{dz} + g(z)\dfrac{df(z)}{dz}$.

    (3) $\dfrac{d}{dz}[\dfrac{f(z)}{g(z)}] = \dfrac{g(z)\dfrac{df(z)}{dz} - f(z)\dfrac{dg(z)}{dz}}{g^2(z)}$, where $g(z) \neq 0$.

    (4) $w = f(\xi)$, $\xi = g(z)$, then $\dfrac{dw}{dz} = \dfrac{dw}{d\xi}\dfrac{d\xi}{dz} = f'(\xi)\dfrac{d\xi}{dz} = f'(\xi)g'(z)$.

    (5) $w = f(z)$, then $z = f^{-1}(w) \Rightarrow \dfrac{dw}{dz} = \dfrac{1}{\dfrac{dz}{dw}}$.

2. **L'Hôpital's Rule**

    In calculus of real functions, when encountering an indeterminate form involving $\dfrac{0}{0}$, L'Hôpital's rule is often applied, this rule continues to hold in complex function calculus. Let $f(z)$, $g(z)$ have derivatives of all orders within a region R that includes $z = z_0$, and $f(z_0) = g(z_0) = 0$, but $g'(z_0) \neq 0$, then

$$\lim_{z \to z_0} \frac{f(z)}{g(z)} = \lim_{z \to z_0} \frac{f'(z)}{g'(z)} = \frac{f'(z_0)}{g'(z_0)}$$

Example 1

Find $\lim\limits_{z \to 0} \dfrac{\bar{z}}{z}$.

Solution

$$\lim_{z \to 0} \frac{\bar{z}}{z} = \lim_{(x, y) \to (0, 0)} \frac{x - iy}{x + iy}.$$

① Approach first from the x-axis, and then from the y-axis.

$$\lim_{z \to 0} \frac{\bar{z}}{z} = \lim_{y \to 0} \lim_{x \to 0} \frac{x - iy}{x + iy} = \lim_{y \to 0} \frac{-iy}{iy} = -1 ;$$

② Approach first from the y -axis, and then from the x -axis,

$$\lim_{z \to 0} \frac{\bar{z}}{z} = \lim_{x \to 0} \lim_{y \to 0} \frac{x - iy}{x + iy} = \lim_{x \to 0} \frac{x}{x} = 1 ,$$

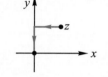

$\therefore \lim\limits_{z \to 0} \dfrac{\bar{z}}{z}$ does not exist.

Q.E.D.

Example 2

Find $\lim\limits_{z \to 0} \dfrac{z^2}{|z|^2}$.

Solution

$$f(z) = \frac{(x + iy)^2}{x^2 + y^2} , \quad \lim_{z \to 0} f(z) = \lim_{\substack{x \to 0 \\ y \to 0}} \frac{(x + iy)^2}{x^2 + y^2} , \text{ let } y = mx,$$

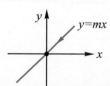

$$\lim_{z \to 0} f(z) = \lim_{x \to 0} \frac{(x + imx)^2}{x^2 + m^2 x^2} = \frac{(1 + im)^2}{1 + m^2} \text{ related to } m$$

$\Rightarrow$ Different slopes m lead to different limit values $\rightarrow$

$\therefore \lim\limits_{z \to 0} \dfrac{z^2}{|z|}$ does not exist.

Q.E.D.

11-3-4 Analytic Function

If $f(z)$ is differentiable at every point within region R, then $f(z)$ is said to be analytic within R. Specifically, if there exist $\delta > 0$, and $\delta \to 0^+$, causes $f(z)$ is differentiable within $|z - z_0| < \delta$, then $f(z)$ is said to be analytic at z_0, as illustrated in Figure 11-16. If the limit of a complex function on the complex plane exists, it is independent of the path taken to approach the differentiable points. This independence of path leads to a set of equations that serve as tools for determining whether a function is analytic or not.

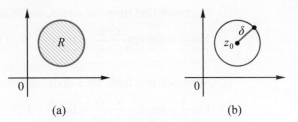

(a)   (b)

Figure 11-16   Diagram of analytic region

▶ **Theorem 11-3-4**

Cauchy-Riemann Equation

Let $f(z) = u(x, y) + iv(x, y)$ is continuous within a neighborhood of δ at $z = z_0 \equiv x_0 + iy_0$,

then $f'(z_0)$ exists if and only if
$$\begin{cases} \dfrac{\partial u}{\partial x} = \dfrac{\partial v}{\partial y} \\[2mm] \dfrac{\partial u}{\partial y} = -\dfrac{\partial v}{\partial x} \end{cases}.$$

The system of equations described above is known as the **Cauchy-Riemann equations**.

Proof

According to the definition of differentiation, we have the following equation:

$$f'(z_0) = \lim_{\substack{\Delta x \to 0 \\ \Delta y \to 0}} \frac{[u(x_0 + \Delta x, y_0 + \Delta y) - u(x_0, y_0)] + i[v(x_0 + \Delta x, y_0 + \Delta y) - v(x_0, y_0)]}{\Delta x + i\Delta y}.$$

Given the premises, this limit exists. Hence, the result remains the same regardless of the direction of approach.

1. First $\Delta y \to 0$, then $\Delta x \to 0$:

$$f'(z_0) = \lim_{\Delta x \to 0} \left\{ \frac{u(x_0 + \Delta x, y_0) - u(x_0, y_0)}{\Delta x} + i\frac{v(x_0 + \Delta x, y_0) - v(x_0, y_0)}{\Delta x} \right\}$$

$$= \frac{\partial u}{\partial x}\bigg|_{(x_0, y_0)} + i\frac{\partial v}{\partial x}\bigg|_{(x_0, y_0)}.$$

2. First $\Delta x \to 0$, then $\Delta y \to 0$:

$$f'(z_0) = \lim_{\Delta y \to 0} \frac{u(x_0, y_0 + \Delta y) - u(x_0, y_0)}{i\Delta y} + i \frac{v(x_0, y_0 + \Delta y) - v(x_0, y_0)}{i\Delta y}$$

$$= \frac{1}{i} \frac{\partial u}{\partial y}\bigg|_{(x_0, y_0)} + \frac{\partial v}{\partial y}\bigg|_{(x_0, y_0)} = \frac{\partial v}{\partial y}\bigg|_{(x_0, y_0)} - i \frac{\partial u}{\partial y}\bigg|_{(x_0, y_0)}.$$

therefore $\begin{cases} \dfrac{\partial u}{\partial x} = \dfrac{\partial v}{\partial y} \\ \dfrac{\partial u}{\partial y} = -\dfrac{\partial v}{\partial x} \end{cases}$, the proof is certified.   ◀

Next are some commonly used theorems, most of which are variations of the Cauchy-Riemann equations. Readers are encouraged to practice proving these theorems themselves.

▶ **Theorem 11-3-5**

Analyticity

$f(z) = u(x, y) + iv(x, y)$ at every point in $z \in R$, if the first-order partial derivatives exist and are continuous, if and only if f is analytic on R (as seen in Theorems 11-3-7~11-3-8).   ◀

▶ **Theorem 11-3-6**

Condition for Analyticity at a Point

Let $f(z) = u(x, y) + iv(x, y)$ and $z = z_0 = x_0 + iy_0$,

then $f(z) = u + iv$ is differentiable at $z = z_0 \Leftrightarrow \begin{cases} \dfrac{\partial u}{\partial x} = \dfrac{\partial v}{\partial y} \\ \dfrac{\partial u}{\partial y} = -\dfrac{\partial v}{\partial x} \end{cases}$.   ◀

▶ **Theorem 11-3-7**

Condition for Function to Be Analytic

Let $f(z) = u(x, y) + iv(x, y)$, the first-order partial derivatives exist and are continuous within a region R in the z-plane,

then $f(z)$ is an analytic function within $R \Leftrightarrow \begin{cases} \dfrac{\partial u}{\partial x} = \dfrac{\partial v}{\partial y} \\ \dfrac{\partial u}{\partial y} = -\dfrac{\partial v}{\partial x} \end{cases}$.   ◀

▶ **Theorem 11-3-8**

Constant Functions

Let $f(z)$ is analytic within R and for every z in R has $f'(z) = 0$, then $f(z)$ is a constant function within R.

Proof

Let $f(z) = u(x, y) + iv(x, y)$, $\because f(z)$ is analytic within R, $\therefore$ it follows that for every point within R, $f'(z)$ always exists, and

$$f'(z) = \frac{\partial u}{\partial x} + i\frac{\partial v}{\partial x} = \frac{\partial v}{\partial y} - i\frac{\partial u}{\partial y} = 0 \Rightarrow \frac{\partial u}{\partial x} = \frac{\partial v}{\partial x} = \frac{\partial v}{\partial y} = \frac{\partial u}{\partial y} = 0,$$

thus, both u, v are constants. $\Rightarrow f(z)$ is a constant function. ◀

▶ **Theorem 11-3-9**

The Cauchy-Riemann Equations in Polar Coordinates

Let $f(z) = u(r, \theta) + iv(r, \theta) \in C'$ at $z = z_0 \neq 0$, then $f'(z_0)$ exists $\Leftrightarrow \begin{cases} \dfrac{\partial u}{\partial r} = \dfrac{1}{r}\dfrac{\partial v}{\partial \theta} \\ \dfrac{\partial v}{\partial r} = -\dfrac{1}{r}\dfrac{\partial u}{\partial \theta} \end{cases}$.

The above equations are referred to as the polar coordinate form of the Cauchy-Riemann equations. ◀

Example 3

Let $f(z) = \begin{cases} \dfrac{(\bar{z})^2}{z} - 0, z \neq 0 \\ 0, z = 0 \end{cases}$, prove that f is not differentiable at $z = 0$.

Solution

$$f'(0) = \lim_{z \to 0}\frac{f(z) - f(0)}{z} = \lim_{z \to 0}\frac{\frac{(\bar{z})^2}{z} - 0}{z} = \lim_{z \to 0}\frac{(\bar{z})^2}{z^2} = \lim_{\substack{x \to 0 \\ y \to 0}}\frac{(x - iy)^2}{(x + iy)^2},$$

we take $y = mx$, then $f'(0) = \lim_{x \to 0}\frac{(1 - im)^2}{(1 + im)^2} = \frac{(1 - im)^2}{(1 + im)^2}$, which is related to the value of

$m \Rightarrow f'(z)\big|_{z=0}$ does not exist $\Rightarrow f(z)$ is not differentiable at $z = 0$. **Q.E.D.**

Example 4

Assuming that $f(z) = e^x(x\cos y - y\sin y) + ie^x(y\cos y + x\sin y)$.

(1) Prove that $f(z)$ is analytic.

(2) Determine the points where $f'(z)$ exists and find its value.

Solution

(1) $\begin{cases} u = e^x(x\cos y - y\sin y) \\ v = e^x(y\cos y + x\sin y) \end{cases}$, $f(z) = u + iv$, then the first-order partial derivatives of u, v

exist and are continuous on the complex plane,

also $\dfrac{\partial u}{\partial x} = e^x(x\cos y - y\sin y + \cos y)$; $\dfrac{\partial v}{\partial y} = e^x(\cos y - y\sin y + x\cos y)$;

$\dfrac{\partial u}{\partial y} = e^x(-x\sin y - \sin y - y\cos y)$; $\dfrac{\partial v}{\partial x} = e^x(y\cos y + x\sin y + \sin y)$;

we obtain $\dfrac{\partial u}{\partial x} = \dfrac{\partial v}{\partial y}$, $\dfrac{\partial u}{\partial y} = -\dfrac{\partial v}{\partial x}$, $\therefore f(z)$ is analytic at the complex plane.

(2) $f'(z) = \dfrac{\partial u}{\partial x} + i\dfrac{\partial v}{\partial x} = e^x(x\cos y - y\sin y + \cos y) + ie^x(x\sin y + \sin y + y\cos y)$

$\qquad = e^x[x(\cos y + i\sin y) + (\cos y + i\sin y) + y(i\cos y - \sin y)]$

$\qquad = e^x[x \cdot e^{iy} + e^{iy} + iy \cdot e^{iy}]$

$\qquad = e^x \cdot e^{iy}[x + iy + 1] = e^{x+iy}(x + iy + 1)$

$\qquad = e^z(z+1)$.

Q.E.D.

The concept of analytic functions, unique to complex function theory, sets it apart from the general definitions in calculus. Here, we will introduce several common applications of this concept.

1. Entire Function

$f(z)$ is analytic at every point in C, then $f(z)$ is called an entire function, for instance, $\sin z$, z^2, $e^z, \cdots$ are all entire functions. Liouville theorem states that: if an entire function is bounded, then it must be a constant function. Additionally, the following theorems are also highly representative:

▶ **Theorem 11-3-10**

Harmonic Function

Let the second-order partial derivatives of $\varphi(x, y)$ exist and are continuous in R^2, and

$\nabla^2\varphi = \dfrac{\partial^2\varphi}{\partial x^2} + \dfrac{\partial^2\varphi}{\partial y^2} = 0$ then $\varphi(x, y)$ is called a harmonic function within R^2. Furthermore,

let $f(z) = u(x, y) + iv(x, y)$ is analytic within C, then u, v must be harmonic functions within R^2, which implies

$$\nabla^2 u = \frac{\partial^2 u}{\partial x^2} + \frac{\partial^2 u}{\partial y^2} = 0, \quad \nabla^2 v = \frac{\partial^2 v}{\partial x^2} + \frac{\partial^2 v}{\partial y^2} = 0.$$

Proof

Because $f(z) = u(x, y) + iv(x, y)$ is analytic within R^2, the Cauchy-Riemann

equations state that $\begin{cases} \dfrac{\partial u}{\partial x} = \dfrac{\partial v}{\partial y} \\ \dfrac{\partial u}{\partial y} = -\dfrac{\partial v}{\partial x} \end{cases}$, then partial differentiating both equations

individually, we obtain $\dfrac{\partial^2 u}{\partial x^2} = \dfrac{\partial^2 u}{\partial x\partial y}$, $\dfrac{\partial^2 u}{\partial y^2} = -\dfrac{\partial^2 u}{\partial y\partial x}$, also $u \in C(R^2)^2$ at R^2

$\Rightarrow \dfrac{\partial^2 u}{\partial x\partial y} = \dfrac{\partial^2 u}{\partial y\partial x} \Rightarrow \dfrac{\partial^2 u}{\partial x^2} + \dfrac{\partial^2 u}{\partial y^2} = 0 \Rightarrow \nabla^2 u = 0$, similarly, $\nabla^2 v = 0$.[3]   ◀

In other words, the real and imaginary parts of an analytic function serve as solutions to the Laplace's equation $\dfrac{\partial^2\varphi}{\partial x^2} + \dfrac{\partial^2\varphi}{\partial y^2} = 0$.

2. **Orthogonal of Families of Curves**

Let $f(z) = u(x, y) + iv(x, y)$ be a single-valued function. Then, the families of curves on the z-plane, $u(x, y) = c$ and $v(x, y) = d$ are referred to as the level curves of $f(z)$ (the contour lines of u, v). Interestingly, the real and imaginary parts of an analytic function are mutually orthogonal, as demonstrated by Theorem 11-3-11 below.

[3] Let u, v be harmonic functions within R and satisfy the Cauchy-Riemann equations, that is $\begin{cases} \dfrac{\partial u}{\partial x} = \dfrac{\partial v}{\partial y} \\ \dfrac{\partial u}{\partial y} = -\dfrac{\partial v}{\partial x} \end{cases}$, then v is referred to as the harmonic conjugate of u within R.

▶ Theorem 11-3-11

Orthogonal Curves

Let $f(z) = u(x, y) + iv(x, y)$ be an analytic function. Then, the level curves $u(x, y) = c$ and $v(x, y) = d$ at $f'(z) \neq 0$ are orthogonal families of curves, as illustrated in Figure 11-17.

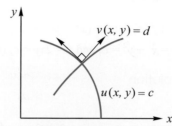

Figure 11-17   Orthogonal Level Curves Illustration

Proof

Considering y as a function of x, differentiate both ends of $u(x, y) = c$ with respect to x,

resulting in: $\dfrac{\partial u}{\partial x} + \dfrac{\partial u}{\partial y}\dfrac{dy}{dx} = 0$, we obtain the slope $m_1 = \dfrac{dy}{dx} = -\dfrac{\dfrac{\partial u}{\partial x}}{\dfrac{\partial u}{\partial y}}$; Similarly, for

$v(x, y) = d$, the slope is $m_2 = \dfrac{dy}{dx} = -\dfrac{\dfrac{\partial v}{\partial x}}{\dfrac{\partial v}{\partial y}}$, on the other hand, from the Cauchy-

Riemann equations: $\Rightarrow u_x = v_y,\ v_x = -u_y,\ \therefore m_1 m_2 = -1$, so the two families of curves are orthogonal.   ◀

▶ Theorem 11-3-12

Relationship Between Analytic Functions and $\bar{z}$

Let w be an analytic function of z, then $\dfrac{\partial w}{\partial \bar{z}} = 0$ (w is not dependent on $\bar z$), meaning

that w is always an explicit function of z.[4]   ◀

---

[4] If the first-order partial derivatives of $w(z, \bar{z})$ are continuous in the region R, then $\dfrac{\partial w}{\partial \bar{z}} = 0 \Leftrightarrow \begin{cases} \dfrac{\partial u}{\partial x} = \dfrac{\partial v}{\partial y} \\ \dfrac{\partial u}{\partial y} = -\dfrac{\partial v}{\partial x} \end{cases} \Leftrightarrow w(z, \bar{z})$ is analytic within R.

The following examples will make use of the application concepts of analytic functions discussed earlier, which will aid readers in understanding the distinctive features of analytic functions.

Example 5

Let $f(z) = z^2 = (x^2 - y^2) + i2xy$, prove that the level curves of $f(z)$ defined by $u = x^2 - y^2 = c_1$ and $v = 2xy = c_2$ (where $c_1, c_2 \neq 0$) are orthogonal.

Solution

On the curve of $u = x^2 - y^2 = c_1$, $\dfrac{dy}{dx} = -\dfrac{\dfrac{\partial u}{\partial x}}{\dfrac{\partial u}{\partial y}} = \dfrac{x}{y}$; on the curve defined by

$v = 2xy = c_2$, $\dfrac{dy}{dx} = -\dfrac{\dfrac{\partial v}{\partial x}}{\dfrac{\partial v}{\partial y}} = -\dfrac{y}{x}$, $\therefore$ at points $x \neq 0$, $y \neq 0$, we have

$\left(\dfrac{dy}{dx}\Big|_{u=c_1}\right) \cdot \left(\dfrac{dy}{dx}\Big|_{v=c_2}\right) = -1$, so $u = x^2 - y^2 = c_1$ and $v = 2xy = c_2$ are orthogonal.

Q.E.D.

Example 6

(1) Please verify that $u = x^2 - y^2 - y$ is a harmonic function.

(2) If u is a harmonic function, please find its conjugate harmonic function.

Solution

(1) Because $\nabla^2 u = \dfrac{\partial^2 u}{\partial x^2} + \dfrac{\partial^2 u}{\partial y^2} = 2 - 2 = 0$, $u(x, y)$ is a harmonic function.

(2) Let $v(x, y)$ be the conjugate harmonic function of $u(x, y)$,

then $\begin{cases} \dfrac{\partial v}{\partial y} = \dfrac{\partial u}{\partial x} = 2x \\ \dfrac{\partial v}{\partial x} = \dfrac{-\partial u}{\partial y} = 2y+1 \end{cases} \Rightarrow \begin{cases} v(x, y) = 2xy + f(x) \\ v(x, y) = 2xy + x + g(y) \end{cases}$, then by comparison, we

obtain $v(x, y) = 2xy + x + c$, where c is an arbitrary constant.

Q.E.D.

Example 7

Please verify if $f(z) = z^*$ is an analytic function, where z^* is the complex conjugate of z.

Solution

$\dfrac{\partial f}{\partial z^*} = 1 \neq 0$, so $f(z)$ is a non-analytic function.

`Q.E.D.`

11-3-5 Singularity

If $\lim\limits_{z \to z_0} \dfrac{f(z) - f(z_0)}{z - z_0}$ does not exist, then z_0 is referred to as a **singularity** of $f(z)$. If within a certain circle (such as $|z - z_0| < \delta$), $f(z)$ is analytic everywhere except at z_0, then z_0 is called an **isolated singularity point** of $f(z)$, for example: $f(z) = \dfrac{1}{z-1}$, then $z = 1$ is an **isolated singularity point** of $f(z)$, as depicted in Figure 11-18. Typically, the singularities we often deal with are isolated.

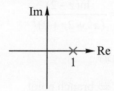

Figure 11-18   Isolated Singular Point $z = 1$

1. **Pole**

   If a singularity can be removed from the original function $f(z)$ by multiplying it with a polynomial, it is called a **pole**. Specifically, if there exists a positive integer m such that $\lim\limits_{z \to a}(z-a)^m f(z)$ exists and is nonzero, then $z = a$ is referred to as a **pole of order m for $f(z)$**. Poles are particularly crucial in the calculation of complex variable integrals and require special attention.

2. **Zero**

   Suppose $f(z) = (z-a)^m g(z)$, and $g(z)$ at $z = a$ is analytic also $g(a) \neq 0$, then $z = a$ is **zero of multiplicites m** of $f(z)$. When the zero of multiplicites m, $z = a$ appears in the denominator of the simplest rational expression for $f(z) = \dfrac{P(z)}{Q(z)}$, by definition, $z = a$ becomes an m-order pole of $f(z)$; while the zeros of the numerator $P(z)$ become the zero of $f(z)$.

Example 8

Try classifying the singular points and poles in $f(z) = \dfrac{z^3 + 2}{z(z-1)^2(z^2+1)^3}$.

Solution

$z = 0 \Rightarrow$ 1st order pole (isolated).
$z = 1 \Rightarrow$ 2nd order pole (isolated).
$z = \pm i \Rightarrow$ 3rd order pole (isolated).

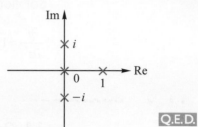

Q.E.D.

3. Branch Point

In the singular points of a multivalued function, the critical points that give rise to multiple values are referred to as branch points.

Example 9

$f(z) = \dfrac{\ln(z-2)}{(z^2+2z+2)^4}$

Solution

$\begin{cases} z = 2 \Rightarrow \text{branch point} \\ z = -1 \pm i \Rightarrow \text{4th order pole} \end{cases}$.

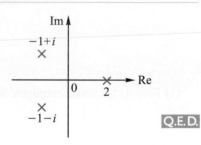

Q.E.D.

4. Removable Singularity

Suppose $f(a)$ does not exist, but $\lim\limits_{z \to a} f(z)$ exists, then $z = a$ is called a removable singularity of $f(z)$.

Example 10

Prove that $z = 0$ is a removable singularity of $f(z) = \dfrac{\sin z}{z}$.

Solution

$\because f(0)$ does not exist, but $\lim\limits_{z \to 0} \dfrac{\sin z}{z} = 1$,

$\therefore z = 0$ is a removable singularity of $f(z)$.

Q.E.D.

5. **Essential Singularity**

Let $z = a$ be a singularity of $f(z)$, that is neither a pole, nor a branch point, nor a removable singularity, then $z = a$ is referred to as an essential singularity of $f(z)$. For example, within $f(z) = e^{\frac{1}{z}}$, it is impossible to multiply any polynomial of the form z^m that would make the limit $\lim\limits_{z \to 0} z^m e^{\frac{1}{z}}$ exist, simultaneously, it cannot be a branch point or a removable singularity. Thus, $z = 0$ is an essential singularity of $f(z)$.

11-3   Exercises

Basic questions

Please identify the location and type of singular points for the given functions in questions 1~3.

1. $f(z) = \dfrac{1}{(z-1)(z+2)}$.

2. $f(z) = \dfrac{1}{(z+3)^3 (z+4)^2}$.

3. $f(z) = \dfrac{1}{z^2 - 1}$.

4. If $u(x, y) = 2x - x^3 + 3xy^2$ is a harmonic function,
   (1) Find its harmonic conjugate function $v(x, y)$.
   (2) For the analytic function
   $f(z) = u(x, y) + iv(x, y)$, find $f'(z)$.

5. If $u(x, y) = e^x \cos y$ is a harmonic function, find a function $v(x, y)$ that causes $f(z) = u + iv$ to be an analytic function.

6. If $u - 3xy^2 - r^3$ is a harmonic function, find a function $v(x, y)$ that causes $f(z) = u + iv$ to be an analytic function.

7. Please verify $f(z) = (2x^2 + y) + i(y^2 - x)$ where $z = x + iy$ is a complex, is non-analytic at every point in the complex plane.

8. Please verify whether the following functions are analytic in the complex plane. ($z = x + iy$)
   (1) $f(z) = x^2 + y^2$.
   (2) $f(z) = e^{x - iy}$.
   (3) $f(z) = (x^2 - y^2) + i2xy$.

Advanced questions

1. Please explain the locations and types of singular points for the following functions.

   (1) $f(z) = \dfrac{2z}{(z^2 - 4)^2}$.

   (2) $f(z) = \dfrac{\ln(z-2)}{(z^2 - 4) \cdot z^2}$.

   (3) $f(z) = e^{\frac{1}{z-3}}$.

2. Please explain the locations and types of singular points for the following functions.

   (1) $f(z) = \dfrac{1}{z - \sin z}$.   (2) $f(z) = \dfrac{z}{1 - \cos z}$.

3. If $f(z) = u(x, y) + iv(x, y)$ is an analytic function, please provide the missing parts for the following subquestions.

   (1) $u(x, y) = x^3 - 3xy^2$, $v(x, y) = ?$ $f(z) = ?$

   (2) $v(x, y) = e^x \sin y$, $u(x, y) = ?$ $f(z) = ?$

4. Determine where
   $f(z) = 2x - x^3 - xy^2 + i(x^2 y + y^3 - 2y)$
   is differentiable and where it is analytic.

5. Determine the singular points of the following functions and classify them.

   (1) $f(z) = \dfrac{z+1}{(z-2)(z^2 + 1)}$.

   (2) $f(z) = \dfrac{1}{\sin(\frac{1}{z})}$.   (3) $f(z) = \dfrac{\sin \sqrt{z}}{\sqrt{z}}$.

11-4 Integration of Complex Functions

In this section, we will introduce the integration of complex functions, a concept similar to line integrals and double integrals in vector calculus.

11-4-1 Line Integral in the Complex Plane

We define the integral of a real-valued function $f(x)$ using Riemann sums. In the complex plane, we adopt a similar concept, albeit with the domain being the complex plane. Consequently, we will have a form resembling that used in defining line integrals.

▶ **Definition 11-4-1**

The Definition of Integration

Let $f(z)$ be a continuous function along a smooth curve C,

then $\int_C f(z)dz = \lim\limits_{\substack{n\to\infty \\ \max|\Delta z_i|\to 0}} \sum\limits_{i=1}^{n} f(\xi_i)\Delta z_i$, as shown in Figure 11-19.

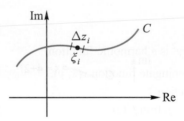

Figure 11-19 Line Integral in the Complex Plane ◀

Based on the definition, we can know: if $f(z) = u(x, y) + iv(x, y)$, then by $dz = dx + idy$, we obtain

$$\int_C f(z)dz = \int_C (u+iv)(dx+idy) = \int_C (udx - vdy) + i\int_C (vdx + udy).$$

The following properties derived directly from the definition:

The Operations of Integral

(1) $\int_C [\alpha f(z) + \beta g(z)]dz = \alpha \int_C f(z)dz + \beta \int_C g(z)dz$.

(2) $\int_A^B f(z)dz = -\int_B^A f(z)dz$, here, A, B represent the two endpoints of a certain curve C.

(3) $\int_{C_1+C_2} f(z)dz = \int_{C_1} f(z)dz + \int_{C_2} f(z)dz$.

▶ **Theorem 11-4-1**

ML Theorem

Assume that $|f(z)| \leq M$ is on the curve C, and the length of curve C is denoted as L, then $\left| \int_C f(z)dz \right| \leq ML$.

Proof

$$\left| \int_C f(z)dz \right| = \lim_{\substack{n \to \infty \\ \max\{\Delta z_i\} \to 0}} \left| \sum_{i=1}^{n} f(\xi_i)\Delta z_i \right| \leq \lim_{\substack{n \to \infty \\ \max\{\Delta z_i\} \to 0}} \sum_{i=1}^{n} |f(\xi_i)| \, |\Delta z_i|$$

$$= \int_C |f(z)| \, |dz| \leq M \int_C |dz| = ML,$$

where $|dz| = |dx + idy| = \sqrt{(dx)^2 + (dy)^2} = ds \Rightarrow \int_C |dz| = L$.   ◀

Example 1

Compute the line integral $\int_C \bar{z}\,dz$ from $z = 0$ to $z = 4 + 2i$ along the curves C provided in (1) and (2),

(1) $C: \ z = t^2 + it$.

(2) C is a straight line from $z = 0$ to $z = 2i$. then from $z = 2i$ to $z = 4 + 2i$.

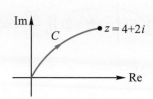

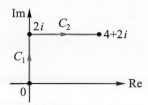

Solution

(1) $z = t^2 + it \Rightarrow \begin{cases} x = t^2 \\ y = t \end{cases}, t \in [0, 2],$

$$\int_C \bar{z}\,dz = \int_C (x - iy)(dx + idy) = \int_C (xdx + ydy) + i(xdy - ydx)$$

$$= \int_0^2 (2t^3 + t)dt + i\int_0^2 (t^2 - 2t^2)dt = 10 - \frac{8}{3}i.$$

(2) $C_1 : \begin{cases} x = 0 \\ y = t \end{cases}, t \in [0, 2]$; $C_2 : \begin{cases} x = t \\ y = 2 \end{cases}, t \in [0, 4]$, $\int_{C_1} \bar{z} \, dz = \int_0^2 (-it) \cdot (idt) = \int_0^2 t \, dt = 2$;

$\int_{C_2} \bar{z} \, dz = \int_{C_2} (t - 2i) \cdot dt = \int_0^4 (t - 2i) dt = 8 - 8i$,

$\therefore \int_C \bar{z} \, dz = \int_{C_1} \bar{z} \, dz + \int_{C_2} \bar{z} \, dz = 10 - 8i$.[5]

Q.E.D.

11-4-2 The Complex Form of Green's Theorem

Recalling the Green's theorem for real-valued functions mentioned in Chapter 8: let $P(x, y)$, $Q(x, y)$ exist and have continuous first partial derivatives in a region R and along its boundary C, then: $\iint_R [\frac{\partial Q}{\partial x} - \frac{\partial P}{\partial y}] dx dy = \oint_C P dx + Q dy$,

where R is a singly (multiply) connected region in the xy-plane, C is the boundary of R, and C is oriented counterclockwise with respect to R, as shown in Figure 11-20.

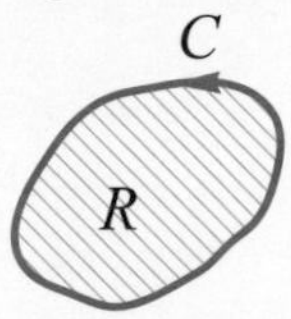

Figure 11-20

Complex Version of Green's Theorem

Let $F(z, \bar{z})$ be continuous defined in a certain region R and its boundary C in the z-plane, and first partial derivatives $\frac{\partial F}{\partial z}$, $\frac{\partial F}{\partial \bar{z}} \in C$. Then, the theorem states:

$$\oint_C F(z, \bar{z}) dz = 2i \iint_R \frac{\partial F}{\partial \bar{z}} dx dy$$

This is illustrated in Figure 11-21.

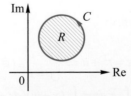

Figure 11-21   Region R in the complex plane

---

[5] From Example 1, it is evident that the line integral of a complex function is generally dependent on the chosen path.

Example 2

Assuming the closed curve C is the boundary of a square with vertices at $z = 0$, 1, $1 + i$, i and if the curve C is oriented counterclockwise, calculate $\oint_C \pi \cdot e^{(\pi z)} dz$.

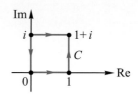

Solution

$$F(z, \bar{z}) = \pi \cdot e^{\pi \bar{z}}, \text{ then } \frac{\partial F}{\partial \bar{z}} = \pi^2 e^{\pi \bar{z}},$$

$$\therefore \oint_C \pi e^{\pi \bar{z}} dz = 2i \int_0^1 \int_0^1 \pi^2 e^{\pi \bar{z}} dxdy = 2i\pi^2 \int_0^1 \int_0^1 e^{\pi x} \cdot e^{-i\pi y} dxdy$$

$$= 2i\pi \int_0^1 e^{-i\pi y} (e^\pi - 1) dy = 2i\pi (e^\pi - 1) \cdot \frac{1}{-i\pi} e^{-i\pi y} \Big|_0^1$$

$$= -2(e^\pi - 1) \cdot [e^{-i\pi} - e^0] = -2(e^\pi - 1) \cdot (-1 - 1) = 4(e^\pi - 1) . \quad \boxed{\text{Q.E.D.}}$$

Example 3

Find $\oint_C \operatorname{Re}(z) dz$, where C is the boundary of the right half-circle with a radius of 1 on the complex plane as depicted in the figure. 〔Hint: $\operatorname{Re}(z) = \frac{1}{2}(z + \bar{z})$ 〕

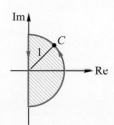

Solution

$$\oint_C \operatorname{Re}(z) dz = \oint_c \frac{1}{2}(z + \bar{z}) dz = 2i \iint_R \frac{\partial}{\partial \bar{z}} [\frac{1}{2}(z + \bar{z})] dxdy = i \iint_R 1 \cdot dxdy = \frac{\pi}{2} i . \quad \boxed{\text{Q.E.D.}}$$

11-4-3 Cauchy's Integral Theorem

Next, let's discuss what can be considered one of the most important theorems in the theory of complex functions—the Cauchy's integral theorem. Aside from directly deriving the values of complex functions at singular points (residue theorem), complex versions of Taylor series and Laurent series are also major implications of Cauchy's integral theorem. A true understanding of complex functions, as well as applications like improper integral of rational functions, all start with Cauchy's theorem. It's crucial for readers to familiar with this theorem.

▶ **Theorem 11-4-2**

Cauchy's Integral Theorem

Suppose $f(z)$ is analytic within a simple closed curve C and its interior region R (simply connected), and $f'(z)$ is continuous. Then, $\oint_C f(z)dz = 0$, as shown in Figure 11-22.

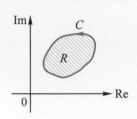

Figure 11-22

Here, we can make a small observation: since $f(z)$ is an analytic function, so $\dfrac{\partial f}{\partial \bar{z}} = 0$, then Green's theorem tells us that $\oint_C f(z)\,dz = 2i \iint_R (\dfrac{\partial f}{\partial \bar{z}})\,dxdy = 0$; on the other hand, if we remove the condition of $f'(z)$ being continuous within R, then $\oint_C f(z)dz = 0$, and this is referred to as Cauchy-Goursat theorem.

When discussing Green's theorem on the real plane, if there exists a function F such that $Pdx + Qdy = dF$, then $\oint_C Pdx + Qdy = \oint_C dF$ becomes independent of the shape of C, simplifying the computation. A similar phenomenon exists in the complex plane:

▶ **Theorem 11-4-3**

The Important Inference of Cauchy's Integral Theorem

As shown in Figure 11-23, let D be a connected domain. The following four statements are equivalent when $f(z)$ is continuous within D.

(1) For any simple closed contour C within D, we have $\oint_C f(z)dz = 0$.

(2) For any two points z_1, z_2 within D,

$\int_{z_1}^{z_2} f(z)dz$ is unrelated to the path.

(3) There exists a function $F(z)$ within D such that

$$F'(z) = f(z) \Rightarrow \int_{z_1}^{z_2} f(z)dz = F(z)\big|_{z_1}^{z_2}.$$

(4) $f(z)$ is an analytic function within D.

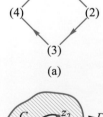

(a)

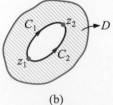

(b)

Figure 11-23 ◀

Example 4

Calculate $\int_C f(z)dz$, where $f(z)=e^z$,

C is a straight line from 1 to $1+i\dfrac{\pi}{2}$.

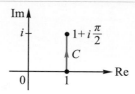

Solution

$\because e^z$ is an entire function so it is definitely analytic, then $(e^z)'=e^z$,

$\therefore \int_C f(z)dz = \int_1^{1+i\frac{\pi}{2}} e^z dz = e^z \Big|_1^{1+i\frac{\pi}{2}} = e^{1+i\frac{\pi}{2}} - e = e(i-1)$. 　　　Q.E.D.

Integration over Multiply Connected Regions

When dealing with integration over complex regions that have "holes," the Cauchy's integral theorem informs us that the integral value over the outer boundary is equal to the sum of integral values over the inner boundaries. In other words, to compute an integral over a large complex multiply connected region, we can focus on the integral values along the boundaries of the "holes." Let's begin by considering the case where there is only one "hole":

▶ Theorem 11-4-4

Cauchy's Integral Theorem for Multiply Connected Regions

Let $f(z)$ be analytic on two non-intersecting curves C and C' also its surrounded region R, then $\oint_C f(z)dz + \oint_{C'} f(z)dz = 0$ where C, C' are traversed in the positive direction relative to R (when the slanted region is on the left side during traversal, it is referred to as a positive direction, curve C is traversed counterclockwise, while curve C' is traversed clockwise), as shown in Figure 11-24.

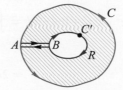

Figure 11-24　multiply connected regions

Proof

$\oint_{C+\overline{AB}+C'+\overline{BA}} f(z)dz = 0$

$\Rightarrow \int_C f(z)dz + \int_{\overline{AB}} f(z)dz + \int_{C'} f(z)dz + \int_{\overline{BA}} f(z)dz = 0$

$\Rightarrow \int_C f(z)dz + \int_{C'} f(z)dz = 0$.　　　◀

As shown in the proof, if curve C' is traversed counterclockwise, then

$$\oint_C f(z)dz = \oint_{C'} f(z)dz \ .$$

▶ **Theorem 11-4-5**

The Principle of Contour Deformation

Suppose $f(z)$ is analytic within the regions enclosed by non-intersecting contours C, C_1, $C_2, \cdots$, C_k, and their corresponding domain R, then $\oint_C f(z)dz + \sum_{i=1}^{k} \int_{C_i} f(z)dz = 0$, where

C, C_1, $C_2, \cdots$, C_k are traversed in the positive direction relative to R. Furthermore, if C_1, $C_2, \cdots$, C_k are all oriented counterclockwise, then

$$\oint_C f(z)dz = \oint_{C_1} f(z)dz + \oint_{C_2} f(z)dz + \cdots + \oint_{C_k} f(z)dz \ .$$

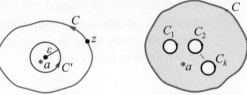

Figure 11-25   The principle of contour deformation

Example 5

Calculate $\oint_C [z^2 + 2z^5 + \text{Im}(z)]dz$, where C is the square curve with vertices at $0, -2i,$
$2 - 2i, 2.$

Solution

$\because z^2 + 2z^5$ is an entire function so it can be analytic, then $\oint_C (z^2 + 2z^5) = 0$, so

$$\oint_C [z^2 + 2z^5 + \text{Im}(z)]dz = \oint_C \text{Im}(z)dz = 2i \iint_R (\frac{\partial \text{Im}(z)}{\partial \bar{z}})dxdy$$

$$= 2i \iint_R (\frac{\partial(\frac{z-\bar{z}}{2i})}{\partial \bar{z}})dxdy = -2^2 = -4 \ .$$

【another method】

$$\oint_C [z^2 + 2z^5 + \text{Im}(z)]dz = \oint_C \text{Im}(z)dz = \oint_C y(dx+idy) = \oint_C ydx + i\oint_C ydy$$

$$= 0 + \int_0^2 (-2)dx + 0 + \int_2^0 0dx + i[\int_{y=0}^{-2} ydy + 0 + \int_{y=-2}^{0} ydy + 0]$$

$$= -4 + i[\frac{1}{2}y^2 \Big|_0^{-2} + \frac{1}{2}y^2 \Big|_{-2}^{0}] = -4 \ .$$

Q.E.D.

Example 6

Prove that $\displaystyle\oint_C (z-a)^n\,dz = \begin{cases} 0,\, n \neq -1 \\ 2\pi i,\, n = -1 \end{cases}$,

where C is any simple closed curve that contains
the point $z = a$ in its interior.

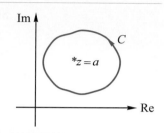

Solution

① $n = 0, 1, 2, \cdots$, $\because (z-a)^n$ is an nth-degree polynomial, $\therefore$ it is must analytic both on
C and within the interior. Therefore, by the Cauchy integral theorem
$$\Rightarrow \oint_C (z-a)^n = 0.$$

② $n = -1, -2, \cdots$, $\because (z-a)^n$ at $z = a$, it is not analytic $\therefore$ let C' be a circle inside C
within $|z-a| = \varepsilon$, then $(z-a)^n$ is analytic on C, C' and within the region enclosed
by them. According to the principle of contour deformation, we know
$$\oint_C (z-a)^n dz = \oint_{C'} (z-a)^n dz, \text{ on } C', \ z = a + \varepsilon \cdot e^{i\theta}, \ \theta \in [0, 2\pi], \ dz = i\varepsilon e^{i\theta} d\theta,$$
$$\therefore \oint_{C'} (z-a)^n dz = \int_0^{2\pi} (\varepsilon e^{i\theta})^n i\varepsilon e^{i\theta} d\theta = i\varepsilon^{n+1} \int_0^{2\pi} e^{i(1+n)\theta} d\theta$$

$$= \begin{cases} \dfrac{\varepsilon^{1+n}}{(1+n)} e^{i(1+n)\theta} \Big|_0^{2\pi} ,\, n = -2, -3, \cdots \\ 2\pi i,\, n = -1 \end{cases}$$

$$= \begin{cases} 0,\, n = -2, -3, \cdots \\ 2\pi i,\, n = -1 \end{cases}.$$

$$\therefore \oint_C (z-a)^n dz = \begin{cases} 0,\, n \neq -1 \\ 2\pi i,\, n = -1 \end{cases}.$$

Q.E.D.

It is important to note that the converse statement of the Cauchy theorem is not true:
$\oint_C f(z)dz = 0$ cannot be inferred that $f(z)$ is analytic within the interior of C. For

example, $f(z) = \dfrac{1}{(z-a)^2}$, even though $f(z)$ is not analytic at $z = 0$, but $\oint_C f(z)dz = 0$.

To ensure analytic, it is necessary for $f(z)$ to be continuous on C and its interior.

11-4-4 The Cauchy Integral Formula

▶ **Theorem 11-4-6**

The Cauchy Integral Formula

Suppose that $f(z)$ is analytic both on and inside a simple closed contour C, and $z = a$ is a
point within C, then $\displaystyle\oint_C \frac{f(z)}{(z-a)}\,dz = 2\pi i f(a)$.

Proof

Since $\dfrac{f(z)}{z-a}$ inside C only contains one singularity at $z = a$, let C' be a circle within C,

centered at $z = a$ with a radius of $\varepsilon \to 0^+$, then $\dfrac{f(z)}{z-a}$ is analytic on both C and C', as

well as within the intermediate region enclosed by the two contours. As established by the principle of contour deformation, we know

$$\oint_C \frac{f(z)}{z-a}\,dz = \oint_{C'} \frac{f(z)}{z-a}\,dz \;.$$

Hence, we focus on the integration on C': let $z = a + \varepsilon e^{i\theta}$, $\theta \in [0, 2\pi]$, $dz = i\varepsilon e^{i\theta}\,d\theta$,

then $\displaystyle\oint_{C'} \frac{f(z)}{z-a}\,dz = \int_0^{2\pi} \frac{f(a+\varepsilon e^{i\theta})}{\varepsilon e^{i\theta}} i\varepsilon e^{i\theta}\,d\theta = i\int_0^{2\pi} f(a+\varepsilon e^{i\theta})\,d\theta\,$,

so $\displaystyle\oint_C \frac{f(z)}{z-a}\,dz = i\int_0^{2\pi} f(a+\varepsilon e^{i\theta})\,d\theta = i\lim_{\varepsilon \to 0}\int_0^{2\pi} f(a+\varepsilon e^{i\theta})\,d\theta$

$$= i\int_0^{2\pi} f(a)\,d\theta = 2\pi i f(a)\;.$$

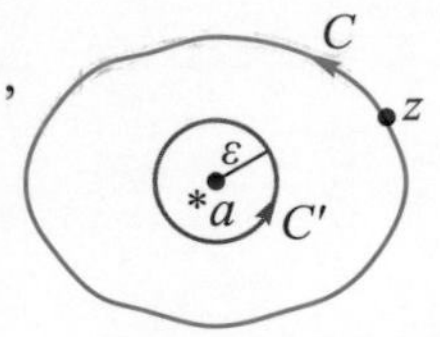

Figure 11-26 ◄

► **Theorem 11-4-7**

The Generalized Cauchy Integral Formula

If $f(z)$ on the complex plane, it is analytic on the simple closed contour C, and within its

interior, and $z = a$ is a point within C, then $f^{(n)}(a) = \dfrac{n!}{2\pi i}\displaystyle\oint_C \frac{f(z)}{(z-a)^{n+1}}\,dz$; $n = 0, 1, 2, \cdots$. ◄

The generalized Cauchy integral formula informs us that if $f(z)$ is analytic within a region R, then all its derivatives of every order are also analytic within R. Similarly, if $f(z)$ is analytic at a point z_0, then the derivatives of every order of $f(z)$ at z_0 are also analytic.[6]

Example 7

In the following curves (1)~(2), use the Cauchy integral formula, calculate

$$\int_C \frac{z^2}{(z-2)(z-6)}\,dz\;.$$

(1) When $C: |z| = 1$, C is traversed in any direction.

(2) When C encircles $|z| = 4$ in the positive direction and $|z| = 3$ in the negative direction.

(3) When $C: |z-2| = 1$ is traversed in the positive direction.

---

[6] 「$f(z)$ exists $\Leftrightarrow f^{(n)}(z)$ exists」 in the context of real-variable functions, this does not necessarily hold true.

Solution

(1) Inside $C: |z| = 1$, $\dfrac{z^2}{(z-2)(z-6)}$ is analytic,

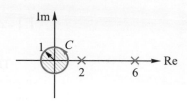

$$\therefore \int_C \frac{z^2}{(z-2)(z-6)} = 0 .$$

(2) No singular points within C,

$$\therefore \int_C \frac{z^2}{(z-2)(z-6)} = 0 .$$

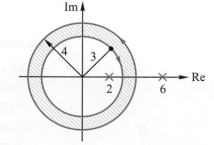

(3) Transform it to a form that can be utilized with the Cauchy integral formula. Within C, there is only one singular point, $z = 2$,

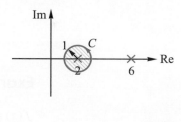

$$\therefore \int_C \frac{z^2}{(z-2)(z-6)} dz = \int_C \frac{\frac{z^2}{z-6}}{(z-2)} dz$$

$$= 2\pi i \cdot \left(\frac{z^2}{z-6} \right) \Big|_{z=2}$$

$$= 2\pi i \cdot \frac{4}{-4}$$

$$= -2\pi i .$$

Q.E.D.

Example 8

Calculate $\displaystyle\oint_C \frac{e^{2z}}{(z+1)^4} dz$, $C: |z| = 3$ is traversed in the positive direction.

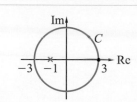

Solution

From the Cauchy integral formula, let $f(z) = e^{2z}$ be a holomorphic function,

then $\displaystyle\oint_C \frac{f(z)}{(z-a)^{n+1}} dz = \frac{2\pi i}{n!} f^{(n)}(a)$, where $n = 3$, $a = -1$, $f'''(a) = 8e^{2z}\big|_{z=-1} = 8e^{-2}$,

$$\therefore \oint_C \frac{e^{2z}}{(z+1)^4} dz = \frac{2\pi i}{3!} 8e^{-2} = \frac{8\pi i}{3} e^{-2} .$$

11-4-5 Relevant Theorem (Selected Reading)

The Cauchy integral formula provides us with a method for evaluating values at singular points. In fact, the open ball containing the singular point itself is also a connected region. The union of these two forms a complete connected region. This "digging a hole" approach then informs us about where the extreme values of a function lie on a connected region.

1. **Maximum and Minimum Modulus Theorem**

   (1) Maximum Modulus Theorem:
   Suppose $f(z)$ is analytic both on and within a simple closed contour C, and $f(z)$ is not a constant function. Then, the maximum value of $|f(z)|$ is attained on C.

   (2) Minimum Modulus Theorem:
   Suppose $f(z)$ is analytic both on and within a simple closed contour C, and $f(z) \neq 0$ inside C, then the minimum value of $|f(z)|$ is attained on C.

Example 9

$f(z) = e^{1-2z}$, find the maximum and minimum value of $|f(z)|$ in the region D: $|\text{Re}(z)| + |\text{Im}(z)| \leq 4$.

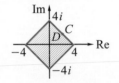

Solution

$f(z) = e^{1-2z}$ is neither a constant nor zero, according to the maximum and minimum modulus theorems, we know that the maximum and minimum values of $|f(z)|$ must occur on the boundary C of the region D, also $z = 4, |f(z)| = |e^{1-2z}| = e^{1-2\text{Re}(z)} = e^{-7}$ is the minimum on C and $z = -4, |f(z)| = |e^{1-2z}| = e^{1-2\text{Re}(z)} = e^{9}$ is the maximum on C.

2. **Argument Theorem**
   Complex functions can be applied in the field of automatic control theory, and one particularly useful principle is the argument principle. This principle serves as a theoretical basis for Nyquist stability criteria. The theorem for this principle is as follows.

▶ **Theorem 11-4-8**

Argument Theorem

Suppose $f(z)$ is analytic on the simple closed contour C, and within C it is also analytic except for several poles $a_1, a_2, \cdots, a_m$, then $\dfrac{1}{2\pi i} \oint_C \dfrac{f'(z)}{f(z)} dz = N - P$, which N represents the sum of the orders of the zeros of $f(z)$ within C; P represents the sum of the orders of the poles of $f(z)$ within C.

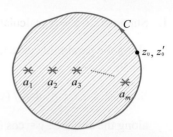

Figure 11-27   ◀

Example 10

Suppose that $f(z) = \dfrac{(z^2+1)^3}{(z^2+2z+2)^2}$,

find $\oint_C \dfrac{f'(z)}{f(z)} dz$, $C: |z| = 4$.

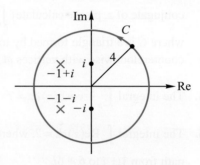

Solution

$z = \pm i$ are both third-order zeros of $f(z)$,

$z = -1 \pm i$ are both second-order poles of $f(z)$,

【Method 1】: By the argument principle, we know that the original expression is

$$2\pi i \cdot (N - P) = 2\pi i (6 - 4) = 4\pi i .$$

【Method 2】: $f(z) = \dfrac{(z+i)^3 (z-i)^3}{(z+1-i)^2 (z+1+i)^2}$, by applying the logarithmic derivative to both sides of the equation, we get

$$\frac{f'(z)}{f(z)} = \frac{3}{z+i} + \frac{3}{z-i} - \frac{2}{z+1-i} - \frac{2}{z+1+i} ,$$

then $\oint_c \dfrac{f'(z)}{f(z)} dz = 3 \cdot 2\pi i + 3 \cdot 2\pi i - 2 \cdot 2\pi i - 2 \cdot 2\pi i$

$$= (3+3-2-2) \cdot 2\pi i = 4\pi i .$$

Q.E.D.

11-4 Exercises

Basic questions

1. Suppose $z = x + iy$, calculate the integral $\int_C \bar{z}\,dz$ along the parabolic curve C: $y = x^2$, z from 0 to $3 + 9i$.

2. Suppose $z = x + iy$, and $f(z) = x^2 + iy^2$, please along the curve C: $y = \cos(x)$, x from 0 to $\dfrac{\pi}{2}$, calculate the integral $\int_C f(z)\,dz$.

3. Suppose $z = x + iy$, and $\bar{z} = x - iy$ is the complex conjugate of z, please calculate $\left| \int_C (e^z - \bar{z})\,dz \right|$, where C is a triangle formed by traversing counterclockwise with vertices at $3i, -4, 0$.

4. The integral $\int_{-\pi}^{1+\frac{\pi}{2}i} \cosh(z)\,dz = ?$

5. The integral $\int_C \text{Re}(z)\,dz = ?$, where C is the shortest path from $1 + i$ to $6 + 6i$.

6. Calculate the integral $\int_C [z - \text{Re}(z)]\,dz$, where C is the positively oriented circle centered at $z = 0$ with a radius of 2.

7. Suppose $z = x + iy$, and $\bar{z} = x - iy$ is the complex conjugate of z, please calculate the integral $\int_C [z^2 + \text{Im}(z)]\,dz$, where C is the positively oriented circle centered at $z = 0$ with a radius of 1.

8. Using the Cauchy integral formula, calculate $\oint_C \dfrac{\cos z}{z^3}\,dz$, where C is a circle traversed counterclockwise with $z = 0$ as its center and a radius of 1.

Advanced questions

1. Using the Cauchy integral formula, calculate $\oint_C \dfrac{\sin \pi z^2 + \cos \pi z^2}{(z-1)(z-2)}\,dz$, where C is a circle traversed counterclockwise with $z = 0$ as its center and a radius of 3.

2. Using the Cauchy integral formula, calculate $\oint_C \dfrac{1 - e^{2z}}{z^2}\,dz$, where C is a counterclockwise-oriented circle as shown below.
(1) $|z| = 1$.
(2) $|z| = 2$.
(3) $|z - 2| = 1$.

3. $f(z) = z^2 - z$, find the maximum and minimum value of $|f(z)|$ in the region D: $|z| \leq 1$.

4. C is the counterclockwise-oriented circle $|z| = 2$, using the Cauchy integral formula, calculate the following integral.
(1) $\oint_C \dfrac{1}{z^2 - 4z + 3}\,dz$.
(2) $\oint_C \dfrac{1}{z^2 - 1}\,dz$.

5. Please prove the complex form of the Green's theorem: $\oint_C F(z, \bar{z})\,dz = 2i \iint_R \dfrac{\partial F}{\partial \bar{z}}\,dxdy$.

6. Prove the Argument Theorem: $\dfrac{1}{2\pi i} \oint_C \dfrac{f'(z)}{f(z)}\,dz = N - P$, where $f(z)$ is analytic within C except for several poles, and $\begin{cases} N: \text{the zeros of } f(z) \text{ inside } C \\ P: \text{the poles of } f(z) \text{ inside } C \end{cases}$.

11-5 Taylor Series Expansion and Laurent Series Expansion

In calculus, for points where a function is analytic, it can be expanded using a Taylor series. Similarly, for complex functions, a Taylor series expansion can be done at analytic points, while at singular points, a Laurent series expansion can be carried out (which can be thought of as a Taylor expansion involving negative powers). The existence of these series is primarily established through the Cauchy integral formula and geometric series, concepts typically encountered in secondary education.

11-5-1   Taylor Expansion (Can Be Performed Only at Regular Points)

▶ **Theorem 11-5-1**

Taylor Expansion

As shown in Figure 11-28, let $f(z)$ be analytic within and on the circle C with center $z = a$ and radius R. Then, for any point z within C, $|z-a| < R$ always holds

$$f(z) = \sum_{n=0}^{\infty} a_n (z-a)^n , \text{ where } a_n = \frac{1}{2\pi i} \oint_C \frac{f(w)}{(w-a)^{n+1}} dw = \frac{f^{(n)}(a)}{n!}.$$

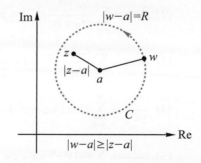

Figure 11-28   The region of Taylor expansion

Proof

First, the Cauchy integral formula implies that $\dfrac{1}{2\pi i} \oint_C \dfrac{f(w)}{(w-a)^{n+1}} dw = \dfrac{f^{(n)}(a)}{n!}$. In the

derivation of the formula, we start with $n = 1$, $f(z) = \dfrac{1}{2\pi i} \oint_C \dfrac{f(w)}{w-z} dw$, where

$C: |w-a| = R$. Now, since the common ratio is $\left| \dfrac{z-a}{w-a} \right| < 1$, the integrand can be

rewritten using a geometric series as follows:

$$\frac{1}{w-z} = \frac{1}{(w-a)-(z-a)} = \frac{1}{w-a}[\frac{1}{1-(\frac{z-a}{w-a})}] = \frac{1}{w-a}\sum_{n=0}^{\infty}\frac{(z-a)^n}{(w-a)^n} = \sum_{n=0}^{\infty}\frac{(z-a)^n}{(w-a)^{n+1}},$$

upon substituting into the Cauchy integral formula, we obtain:

$$f(z) = \frac{1}{2\pi i}\oint_C f(w)\{\sum_{n=0}^{\infty}\frac{(z-a)^n}{(w-a)^{n+1}}\}dw = \sum_{n=0}^{\infty}[\frac{1}{2\pi i}\oint_C \frac{f(w)}{(w-a)^{n+1}}dw](z-a)^n$$

we take $a_n = \frac{1}{2\pi i}\oint_C \frac{f(w)}{(w-a)^{n+1}}dw$, this yields the desired expansion, which also

informs us that the Taylor expansion is unique.   ◀

1. **Taylor Expansion Within the Neighborhood of Infinity**

    If there exists a positive number $R > 0$ which causes $f(z)$ is differentiable at every point within the region $|z| > R$ (including $z = \infty$), then $f(z)$ is said to be analytic within a neighborhood of infinity. In such cases, through the variable transformation $z = \frac{1}{t}$; similarly, taking $f(t) = f(\frac{1}{z})$, then $g(t)$ returns to a familiar situation for us.

2. **Common Maclaurin Series**

    Based on the earlier Taylor series expansion, we can obtain the Taylor expansion for common functions at $z = 0$, which is called Maclaurin series as follows:

    (1) $e^z = \sum_{n=0}^{\infty}\frac{z^n}{n!} = 1 + \frac{z}{1!} + \frac{z^2}{2!} + \cdots, \quad \forall |z| < \infty$.

    (2) $\sin z = \sum_{n=0}^{\infty}\frac{(-1)^n}{(2n+1)!}z^{2n+1} = z - \frac{z^3}{3!} + \cdots, \quad \forall |z| < \infty$.

    (3) $\cos z = \sum_{n=0}^{\infty}\frac{(-1)^n}{2n!}z^{2n} = 1 - \frac{z^2}{2!} + \frac{z^4}{4!}\cdots, \quad \forall |z| < \infty$.

    (4) $\frac{1}{1-z} = \sum_{n=0}^{\infty}z^n = 1 + z + z^2 + \cdots, \quad \forall |z| < 1$.

    (5) $\frac{1}{1+z} = \sum_{n=0}^{\infty}(-1)^n z^n = 1 - z + z^2 + \cdots + (-1)^n z^n + \cdots, \quad \forall |z| < 1$.

    (6) $\ln(1+z) = \sum_{n=0}^{\infty}\frac{(-1)^n}{n+1}z^{n+1} = z - \frac{1}{2}z^2 + \frac{1}{3}z^3 \cdots, \quad \forall |z| < 1$.

    Since Taylor expansions are unique, in many cases, we don't necessarily need to calculate coefficients through integration. If conditions permit, we can directly write down the expansion using geometric series. This simplifies the Taylor expansion, as illustrated in the following example.

Example 1

Suppose $f(z) = \dfrac{z-1}{z+1}$, find the Taylor series expansion at the following points:

(1) for $z = 0$ expand,   (2) for $z = 1$ expand.

Solution

(1) Expand using a geometric series a $z = 0$:

$$f(z) = \frac{z-1}{z+1} = 1 - \frac{2}{z+1} , \text{ at } |z| < 1, \text{ we have}$$

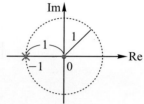

$$\frac{1}{1+z} = \sum_{n=0}^{\infty} (-1)^n z^n = 1 - z + z^2 - z^3 + \cdots,$$

$$\therefore f(z) = 1 - 2\sum_{n=0}^{\infty} (-1)^n z^n = -1 - 2\sum_{n=1}^{\infty} (-1)^n z^n , \text{ where } |z| < 1.$$

(2) Expand at $z = 1$: let $u = z - 1$, $f(u) = \dfrac{u}{2+u} = 1 - \dfrac{2}{2+u} = 1 - \dfrac{1}{1+\dfrac{u}{2}}$,

at $|\dfrac{u}{2}| < 1 \Rightarrow |u| < 2 \Rightarrow |z-1| < 2$,

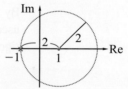

now $f(u) = 1 - \sum_{n=0}^{\infty} (-1)^n (\dfrac{u}{2})^n = 1 - \sum_{n=0}^{\infty} \dfrac{(-1)^n}{2^n} u^n$,

$$\therefore f(z) = 1 - \sum_{n=0}^{\infty} \frac{(-1)^n}{2^n} (z-1)^n , \text{ where } |z-1| < 2.$$

Q.E.D.

Example 2

Find $f(z) = \dfrac{1}{5-3z}$, Taylor series expansion centered at $z = 1$, and discuss the convergence range.

Solution

(1) Let $t = z - 1$, then $f(z) = \dfrac{1}{5-3(t+1)} = \dfrac{1}{2-3t} = \dfrac{1}{2(1-\dfrac{3}{2}t)} = \dfrac{1}{2} \cdot \dfrac{1}{1-\dfrac{3}{2}t}$,

at $|\dfrac{3}{2}t| < 1 \Rightarrow |t| < \dfrac{2}{3}$, therefore,

$$f(z) = \frac{1}{2} \cdot \sum_{n=0}^{\infty} (\frac{3}{2}t)^n = \frac{1}{2}[1 + \frac{3}{2}t + (\frac{3t}{2})^2 + \cdots]$$

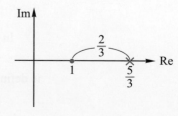

$$= \frac{1}{2}\{1 + \frac{3}{2}(z-1) + [\frac{3}{2}(z-1)]^2 + \cdots\}$$

$$= \frac{1}{2} + \frac{3}{4}(z-1) + \frac{9}{8}(z-1)^2 + \cdots,$$

where, $|z-1| < \dfrac{2}{3}$.

(2) By the ratio test for convergence,

$$\lim_{n \to \infty} \left| \frac{\frac{1}{2}[\frac{3}{2}(z-1)]^{n+1}}{\frac{1}{2}[\frac{3}{2}(z-1)]^{n}} \right| = \left| \frac{3}{2}(z-1) \right| < 1 \Rightarrow |z-1| < \frac{2}{3} ,$$

is its convergence range.

Q.E.D.

11-5-2 Laurent Expansion

When the domain of a complex function has a "hole," we can utilize "branch cuts" to transform the situation back to one without the hole. In this context, the Cauchy integral formula informs us

$$f(a) = \frac{1}{2\pi i} \oint_{C_1} \frac{f(w)}{w-a} dw - \frac{1}{2\pi i} \oint_{C_2} \frac{f(w)}{w-a} dw$$

By rewriting each of the two integral terms in a manner similar to the geometric series used in deriving Taylor expansions, a more general Laurent series can be obtained. For a rigorous proof, readers can refer to specialized books on complex analysis.

▶ **Theorem 11-5-2**

Laurent Expansion

Suppose $f(z)$ is analytic within and on the circles C_1, C_2 centered at $z = a$ with radius R_1, R_2 respectively, and also within the enclosed intermediate region D. Then, for any point z within D, it always holds that

$$f(z) = \sum_{n=-\infty}^{-1} a_n(z-a)^n + \sum_{n=0}^{\infty} a_n(z-a)^n , \quad R_1 < |z-a| < R_2 ,$$

where $a_n = \frac{1}{2\pi i} \oint_C \frac{f(z)}{(z-a)^{n+1}} dz$.

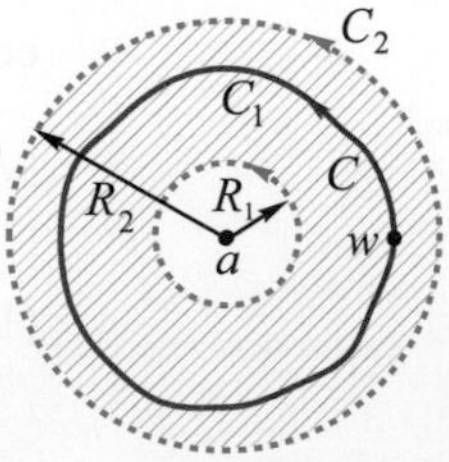

Figure 11-29   The region of Laurent expansion ◀

In the Laurent expansion , $f(z) = \sum_{n=-\infty}^{-1} a_n(z-a)^n + \sum_{n=0}^{\infty} a_n(z-a)^n$, the regular part

is defined as $\sum_{n=0}^{\infty} a_n(z-a)^n$, and it converges within the range $|z-a| < R_2$; the principal

part is $\sum_{n=-\infty}^{-1} a_n(z-a)^n$, and it converges within the range $|z-a| > R_1$.

1. **Properties**

   (1) If $f(z)$ is analytic within $|z-a| < R_1$, then the principal part is zero, and in this case, the Laurent series is the same as the Taylor series.

   (2) Within the same expansion range, the Laurent series has uniqueness (for the same function). However, if the expansion ranges are different, the Laurent series will also be different.

   (3) If $f(z)$ is analytic within $|z-a| < R_1$ except at $z = a$, then the convergence range of the Laurent series can be extended to: $0 < |z-a| < R_2$.

Example 3

If $f(z) = \dfrac{1}{(z-1)(z-3)}$, find the Laurent series of $f(z)$ for

$z = 0$, in the following range (1)~(4).

(1) $|z| = 1$.   (2) $1 < |z| < 3$.

(3) $3 < |z| < \infty$.   (4) $0 < |z-3| < 2$.

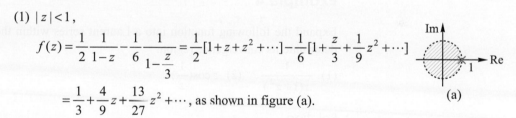

Solution

$$f(z) = -\frac{1}{2}\frac{1}{z-1} + \frac{1}{2}\frac{1}{z-3},$$

(1) $|z| < 1$,

$$f(z) = \frac{1}{2}\frac{1}{1-z} - \frac{1}{6}\frac{1}{1-\dfrac{z}{3}} = \frac{1}{2}[1+z+z^2+\cdots] - \frac{1}{6}[1+\frac{z}{3}+\frac{1}{9}z^2+\cdots]$$

$$= \frac{1}{3} + \frac{4}{9}z + \frac{13}{27}z^2 + \cdots, \text{ as shown in figure (a).}$$

(2) $1 < |z| < 3$,

$$f(z) = -\frac{1}{2z}\frac{1}{1-\dfrac{1}{z}} + \frac{1}{-6}\frac{1}{1-\dfrac{z}{3}} = -\frac{1}{2z}(1+\frac{1}{z}+\frac{1}{z^2}+\cdots) - \frac{1}{6}[1+\frac{1}{3}z+\frac{1}{9}z^2+\cdots]$$

$$= -\frac{1}{2z}\sum_{n=0}^{\infty}(\frac{1}{z})^n - \frac{1}{6}\sum_{n=0}^{\infty}(\frac{1}{3}z)^n$$

$$= -\frac{1}{2}\sum_{n=0}^{\infty}(\frac{1}{z})^{n+1} - \frac{1}{6}\sum_{n=0}^{\infty}(\frac{1}{3}z)^n,$$

as shown in figure (b).

(3) $3 < |z| < \infty$,

$$f(z) = -\frac{1}{2z} \frac{1}{1-\dfrac{1}{z}} + \frac{1}{2z} \frac{1}{1-\dfrac{3}{z}}$$

$$= -\frac{1}{2z}\left(1 + \frac{1}{z} + \frac{1}{z^2} + \cdots\right) + \frac{1}{2z}\left(1 + \frac{3}{z} + \frac{9}{z^2} + \cdots\right)$$

$$= -\frac{1}{2z}\sum_{n=0}^{\infty}\left(\frac{1}{z}\right)^n + \frac{1}{2z}\sum_{n=0}^{\infty}\left(\frac{3}{z}\right)^n, \text{ as shown in figure (c).}$$

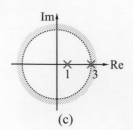

(c)

(4) $0 < |z-3| < 2$, let $t = z - 3 \Rightarrow 0 < |t| < 2$,

$$f(z) = \frac{-1}{2}\frac{1}{t+2} + \frac{1}{2}\frac{1}{t} = \frac{1}{2t} - \frac{1}{4}\frac{1}{1+\dfrac{t}{2}} = \frac{1}{2t} - \frac{1}{4}\sum_{n=0}^{\infty}(-1)^n\left(\frac{t}{2}\right)^n$$

$$= \frac{1}{2t} - \frac{1}{4}\left[1 - \frac{t}{2} + \frac{t^2}{4} - \frac{t^3}{8} + \cdots\right]$$

$$= \frac{1}{2}\cdot\frac{1}{z-3} - \frac{1}{4}\left[1 - \frac{1}{2}(z-3) + \frac{1}{4}(z-3)^2 - \frac{1}{8}(z-3)^3 + \cdots\right]$$

$$= \frac{1}{2}\frac{1}{z-3} - \frac{1}{4}\sum_{n=0}^{\infty}(-1)^n\left(\frac{z-3}{2}\right)^n, \text{ as shown in figure (d).}$$

Im, Re (b)

Q.E.D.

Example 4

Expand the following function into a Laurent series within the range $0 < |z| < R$, and determine its convergence region.

(1) $\dfrac{1}{z(1+z^2)}$. (2) $z\cos\left(\dfrac{1}{z}\right)$.

Solution

(1) $\dfrac{1}{z(z^2+1)} = \dfrac{1}{z} + \dfrac{-\dfrac{1}{2}}{z-i} + \dfrac{-\dfrac{1}{2}}{z+i} = \dfrac{1}{z} - \dfrac{1}{2}\cdot\dfrac{1}{i}\dfrac{1}{\dfrac{z}{i}-1} - \dfrac{1}{2i}\dfrac{1}{1+\dfrac{z}{i}}$

$= \dfrac{1}{z} + \dfrac{1}{2i}\dfrac{1}{1-\dfrac{z}{i}} - \dfrac{1}{2i}\dfrac{1}{1+\dfrac{z}{i}}$

$= \dfrac{1}{z} + \dfrac{1}{2i}\left\{1 + \dfrac{z}{i} + \left(\dfrac{z}{i}\right)^2 + \cdots\right\} - \dfrac{1}{2i}\left\{1 - \dfrac{z}{i} + \left(\dfrac{z}{i}\right)^2 \cdots\right\}$

$= \dfrac{1}{z} + \dfrac{1}{i^2}\times z + \dfrac{1}{i^4}\times z^3 + \dfrac{1}{i^6}\times z^5 + \cdots,$

thus, when $0 < |z| < 1$, the series converges.

【Another method】

$$\frac{1}{z(z^2+1)} = \frac{1}{z}\cdot\frac{1}{1+z^2} = \frac{1}{z}\cdot[1 - z^2 + z^4 - z^6 + \cdots] = \frac{1}{z} - z + z^3 - z^5 + \cdots.$$

(2) $z \cdot \cos\dfrac{1}{z} = z \cdot \{1 - \dfrac{1}{2!}\dfrac{1}{z^2} + \dfrac{1}{4!}\dfrac{1}{z^4} - \dfrac{1}{6!}\dfrac{1}{z^6} + \cdots\} = z - \dfrac{1}{2!}\dfrac{1}{z} + \dfrac{1}{4!}\dfrac{1}{z^3} - \dfrac{1}{6!}\dfrac{1}{z^5} + \cdots$,

hence, the series converges when $0 < |z| < \infty$.

Q.E.D.

2.　The Relationship Between Isolated Singularity and Laurent Series

In Section 11-3, we defined various types of singularities. In fact, these singularities can also be defined using Laurent series as follows:

▶ **Definition 11-5-1**

Classification of Singularities

If $f(z)$ has a Laurent series expansion on $0 < |z - \alpha| < \rho$:

$f(z) = \cdots + \dfrac{a_{-1}}{z - \alpha} + a_0 + a_1(z - \alpha) + \cdots, \quad 0 < |z - \alpha| < \rho$, then α is called an isolated

singular point of $f(z)$, based on the conditions of the principal part, the following classifications are made:

1.　If the principal part is zero, that is $f(z) = a_0 + a_1(z - \alpha) + a_2(z - \alpha)^2 + \cdots$,
　　 $0 < |z - \alpha| < \rho$, then $z = \alpha$ is termed a **removable singularity** of $f(z)$.

2.　If the highest power in the principal part is n, which means if
　　 $f(z) = \dfrac{a_{-n}}{(z - \alpha)^n} + \cdots + \dfrac{a_{-1}}{z - \alpha} + a_0 + \cdots$, then $z = \alpha$ is termed **a pole of order n** for
　　 $f(z)$.

3.　If the principal part involves an infinite series, that is
　　 $f(z) = \cdots + \dfrac{a_{-n}}{(z - \alpha)^n} + \cdots + \dfrac{a_{-1}}{(z - \alpha)} + a_0 + \cdots$, then $z = \alpha$ is termed an **essential**
　　 singularity of $f(z)$.　　　　　　　　　　　　　　　　　　　　◀

Example 5

Find the Laurent series for the following function and explain the type of singularity.

(1) $\dfrac{e^{2(z-2)}}{(z-4)^2}$, for $z = 4$.　(2) $\dfrac{1 - \cos 2z}{z^3}$, for $z = 0$.

Solution

(1) Let $t = z - 4$,

$$\therefore f(z) = \frac{e^{2(t+2)}}{t^2} = \frac{e^4}{t^2} \cdot e^{2t} = \frac{e^4}{t^2}[1 + 2t + \frac{1}{2!}(2t)^2 + \cdots] = \frac{e^4}{t^2} + \frac{2e^4}{t} + \cdots$$

$$= \frac{e^4}{(z-4)^2} + \frac{2e^4}{(z-4)} + \cdots, \quad \therefore z = 4 \text{ is a second-order pole.}$$

(2) $f(z) = \dfrac{1 - \cos 2z}{z^3} = \dfrac{1}{z^3}[1 - (1 - \dfrac{1}{2!}(2z)^2 + \dfrac{1}{4!}(2z)^4 \cdots)] = \dfrac{2}{z} - \dfrac{16z}{4!} + \cdots,$

$\therefore z = 0$ is a first-order pole.

Q.E.D.

Example 6

Please determine the type of singularity for the following functions.

(1) $f(z) = \dfrac{\sin z}{z}$.   (2) $e^{\frac{1}{z}}$.

Solution

(1) $z = 0$ is a singularity of $f(z)$, also

$$f(z) = \frac{\sin z}{z} = \frac{z - \dfrac{z^3}{3!} + \dfrac{z^5}{5!} - \dfrac{z^7}{7!} + \cdots}{z} = 1 - \frac{1}{3!}z^2 + \frac{1}{5!}z^4 - \frac{1}{7!}z^6 + \cdots,$$

with no principal part in the Laurent series, thus $z = 0$ is a removable singularity.

(2) $z = 0$ is a singularity of $f(z)$, also

$$f(z) = e^{\frac{1}{z}} = 1 + \frac{1}{1!}\frac{1}{z} + \frac{1}{2!}(\frac{1}{z})^2 + \frac{1}{3!}(\frac{1}{z})^3 + \cdots,$$

with an infinite number of terms in the principal part of the Laurent series, thus $z = 0$ is an essential singularity.

Q.E.D.

11-5   Exercises

Basic questions

1. Let $f(z) = \dfrac{1}{(z-1)(z-2)}$, and find the Laurent

series for this function in the given ranges(1)~(2).
(1) $|z| < 1$.
(2) $|z| > 1$.

2. Let $f(z) = \dfrac{1}{z(z-1)}$, and find the Laurent series for

this function in the given ranges(1)~(2).
(1) $|z| > 1$.
(2) $0 < |z-1| < 1$.

3. Let $f(z) = \dfrac{1}{z^2(z+2i)}$, and find the Laurent series

for this function in the given ranges $0 < |z| < 2$.

4. Let $f(z) = \dfrac{5}{(z+2)(z-3)}$, and find the Laurent

series and the convergence region for this function
at $z = 3$.

5. Let $f(z) = \dfrac{1}{z(z-1)(z-2)}$, find the Laurent series

for this function at $z = 0$ with a convergence region
of $1 < |z| < 2$.

4. Let $f(z) = \dfrac{1}{z - \sin z}$. What is the singularity

property of $z = 0$?

5. Indicate the type of singularity that

$$f(z) = \dfrac{1}{z(e^z - 1)} \text{ holds.}$$

Advanced questions

1. Let $f(z) = \dfrac{z^2 - 2z + 2}{(z-2)}$, find the Laurent series for

this function at $|z - 1| > 1$.

2. Let $f(z) = \dfrac{(\sin z) \cdot (\cos 2z)}{z^3}$, find the Laurent series

of the first three non-zero terms for this function at
$z = 0$, and determine the singularity property at
$z = 0$.

3. Let $f(z) = \dfrac{2i}{4 + iz}$, find the Laurent series for this

function at $z = -3i$.

11-6 Residue Theorem

When performing a contour integral of a complex function's Laurent series, only certain terms will contribute. This concept is extensively employed in the evaluation of contour integrals involving complex functions. Next, we will systematically elucidate this phenomenon through the concepts of poles and other aspects introduced in Section 11-5.

▶ **Definition 11-6-1**

Residue

If a single-valued function $f(z)$ has an isolated singularity at $z = a$, then $\dfrac{1}{2\pi i} \oint_C f(z)\,dz$ is called the residue of $f(z)$ at $z = a$, denoted by the symbol $\operatorname{Res} f(a)$.   ◀

1. **The Calculation of Residue**

   For the neighborhood of the isolated singularity $z = a$ of $f(z)$, where $0 < |z - a| < R$, the Laurent series can be written as:

   $f(z) = \displaystyle\sum_{n=-\infty}^{\infty} a_n (z-a)^n$, $0 < |z - a| < R$, by integrating both sides and referring to Example 6 in Section 11-4, we can deduce that:

   $$\oint_C f(z)\,dz = \oint_C \left[\cdots + \frac{a^2}{(z-a)^2} + \frac{a_{-1}}{z-a} + a_0 + a_1(z-a) + \cdots\right]dz = 2\pi i \cdot a_{-1}$$

   Therefore, the calculation of residues corresponds to the following algorithms:

▶ **Theorem 11-6-1**

The Solution of Residue

---

(1) If $z = a$ is a removable singularity of $f(z)$, then $\operatorname{Res} f(a) = 0$.

(2) If $z = a$ is an essential singularity of $f(z)$, then $\operatorname{Res} f(a) = a_{-1}$, expanding in a Laurent series at $z = a$, yielding a_{-1}.

(3) Let $z = a$ be an m-th order pole of $f(z)$:

   ① When the order m is larger, expand at $z = a$ in a Laurent series to obtain a_{-1}.

   ② For orders $m = 1, 2$, substitute the formula $\displaystyle\lim_{z \to a} \frac{1}{(m-1)!} \frac{d^{m-1}}{dz^{m-1}} (z-a)^m f(z)$.   ◀

Example 1

Find the residues of following functions at $z = 0$.

(1) $\dfrac{z - \sin z}{z}$. (2) $\dfrac{\cot z}{z^4}$. (3) $\dfrac{\sinh z}{z^4(1-z^2)}$. (4) $z^2 e^{\frac{1}{z}}$.

Solution

(1) $f(z) = \dfrac{1}{z}(z - \sin z) = \dfrac{1}{z}[z - (z - \dfrac{1}{3!}z^3 + \dfrac{1}{5!}z^5 \cdots)] = \dfrac{1}{6}z^2 - \dfrac{1}{120}z^4 + \cdots$,

$\therefore \operatorname{Res} f(0) = 0$.

(2) $f(z) = \dfrac{1}{z^4}\dfrac{\cos z}{\sin z} = \dfrac{1}{z^4} \cdot \dfrac{1 - \dfrac{1}{2}z^2 + \dfrac{z^4}{4!}\cdots}{z - \dfrac{1}{6}z^3 + \dfrac{1}{120}z^5 \cdots}$ by performing long division, we

obtain $f(z) = \dfrac{1}{z^4}[\dfrac{1}{z} - \dfrac{1}{3}z - \dfrac{1}{45}z^3 + \cdots] = \dfrac{1}{z^5} - \dfrac{1}{3}\dfrac{1}{z^3} - \dfrac{1}{45}\dfrac{1}{z} + \cdots$,

$\therefore \operatorname{Res} f(0) = -\dfrac{1}{45}$.

(3) $\sin(iz) = i \sinh z \Rightarrow \sinh z = \dfrac{1}{i}\sin(iz) = \dfrac{1}{i}[(iz) - \dfrac{1}{6}(iz)^3 + \dfrac{1}{120}(iz)^5 \cdots]$

$= z + \dfrac{1}{6}z^3 + \dfrac{1}{120}z^5 + \cdots$;

$f(z) = \dfrac{\sinh z}{z^4(1-z^2)} = \dfrac{1}{z^4}\dfrac{z + \dfrac{1}{6}z^3 + \dfrac{1}{120}z^5 + \cdots}{1 - z^2}$

by performing long division, we obtain $f(z) = \dfrac{1}{z^4}[z + \dfrac{7}{6}z^3 + \cdots] = \dfrac{1}{z^3} + \dfrac{7}{6}\dfrac{1}{z} + \cdots$,

$\therefore \operatorname{Res} f(0) = \dfrac{7}{6}$.

(4) $f(z) = z^2 e^{\frac{1}{z}} = z^2(1 + \dfrac{1}{z} + \dfrac{1}{2}\dfrac{1}{z^2} + \dfrac{1}{6}\dfrac{1}{z^3} + \cdots) = z^2 + z + \dfrac{1}{2} + \dfrac{1}{6}\dfrac{1}{z} + \cdots$,

$\therefore \operatorname{Res} f(0) = \dfrac{1}{6}$.

Q.E.D.

2. The Residue at Infinity

To establish a reasonable definition for the residue at infinity, let's consider a degenerate case where C_R is $|z| = R(R \to \infty)$ and oriented clockwise, as shown in Figure 11-30. Once again, through the Cauchy's theorem, we deduce that

$$\oint_{C_R} z^n dz = \begin{cases} 0, n \neq -1 \\ -2\pi i, n = -1 \end{cases}.$$ Let $z = \dfrac{1}{w}$ and substitute it into the Laurent series of $f(z)$,

expanding $f(z) = a_0 + a_1 \dfrac{1}{w} + a_2 \dfrac{1}{w^2} + \cdots + a_{-1}w + a_{-2}w^2 + \cdots$ then we obtain the

Laurent series of $f(z)$ at $z = \infty$. Therefore,

$$\int_{C_R} f(z)dz = \oint_{C_R} (a_0 + a_1 z + \cdots + \frac{a_{-1}}{z} + \frac{a_{-2}}{z^2} + \cdots) = \oint_{C_R} \frac{a_{-1}}{z} dz = -2\pi i a_{-1} ,$$ in other

words:

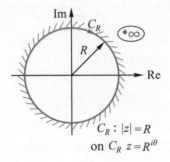

Figure 11-30　Diagram illustrating the residue at infinity

▶ Definition 11-6-2

The Residue at Infinity

The residue at infinity $\dfrac{1}{2\pi i} \oint_{C_R} f(z)dz = -a_{-1}$. ◀

Example 2

Find the residue of function $f(z)$ at $z = \infty$, where $f(z) = \dfrac{1}{\sin(\dfrac{1}{z})}$.

Solution

Let $z = \dfrac{1}{w}$, then $z \to \infty$, $w \to 0$,

$$f(z) = f(\frac{1}{w}) = \frac{1}{\sin w} = \frac{1}{w - \dfrac{1}{3!}w^3 + \cdots} = \frac{1}{w} + \frac{1}{6}w + \frac{7}{360}w^3 + \cdots$$

$$= z + \frac{1}{6z} + \frac{7}{360z^3} + \cdots,$$

$$\therefore \operatorname{Res} f(\infty) = -a_{-1} = -\frac{1}{6}.$$

Q.E.D.

▶ **Theorem 11-6-2**

Residue Theorem

In a simply connected domain, suppose $f(z)$ is analytic on a simple closed curve C, and within C it is also analytic except at several isolated singular points a_1, a_2, $\cdots$, a_m. In this case, the relationship between integration and residues can be expressed using the following identity:

$$\oint_C f(z)dz = 2\pi i \sum_{k=1}^{m} \operatorname{Res} f(a_k).$$

Proof

Let C_k be a simple closed curve within C that contains only one singular point a_k. Then, $f(z)$ is analytic in the regions enclosed by C_1, C_2, $\cdots$, C_m and C.

$\therefore$ Therefore, according to the principle of contour deformation,

we know $\oint_C f(z)dz = \sum_{k=1}^{m} \oint_{C_k} f(z)dz = 2\pi i \sum_{k=1}^{m} \operatorname{Res} f(a_k)$.

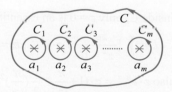

Figure 11-31   The residue theorem of simply connected domain   ◀

3. **Promotion**

Extending the above, if $f(z)$ on C has multiple first-order poles b_1, b_2, $\cdots$, b_l, then

$$\oint_C f(z)dz = 2\pi i \sum_{k=1}^{m} \operatorname{Res} f(a_k) + 2\pi i \sum_{k=1}^{l} \operatorname{Res} f(b_k)$$

Example 3

Along the following paths, find $I = \oint_C \dfrac{\cos z}{z^2(z-2)} dz$.

(1) $C: |z| = 1$.   (2) $C: |z| = 3$.

Solution

(1) $C: |z| = 1$, there is only one second-order pole at $z = 0$ within C.

$$\therefore \operatorname{Res} f(0) = \lim_{z \to 0} \frac{d}{dz} [z^2 \frac{\cos z}{z^2(z-2)}]$$

$$= \lim_{z \to 0} \frac{-\sin z \cdot (z-2) - \cos z}{(z-2)^2} = -\frac{1}{4} ,$$

$$\therefore I = 2\pi i \cdot (-\frac{1}{4}) = -\frac{\pi}{2} i .$$

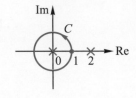

(2) $C: |z| = 3$, there are two singularities $z = 0, 2$ within C, and $z = 0$ is a second-order pole, $z = 2$ is a first-order pole.

also $\operatorname{Res} f(2) = \lim\limits_{z \to 2} (z-2) \dfrac{\cos z}{z^2(z-2)} = \dfrac{1}{4} \cos 2$,

$$\therefore I = 2\pi i [\operatorname{Res} f(0) + \operatorname{Res} f(2)]$$

$$= 2\pi i (-\frac{1}{4} + \frac{1}{4} \cos 2) = \frac{1}{2} \pi i (\cos 2 - 1) .$$

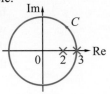

Q.E.D.

Example 4

$C: \ 9x^2 + y^2 = 9$ (counterclockwise direction around)

find $I = \oint_C [\dfrac{z \cdot e^{\pi z}}{z^2 - 16} + z \cdot e^{\pi/z}] dz$; where $z = x + iy$.

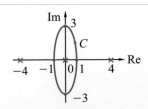

Solution

Inside C, only $z = 0$ is an essential singularity,

$$\therefore I = \oint_C [\frac{z e^{\pi z}}{z^2 - 16} + z \cdot e^{\pi/z}] dz = \oint_C z \cdot e^{\pi/z} dz ,$$

also $z \cdot e^{\pi/z} = z \cdot (1 + \dfrac{\pi}{z} + \dfrac{1}{2} (\dfrac{\pi}{z})^2 + \cdots) = z + \pi + \dfrac{1}{2} \pi^2 \dfrac{1}{z} + \cdots$; $a_{-1} = \dfrac{1}{2} \pi^2$,

$$\therefore I = 2\pi i \cdot \frac{1}{2} \pi^2 = i\pi^3 .$$

Q.E.D.

4. **The Relationship Between** $\operatorname{Res} f(\infty)$

Suppose $f(z)$ is analytic on $C: |z| = R$, with n singularities inside C and m singularities outside C, where $n >> m$. By considering the integration region outside C, we have

$$\oint_C f(z)dz = 2\pi i[\sum_{\text{inside } C} \operatorname{Res} f(z)] = -2\pi i[\sum_{\text{outside } C} \operatorname{Res} f(z)]$$

$$= -2\pi i[\operatorname{Res} f(\infty) + \sum_{\text{outside } C} \operatorname{Res} f(z)]$$

as illustrated in Figure 11-32.

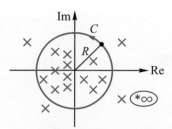

Figure 11-32   Relation with the Residue at Infinity

Example 5

Calculate $\dfrac{1}{2\pi i} \displaystyle\int_C \dfrac{dz}{(z^{100}+1)(z-4)}$, where C is the path among the following (1)~(2).

(1) $C: |z| = \infty$ (counterclockwise).   (2) $C: |z| = 3$ (counterclockwise).

Solution

(1) Suppose C_R is the circle of $|z| = R$, when $R \to \infty$, $C_R \to C$,

on C_R, $z = \operatorname{Re}^{i\theta}$, $dz = i\operatorname{Re}^{i\theta}d\theta$, $\theta: 0 \sim 2\pi$,

$$\therefore \int_C \frac{dz}{(z^{100}+1)(z-4)} = \lim_{R\to\infty}\int_{C_R} \frac{i\operatorname{Re}^{i\theta}d\theta}{(R^{100}e^{i100\theta}+1)(\operatorname{Re}^{i\theta}-4)} \ ,$$

$$\because \lim_{R\to\infty}|\int_0^{2\pi}\frac{i\operatorname{Re}^{i\theta}d\theta}{(R^{100}e^{i100\theta}+1)(\operatorname{Re}^{i\theta}-4)}| \le \lim_{R\to\infty}\int_0^{2\pi}\frac{|i\operatorname{Re}^{i\theta}|d\theta}{|R^{100}e^{i100\theta}+1||\operatorname{Re}^{i\theta}-4|}$$

$$\le \lim_{R\to\infty}\int_0^{2\pi}\frac{M}{R^{100}}d\theta = \lim_{R\to\infty}\frac{2\pi M}{R^{100}} = 0 \quad (ML \text{ theorem, } M \text{ is a constant}),$$

so $\dfrac{1}{2\pi i}\displaystyle\int_C \dfrac{dz}{(z^{100}+1)(z-4)} = 0$.

【another solution】

$$\frac{1}{2\pi i}\int_{C_R} \frac{dz}{(z^{100}+1)(z-4)} = -\operatorname{Res} f(\infty) ,$$

also $f(z) = \dfrac{1}{(z^{100}+1)(z-4)} = \dfrac{1}{z^{101}-4z^{100}+z-4} = \dfrac{1}{z^{101}} + \dfrac{4}{z^{102}} + \cdots$,

$\operatorname{Res} f(\infty) = 0$, $\dfrac{1}{2\pi i}\displaystyle\oint_C \dfrac{dz}{(z^{100}+1)(z-4)} = -\operatorname{Res} f(\infty) = 0$.

(2) $C: |z| = 3$,

$$\frac{1}{2\pi i} \int_C \frac{dz}{(z^{100}+1)(z-4)} = \frac{1}{2\pi i}[-2\pi i \operatorname{Res} f(4) - 2\pi i \operatorname{Res} f(\infty)],$$

also

$$\operatorname{Res} f(4) = \lim_{z \to 4}(z-4) \cdot \frac{1}{(z^{100}+1)(z-4)} = \frac{1}{4^{100}+1},$$

$$\therefore \frac{1}{2\pi i} \int_C \frac{dz}{(z^{100}+1)(z-4)} = -\frac{1}{4^{100}+1}.$$

Q.E.D.

11-6 Exercises

Basic questions

1. Please calculate the following integral $\oint_C \frac{dz}{z^2+4}$, C represents the circle of $|z-2i| = 1$.

2. Please attempt to compute the integral of

$\oint_C \frac{z+i}{z-3i} dz$, where

(1) $C: \; |z-i| = 1$,

(2) $C: \; |z-i| = 3$.

3. Find $\oint_C \frac{1}{z^2+1} dz$, where

(1) $C: \; |z+i| = 1$, counterclockwise,

(2) $C: \; |z-i| = 1$, counterclockwise.

4. $C: \; |z+i| = 4$, counterclockwise,

find $\oint_C \frac{1}{(z^2+1)(z-2i)^2} dz$.

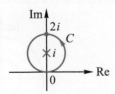

2. $C: \; |z-\frac{1}{2}| = 1$, counterclockwise, find

$\oint_C \frac{1}{z^2 \sin z} dz$.

3. $C: \; |z| = 2$, counterclockwise, find $\oint_C \frac{e^z}{z^4+5z^3} dz$.

4. $C: \; |z| = 1$, clockwise, find $\oint_C \tan \pi z dz$.

5. $C: \; |z| = 3$, counterclockwise, find $\oint_C \frac{z^3 e^{\frac{1}{z}}}{1+z^3} dz$.

6. $C: \; |z| = 0.5$, counterclockwise, find $\oint_C \frac{e^{\frac{1}{z}}}{1+z} dz$.

7. $C: \; |z-i| = 1$, find

(1) $\oint_C \frac{\sin(1+z^2)}{1+z^2} dz$,

(2) $\oint_C \frac{\sin(1+z^2)}{(1+z^2)^2} dz$,

(3) $\oint_C \frac{1}{\sin(\frac{1}{z-i})} dz$.

Advanced questions

1. $C: \; |z-1| = 4$, clockwise, find $\oint_C \frac{2z^3+z^2+4}{z^4+4z^2} dz$.

11-7 **Definite Integral of Real Variable Functions**

One significant application of studying complex functions is to solve certain real definite integrals of real variable function that are difficult to evaluate using calculus methods. The techniques introduced in this section have minor differences in the specific details, but the overall strategies involve classifying singularities and then utilizing the residue theorem. The following are the integrals discussed in this section.

(1) $\int_0^{2\pi} f(\cos\theta, \sin\theta)d\theta$ trigonometric function type.

(2) $\int_{-\infty}^{\infty} F(x)dx$ rational function type.

(3) $\int_{-\infty}^{\infty} F(x)e^{imx}dx$ Fourier transform (integral) type.

(4) Laplace inverse transform type.

11-7-1 Trigonometric Function type

$$I = \int_0^{2\pi} F(\cos\theta, \sin\theta)d\theta$$

The strategy is: convert the problem into the polar coordinates of the complex plane and confine the integration to the unit circle C at $|z| - 1$. As illustrated in Figure 11-33, let $C: |z| = 1$, on C, $z = e^{i\theta}$, $z^{-1} = e^{-i\theta}$, $dz = ie^{i\theta}d\theta = izd\theta$, $\theta : 0 \sim 2\pi$,

simultaneously, we obtain from Euler's formula $\begin{cases} \cos\theta = \dfrac{e^{i\theta} + e^{-i\theta}}{2} = \dfrac{z + z^{-1}}{2} \\ \sin\theta = \dfrac{e^{i\theta} - e^{-i\theta}}{2i} = \dfrac{z - z^{-1}}{2i} \end{cases}$. Hence,

we obtain the change of variable from the original equation.

$$\int_c f(z)dz = \int_0^{2\pi} F(\cos\theta, \sin\theta)d\theta = \oint_c F(z, z^{-1})\frac{dz}{iz} \text{ , consequently, we formally}$$

obtain

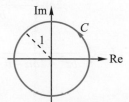

Figure 11-33   the unit circle of complex plane

▶ **Theorem 11-7-1**

The integral of a complex function $f(z)$

$$\int_c f(z)dz = 2\pi i \sum_{\text{inside } C} \text{Res } f(z) + \pi i \sum_{\text{on } C} \text{Res } f(z), \text{ where}$$

$f(z) = \dfrac{F(z, z^{-1})}{iz}$, $\cos\theta = \dfrac{z + z^{-1}}{2}$, $\sin\theta = \dfrac{z - z^{-1}}{2i}$ and the singularities of function

$f(z)$ on the unit circle C can only be simple poles (first-order poles).   ◀

Example 1

Find $\displaystyle\int_0^{2\pi} \dfrac{d\theta}{a + b\sin\theta}$, where $a > |b|$.

Solution

(1) Let $C: |z| = 1$, then on the unit circle C,

$z = e^{i\theta}$, $z^{-1} = e^{-i\theta}$, $dz = ie^{i\theta}d\theta = izd\theta$,

$\theta : 0 \sim 2\pi$, $\sin\theta = \dfrac{z - z^{-1}}{2i}$.

(2) $I = \displaystyle\oint_C \dfrac{\dfrac{1}{iz}dz}{a + b \cdot (\dfrac{z - z^{-1}}{2i})} = \oint_C \dfrac{2dz}{bz^2 + 2iaz - b}$.

(3) By $bz^2 + 2aiz - b = 0 \Rightarrow z = \dfrac{-a \pm \sqrt{a^2 - b^2}}{b}i$, only has one singularity

$z = \alpha = \dfrac{-a + \sqrt{a^2 - b^2}}{b}i$, and

$\text{Res } f(\alpha) = \displaystyle\lim_{z \to \alpha}(z - \alpha)\dfrac{2}{bz^2 + 2iaz - b} = \lim_{z \to \frac{-a + \sqrt{a^2 - b^2}}{b}i} \dfrac{2}{2bz + 2ia}$

$= \dfrac{2}{2b(\dfrac{-a + \sqrt{a^2 - b^2}}{b}i) + 2ia} = \dfrac{1}{i\sqrt{a^2 - b^2}}$.

(4) $\therefore I = \displaystyle\oint_C \dfrac{2dz}{bz^2 + 2aiz - b} = 2\pi i \cdot \dfrac{1}{i\sqrt{a^2 - b^2}} = \dfrac{2\pi}{\sqrt{a^2 - b^2}}$.[7]   Q.E.D.

---

[7] (1) $a > |b|$, $a > 0$, and $a^2 > b^2$, $\therefore \sqrt{a^2 - b^2} > 0$, thus, $\left|\dfrac{-a - \sqrt{a^2 - b^2}}{b}\right| > 1$ is outside C.

(2) $\displaystyle\int_0^{2\pi} \dfrac{1}{5 + 3\sin\theta}d\theta = \dfrac{2\pi}{\sqrt{5^2 - 3^2}} = \dfrac{\pi}{2}$.

11-7-2 The Improper Integral of Rational Function: $I \equiv \int_{-\infty}^{\infty} F(x)dx$

In complex analysis, we refer to the improper integral $\int_{-\infty}^{\infty} F(x)dx = \lim_{R \to \infty} \int_{-R}^{R} F(x)dx$ as the Cauchy principal value. However, this definition naturally raises questions about its existence. Therefore, we define the existence of the Cauchy principal value if both

$\int_{-\infty}^{0} F(x)dx = \lim_{R \to \infty} \int_{-R}^{0} F(x)dx$ and $\int_{0}^{\infty} F(x)dx = \lim_{R \to \infty} \int_{0}^{R} F(x)dx$ simultaneously exist.

When the integrand is a rational function, the key issue in integration is the location of the zero roots of the denominator, which cause discontinuities in the rational function and consequently lead to integrals becoming improper. Now, from the perspective of Laurent series, these points of discontinuity on the complex plane become singularities of the rational function. This approach provides room for the Cauchy integral formula to come into play, allowing us to systematically handle improper integrals through the calculation of residues.

1. Calculation of the Principal Value (P.V.) with no Singularities on the Real Axis

▶ **Theorem 11-7-2**

If the Function Is Bounded, Then the Improper Integral Is Zero

If C_R is the circular arc with center at $z = 0$ and radius R, as shown in Figure 11-34. If on C_R, $|F(z)| \le \dfrac{M}{R^k}$ (where $k > 1$, M is a constant), then $\lim_{R \to \infty} \int_{C_R} F(z)dz = 0$.

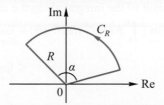

Figure 11-34   The circular arc on the complex plane

Proof

According to the ML theorem, it is known that $|\int_{C_R} F(z)dz| \le \dfrac{M}{R^k} \cdot \alpha R = \dfrac{\alpha M}{R^{k-1}}$, where α represents the angle subtended by the arc of C_R. Taking the limit, we obtain

$\lim_{R \to \infty} |\int_{C_R} F(z)dz| \le \lim_{R \to \infty} \dfrac{\alpha M}{R^{k-1}} = 0$ ($\because k - 1 > 0$), which completes the proof.   ◀

▶ **Theorem 11-7-3**

The Application of Residue

Suppose $F(x) = \dfrac{P(x)}{Q(x)}$ be rational functions of x, if the degree of $Q(x)$ is more than one degree greater than the degree of $P(x)$, and $Q(x) = 0$ has no real roots. Then

$$\int_{-\infty}^{\infty} F(x)dx = 2\pi i \sum_{\mathrm{im}(z)>0} \mathrm{Res}[F(z)] = -2\pi i \sum_{\mathrm{im}(z)<0} \mathrm{Res}[F(z)], \text{ where Im }(z) > 0$$

represents the upper half-plane of the complex plane, while Im $(z) < 0$ represents the lower half-plane.

Proof

In order to utilize residues, we first construct an integration path that encloses the singularities $C = C_R + \Gamma$, where

C_R: $z = Re^{i\theta}$, $\Gamma:[-R, R]$, as shown in Figure 11-35.

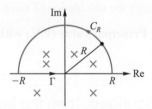

Figure 11-35　Upper half-plane of the complex plane with an infinite semicircle

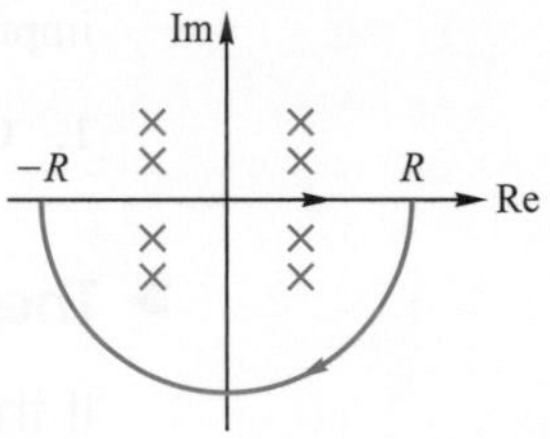

Figure 11-36　Lower half-plane of the complex plane with an infinite semicircle

Then $dz = iRe^{i\theta}d\theta$, $\theta \in [0, \pi]$, Γ: $z = x$, $dz = dx$, $x \in [-R, R]$. First, observe the result of the integral along the arc as R approaches infinity:

$$\lim_{R\to\infty} |\int_0^\pi F(Re^{i\theta})iRe^{i\theta}d\theta| \leq \lim_{R\to\infty}\int_0^\pi |F(Re^{i\theta})||iRe^{i\theta}|\,d\theta$$

$$\leq \lim_{R\to\infty}\int_0^\pi |\frac{P(Re^{i\theta})}{Q(Re^{i\theta})}|\,Rd\theta \leq \lim_{R\to\infty}\int_0^\pi \frac{M}{R^{k+1}}Rd\theta \quad (k > 0)$$

$$= \lim_{R\to\infty}\frac{M\pi}{R^k} = 0.$$

Therefore, $\lim\limits_{R\to\infty}\oint_C f(z)dz = \lim\limits_{R\to\infty}\int_0^\pi F(Re^{i\theta})iRe^{i\theta}d\theta + \int_{-\infty}^{\infty} F(x)dx = \int_{-\infty}^{\infty} F(x)dx$. Now,

by the residue theorem, $\int_{-\infty}^{\infty} F(x)dx = 2\pi i \sum\limits_{\mathrm{im}(z)>0} \mathrm{Res}F(z)$, which completes the proof.

Similarly, the formula can be derived in the lower half-plane.[8]

---

[8] (1) $\deg[Q(x)] > \deg[P(x)] + 1 \Rightarrow \dfrac{P(x)}{Q(x)} = \dfrac{a_m x^m + a_{m-1}x^{m-1} + \cdots}{x^{n-m}(b_m x^m + b_{m-1}x^{m-1} + \cdots)} = \dfrac{a_m + a_{m-1}x^{-1} + \cdots}{x^{n-m}(b_m + b_{m-1}x^{-1} + \cdots)} \Rightarrow |F(z)| \leq \dfrac{M}{R^{k+1}}$, $k > 0$, when $R \to \infty$.

(2) The curve C can also be changed to a lower semicircle, that is obtained $\int_{-\infty}^{\infty} F(x)dx = -2\pi i \sum \mathrm{Res}$ in the lower half-plane.

Example 2

Calculate the integral $\displaystyle\int_0^\infty \frac{dx}{1+x^4}$ by using the residue (residue value) theorem.

Solution

(1) $I = \displaystyle\int_0^\infty \frac{dx}{1+x^4} = \frac{1}{2}\int_{-\infty}^\infty \frac{1}{1+x^4}dx$ $\left(\because \frac{1}{1+z^4}\text{ is even function}\right)$, let $F(z) = \dfrac{1}{1+z^4}$,

from $1+z^4 = 0 \Rightarrow z = e^{i\frac{\pi}{4}},\ e^{i\frac{3\pi}{4}},\ e^{i\frac{5\pi}{4}},\ e^{i\frac{7\pi}{4}}$ are first-order poles, with only

$z = e^{i\frac{\pi}{4}},\ e^{i\frac{3\pi}{4}}$ located in the upper half-plane.

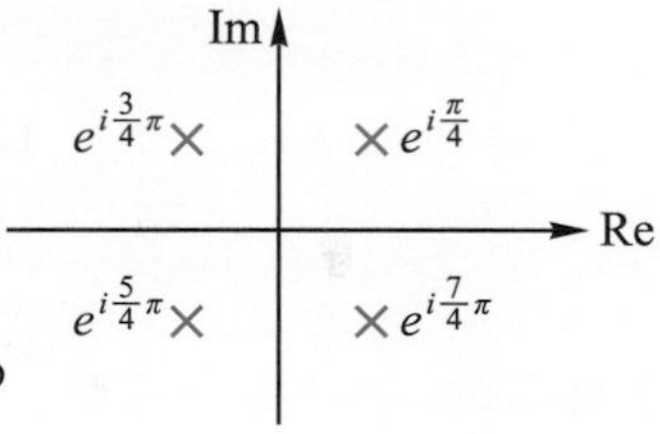

(2) $\operatorname{Res} f(e^{i\frac{\pi}{4}}) = \displaystyle\lim_{z\to e^{i\frac{\pi}{4}}} (z - e^{i\frac{\pi}{4}})\cdot \frac{1}{(z^4+1)} = \frac{1}{4}e^{-i\frac{3\pi}{4}}$,

$\operatorname{Res} f(e^{i\frac{3\pi}{4}}) = \displaystyle\lim_{z\to e^{i\frac{3\pi}{4}}} (z - e^{i\frac{3\pi}{4}})\cdot \frac{1}{(z^4+1)} = \frac{1}{4}e^{-i\frac{9\pi}{4}}$, so

$\displaystyle\int_0^\infty \frac{dx}{1+x^4} = \frac{1}{2}\cdot 2\pi i\cdot\left(\frac{1}{4}e^{-i\frac{3\pi}{4}} + \frac{1}{4}e^{-i\frac{9\pi}{4}}\right) = \frac{\pi i}{4}\left(e^{-i\frac{1\pi}{4}} - e^{i\frac{\pi}{4}}\right) = \frac{\pi i}{4}\cdot 2i\cdot \frac{e^{-i\frac{1\pi}{4}} - e^{i\frac{\pi}{4}}}{2i}$

$\qquad = \frac{\pi}{2}\cdot \sin\frac{\pi}{4} = \frac{\sqrt{2}}{2}\times\frac{\pi}{2} = \frac{\sqrt{2}}{4}\pi$.

Q.E.D.

Im axis diagram: $e^{i\frac{3}{4}\pi}\times$ and $\times e^{i\frac{\pi}{4}}$ in upper half-plane; $e^{i\frac{5}{4}\pi}\times$ and $\times e^{i\frac{7}{4}\pi}$ in lower half-plane; with Re and Im axes.

2. **There Are Singularities on the Real Axis**

Suppose $f(x)$ has several singularities $x_1, x_2, \cdots, x_n$ (distinct) on $x \in (a, b)$ and $a < x_1 < x_2 < \cdots < x_n < b$, as shown in Figure 11-37. Then the Cauchy principal value (abbreviated as P.V.) of $f(x)$ on $[a, b]$ is defined as:

$$\text{P.V.} \int_{-\infty}^\infty f(x)dx = \lim_{\varepsilon\to 0}\left[\int_a^{x_1-\varepsilon} f(x)dx + \int_{x_1+\varepsilon}^{x_2-\varepsilon} f(x)dx + \cdots + \int_{x_n+\varepsilon}^b f(x)dx\right].$$

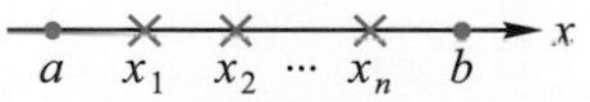

Figure 11-37   The path circumventing the singularities

3. **Combining the Calculation of Residues on the Real Axis**

If $F(z)$ has several first-order poles on the real axis, then the Cauchy principal value is:

$$\text{P.V.} \int_{-\infty}^\infty F(x)dx = 2\pi i\left(\sum \operatorname{Res} \text{ on the upper half-plane}\right) + \left(\pi i\sum \operatorname{Res} \text{ on } x\text{-axis}\right)$$

$$= -2\pi i\left(\sum \operatorname{Res} \text{ on the lower half-plane}\right) - \pi i\left(\sum \operatorname{Res} \text{ on } x\text{-axis}\right).$$

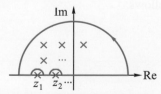

Figure 11-38   The path circumventing the singularities

Example 3

Calculate $\displaystyle\int_{-\infty}^{\infty}\frac{3x+2}{x(x-4)(x^2+9)}dx$.

Solution

(1) Let $f(z)=\dfrac{3z+2}{z(z-4)(z^2+9)}$ then taking the

integration path as shown in the diagram on the right. $C=C_R+\Gamma_1+C_{\varepsilon_1}+\Gamma_2+C_{\varepsilon_2}+\Gamma_3$.

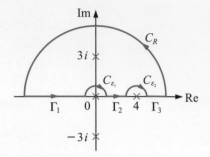

(2) Obtained from the residue theorem,

$$\int_{-\infty}^{\infty}\frac{3x+2}{x(x-4)(x^2+9)}dx=2\pi i\operatorname{Res}f(3i)+\pi i\left[\operatorname{Res}f(0)+\operatorname{Res}f(4)\right],$$

also $\operatorname{Res}f(3i)=\lim\limits_{z\to 3i}(z-3i)\cdot\dfrac{3z+2}{z(z-4)(z^2+9)}=\dfrac{9i+2}{3i(3i-4)\cdot 6i}$

$$=-\frac{1}{18}\cdot\frac{(9i+2)(3i+4)}{-25}=\frac{1}{450}(-27+36i+6i+8)=-\frac{19}{450}+\frac{7}{75}i\,,$$

$$\operatorname{Res}f(0)=-\frac{2}{36}=-\frac{1}{18}\,,\quad \operatorname{Res}f(4)=\frac{14}{4\cdot 25}=\frac{7}{50}\,.$$

(3) $PV\displaystyle\int_{-\infty}^{\infty}\frac{3x+2}{x(x-4)(x^2+9)}dx=2\pi i(-\frac{19}{450}+\frac{7}{75}i)+\pi i(-\frac{1}{18}+\frac{7}{50})$

$$=-\frac{38\pi i}{450}-\frac{14\pi}{75}-\frac{\pi i}{18}+\frac{7\pi i}{50}$$

$$=\frac{-76\pi i}{900}-\frac{168\pi}{900}-\frac{50\pi i}{900}+\frac{126\pi i}{900}=\frac{-14\pi}{75}\pi\,.\quad \boxed{\text{Q.E.D.}}$$

11-7-3 Fourier Transform (Integrals) Type

$$\int_{-\infty}^{\infty}F(x)e^{imx}\,dx$$

In Chapter 9, we learned various calculations related to Fourier integrals and transformations. Many of these integrals are not easy to solve directly, but they can be evaluated using the residue theorem from complex analysis. The approach is outlined as follows.

▶ **Theorem 11-7-4**

Calculation of Fourier Transform (Integrals)

Suppose $F(x) = \dfrac{P(x)}{Q(x)}$ be rational functions, and $P(x)$, $Q(x)$ lack real roots. If $\deg[Q(x)] > \deg[P(x)]$,

then $\begin{cases} (1)\, m > 0 \Rightarrow \displaystyle\int_{-\infty}^{\infty} F(x)e^{imx}\, dx = 2\pi i (\sum \text{Res on the upper half-plane}) \\ (2)\, m < 0 \Rightarrow \displaystyle\int_{-\infty}^{\infty} F(x)e^{imx}\, dx = -2\pi i (\sum \text{Res on the lower half-plane}) \end{cases}$.

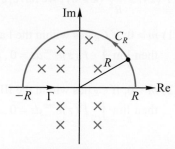

Figure 11-39   Diagram of an infinite semicircle on the upper half-plane

Proof

Taking the integration path $C = C_R + \Gamma$, as shown in Figure 11-39. Then

① on the circular arc C_R: $z = Re^{i\theta}$, $dz = iRe^{i\theta}d\theta$, $\theta \in [0, 2\pi]$;

② in the diameter Γ: $z = x$, $dz = dx$, $x \in [-R, R]$.

Therefore, the line integral can be decomposed as
$$I = \oint_C F(z)e^{imz}\,dz = \int_{C_R} F(z)e^{imz}\,dz + \int_{\Gamma} F(z)e^{imz}\,dz .$$

Let's first examine the behavior of the integral along the circular arc C_R as R approaches infinity:

$$\lim_{R \to \infty} \int_{C_R} F(Re^{i\theta})e^{imRe^{i\theta}} iRe^{i\theta}\,d\theta \le \lim_{R \to \infty} |\int_0^{\pi} e^{imRe^{i\theta}} F(Re^{i\theta}) iRe^{i\theta}\,d\theta |$$

$$\le \lim_{R \to \infty} \int_0^{\pi} |e^{imRe^{i\theta}}| \, \| F(Re^{i\theta}) \| \, | iRe^{i\theta} | \, d\theta$$

$$\le \lim_{R \to \infty} \frac{M}{R^{k-1}} \int_0^{\pi} e^{-mR\sin\theta}\,d\theta \quad (k \ge 0)$$

$$= \lim_{R \to \infty} \frac{2M}{R^{k-1}} \int_0^{\pi/2} e^{-mR\sin\theta}\,d\theta \le \lim_{R \to \infty} \frac{2M}{R^{k-1}} \int_0^{\pi/2} e^{-mR(\frac{2\theta}{\pi})}\,d\theta$$

$$= \lim_{R \to \infty} \frac{\pi M}{mR^k} (-e^{-mR\frac{2\theta}{\pi}} \Big|_0^{\pi/2}) = \lim_{R \to \infty} \frac{\pi M}{mR^k}(1 - e^{-mR}) = 0 ,$$

$$\therefore \lim_{R\to\infty} \int_0^\pi e^{im\,Re^{i\theta}} F(Re^{i\theta}) iRe^{i\theta}\, d\theta = 0 \text{, therefore,}$$

$$\lim_{R\to\infty} \oint_c F(z)e^{imz}\, dz = \lim_{R\to\infty} \int_{C_R} F(Re^{i\theta})e^{imz} iRe^{i\theta}\, d\theta + \int_\Gamma F(x)e^{imx}\, dx = \int_{-\infty}^\infty F(x)e^{imx}\, dx \,,$$

hence, by the residue theorem: $\displaystyle\lim_{R\to\infty} \int_{-R}^R F(x)e^{imx}\, dx = \lim_{R\to\infty} I = 2\pi i \sum \mathrm{Res}$ on the upper half-plane. ◄

In this theorem, C_R is the circular arc with center at $z = 0$ and radius R. Since $|F(z)| \le \dfrac{M}{R^k}(k > 0)$, we have $|e^{imz}| = e^{-mR\sin\theta}$, which indicates:

(1) $m > 0$; if C_R lies within the I and II quadrants ($0 \le \theta \le \pi$),
then $\displaystyle\lim_{R\to\infty} \int_{C_R} F(z)e^{imz}\, dz = 0$, as shown in Figure (a).

(2) $m < 0$; if C_R lies within the III and IV quadrants ($\pi \le \theta \le 2\pi$),
then $\displaystyle\lim_{R\to\infty} \int_{C_R} F(z)e^{imz}\, dz = 0$, as shown in Figure (b).

(3) $m > 0$; if C_R lies within the II and III quadrants ($\dfrac{\pi}{2} \le \theta \le \dfrac{3\pi}{2}$),
then $\displaystyle\lim_{R\to\infty} \int_{C_R} e^{mz} F(z)\, dz = 0$, as shown in Figure (c).

(4) $m < 0$; if C_R lies within the IV and I quadrants ($-\dfrac{\pi}{2} \le \theta \le \dfrac{\pi}{2}$),
then $\displaystyle\lim_{R\to\infty} \int_{C_R} e^{mz} F(z)\, dz = 0$, as shown in Figure (d).

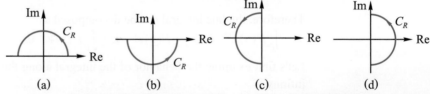

| (a) | (b) | (c) | (d) |

Figure 11-40　　Upper, lower, left, and right infinite semicircles

▶ **Theorem 11-7-5**

Calculation of the Fourier Transform (Integral) with a Simple Pole on the Real Axis

Following Theorem 1, if $F(z)$ has several first-order poles on the x-axis, then

(1) $m > 0$, P.V. $\displaystyle\int_{-\infty}^{\infty} F(x)e^{imx}dx = 2\pi i\sum \mathrm{Res}$ on the upper half-plane $+\pi i\sum \mathrm{Res}$ on the x-axis.

(2) $m < 0$, P.V. $\displaystyle\int_{-\infty}^{\infty} F(x)e^{imx}dx = -2\pi i\sum \mathrm{Res}$ on the lower half-plane $-\pi i\sum \mathrm{Res}$ on the x-axis.   ◀

Furthermore, due to $\mathrm{Re}[e^{imx}] = \cos mx$, $\mathrm{Im}[e^{imx}] = \sin mx$, we can extend the following theorem.

▶ **Theorem 11-7-6**

Calculation of the Fourier Sine (Cosine) Transform (Integral)

(1) $\displaystyle\int_{-\infty}^{\infty} F(x)\cdot \cos mx dx = \mathrm{Re}[\int_{-\infty}^{\infty} e^{imx}F(x)dx]$.

(2) $\displaystyle\int_{-\infty}^{\infty} F(x)\cdot \sin mx dx = \mathrm{Im}[\int_{-\infty}^{\infty} e^{imx}F(x)dx]$.   ◀

Example 4

Calculate $\displaystyle\int_{0}^{\infty} \frac{\cos 2x}{4x^4 +13x^2 +9} dx$

Solution

(1) $\displaystyle\int_{0}^{\infty} \frac{\cos 2x}{4x^4 +13x^2 +9} dx = \frac{1}{2}\int_{-\infty}^{\infty} \frac{\cos 2x}{4x^4 +13x^2 +9} dx = \frac{1}{2}\mathrm{Re}\int_{-\infty}^{\infty} \frac{e^{i2x}}{4x^4 +13x^2 +9} dx$,

let $f(z) = \dfrac{e^{i2z}}{4z^4 +13z^2 +9}$, considering the integration path as shown in the diagram on the right, where there are only $z = i$, $\dfrac{3}{2}i$, two first-order poles inside C and $C = C_R + \Gamma$.

(2) From the residue theorem, we know that:

$$\int_{-\infty}^{\infty} \frac{e^{i2x}}{4x^4 + 13x^2 + 9} dx = 2\pi i [\operatorname{Res} f(i) + \operatorname{Res} f(\frac{3}{2}i)],$$

also $\operatorname{Res} f(i) = \lim_{z \to i}(z - i) \cdot \dfrac{e^{i2z}}{4z^4 + 13z^2 + 9} = \dfrac{e^{-2}}{10i}$;

$$\operatorname{Res} f(\frac{3}{2}i) = \lim_{z \to \frac{3}{2}i}(z - \frac{3}{2}i) \cdot \frac{e^{i2z}}{4z^4 + 13z^2 + 9} = -\frac{e^{-3}}{15i},$$

$$\therefore \int_{-\infty}^{\infty} \frac{e^{i2x}}{4x^4 + 13x^2 + 9} dx = 2\pi i(\frac{e^{-2}}{10i} - \frac{e^{-3}}{15i}) = 2\pi(\frac{e^{-2}}{10} - \frac{e^{-3}}{15}).$$

(3) $\therefore \displaystyle\int_0^{\infty} \frac{\cos 2x}{4x^4 + 13x^2 + 9} = \frac{1}{2}\operatorname{Re}\int_{-\infty}^{\infty} \frac{e^{i2x}}{4x^4 + 13x^2 + 9} dx = \pi[\frac{e^{-2}}{10} - \frac{e^{-3}}{15}].$ Q.E.D.

Example 5

Calculate the Cauchy principal value of $\displaystyle\int_{-\infty}^{\infty} \frac{\sin x}{x(x^2 - 2x + 2)} dx$.

Solution

(1) P.V. $\displaystyle\int_{-\infty}^{\infty} \frac{\sin x}{x(x^2 - 2x + 2)} dx = \operatorname{Im}[\int_{-\infty}^{\infty} \frac{e^{ix}}{x(x^2 - 2x + 2)} dx],$

let $f(z) = \dfrac{e^{iz}}{z(z^2 - 2z + 2)}$, considering the

integration path as shown in the diagram on the right, where only $1 + i$ is a first-order pole inside C, while $z = 0$ is a simple pole on C.

(2) From the residue theorem, we know that:

$$\int_{-\infty}^{\infty} \frac{e^{ix} dx}{x(x^2 - 2x + 2)} = 2\pi i \operatorname{Res} f(1 + i) + \pi i \operatorname{Res} f(0),$$

$$\operatorname{Res} f(1 + i) = \lim_{z \to 1+i} [z - (1 + i)]\frac{e^{iz}}{z(z^2 - 2z + 2)} = \frac{e^{i(1+i)}}{(1 + i)2i}$$

$$= \frac{e^{-1}(\cos 1 + i \sin 1)(1 - i)}{4i} = \frac{e^{-1}[(\cos 1 + \sin 1) + i(\sin 1 - \cos 1)]}{4i},$$

$$\operatorname{Res} f(0) = \frac{1}{2},$$

$$\therefore \int_{-\infty}^{\infty} \frac{e^{ix}}{x(x^2 - 2x + 2)} dx = \frac{\pi}{2}e^{-1}[(\cos 1 + \sin 1) + i(\sin 1 - \cos 1)] + \frac{\pi}{2}i$$

$$= \frac{\pi}{2}e^{-1}(\cos 1 + \sin 1) + i\frac{\pi}{2}e^{-1}(\sin 1 - \cos 1) + \frac{\pi}{2}i.$$

(3) P.V. $\displaystyle\int_{-\infty}^{\infty} \frac{\sin x}{x(x^2 - 2x + 2)} dx = \frac{\pi}{2}e^{-1}(\sin 1 - \cos 1) + \frac{\pi}{2}.$ Q.E.D.

11-7-4 Laplace Inverse Type

Recalling from Laplace transformations, if $F(s) = \mathcal{L}\{f(t)\} = \int_0^\infty e^{-st} f(t) dt$

then $f(t) = \dfrac{1}{2\pi i} \int_{a-i\infty}^{a+i\infty} e^{st} F(s) ds$, $t > 0$, as shown in Figure 11-41.

The choice of the real number a is such that all singularities of $F(z)$ on the complex plane remain to the left of $z = a$, and the integration path is taken as $C = C_R + C_1$. Let's illustrate this principle with an example.

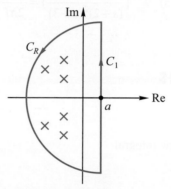

Figure 11-41   Laplace inverse transform integration path diagram.

Example 6

Use the residue theorem to calculate $\mathcal{L}^{-1}\{\dfrac{1}{(s+2)^2(s+3)}\}$.

Solution

(1) $\mathcal{L}^{-1}\{\dfrac{1}{(s+2)^2(s+3)}\} = \dfrac{1}{2\pi i} \int_{a-i\infty}^{a+i\infty} \dfrac{e^{st}}{(s+2)^2(s+3)} ds$,

let $F(z) = \dfrac{e^{zt}}{(z+2)^2(z+3)}$,

then $\oint_C F(z) dz = \int_{C_1} F(z) dz + \int_{C_R} F(z) dz$,

where $\displaystyle \int_{C_R} F(z) dz = \lim_{R\to\infty} \int_{\theta_0}^{2\pi-\theta_0} \dfrac{e^{Re^{i\theta}t}}{(Re^{i\theta}+2)^2(Re^{i\theta}+3)} iRe^{i\theta} d\theta$

$\displaystyle \le \lim_{R\to\infty} \int_{\theta_0}^{2\pi-\theta_0} \dfrac{|e^{Re^{i\theta}t}|}{R^2} M d\theta = 0$,

$\therefore \oint_C F(z) dz = \int_{C_1} F(z) dz = \int_{a-i\infty}^{a+i\infty} \dfrac{e^{zt}}{(z+2)^2(z+3)} dz$.

(2) From the residue theorem: $\oint_C F(z)dz = 2\pi i[\operatorname{Re} sF(-3) + \operatorname{Re} sF(-2)]$,

$$\operatorname{Re} sF(-3) = \lim_{z \to -3}(z+3)\frac{e^{zt}}{(z+2)^2(z+3)} = e^{-3t},$$

$$\operatorname{Re} sF(-2) = \lim_{z \to -2}\frac{d}{dz}[(z+2)^2\frac{e^{zt}}{(z+2)^2(z+3)}] = te^{-2t} - e^{-2t},$$

$$\therefore \int_{a-i\infty}^{a+i\infty}\frac{e^{zt}}{(z+2)^2(z+3)} = 2\pi i(te^{-2t} - e^{-2t} + e^{-3t}).$$

(3) $\mathscr{L}^{-1}\{\frac{1}{(s+2)^2(s+3)}\} = \frac{1}{2\pi i}\cdot 2\pi i(te^{-2t} - e^{-2t} + e^{-3t}) = te^{-2t} - e^{-2t} + e^{-3t}.$ Q.E.D.

11-7 Exercises

Basic questions

1. Calculate the trigonometric integral
$$\int_0^{2\pi}\frac{d\theta}{5+3\sin\theta}.$$

2. Calculate the trigonometric integral
$$\int_0^{\pi}\frac{d\theta}{5+3\cos\theta}.$$

3. Calculate the integral $\int_{-\infty}^{\infty}\frac{dx}{(x-1)(x^2+3)}.$

4. Calculate the principal value of the integral
$$\int_{-\infty}^{\infty}\frac{dx}{x^4-1}.$$

5. Find $\int_{-\infty}^{\infty}\frac{x\cdot\cos x}{x^2-3x+2}dx.$

Advanced questions

1. Calculate the trigonometric integral $\int_0^{2\pi}\frac{\cos\theta d\theta}{3+\sin\theta}.$

2. Calculate the trigonometric integral
$$\int_0^{\pi}\frac{2\sin^2\theta d\theta}{5-4\cos\theta}.$$

3. Calculate the trigonometric integral
$$\int_0^{2\pi}\frac{1d\theta}{(2+\cos\theta)^2}.$$

4. Calculate the trigonometric integral
$$I = \int_0^{\pi}\frac{\cos\theta}{1-2a\cos\theta+a^2}d\theta, a \in R \text{ and } |a| \neq 1.$$

5. Calculate the integral $\int_{-\infty}^{\infty}\frac{dx}{(x^2+1)(x^2+9)}.$

6. Calculate the integral $\int_0^{\infty}\frac{x^2}{x^6+1}dx.$

7. Calculate the principal value of the integral
$$\int_{-\infty}^{\infty}\frac{dx}{(x^2-3x+2)(x^2+1)}.$$

8. Calculate the principal value of the integral
$$\int_{-\infty}^{\infty}\frac{\sin x}{x(x^2+1)}dx.$$

9. Calculate the principal value of the integral
$$\int_{-\infty}^{\infty}\frac{\cos mx}{x^4-1}dx.$$

10. Calculate the principal value of the integral
$$\int_{-\infty}^{\infty}\frac{1}{x(x^2-4x+5)}dx.$$

11. Find $\int_0^{\infty}\frac{x\cdot\sin x}{x^2+4}dx.$

12. Find $\int_{-\infty}^{\infty}\frac{\cos 3x}{x^2+9}dx.$

13. Find $\int_{-\infty}^{\infty}\frac{x\sin ax}{x^4+4}dx, a > 0.$

Appendix

Appendix 1　　　　　**References**

1. Erwin Kreyszig, Advanced Engineering Mathematics. 8th Edition, John Wiley & Sons. Inc., 1999.

2. Peter V. O'Nell. Advanced Engineering Mathematics, 5th Edition, Brooks/Cole-Thomson Learning, Inc., 2003.

3. Pennis G. Zill & Warren S. Wright, Advanced Enginearing Mathematics. 5th Edition, Jones & Bartlett Karning, October 1, 2012.

4. Michael D. Greenberg, Advanced Engineering Mathematics, second Edition, Prentice-Hall, Inc., 1998.

5. C. Ray Wylie, Advanced Engineering Mathematics, 6th Edition, McGraw-Hill, Inc., 1995.

6. Dennis G. Zill & Micharel R. Cullen, Differential Equation with Boundary Value Problems. 4th edition, Brooks/Cole-Thomson Learning, Inc., 1997.

7. Mary L. Boas, Mathematical Methods in the Physical Sciences, 2nd edition, John Wiley & Sons, Inc., 1983.

8. William E. Boyce & Richard C. DiPrima, Elementary Differential Equation and Boundary Value Problems, 5th edition, John Wiley & Sons, Inc., 1992.

9. R. Kent Nagle & Edward B. Saff, Fundamentals of Differential Equation, Benjarnin/Cummings Publishing Company, Inc., 1986.

10. Murray R. Spiegel, Schaum's Outline Series of Theory and Problems of Advanced Mathematics for Engineers and Scientists, McGraw-Hill, Inc., 1971.

11. C. H. Edwards, Jr. & David E. Penney, Elementary Differential Equation with Boundary Value Problems, Prentice-Hall. Inc., 1993.

12. D. V. Widder, The Laplace Transform, Princeton University Press, Princeton, N, J., 1941.

13. Grossman Derrick, 廖東成、吳嘉祥譯, 高等工程數學(上)、(下), 台北圖書有限公司, 1990。

14. 圖立編譯館部定大學用書編審委員會主編, 朱越生編著, 部定大學用書工程數學(上)、(下), 圖立編譯館主編, 正中書局印行, 民國 61 年。

Appendix 2

Appendix 3

Appendix 4

Appendix 5

Index